Developments in Mathematics

VOLUME 38

More information about this series at http://www.springer.com/series/5834

Geoffrey Mason • Ivan Penkov • Joseph A. Wolf
Editors

Developments and Retrospectives in Lie Theory

Algebraic Methods

Springer

Editors
Geoffrey Mason
Department of Mathematics
University of California, Santa Cruz
Santa Cruz, CA, USA

Ivan Penkov
Jacobs University
Bremen, Germany

Joseph A. Wolf
Department of Mathematics
University of California, Berkeley
Berkeley, CA, USA

ISSN 1389-2177 ISSN 2197-795X (electronic)
ISBN 978-3-319-37820-6 ISBN 978-3-319-09804-3 (eBook)
DOI 10.1007/978-3-319-09804-3
Springer Cham Heidelberg New York Dordrecht London

Printed on acid-free paper

Springer is part of Springer Science+Business Media (www.springer.com)

*** Please Post *** *** Please Post *** *** Please Post ***

Lie Groups, Lie Algebras and their Representations

Third Announcement -- October 1991

This is to announce a new program ***Lie Groups, Lie Algebras and Their Representations*** starting in the academic year 1991-92. We plan to meet at various University of California campuses for one weekend every other month during the year. The purpose of the program is to communicate results and ideas rather than to deliver polished presentations. The tentative format of the meetings is

Saturday: one late morning talk and two afternoon talks; these scheduled in advance.
Sunday: two morning talks and one early afternoon talk; some may be scheduled in advance and the rest will be scheduled Saturday.

The meetings are scheduled as follows:

Location	Dates	Local Organizer	e-mail address
Berkeley	October 19, 20	Joseph Wolf	jawolf@cartan.berkeley.edu
Riverside	December 7, 8	Ivan Penkov	penkov@ucrmath.ucr.edu
Davis	February 8, 9 ??	Alice Fialowski	fialowski@bolyai.ucdavis.edu
UCLA	April 4, 5	Robert Blattner	blattner@math.ucla.edu
UC Santa Cruz	May 30, 31	Daniel Goldstein	danny@cats.ucsc.edu

The program for the first meeting, Saturday October 19 and Sunday October 20, is

time	speaker	title
Sat 11:00	Joseph Wolf	Lie structures on direct limit groups and their completions (joint work with L. Natarajan and E. Rodríguez-Carrington)
Sat 02:00	Dmitri Fuchs	On the cohomology of Lie algebras of vector fields in the line and of Hamiltonian vector fields in the plane
Sat 04:00	Nicolai Reshetikhin	q-holonomic systems and quantum affine Lie algebras, I
Sun 09:00	Ivan Penkov	Infinite dimensional irreducible modules over finite dimensional Lie superalgebras
Sun 10:30	Yan Soibelman	Algebras of functions on compact quantum groups
Sun 01:00	Nicolai Reshetikhin	q-holonomic systems and quantum affine Lie algebras, II

These talks will take place in 9 Evans Hall on the Berkeley campus. The one hour scheduled time of any talk will be extended by the informal discussion that takes place during that talk.

There are no registration fees. No support is currently available. We hope to minimize participants' expenses by car-pooling and various informal arrangements for staying over Saturday night.

People interested in participating, or speaking and participating, at a meeting, should contact the local organizer for that meeting. People interested in helping with the local organization should contact both the local organizer and Joseph Wolf. If you wish to be on the mailing list for further announcements, and for meeting programs/schedules, write to Joseph Wolf.

Announcement of the First Conference, October 1991

Preface

The West Coast of the United States has a longstanding tradition in Lie theory, although before 1991 there had been no systematic cooperation between the various strongholds. In 1991, partially inspired by the arrival of some well–known Lie theorists from eastern Europe, the situation changed. A new structure emerged: a seminar that would meet at various University of California campuses three or four times a year. The purpose of the seminar was to foster contacts between researchers and graduate students at the various campuses by facilitating the sharing of ideas prior to formal publication. This idea quickly gained momentum, and became a great success. It was enthusiastically supported by graduate students. A crucial feature of the entire endeavor was the feeling of genuine interest for the work of colleagues and the strong desire to collaborate.

The first meeting of the new seminar "Lie Groups, Lie Algebras and their Representations" was held in Berkeley on October 19 and 20, 1991. On the second day of the seminar, excitement was made even more memorable by the historic Berkeley–Oakland fire, which we observed from Evans Hall. The original announcement for that meeting is reprinted here on page v. The phrase "The purpose of the program is to communicate results and ideas rather than to deliver polished presentations" quickly became, and still is, the guiding principle of the seminar. We never restricted ourselves to Lie Theory per se, and speakers from geometry, algebra, complex analysis, and other adjacent areas were often invited.

NSF travel grants were crucial to the success of the seminar series. These grants funded travel for speakers, graduate students and postdoctoral researchers, and we thank the National Science Foundation for its continued support.

Over the years our idea became widely popular. On occasion the seminar took place in Salt Lake City, in Stillwater, Oklahoma, and in Eugene, Oregon. In addition, colleagues from other regional centers of Lie theory and related areas picked up on our idea and created their own meeting series. This is how the "Midwest Lie Theory

Seminars", the "Midwest Group Theory Seminar", the "Southeastern Lie Theory Workshops", and other regional series emerged.

The California Lie Theory Seminar has now been alive and well for 23 years. Joe Wolf has always played a central role in the seminar, along with Geoff Mason and Ivan Penkov. When Ivan left for Germany in 2004, Susan Montgomery and Milen Yakimov joined the team of organizers.

Over the course of these 23 years, at about 20 talks per year, some 450 talks have been hosted by the seminar. It seemed unrealistic to give an overview of all of the topics covered over all these years, and similarly unrealistic to try to publish a comprehensive set of volumes. Rather, we settled on two retrospective volumes containing work representative of the seminar as a whole. We started with a list of participants who spoke more than once in the seminar, and invited them to submit work relevant to their seminar talks. For obvious reasons we did not hear from everyone. Nevertheless, there was a strong response, and the reader of these Volumes will find 26 research papers, all of which received strong referee reports, in the greater area of Lie Theory. We decided to split the papers into two volumes: "Algebraic Methods" and "Geometric/Analytic Methods". We thank Springer, the publisher of these volumes, and especially Ann Kostant and Elizabeth Loew, for their cooperation and assistance in this project.

This is the Algebraic Methods volume.

Santa Cruz, USA — Geoffrey Mason
Bremen, Germany — Ivan Penkov
Berkeley, USA — Joseph A. Wolf

Contents

Group Gradings on Lie Algebras, with Applications to Geometry. I

Yuri Bahturin, Michel Goze, and Elisabeth Remm

Abstract In this article, which is the first part of a sequence of two, we discuss modern approaches to the classification of group gradings on simple and nilpotent Lie algebras. In the second article we discuss applications and related topics in differential geometry.

Key words Lie and associative algebras of linear transformations • Graded algebras • Hopf algebras • Automorphisms • Involution gradings

Mathematics Subject Classification (2010): 16K20, 16T05, 16W10, 16W20, 16W50, 17B65, 17B70, 17B40.

1 Introduction

This article is the first is a sequence of two articles in which we discuss several topics related to the first author's presentations given at the workshops "Lie Groups, Lie Algebras and their Representations" at the University of Southern California and several campuses of the University of California. The main topics covered were the theory and representations of simple locally finite Lie algebras, Lie superalgebras, and group gradings on simple Lie algebras, with applications to differential geometry. The most recent talk, in 2010 at the University of Southern California, was devoted to a more abstract topic: the study of the distortion arising when one Lie algebra is embedded in another as a subalgebra.

At least in one of the areas, namely, in the theory of group gradings on simple Lie algebras, the development that followed was truly spectacular. Thanks to the efforts

Y. Bahturin (✉)
Department of Mathematics and Statistics, Memorial University of Newfoundland, St. Johns, NL A1C5S7, Canada
e-mail: bahturin@mun.ca

M. Goze • E. Remm
Université de Haute Alsace, LMIA, 4 rue des Frères Lumière, 68093 Mulhouse, France
e-mail: michel.goze@uha.fr; elisabeth.remm@uha.fr

G. Mason et al. (eds.), *Developments and Retrospectives in Lie Theory: Algebraic Methods*, Developments in Mathematics 38,
DOI 10.1007/978-3-319-09804-3_1

of several researchers in several countries, now we have a complete classification of abelian group gradings on all classical simple Lie algebras over an algebraically closed fields of characteristic different from 2. The detailed exposition of this theory can now be found in a monograph [19], which has incorporated the results of [6], where the above classification, up to the isomorphism, has been completed, with the exception of D_4 and $\mathfrak{psl}_3$. In the same monograph [19], the authors present the classification of group gradings on almost all exceptional algebras, which heavily depends on the study of various classes of nonassociative algebras, such as octonions, Albert algebra, and so on. In the case char $F = 0$, a description of fine gradings on simple Lie algebras, with the same exception, as above, has appeared in [22] and a description of all group gradings was obtained in [9] and [11].

Lately, the study of gradings on Lie algebras went beyond the boundaries of the area of classical or exceptional simple finite-dimensional algebras. In the paper [3] the authors classify up to isomorphism all abelian group gradings on finitary simple Lie algebras of linear transformations. In the paper [7] the authors classify all abelian group gradings on two types of restricted simple Lie algebras of Cartan type over algebraically closed field of characteristic $p > 3$.

These results became possible thanks to the extensive use of the methods involving the techniques of the theory of Hopf algebras. Probably the first papers where this approach was successfully implemented were [5] and [8]. This is a well-known fact that an algebra A over a field F graded by an abelian group G bears a canonical structure of the (right) comodule algebra over the group algebra $H = FG$ (see [24]). Conversely, the H-comodule algebra canonically becomes a G-graded algebra. If A is an associative envelope of a Lie algebra L then in some cases of interest to us, the H-comodule structure of L can be extended to the H-comodule structure of A. Then A becomes G-graded and the grading of L becomes the restriction of a G-grading of an associative algebra A. It is a usual pattern that the gradings of simple associative algebras are easier to determine.

The problem of extending the H-comodule structure from Lie to associative algebras in the case of finite-dimensional algebras is approached using the technique of affine group schemes [32]. These techniques, however, do not apply in the case of infinite-dimensional algebras. In the case where the algebras are infinite-dimensional (or just of sufficiently great dimension) quite another approach works. As it turns out, the question about the possibility of extending the H-comodule structure from a Lie to an associative algebra is analogous to the problem of extending automorphisms of Lie algebras to the automorphisms or negatives of antiautomorphisms of their associative envelopes. This latter problem, named after I. Herstein, was successfully solved in the 1990s using so-called *functional identities* (see [18]). An appropriate adaptation of the methods of this book made it possible to classify in [3] all abelian group gradings on simple finitary Lie algebras over algebraically closed fields of characteristic other than 2 and 3. Note that these results are also related to the theory of locally finite simple Lie algebras, which was one of the topics of research of several people related to the West Coast Lie Theory workshop "Lie Groups, Lie Algebras and their Representations".

Since Cartan decompositions of semisimple Lie algebras are a special case of fine gradings, it is our strong belief that the classification of all gradings on simple Lie algebras should be important for the study of their structure, representations and other properties.

As examples, the study of graded identical relations of algebras lately became an important branch of the theory of algebras with polynomial identities (PI-algebras). Graded polynomial identities of algebras are often much easier to study, yet they define the ordinary ones.

2 Group Gradings of Lie Algebras

In what follows $\mathbb{K}$ is an algebraically closed field of characteristic other than 2. For more details on group gradings of algebras, see a recent monograph [19]. These definitions apply to other classes of algebras, with an appropriate change of notation.

2.1 Definition

Let L be a Lie algebra over $\mathbb{K}$ and S a set.

Definition 2.1. An S-grading Γ of L with support S is a decomposition

$$\Gamma : L = \bigoplus_{s \in S} L_s$$

of L as the sum on nonzero vector subspaces L_s satisfying the following condition. For any $s_1, s_2 \in S$ such that $[L_{s_1}, L_{s_2}] \neq 0$ there is $\mu(s_1, s_2) \in S$ such that $[L_{s_1}, L_{s_2}] \subset L_{\mu(s_1,s_2)}$.

If S is a subset of a group G such that $\mu(s_1, s_2) = s_1 s_2$ in G, we say that Γ is a *group grading by the group G with support S*. Such a group G is not defined uniquely, but for any group grading there is a universal grading group $U(\Gamma)$ such that any other grading group G of Γ is a factor group of $U(\Gamma)$. The universal group is given in terms of generators and defining relations if one chooses S as the set of generators and all the above equations $s_1 s_2 = \mu(s_1, s_2)$ as the set of defining relations.

2.2 Equivalent and Isomorphic Gradings

If $\Gamma' : L = \bigoplus_{s' \in S'} L'_{s'}$ is another grading of L, we say that Γ is *equivalent* to Γ' if there is an automorphism $\varphi \in \operatorname{Aut} L$ and a bijection $\sigma : S \to S'$ such that $\varphi(L_s) = L'_{\sigma(s)}$. It follows that $\sigma(\mu(s_1, s_2)) = \mu(\sigma(s_1), \sigma(s_2))$.

Two group gradings Γ and Γ' of a Lie algebra L by groups G and G' with supports S and S' are called *weakly isomorphic* if they are equivalent, as above, and the map $\sigma : S \to S'$ extends to the isomorphism of groups G to G'. The strongest relation is the *isomorphism* of G-gradings. In this case both Γ and Γ' are gradings by the same group; they have the same support and the isomorphism of groups σ is identity. As a result, we have $\varphi(L_g) = L'_g$, for any $g \in G$. In the case of group G-gradings with support S we will write $\Gamma : L = \bigoplus_{g\in G} L_g$, assuming that $L_g = \{0\}$ if $g \in G \setminus S$.

2.3 Refinements, Coarsenings, Fine Gradings

Definition 2.2. Let Γ, as above, and $\Gamma' : L = \bigoplus_{t\in T} L'_t$ be two gradings on L with supports S and T, respectively. We say that Γ is a *refinement* of Γ' or Γ' is a *coarsening* of Γ if for any $s \in S$ there exists $t \in T$ such that $L_s \subset L'_t$.

Coarsening and refinements often arise as follow. Let $\Gamma : L = \bigoplus_{g\in G} L_g$ and let $\varepsilon : G \to K$ be a homomorphism of groups. We set $L'_k = \bigoplus_{\varepsilon(g)=k} L_g$, for any $k \in K$. Then we obtain the image-grading $\varepsilon(\Gamma) = \bigoplus_{k\in K} L'_k$. If ε is an onto homomorphism, we say that $\varepsilon(\Gamma)$ is a *factor-grading* of the grading Γ. Clearly, $\varepsilon(\Gamma)$ a coarsening of Γ and Γ is a refinement of $\varepsilon(\Gamma)$.

A group grading Γ of L is called a *fine grading* if doesn't admit proper group refinements.

Remark 2.1. If Γ' is a refinement of Γ, then Γ viewed as a $U(\Gamma)$-grading is a factor grading of Γ' viewed as a $U(\Gamma')$-grading.

2.4 Abelian Groups Gradings

In the case of Lie algebras it is natural to assume that all groups involved in the group gradings are abelian. In fact, for many Lie algebras, like finite-dimensional simple ones, this is satisfied (see [11]) in the sense that the partial function $\mu : S \times S \to S$ appearing in the definition of the grading is symmetric or that the elements of the support in the group grading commute. So in what follows we always deal with gradings by abelian groups. In addition, when we study finite-dimensional algebras, the supports of the gradings are finite sets, so our groups are finitely generated.

Now given a finitely generated abelian group G, we denote by $\hat{G}$ the group of (1-dimensional) characters of G, that is, the group of all homomorphisms $\chi : G \to F^{\times}$ where $F^{\times}$ is the multiplicative group of the field F. If $\Gamma : L = \bigoplus_{g\in G} L_g$ is a grading of a Lie algebra L with a grading group G, there is an action of $\hat{G}$ by semisimple automorphisms of L given on the homogeneous elements of L by $\chi * x = \chi(g)x$ where $\chi \in \hat{G}$ and $x \in L_g$. If G is generated by the support

S of Γ, different characters act differently. Indeed, assume $\chi_1, \chi_2 \in \hat{G}$ are such that $\chi_1 * x = \chi_2 * x$, for any $x \in L$. Choose any $s \in S$ and $0 \neq x \in L_s$. In this case $\chi_1(s)x = \chi_1 * x = \chi_2 * x = \chi_2(s)x$. As a result, $\chi_1(s) = \chi_2(s)$, for any $s \in S$. Since χ_1 and χ_2 are homomorphisms and G is generated by S, we have $\chi_1 = \chi_2$, as claimed. This allows us, in this important case, to view $\hat{G}$ as a subgroup of Aut L. We will view Aut L as an algebraic group. When we study finite-dimensional algebras, then G is finitely generated abelian and so $\hat{G}$ is the group of characters of a finitely generated abelian group. If $G \cong \mathbb{Z}^m \times A$, where m is an integer, $m \geq 0$, and A a finite abelian group, then $\hat{G} \cong (F^\times)^m \times \hat{A}$. Such subgroups of algebraic groups, consisting of semisimple elements, are called *quasitori*.

A quasitorus is a generalization of the notion of a torus, which is an algebraic subgroup of Aut L isomorphic to $\hat{G} \cong (F^\times)^m$, for some m. A torus, which is not contained in a larger torus is called *maximal*. The following result is classical.

Theorem 2.1. *In any algebraic group any two maximal tori are conjugate by an inner automorphism.*

Another well-known result is often attributed to Platonov [27] (but see also [28]).

Theorem 2.2. *Any quasitorus is isomorphic to a subgroup in the normalizer of a maximal torus.*

Thus, if we find a maximal torus D in Aut L equal to its normalizer in Aut L, then for any grading of L by a finitely generated abelian group G, there is $\varphi \in$ Aut L such that $\varphi\hat{G}\varphi^{-1} \subset D$.

Every time we have a quasitorus Q in Aut L there is root space decomposition of L with roots from the group of (algebraic) characters $\mathfrak{X}(Q)$ for the group Q, the root subspace for $\lambda \in \mathfrak{X}(Q)$ given by

$$L_\lambda = \{x \in L \mid \alpha(x) = \lambda(\alpha)x \text{ for any } \alpha \in Q\}.$$

If $Q \subset D$ then, by duality, $\mathfrak{X}(Q)$ is a factor group of $\mathfrak{X}(D) \cong \mathbb{Z}^m$ where $m = \dim D$. The root space decomposition by D is the refinement of the root space decomposition by Q and so the grading by $\mathfrak{X}(Q)$ is a coarsening of a grading by $\mathfrak{X}(D) \cong \mathbb{Z}^m$.

Now assume that we deal with a grading of L by a finitely generated abelian group G, $\Gamma : L = \bigoplus_{g\in G} L_g$. Assume that $p = \operatorname{char} F$ and write $G = G_p \times G_{p'}$, where G_p is the Sylow p-subgroup and $G_{p'}$ its complement in G that has no elements of order p. If char $F = 0$, then $G = G_{p'}$. Let us consider the quasitorus $\hat{G} \subset$ Aut L. Then there is $\varphi \in$ Aut L such that $\varphi\hat{G}\varphi^{-1} \subset D$. Let us switch to another G-grading $\varphi(\Gamma) : L = \bigoplus_{g\in G} \varphi(L_g)$. The action of $\hat{G}$ on L induced by $\varphi(\Gamma)$ gives rise to another copy of $\hat{G}$ in Aut L, namely, $\varphi\hat{G}\varphi^{-1}$ and now this subgroup is a subgroup in D. Replacing D by another maximal torus $\varphi^{-1}D\varphi$ we may assume from the very beginning that $\hat{G} \subset D$. Then the root decomposition by D is a refinement of the root decomposition under the action of $\hat{G}$. Thus the grading

$\Gamma' : L = \bigoplus_{\lambda\in\mathfrak{X}(\hat{G})} L_\lambda$ induced by this root decomposition resulting from the action of $\hat{G}$ is a coarsening of the $\mathfrak{X}(D) \cong \mathbb{Z}^m$ grading of L induced by the action of the torus D.

In the case where G has no elements of order $p = \operatorname{char} F$, we have the original grading being a coarsening of the grading induced from the action of a maximal torus. Usually, this is some "standard" torus, and the grading induced by its action is also called "standard."

We summarize the above discussion as follows.

Theorem 2.3. *Let* $\Gamma : L = \bigoplus_{g\in G} L_g$ *be a grading of a finite-dimensional algebra* L *over an algebraically closed field* $\mathbb{K}$ *by a finitely generated abelian group* G*. If* $\operatorname{char}\mathbb{K} = p > 0$*, let* G_p *denote the Sylow* p*-subgroup of* G*. Consider the automorphism group* $A = \operatorname{Aut} L$ *of* L *and assume* D *is a maximal torus of* A*, of dimension* m*, equal to its normalizer in* A*. Then the factor-grading* Γ/G_p *is isomorphic to a factor-grading of the standard* $\mathbb{Z}^m$*-grading of* L *induced by the action of* D *on* L*.*

An important particular case is the following.

Theorem 2.4. *Let* $\Gamma : L = \bigoplus_{g\in G} L_g$ *be a grading of a finite-dimensional Lie algebra* L *over an algebraically closed field* $\mathbb{K}$ *by a finitely generated abelian group* G*. If* $\operatorname{char}\mathbb{K} = p > 0$*, assume* G *has no elements of order* p*. Consider the automorphism group* $A = \operatorname{Aut} L$ *of* L *and assume* D *is a maximal torus of* A*, of dimension* m*, equal to its normalizer in* A*. Then* Γ *is isomorphic to a factor-grading of the standard* $\mathbb{Z}^m$*-grading of* L *induced by the action of* D *on* L*.*

Even more special is the following.

Theorem 2.5. *Let* $\Gamma : L = \bigoplus_{g\in G} L_g$ *be a grading of a finite-dimensional algebra* L *over an algebraically closed field of characteristic zero* $\mathbb{K}$ *by a finitely generated abelian group* G*. Consider the automorphism group* $A = \operatorname{Aut} L$ *of* L *and assume* D *is a maximal torus of* A*, of dimension* m*, equal to its normalizer in* A*. Then* Γ *is isomorphic to a factor-grading of the standard* $\mathbb{Z}^m$*-grading of* L *induced by the action of* D *on* L*.*

2.5 Automorphism Group of a Grading

Since we completely classify gradings up to equivalence for certain classes of algebras, we quote some more results from [19]. Given a grading $\Gamma : L = \bigoplus_{s\in S} L_s$ of an algebra L, the subgroup of the group $\operatorname{Aut} L$ permuting the components of Γ is called the *automorphism group* of the grading Γ and denoted by $\operatorname{Aut}\Gamma$. Each $\varphi \in \operatorname{Aut}\Gamma$ defines a bijection on the support S of the grading: if $\varphi(L_s) = L_{s'}$, then $s \mapsto s'$ is the desired permutation $\sigma(\varphi)$, an element of the symmetric group $\operatorname{Sym} S$. The kernel of the homomorphism $\varphi \mapsto \sigma(\varphi)$ is called the *stabilizer* of the grading

Γ, denoted by Stab Γ. Finally a subgroup of Stab Γ, whose elements are scalar maps on each graded component of Γ is called the *diagonal group* of Γ and denoted by Diag Γ.

Definition 2.3. Let $Q \subset$ Aut Γ be a quasitorus. Let Γ be the eigenspace decomposition of L with respect to Q. Then the quasitorus Diag Γ in Aut Γ is called the saturation of Q. We always have $Q \subset$ Diag Γ. If $Q =$ Diag Γ, then we say that Q is a *saturated* quasitorus.

A quasitorus Q is saturated if and only if the group $\mathscr{X}(Q)$ of algebraic characters of Q is $U(\Gamma)$, the universal group of Γ.

Proposition 2.1. *The equivalence classes of gradings on L are in one-to-one correspondence with the conjugacy classes of saturated quasitori in* Aut L.

Notice that if we already know that every quasitorus is conjugate to a subgroup of a fixed maximal torus and that two subgroups of the maximal torus are conjugate if and only if they are equal, we can say that the *equivalence classes of gradings are in one-to-one correspondence with the saturated subquasitori of a fixed maximal torus.*

2.6 Group Gradings and Actions of Hopf Algebras

In the case where the action of $\hat{\Gamma}$ by automorphisms does not completely reflect the G-grading, one can still find the transformations which are responsible for the gradings. For this one need to consider the group algebra $H = FG$ of the group G. The Hopf algebra structure on H, that is, the coproduct Δ, the counit ε and the antipode S are given as follows: $\Delta(g) = g \otimes g$, $\varepsilon(g) = 1$ and $S(g) = g^{-1}$, for any $g \in G$. Now let us consider the *finite dual* $K = H^\circ$ of H, which is just the ordinary dual H^* in the case where $|G| < \infty$, and the set of linear functions $f : H \to \mathbb{K}$ such that Ker f contains a two-sided ideal of finite codimension in H. All operations on K are defined by duality. The action of $f \in K$ on $A = \bigoplus_{g \in G} A_g$ is defined by setting $f * a = f(g)a$ if $a \in A_g$. If G is a finitely generated abelian group (or, more generally, a residually linear group), then K separates points of G and the G-grading can be recovered from the K-action. One has to set $A_g = \{x \in A \mid f * x = f(g)x$ for all $f \in K\}$.

If G is a finite group of order coprime to the characteristic of the base field, the basis of K is formed by the characters of G; they act as automorphisms and so we are bounced back to the situation described earlier. In the case where char $F = p > 0$ and G is an elementary abelian group of order p^n, it is known that K is a restricted enveloping algebra for an abelian p-Lie algebra spanned by n semisimple commuting derivations. In this case the study of gradings on an algebra is reduced to the study of action by derivations. In all other cases, although K has a very simple structure as an algebra, just the direct sum of the copies of the ground field, the action of the elements of K on the products of element of A can be extremely complex. However, this approach was successfully applied to the study of gradings on algebras of type A in characteristic $p > 0$ [5].

3 Transfer

If A is an associative algebra over a field $\mathbb{K}$, then we denote by $A^{(-)}$ the Lie algebra on the same vector space A under the bracket operation $[a, b] = ab - ba$, for all $a, b \in A$. By the Poincaré–Birkhoff–Witt Theorem, for any Lie algebra L there is an associative algebra A such that L is (isomorphic to) a subalgebra of $A^{(-)}$. If A is generated by (the image of) L, we say that A is an *associative envelope* of L. For any L one can choose the *universal enveloping algebra* $U(L)$ as an associative envelope, the drawback being that $U(L)$ is an infinite-dimensional algebra, for any nonzero L. In many cases, however, there are more manageable associative envelopes. For example, by the Ado–Iwasawa Theorem, every finite-dimensional Lie algebra has a finite-dimensional associative envelope. A Lie algebra consisting of linear transformations of a vector space V is a subalgebra of the associative algebra End (V), etc.

It was noted a while ago [26] that the study of group gradings on simple Lie algebras is closely related to the study of gradings on their associative envelopes. This is especially true in the case of classical simple Lie algebras over algebraically closed fields of characteristic zero for two reasons. First, because, as we saw, the study of gradings is equivalent to the study of quasitori in the automorphism groups of these algebras. Second, because all these algebras have matrix realizations and, with the exception of D_4, their automorphisms extend to the automorphisms or negatives of the automorphisms of the respective matrix algebras. The situation is especially benign in the case of algebras of the types B, C, D, except D_4. In this case, every automorphism of a Lie algebra of one of these types is given by a matrix conjugation. As a result, any G-grading of a Lie algebra, of the type $so(n)$ or $sp(n)$, n even, can be obtained by restriction from a G-grading of the matrix algebra M_n.

3.1 Affine Group Schemes

An object that most fully reflects the structure of gradings on a finite-dimensional algebra A by abelian groups is called the *automorphism group scheme* of A over a field $\mathbb{K}$, denoted by **Aut** A. This is a *representable* functor $\mathscr{F}$ from the category *Comm* of commutative associative unital algebras over $\mathbb{K}$ to the category *Ab* of abelian groups, which associates with each commutative associative unital algebra R the group $\mathrm{Aut}_R(A \otimes R)$. The value $\mathscr{F}(\mathbb{K})$ is just the ordinary Aut A. Being representable for a functor $\mathcal{G}$ means that there is there is a finitely generated Hopf algebra H such that $\mathcal{G}(R) = \mathrm{Alg}_{\mathbb{K}}(H, R)$, the group of algebra homomorphisms from H to R under the convolution product $(f * g)(h) = \sum_h f(h_1)g(h_2)$, where $\Delta(h) = \sum_h h_1 \otimes h_2$, h being an arbitrary element of H. Given a finitely generated abelian group G, the affine group scheme represented by the group algebra $H = \mathbb{K}G$ is denoted by G^D.

We quote the following results from [19, Section 1.4].

Proposition 3.1. *The G-gradings on an algebra A are in one-to-one correspondence with the morphisms of affine group schemes $G^D \to$ **Aut** A. Two G-gradings are isomorphic if and only if the corresponding morphisms are conjugate by an element of* Aut A.

Theorem 3.1. *Let A and B be finite-dimensional algebras. Assume that we have a morphism θ :* **Aut** *$A \to$* **Aut** *B. Then, for any abelian group G, we have a mapping $\Gamma \to \theta(\Gamma)$ from G-gradings of A to G-gradings of B. If $\Gamma \cong \Gamma'$, then $\theta(\Gamma) \cong \theta(\Gamma')$. If θ is an isomorphism and G is the universal grading group of a fine grading Γ, then $\theta(\Gamma)$ is a fine grading with universal group G.*

Using this theorem and the results about the gradings of nonassociative algebras, such as octonions or the Albert algebra, made it possible to describe abelian group gradings on simple Lie algebras of the types G_2 and F_4 in an amazing generality of arbitrary algebraically closed fields of characteristic not 2 or 3 in the case of G_2 and not 2, in the case of F_4. The group scheme **Aut** $L \cong$ **Aut O**, where **O** is the octonion algebra in the case where L is of the type G_2 and **Aut** $L \cong$ **Aut A** where **A** is the Albert algebra in the case where L is of the type F_4. For details see [19, Chapters 4, 5]. In both cases, the Lie algebra L is the derivation algebra of the respective nonassociative algebra $\mathcal{C}$ and the isomorphism of the group schemes is given by the adjoint map (see below).

The approach via group schemes paved the way for the classification of abelian group gradings on simple Cartan Lie algebras. These algebras arise as subalgebras of the derivation algebras of so-called *divided power algebras* (Witt algebras, type W; Special algebras, type S; Hamiltonian algebras, type H; Contact algebras, type K). For the classification of Cartan type Lie algebras see [29, 30]. A fairly recent paper [7] is devoted to the classification of all abelian group gradings on the restricted Lie algebras of the types W and S, with important information in the case S. In the case of restricted algebras, we only need to consider the derivation algebras of so the *truncated polynomial algebras*, that is, the algebras of the form $\mathcal{O}_p(n) = \mathbb{K}[x_1, \ldots, x_n]/(x_1^p, \ldots, x_n^p)$, where $n \geq 1$, $p = \mathrm{char}\,\mathbb{K}$. Using the isomorphism of the group schemes Ad : **Aut** $\mathcal{O}_p(n) \to$ **Aut** $W_p(n)$, where Ad : $\varphi \to (D \to (\varphi^{-1} \circ D \circ \varphi))$ where φ is an automorphism of $\mathcal{O}_p(n)$ and D is a derivation of $\mathcal{O}_p(n)$ one easily transfers the gradings from $\mathcal{O}_p(n)$ to $W_p(n)$. For more details see [19, Chapter 7].

3.2 Group Gradings, Comodules and Functional Identities

Suppose a Lie algebra L over a field F is graded by an abelian group G. This is well known [24] to be equivalent to L being a (right) *H-comodule Lie algebra* over the group algebra $H = FG$, that is, to the existence of a Lie homomorphism $\rho : L \to L \otimes H$ such that

(1) $$(\rho \otimes \mathrm{id}_H)\rho = (\mathrm{id}_L \otimes \Delta)\rho$$

and

(2) $$(\mathrm{id}_L \otimes \varepsilon)\rho = \mathrm{id}_L.$$

In the case of a graded algebra, ρ is determined by $\rho(a_g) = a_g \otimes g$ where a_g is a homogeneous element of degree g. If L is a Lie subalgebra generating an associative algebra A and ρ extends to an associative homomorphism $\rho : A \to A \otimes H$, then A also becomes G-graded. One easily checks that (1) and (2) will still be satisfied because L generates A as an associative algebra. Since both L_g and A_g are defined as the sets of elements x in L and A satisfying $\rho(x) = x \otimes g$, we have $L_g = L \cap A_g$. In what follows, we will see that this extension of ρ from L to A indeed happens in the case where L is the Lie algebra of skew-symmetric elements in a central simple associative algebra over any field of characteristic not 2, provided that $\dim L \geq 21$. The case of L being infinite-dimensional is not excluded but is rather welcome! On the other hand, it is clear that some gradings of simple Lie algebras cannot be induced from the associative gradings. This is true for the grading of $\mathfrak{sl}(n)$ by $\mathbb{Z}_2$, which represents every matrix as the sum its of skew-symmetric and symmetric components. To correctly handle the situation arising, let us extend the Lie homomorphism $\rho : L \to L \otimes H$ to $\bar{\rho} : L \otimes H \to L \otimes H$ by setting $\bar{\rho}(x \otimes h) = \rho(x)(1 \otimes h)$. Clearly, this is a surjective Lie homomorphism. We want to extend $\bar{\rho}$ to $A \otimes H$.

Recall that a map σ from an associative algebra $\mathscr{A}'$ to a unital associative algebra $\mathscr{A}''$ is a *direct sum of a homomorphism and the negative of an antihomomorphism* if there exist central idempotent e_1 and e_2 in A'' with $e_1 + e_2 = 1$ such that $x \mapsto e_1\sigma(x)$ is a homomorphism and $x \mapsto e_2\sigma(x)$ is the negative of an antihomomorphism. Maps with this property are clearly Lie homomorphisms; the nontrivial question is whether all Lie homomorphisms can be described in terms of such maps.

The answer to these questions is partially given in the following theorems from [2]. The technique used, so-called *Functional Identities*, was developed in course of solution of the famous Herstein problems about Lie homomorphisms of associative algebras (see [15–17]). At this time, the best source is the monograph by M. Brešar, M. Chebotar' and W. Martindale III [18].

We call a unital associative algebra over a field $\mathbb{K}$ *central* if its center equals $\mathbb{K} \cdot 1$.

Theorem 3.2. *Let A be a central simple algebra such that* $\dim_F A \geq 64$. *Let H be a unital commutative algebra. If A is not unital, then assume that H is finite dimensional. Then every surjective Lie homomorphism $\rho : [A, A] \otimes H \to [A, A] \otimes H$ can be extended to a direct sum of a homomorphism and the negative of an antihomomorphism $\sigma : A \otimes H \to A \otimes H$.*

Theorem 3.3. *Let* $\operatorname{char} F \neq 2$ *and let A be a central simple algebra such that* $\dim_F A \geq 441$. *Suppose that A has an involution and set $K = K_A$. Let H be a unital commutative algebra. If A is not unital, assume that H is finite dimensional. Then every surjective Lie homomorphism $\rho : [K, K] \otimes H \to [K, K] \otimes H$ can be extended to a homomorphism $\sigma : A \otimes H \to A \otimes H$.*

These transfer theorems and the techniques of affine group schemes have been successfully applied and the final classification of abelian group gradings on classical simple Lie algebras of types except D_4 over algebraically closed fields of characteristic other than 2 (in one case, different from 3) was obtained in [4]. The case of D_4 was accomplished by A. Elduque (see the details in [19]).

3.3 Case of Finitary Simple Algebras

Theorems 3.2 and 3.3 work well for infinite-dimensional algebras. For example, one can apply them to the case of Lie algebras L, which are direct limits of algebras $\{L_i = \mathfrak{sl}(n_i, \mathbb{K}) \mid k \in \mathbb{N}\}$, where each L_i is embedded in L_{i+1} by a diagonal embedding $X \to \operatorname{diag}(X, \ldots, X)$. In this case the unital algebra A is the direct limit of $\{A_i = M_{n_i}(\mathbb{K}) \mid k \in \mathbb{N}\}$, with the same embeddings. One could also consider the orthogonal or symplectic Lie algebras of skew-symmetric elements in A with respect to appropriate involutions. One drawback is that these theorems give no answer in the case of gradings of non-unital simple algebras by infinite groups, which is exactly the case when one deals with simple Lie algebras of finitary linear transformations. These algebras are direct limit of algebras of the same kind as before but the embedding is given by setting $X \to \operatorname{diag}(X, 0)$. Let us recall a few definitions.

An infinite-dimensional simple Lie algebra L of linear operators on an infinite-dimensional space over a field $\mathbb{K}$ is called *finitary* if L consists of linear operators of finite rank. These algebras were classified in [13] over any $\mathbb{K}$ with $\operatorname{char} \mathbb{K} = 0$ and in [14] over an algebraically closed $\mathbb{K}$ with $\operatorname{char} \mathbb{K} \neq 2, 3$. Under the latter assumption, finitary simple Lie algebras over $\mathbb{K}$ can be described in the following way.

Let U be an infinite-dimensional vector space over $\mathbb{K}$. Let $\Pi \subset U^*$ be a *total* subspace, i.e., for any $v \neq 0$ in U there is $f \in \Pi$ such that $f(v) \neq 0$. Let $\mathfrak{F}_\Pi(U)$ be the space spanned by the linear operators of the form $v \otimes f$, $v \in U$ and $f \in \Pi$, defined by $(v \otimes f)(u) = f(u)v$ for all $u \in U$. It is known from [23] that $R = \mathfrak{F}_\Pi(U)$ is a (non-unital) simple associative algebra.

The commutator $[R, R]$ is a simple Lie algebra, which is denoted by $\mathfrak{fsl}(U, \Pi)$. The algebra R admits an ($\mathbb{K}$-linear) involution if and only if there is a nondegenerate bilinear form $\Phi\colon U \times U \to \mathbb{K}$ that identifies U with Π and that is either symmetric or skew-symmetric. The set of skew-symmetric elements with respect to this involution, i.e., the set of $r \in R$ satisfying $\Phi(ru, v) + \Phi(u, rv) = 0$ for all $u, v \in U$, is a simple Lie algebra, which is denoted by $\mathfrak{fso}(U, \Phi)$ if Φ is symmetric and $\mathfrak{fsp}(U, \Phi)$ if Φ is skew-symmetric.

In [14] it is shown that if $\mathbb{K}$ is algebraically closed and char $\mathbb{K} \neq 2, 3$, then any finitary simple Lie algebra over $\mathbb{K}$ is isomorphic to one of $\mathfrak{fsl}(U, \Pi)$, $\mathfrak{fso}(U, \Phi)$ or $\mathfrak{fsp}(U, \Phi)$. The most important special case is that of countable (infinite) dimension. Then U has countable dimension and the isomorphism class of the Lie algebra does not depend on Π or Φ. Hence, there are exactly three finitary simple Lie algebras of countable dimension: $\mathfrak{sl}(\infty)$, $\mathfrak{so}(\infty)$ and $\mathfrak{sp}(\infty)$.

In the paper [3] we have given a complete classification of the abelian group gradings on finitary simple Lie algebras. This classification was obtained using the transfer to the case of associative algebras. Since the authors did not want to restrict themselves to the case of finite grading groups, as suggested by Theorems 3.2 or 3.3, they proved the following result.

To state the results, we set $\mathscr{A}^\sharp = \mathscr{A} + \mathbb{K}1$; thus, $\mathscr{A}^\sharp = \mathscr{A}$ if $1 \in \mathscr{A}$. If X is a subset of an associative algebra, then by $\langle X \rangle$ we denote the subalgebra generated by X.

Theorem 3.4. *Let $\mathscr{A}$ be any algebra satisfying $\mathfrak{F}_\Pi(U) \subset \mathscr{A} \subset \operatorname{End}(U)$, where U is infinite-dimensional, and let $\mathscr{L}$ be a noncentral Lie ideal of $\mathscr{A}$. If $\mathscr{H}$ is a unital commutative associative algebra, $\mathscr{M}$ is a Lie ideal of some associative algebra, and $\rho\colon \mathscr{M} \to \mathscr{L} \otimes \mathscr{H}$ is a surjective Lie homomorphism, then there exist a direct sum of a homomorphism and the negative of an antihomomorphism $\sigma\colon \langle \mathscr{M} \rangle \to \mathscr{A}^\sharp \otimes \mathscr{H}$ and a linear map $\tau\colon \mathscr{M} \to 1 \otimes \mathscr{H}$ such that $\tau([\mathscr{M}, \mathscr{M}]) = 0$ and $\rho(x) = \sigma(x) + \tau(x)$ for all $x \in \mathscr{M}$.*

If $\mathscr{A}$ is any associative algebra with an involution φ, then by $\mathscr{K}(\mathscr{A}, \varphi)$ we denote the Lie algebra of skew-symmetric elements in $\mathscr{A}$:

$$\mathscr{K}(\mathscr{A}, \varphi) = \{a \in \mathscr{A} \mid \varphi(a) = -a\}.$$

Theorem 3.5. *Let $\mathscr{A}$ be any algebra satisfying $\mathfrak{F}_\Pi(U) \subset \mathscr{A} \subset \operatorname{End}(U)$, where U is infinite-dimensional and the characteristic of the ground field is other than 2. Assume that $\mathscr{A}$ has an involution φ and let $\mathscr{L}$ be a noncentral Lie ideal of $\mathscr{K}(\mathscr{A}, \varphi)$. If $\mathscr{H}$ is a unital commutative associative algebra, $\mathscr{U}$ is an associative algebra with involution $*$, $\mathscr{M}$ is a Lie ideal of $\mathscr{K}(\mathscr{U}, *)$, and $\rho\colon \mathscr{M} \to \mathscr{L} \otimes \mathscr{H}$ is a surjective Lie homomorphism, then there exist a homomorphism $\sigma\colon \langle \mathscr{M} \rangle \to \mathscr{A}^\sharp \otimes \mathscr{H}$ and a linear map $\tau\colon \mathscr{M} \to 1 \otimes \mathscr{H}$ such that $\tau([\mathscr{M}, \mathscr{M}]) = 0$ and $\rho(x) = \sigma(x) + \tau(x)$ for all $x \in \mathscr{M}$.*

Note that in the case of simple algebras the map τ is always zero.

4 Group Gradings on Algebras of Finitary Linear Transformations

For the transfer approach to work, one needs to classify the gradings by groups on associative simple finitary algebras. In this section the grading group G does not need to be abelian.

Now the basic idea about the gradings in the case of finite-dimensional associative algebras (even graded Artinian algebras) is as follows. Let us call an algebra *graded simple* if it has no nontrivial proper ideals which are graded subspaces. Let us call a G-graded algebra a *graded division algebra* if any nonzero homogeneous element is invertible. An algebra is called graded left Artinian if it satisfies the descending chain condition for its left ideals. A well-known adaptation of the classical Wedderburn–Artin Theorem (see [25]) is the following.

Theorem 4.1. *Every G-graded left Artinian graded simple algebra A is isomorphic to an algebra of endomorphisms* $\mathrm{End}_D(V)$ *of a G-graded right vector space V over a graded division algebra D.*

Basically the same result holds in the case of simple associative algebras or finitary linear transformations, which belong to a wider class of primitive algebras with minimal ideals.

As shown in [23, IV.9], R is a primitive algebra with minimal one-sided ideals if and only if it is isomorphic to a subalgebra of $\mathfrak{L}_\Pi(U)$ containing the ideal $\mathfrak{F}_\Pi(U)$ where U is a right vector space over a division algebra Δ, Π is a total subspace of the left vector space U^* over Δ, $\mathfrak{L}_\Pi(U)$ is the algebra of all continuous Δ-linear operators on U, and $\mathfrak{F}_\Pi(U)$ is the set of all operators in $\mathfrak{L}_\Pi(U)$ whose image has finite dimension over Δ. The term *continuous* refers here to the topology on U with a neighbourhood basis at 0 consisting of the sets of the form $\ker f_1 \cap \ldots \cap \ker f_k$ where $f_1, \ldots, f_k \in \Pi$ and $k \in \mathbb{N}$. A linear operator $\mathscr{A} : U \to U$ is continuous with respect to this topology if and only if the adjoint operator $\mathscr{A}^* : U^* \to U^*$ leaves the subspace Π invariant. (In particular, if Δ is $\mathbb{R}$ and $\mathbb{C}$, U is a Banach space and Π consists of all bounded linear functionals on U, then a linear operator on U is continuous in our sense if and only if it is bounded.)

In our paper [3], we show that if an algebra R as above is given a grading by a group G, then it becomes a graded primitive algebra with minimal one-sided graded ideals. To state the graded analogue of the quoted result from [23, IV.9], let us remind that a linear transformation $f : M \to N$ of G-graded vector spaces is said to be *homogeneous of degree h* if $f(M_g) \subset N_{hg}$, for all $g \in G$. Thus, the set $\mathrm{Hom}^{\mathrm{gr}}(M, N)$ of finite sums of homogeneous maps from M to N is a G-graded vector space. A homomorphism of graded vector spaces is a homogeneous map of degree e.

4.1 Graded Division Algebras

In the case where $\mathbb{K}$ is algebraically closed and we consider a G-grading on $R = \mathfrak{F}_\Pi(U)$, which is a locally finite simple algebra with minimal ideals, we have that in the presentation of $R \cong \mathfrak{F}_W^{gr}(V)$ in Theorem 4.4, the graded division algebra is isomorphic to a matrix algebra $M_\ell(\mathbb{K})$, as an ungraded algebra. If we restrict ourselves to the case where G is abelian (this is all we need when we deal with simple Lie algebras!) such graded division algebras have been described in [10] and classified up to isomorphism [4]. The following statement can be found in [19].

Theorem 4.2. *Let T be a finite abelian group and let $\mathbb{K}$ be an algebraically closed field. There exists a grading on the matrix algebra $M_n(\mathbb{K})$ with support T making $M_n(\mathbb{K})$ a graded division algebra if and only if* char $\mathbb{K}$ *does not divide n and $T \cong \mathbb{Z}_{\ell_1}^2 \times \cdots \times \mathbb{Z}_{\ell_r}^2$ where $\ell_1 \cdots \ell_r = n$. The isomorphism classes of such gradings are in one-to-one correspondence with nondegenerate alternating bicharacters $\beta\colon T \times T \to \mathbb{K}^\times$. All such gradings belong to one equivalence class.* □

A *standard realization* can be obtained as follows.

If R is finite-dimensional or a finitary algebra, then D is finite-dimensional and $T = \operatorname{Supp} D$ is a finite subgroup of G. Every homogeneous component of D is one-dimensional and, one can easily see that D is isomorphic to a group algebra of T twisted by a 2-cocycle $\sigma : T \times T \to \mathbb{K}^\times$, usually denoted by $\mathbb{K}^\sigma T$. This means that we can choose a basis X_t in each D_t, $t \in T$ and then we will have $X_t X_u = \sigma(t, u) X_{tu}$. Let us introduce an alternating bicharacter $\beta = \beta_\sigma$ by setting $\beta(t, u) = \sigma(t, u)\sigma(u, t)^{-1}$. This depends only on the cohomology class of σ, which defines the isomorphism class of $\mathbb{K}^\sigma T$. The simplicity of $\mathbb{K}^\sigma T$ is then equivalent to the nondegeneracy of β.

Suppose there exists a nondegenerate alternating bicharacter β on T. One easily shows that T admits a "symplectic basis", i.e., there exists a decomposition of T as the direct product of cyclic subgroups:

$$T = H_1' \times H_1'' \times \cdots \times H_r' \times H_r'' \tag{3}$$

such that $H_i' \times H_i''$ and $H_j' \times H_j''$ are β-orthogonal for $i \neq j$, and H_i' and H_i'' are in duality by β.

4.2 Pauli Gradings on Matrix Algebras

Given a matrix algebra $D = M_n(\mathbb{K})$ over a field $\mathbb{K}$ possessing a primitive nth root of 1, one can define a division grading by the group $T = \langle a\rangle_n \times \langle b\rangle_n \cong \mathbb{Z}_n^2$ on D, as follows. Let

(4)
$$X = X(n,\varepsilon) = \begin{bmatrix} \varepsilon^{n-1} & 0 & 0 \dots 0\,0 \\ 0 & \varepsilon^{n-2} & 0 \dots 0\,0 \\ \dots & & \\ 0 & 0 & 0 \dots \varepsilon\,0 \\ 0 & 0 & 0 \dots 0\,1 \end{bmatrix} \text{ and } Y = Y(n) = \begin{bmatrix} 0\,1\,0 \dots 0\,0 \\ 0\,0\,1 \dots 0\,0 \\ \dots \\ 0\,0\,0 \dots 0\,1 \\ 1\,0\,0 \dots 0\,0 \end{bmatrix}$$

be called *generalized Pauli matrices*. Clearly, X and Y are periodic matrices of order n each. Moreover, $YX = \varepsilon^{-1}XY$. For any $t = a^k b^l \in T$, we set $X_t = X^k Y^l$. There are n^2 of such different matrices, they form a basis of R and setting $D_t = \mathbb{K}X_t$, for each $t \in T$ turns D into a graded algebra. We call this grading a *Pauli grading* of $D = M_n(\mathbb{K})$ and denote by $\Pi(n,\varepsilon)$. Since for any $t \in T$ every nonzero element of D_t is invertible, D is a graded division algebras.

Using Pauli gradings allows us to describe the classes $[\sigma]$ such that β_σ is nondegenerate and the isomorphisms from $\mathbb{K}^\sigma T$ onto a matrix algebra, as follows. Denote by ℓ_i the order of H_i' and H_i''. (We may assume without loss of generality that ℓ_i are prime powers.) If we pick generators a_i and b_i for H_i' and H_i'', respectively, then $\varepsilon_i = \beta(a_i, b_i)$ is a primitive ℓ_i-th root of unity, and all other values of β on the elements $a_1, b_1, \dots, a_r, b_r$ are 1. We can scale the elements X_{a_i} and X_{b_i} so that $X_{a_i}^{\ell_i} = X_{b_i}^{\ell_i} = 1$. Then we consider the Kronecker product $M_{\ell_1}(\mathbb{K}) \otimes \cdots \otimes M_{\ell_r}(\mathbb{K})$ of matrix algebras $M_{\ell_i}(\mathbb{K})$, each with a Pauli grading $\Pi(\ell_i, \varepsilon_i)$. The degree of the product $X_1 \otimes \cdots \otimes X_r$ is equal to the product of the degrees of the factors. Then we map the generators X_{t_i} of $\mathbb{K}^\sigma T$, $t_i \in T_i$, $i = 1, \dots, r$, as follows:

(5)
$$X_{t_i} \mapsto I \otimes \cdots I \otimes X_{t_i} \otimes I \otimes \cdots I,$$

where the only nonzero factor on the ith position is the generalized Pauli matrix from the definition of the Pauli grading $\Pi(\ell_i, \varepsilon_i)$. Comparing defining relations and the dimensions shows that this is an isomorphism of graded algebras.

If we scale X_{a_i} and X_{b_i}, as above, to have $X_{a_i}^{\ell_i} = X_{b_i}^{\ell_i} = 1$, and set $X_{(a_1^{i_1}, b_1^{j_1}, \dots, a_r^{i_r}, b_r^{j_r})} = X_{a_1}^{i_1} X_{b_1}^{j_1} \cdots X_{a_r}^{i_r} X_{b_r}^{j_r}$, then

$$X_{(a_1^{i_1}, b_1^{j_1}, \dots, a_r^{i_r}, b_r^{j_r})} X_{(a_1^{i_1'}, b_1^{j_1'}, \dots, a_r^{i_r'}, b_r^{j_r'})} = \varepsilon_1^{-j_1 i_1'} \cdots \varepsilon_r^{-j_r i_r'} X_{(a_1^{i_1+i_1'}, b_1^{j_1+j_1'}, \dots, a_r^{i_r+i_r'}, b_r^{j_r+j_r'})}.$$

Hence, with this choice of X_t, we obtain a representative of the cohomology class $[\sigma]$ that is multiplicative in each variable, i.e., it is a *bicharacter* (not alternating unless T is trivial).

Summarizing, we derive the following.

Theorem 4.3. *If a matrix algebra $R = M_n(\mathbb{K})$, $\mathbb{K}$ an algebraically closed field, is turned into a G-graded division algebra, then there are $l_1, \dots, l_r$, $n = l_1 \cdots l_r$, a subgroup $T \cong \mathbb{Z}_{l_1}^2 \oplus \cdots \oplus \mathbb{Z}_{l_r}^2$ and $\varepsilon_1, \dots, \varepsilon_r$, where ε_i is an l_ith root of 1, for* *each $i = 1, \ldots, r$, such that R is isomorphic as a graded algebra to the graded Kronecker product $M_{\ell_1}(\mathbb{K}) \otimes \cdots \otimes M_{\ell_r}(\mathbb{K})$ of matrix algebras, on each of which the grading is Pauli.*

In many arguments, however, it is convenient to simply use that any finite-dimensional G-graded division algebra D is a twisted group algebra of a group T, with basis X_t, $t \in T$, $D_t = \mathbb{K}X_t$ and $X_tX_{t'} = \sigma(t,t')X_{tt'}$ for some 2-cocycle $\sigma\colon T \times T \to \mathbb{K}^\times$. The isomorphism classes of such gradings are in bijection with the pairs (T, β) where $T \subset G$ is a subgroup of order ℓ^2 and $\beta\colon T \times T \to \mathbb{K}^\times$ is a nondegenerate alternating bicharacter. Here T is the support of the grading and β is given by $\beta(t,t') = \sigma(t,t')/\sigma(t',t)$, so we get

$$X_tX_{t'} = \beta(t,t')X_{t'}X_t \quad \text{for all} \quad t, t' \in T.$$

Note that char $\mathbb{K}$ cannot divide the order of T.

4.3 Graded Primitive Algebras with Minimal Graded Left Ideals

As mentioned in the introduction, we want to study gradings on the associative algebras of the form $\mathfrak{F}_\Pi(U)$ where U is a vector space over $\mathbb{K}$ and $\Pi \subset U^*$ is a total subspace. A natural class to which these algebras belong is the primitive algebras with minimal left ideals. In fact, the simple algebras in this class are precisely the algebras $\mathfrak{F}_\Pi(U)$ where U is a right vector space over a division algebra and $\Pi \subset U^*$ is a total subspace. We are now going to develop a graded version of the theory in [23, Chapter IV], which will apply to the setting we are interested in—thanks to the following lemma. All algebras in this section will be associative, but not necessarily unital. Unless indicated otherwise, the ground field $\mathbb{K}$ is arbitrary. (In fact, some results will also apply to rings.)

Lemma 4.1. *Let R be a primitive algebra (or ring) with minimal left ideals. Suppose that R is given a grading by a group G. Then R is graded primitive with minimal graded left ideals.*

Proof. Any nonzero left ideal of R is a faithful module because R is prime. Now consider R as a graded left R-module. According to [25, Proposition 2.7.3], the graded socle $S^{\mathrm{gr}}(R)$ always contains the ordinary socle $S(R)$. Hence R must contain a minimal graded left ideal I. Since I is a faithful graded simple R-module, R is graded primitive. □

4.4 A Structure Theorem

Let R be a G-graded algebra (or ring) and let V be a graded simple faithful left module over R. Let $D = \mathrm{End}_R^{\mathrm{gr}}(V)$. We will follow the standard convention of writing the elements of D on the right. By a version of Schur's Lemma (see e.g.,

[25, Proposition 2.7.1]), D is a graded division algebra, and thus V is a graded right vector space over D. Hence R is isomorphic to a graded subalgebra of $\mathrm{End}_D^{\mathrm{gr}}(V)$.

The *graded dual* $V^{\mathrm{gr}\,*}$ is defined as $\mathrm{Hom}_D^{\mathrm{gr}}(V, D)$. Note that $V^{\mathrm{gr}\,*}$ is a graded left vector space over D, with the action of D defined by $(df)(v) = df(v)$ for all $d \in D$, $v \in V$ and $f \in V^{\mathrm{gr}\,*}$. We will often use the symmetric notation (f, v) for $f(v)$.

Let $W \subset V^{\mathrm{gr}\,*}$ be a total graded subspace, i.e., the restriction of the mapping $(\, , \,)$ is a nondegenerate D-bilinear form $W \times V \to D$. Recall the graded subalgebras $\mathfrak{L}_W^{\mathrm{gr}}(V)$ and $\mathfrak{F}_W^{\mathrm{gr}}(V)$ of $\mathrm{End}_D^{\mathrm{gr}}(V)$ defined in the introduction: $\mathfrak{L}_W^{\mathrm{gr}}(V)$ is the set of all operators in $\mathrm{End}_D^{\mathrm{gr}}(V)$ that are continuous with respect to the topology induced by W and the subalgebra $\mathfrak{F}_W^{\mathrm{gr}}(V) \subset \mathfrak{L}_W^{\mathrm{gr}}(V)$ consists of all finite sums of operators of the form $v \otimes w$ where $v \in V$ and $w \in W$ are homogeneous and $v \otimes w$ acts on V as follows:

$$(v \otimes w)(u) = v(w, u) \quad \text{for all} \quad u, v \in V \text{ and } w \in W.$$

It is easy to see that V is a graded simple module for $\mathfrak{F}_W^{\mathrm{gr}}(V)$ and hence for $\mathfrak{L}_W^{\mathrm{gr}}(V)$ or any graded subalgebra of $\mathrm{End}_D^{\mathrm{gr}}(V)$ containing $\mathfrak{F}_W^{\mathrm{gr}}(V)$.

Suppose that R has a minimal graded left ideal I. Then either $I^2 = 0$ or $I = R\varepsilon$ where ε is a homogeneous idempotent (hence of degree e). Indeed, if $I^2 \neq 0$, then there is a homogeneous $x \in I$ such that $Ix \neq 0$ and hence $Ix = I$. The left annihilator L of x in R is a graded left ideal of R such that $I \not\subset L$ and hence $L \cap I = 0$. Let ε be an element of I such that $\varepsilon x = x$. Since x is homogeneous and every homogeneous component of ε is in I, we may assume that ε is homogeneous of degree e. Now $\varepsilon^2 x = \varepsilon x = x$ and hence $\varepsilon^2 - \varepsilon \in L$. Since $I \cap L = 0$, we conclude that $\varepsilon^2 = \varepsilon$. Since $Ix \neq 0$, it follows that $I\varepsilon \neq 0$ and hence $I\varepsilon = I$. Since we are assuming R graded primitive and hence graded prime, the case $I^2 = 0$ is not possible. Hence $I = R\varepsilon$ is a graded simple R-module.

The following is a graded version of a result in [23, III.5].

Lemma 4.2. *Let R be a graded primitive algebra (or ring) with a minimal graded left ideal I. Let V be a faithful graded simple left R-module. Then there exists $g \in G$ such that V is isomorphic to $I^{[g]}$ which is I, with the grading right shifted by g, as a graded R-module. Hence, all faithful graded simple left R-modules are isomorphic up to a right shift of grading.*

Proof. Since IV is a graded submodule of V, we have either $IV = 0$ or $IV = V$. But V is faithful, so $IV = V$. Pick a homogeneous $v \in V$ such that $Iv \neq 0$ and let $g = \deg v$. Then the map $I \to V$ given by $r \mapsto rv$ is a homomorphism of R-modules and sends I_h to V_{hg}, $h \in G$. By graded simplicity of I and V, this map is an isomorphism of R-modules. Hence $I^{[g]}$ is isomorphic to V as a graded R-module. □

Now we will obtain a graded version of the structure theorem in [23, IV.9].

Theorem 4.4. *Let R be a G-graded algebra (or ring). Then R is graded primitive with minimal graded left ideals if and only if there exists a G-graded division algebra D, a graded right vector space V over D and a total graded subspace $W \subset V^{\mathrm{gr}\,*}$ such that R is isomorphic to a graded subalgebra (subring) of $\mathfrak{L}_W^{\mathrm{gr}}(V)$ containing $\mathfrak{F}_W^{\mathrm{gr}}(V)$. Moreover, $\mathfrak{F}_W^{\mathrm{gr}}(V)$ is the only such subalgebra (subring) that is graded simple.*

4.5 An Isomorphism Theorem

We are going to investigate under what conditions two graded simple algebras described by Theorem 4.4 are isomorphic. By Lemma 4.2, V is determined by R up to isomorphism and shift of grading. If $\varphi \in \mathrm{End}_R^{\mathrm{gr}} V$ is homogeneous of degree t, then φ regarded as a map $V^{[g]} \to V^{[g]}$ will be homogeneous of degree $g^{-1}tg$. Indeed, $(V_h)\varphi \subset V_{ht}$, for all $h \in G$, can be rewritten as $(V_{hg}^{[g]})\varphi \subset V_{htg}^{[g]}$. Hence if $D = \mathrm{End}_R^{\mathrm{gr}}(V)$, then $\mathrm{End}_R^{\mathrm{gr}}(V^{[g]}) = {}^{[g^{-1}]}D^{[g]}$. It remains to include the total graded subspace $W \subset V^{\mathrm{gr}\,*}$ in the picture. It is convenient to introduce the following terminology. Fix a group G.

Definition 4.1. Let D and D' be G-graded division algebras (or rings) and let V and V' be graded right vector spaces over D and D', respectively. Let $\psi_0 \colon D \to D'$ be an isomorphism of G-graded algebras (or rings). A homomorphism $\psi_1 \colon V \to V'$ of G-graded vector spaces over $\mathbb{K}$ (or G-graded abelian groups) is said to be ψ_0-*semilinear* if $\psi_1(vd) = \psi_1(v)\psi_0(d)$ for all $v \in V$ and $d \in D$. The *adjoint* to ψ_1 is the mapping $\psi_1^* : (V')^{\mathrm{gr}\,*} \to V^{\mathrm{gr}\,*}$ defined by $(\psi_1^*(f))(v) = \psi_0^{-1}(f(\psi_1(v)))$ for all $f \in (V')^{\mathrm{gr}\,*}$ and $v \in V$.

One easily checks that ψ_1^* is ψ_0^{-1}-semilinear.

Definition 4.2. Let D and D' be G-graded division algebras (or rings), let V and V' be graded right vector spaces over D and D', respectively, and let W and W' be total graded subspaces of $V^{\mathrm{gr}\,*}$ and $(V')^{\mathrm{gr}\,*}$, respectively. An *isomorphism of triples* from (D, V, W) to (D', V', W') is a triple (ψ_0, ψ_1, ψ_2) where $\psi_0 \colon D \to D'$ is an isomorphism of graded algebras (or rings) while $\psi_1 \colon V \to V'$ and $\psi_2 \colon W \to W'$ are isomorphisms of graded vector spaces over $\mathbb{K}$ (or graded abelian groups) such that $(\psi_2(w), \psi_1(v)) = \psi_0((w, v))$ for all $v \in V$ and $w \in W$.

It follows that ψ_1 and ψ_2 are ψ_0-semilinear. Also, for given isomorphisms ψ_0 and ψ_1 there can exist at most one ψ_2 such that (ψ_0, ψ_1, ψ_2) is an isomorphism of triples. Such ψ_2 exists if and only if ψ_1 is ψ_0-semilinear and $\psi_1^*(W') = W$. Indeed, we can take ψ_2 to be the restriction of $(\psi_1^*)^{-1}$ to W. The condition $\psi_1^*(W') = W$ means that $\psi_1 \colon V \to V'$ is a homeomorphism with respect to the topologies induced by W and W'.

The following is a graded version of the isomorphism theorem in [23, IV.11].

Theorem 4.5. *Let G be a group. Let D and D' be G-graded division algebras (or rings), let V and V' be graded right vector spaces over D and D', respectively, and let W and W' be total graded subspaces of $V^{\mathrm{gr}\,*}$ and $(V')^{\mathrm{gr}\,*}$, respectively. Let R and R' be G-graded algebras (or rings) such that*

$$\mathfrak{F}^{\mathrm{gr}}_W(V) \subset R \subset \mathfrak{L}^{\mathrm{gr}}_W(V) \quad \textit{and} \quad \mathfrak{F}^{\mathrm{gr}}_{W'}(V') \subset R' \subset \mathfrak{L}^{\mathrm{gr}}_{W'}(V').$$

If $\psi : R \to R'$ is an isomorphism of graded algebras, then there exist $g \in G$ and an isomorphism (ψ_0, ψ_1, ψ_2) from $({}^{[g^{-1}]}D^{[g]}, V^{[g]}, {}^{[g^{-1}]}W)$ to (D', V', W') such that

$$\psi(r) = \psi_1 \circ r \circ \psi_1^{-1} \quad \textit{for all} \quad r \in R. \tag{6}$$

If other $g' \in G$ and isomorphism $(\psi_0', \psi_1', \psi_2')$ from $({}^{[(g')^{-1}]}D^{[g']}, V^{[g']}, {}^{[(g')^{-1}]}W)$ to (D', V', W') define ψ as above, then there exists a nonzero homogeneous $d \in D$ such that $g' = g \deg d$, $\psi_0'(x) = d^{-1}\psi_0(x)d$ for all $x \in D$, $\psi_1'(v) = \psi_1(v)d$ for all $v \in V$, and $\psi_2'(w) = d^{-1}\psi_2(w)$ for all $w \in W$.

As a partial converse, if (ψ_0, ψ_1, ψ_2) is an isomorphism of triples as above, then setting $\psi(r) = \psi_1 \circ r \circ \psi_1^{-1}$ defines an isomorphism of G-graded algebras (or rings) $\psi : \mathfrak{L}^{\mathrm{gr}}_W(V) \to \mathfrak{L}^{\mathrm{gr}}_{W'}(V')$ such that $\psi(\mathfrak{F}^{\mathrm{gr}}_W(V)) = \mathfrak{F}^{\mathrm{gr}}_{W'}(V')$.

Note that it follows that any isomorphism of graded algebras $\psi : R \to R'$ extends to an isomorphism $\mathfrak{L}^{\mathrm{gr}}_W(V) \to \mathfrak{L}^{\mathrm{gr}}_{W'}(V')$ and restricts to an isomorphism $\mathfrak{F}^{\mathrm{gr}}_W(V) \to \mathfrak{F}^{\mathrm{gr}}_{W'}(V')$.

4.6 Graded Simple Algebras with Minimal Graded Left Ideals

Fix a grading group G. In view of Theorems 4.4 and 4.5, graded simple algebras (or rings) with minimal graded left ideals are classified by the triples (D, V, W) where D is a graded division algebra (or ring), V is a right graded vector space over D and W is a total graded subspace of $V^{\mathrm{gr}\,*}$. For a fixed D, the triples (D, V, W) can be classified up to isomorphism as follows.

Let T be the support of D and let $\Delta = D_e$. Clearly, T is a subgroup of G and Δ is a division algebra (or ring). Consider the set G/T of left cosets and the set $T\backslash G$ of right cosets of T in G. The map $A \to A^{-1}$ is a bijection between G/T and $T\backslash G$. Clearly, the graded right D-modules ${}^{[g]}D$ and ${}^{[h]}D$ are isomorphic if and only if $gT = hT$ (similarly for graded left D-modules). Any graded right D-module V is a direct sum of modules of this form, which can be grouped into isotypic components. Namely, $V_A = \bigoplus_{a\in A} V_a$ is the isotypic component of V corresponding to ${}^{[g]}D$ where $A = gT$. Note that V_g is a right Δ-module and $V_A \cong V_g \otimes_\Delta {}^{[g]}D$ as graded right D-modules. Select a left transversal S for T, i.e., a set of representatives for the left cosets of T, and let $\tilde{V}_A = V_g$ where g is the unique element of $A \cap S$. Let $\tilde{V} = \bigoplus_{A \in G/T} \tilde{V}_A$. Then $\tilde{V}$ is a right Δ-module and $V \cong \tilde{V} \otimes_\Delta D$ as ungraded right D-modules. We can recover the G-grading on V if we consider $\tilde{V}$ as graded by the set G/T. Similarly, any graded left D-module W can be encoded by a left Δ-module $\tilde{W}$ with a grading by the set $T \backslash G$.

Now observe that since T is the support of D, we have $(W_B, V_A) = 0$ for all $A \in G/T$ and $B \in T \backslash G$ with $BA \neq T$. It follows that $W_{A^{-1}}$ is a total graded subspace of $(V_A)^{\mathrm{gr}\,*}$. Selecting a left transversal S and using S^{-1} as a right transversal, we obtain $\tilde{V}$ and $\tilde{W}$ such that $(\tilde{W}_{A^{-1}}, \tilde{V}_A) \subset \Delta$ and

$$(\tilde{W}_B, \tilde{V}_A) = 0 \quad \text{for all} \quad A \in G/T \text{ and } B \in T \backslash G \text{ with } B \neq A^{-1}. \tag{7}$$

Hence, for each $A \in G/T$, the D-bilinear form $W \times V \to D$ restricts to a nondegenerate Δ-bilinear form $\tilde{W}_{A^{-1}} \times \tilde{V}_A \to \Delta$, which identifies $\tilde{W}_{A^{-1}}$ with a total subspace of the Δ-dual $(\tilde{V}_A)^*$.

Conversely, let $\tilde{V}$ be a right Δ-module that is given a grading by G/T and let $\tilde{W}$ be a left Δ-module that is given a grading by $T \backslash G$. Suppose $\tilde{W}_{A^{-1}}$ is identified with a total subspace of $(\tilde{V}_A)^*$ for each $A \in G/T$ or, equivalently, we have a nondegenerate Δ-bilinear form $\tilde{W} \times \tilde{V} \to \Delta$ that satisfies (7). For each $A \in G/T$, choose $g \in A$ and let $V_A = \tilde{V}_A \otimes_\Delta {}^{[g]}D$. Then V_A is a graded right D-module whose isomorphism class does not depend on the choice of g. Also, using the same g, let $W_{A^{-1}} = D^{[g]} \otimes_\Delta \tilde{W}_{A^{-1}}$. Set $V = \bigoplus_{A \in G/T} V_A$ and $W = \bigoplus_{B \in T \backslash G} W_B$. Extending the Δ-bilinear form $\tilde{W} \times \tilde{V} \to \Delta$ to a D-bilinear form $W \times V \to D$, we can identify W with a total graded subspace of $V^{\mathrm{gr}\,*}$. We will denote the corresponding G-graded algebra $\mathfrak{F}_W^{\mathrm{gr}}(V)$ by $\mathfrak{F}(G, D, \tilde{V}, \tilde{W})$.

Definition 4.3. We will write $(D, \tilde{V}, \tilde{W}) \sim (D', \tilde{V}', \tilde{W}')$ if there is an element $g \in G$ and an isomorphism $\psi_0 \colon {}^{[g^{-1}]}D^{[g]} \to D'$ of graded algebras such that, for any $A \in G/T$, there exists an isomorphism of triples from $(\Delta, \tilde{V}_A, \tilde{W}_{A^{-1}})$ to $(\Delta', \tilde{V}'_{Ag}, \tilde{W}'_{g^{-1}A^{-1}})$ whose component $\Delta \to \Delta'$ is the restriction of ψ_0. (Note that Ag is a left coset for $T' = g^{-1}Tg$.)

Corollary 4.1. *Let G be a group and let R be a G-graded algebra (or ring). If R is graded simple with minimal graded left ideals, then R is isomorphic to some $\mathfrak{F}(G, D, \tilde{V}, \tilde{W})$. Two graded algebras $\mathfrak{F}(G, D, \tilde{V}, \tilde{W})$ and $\mathfrak{F}(G, D', \tilde{V}', \tilde{W}')$ are isomorphic if and only if we have $(D, \tilde{V}, \tilde{W}) \sim (D', \tilde{V}', \tilde{W}')$.*

Proof. The first claim is clear by Theorem 4.4 and the above discussion. Definition 4.4 is set up in such a way that $(D, \tilde{V}, \tilde{W}) \sim (D', \tilde{V}', \tilde{W}')$ if and only if the triples $({}^{[g^{-1}]}D^{[g]}, V^{[g]}, {}^{[g^{-1}]}W)$ and (D', V', W') are isomorphic for some $g \in G$, so the second claim follows from Theorem 4.5. □

An important special case of Corollary 4.1 is where the graded simple algebra R satisfies the descending chain condition on graded left ideals. Then V is finite-dimensional over D, so $W = V^{\mathrm{gr}\,*} = V^*$ and R is isomorphic to the matrix algebra $M_k(D)$ where $k = \dim_D V$. Moreover, $V = V_{A_1} \oplus \cdots \oplus V_{A_s}$ for some distinct $A_1, \ldots, A_s \in G/T$, which can be encoded by two s-tuples: $\kappa = (k_1, \ldots, k_s)$ and $\gamma = (g_1, \ldots, g_s)$ where $k_i = \dim_D V_{A_i}$ are positive integers with $k_1 + \cdots + k_s = k$ and g_i are representatives for the cosets A_i. Therefore, the said algebras R are classified by the triples (D, κ, γ), up to an appropriate equivalence relation. Explicitly, the grading on R can be written as follows. If $\{v_1, \ldots, v_k\}$ is a homogeneous basis of V over D, with $\deg v_i = h_i$, and $\{v^1, \ldots, v^k\}$ is the dual basis of V^*, then $\deg v^i = h_i^{-1}$ and, for any homogeneous $d \in D$, the degree of the operator $v_i d \otimes v^j = v_i \otimes d v^j$ (which is represented by the matrix with d in position (i, j) and zeros elsewhere) equals $h_i (\deg d) h_j^{-1}$. This classification (under the assumption that R is finite-dimensional over $\mathbb{K}$) already appeared in the literature—see [4] and references therein. The G-gradings on $M_k(D)$ defined in this way by k-tuples $(h_1, \ldots, h_k)$ of elements in G will be called *elementary*.

In general, $\mathfrak{F}(G, D, \tilde{V}, \tilde{W})$ can be written as $\tilde{V} \otimes_\Delta D \otimes_\Delta \tilde{W}$ where, for any $\tilde{v} \in \tilde{V}, d \in D, \tilde{w} \in \tilde{W}$, the element $\tilde{v} \otimes d \otimes \tilde{w}$ acts on $V = \tilde{V} \otimes_\Delta D$ as follows: $(\tilde{v} \otimes d \otimes \tilde{w})(u \otimes a) = \tilde{v} \otimes d(w, u)a$ for all $\tilde{u} \in \tilde{V}$ and $a \in D$. Clearly, the multiplication of $\mathfrak{F}(G, D, \tilde{V}, \tilde{W})$ is given by

$$(\tilde{v} \otimes d \otimes \tilde{w})(\tilde{v}' \otimes d' \otimes \tilde{w}') = \tilde{v} \otimes d(\tilde{w}, \tilde{v}')d' \otimes \tilde{w}'. \tag{8}$$

Fixing a left transversal S for T, the G-grading on $\mathfrak{F}(G, D, \tilde{V}, \tilde{W})$ is given by

$$\deg(\tilde{v} \otimes d \otimes \tilde{w}) = \gamma(A)t\gamma(B)^{-1} \quad \text{for all} \quad \tilde{v} \in \tilde{V}_A,\ d \in D_t,\ \tilde{w} \in \tilde{W}_{B^{-1}}, \tag{9}$$

where $t \in T$, $A, B \in G/T$, and $\gamma(A)$ denotes the unique element of $A \cap S$. The isomorphism class of the grading does not depend on the choice of the transversal.

It is known that for any finite-dimensional subspaces $\tilde{V}_1 \subset \tilde{V}$ and $\tilde{W}_1 \subset \tilde{W}$, there exist finite-dimensional subspaces $\tilde{V}_0 \subset \tilde{V}$ and $\tilde{W}_0 \subset \tilde{W}$ such that $\tilde{V}_1 \subset \tilde{V}_0$, $\tilde{W}_1 \subset \tilde{W}_0$, and the restriction of the bilinear form $\tilde{W} \times \tilde{V} \to \Delta$ to $\tilde{W}_0 \times \tilde{V}_0$ is nondegenerate (see e.g., [13, Lemma 5.7]). Selecting dual bases in $\tilde{V}_0$ and $\tilde{W}_0$, we see that $\tilde{V}_0 \otimes_\Delta D \otimes_\Delta \tilde{W}_0$ is a subalgebra of $\mathfrak{F}(G, T, \tilde{V}, \tilde{W})$ isomorphic to $M_k(D)$ where $k = \dim_\Delta \tilde{V}_0 = \dim_\Delta \tilde{W}_0$. Without loss of generality, we may assume that $\tilde{V}_0$ is a graded subspace of $\tilde{V}$ with respect to the grading by G/T and $\tilde{W}_0$ is a graded subspace of $\tilde{W}$ with respect to the grading by $T\backslash G$. Then our subalgebra $\tilde{V}_0 \otimes_\Delta D \otimes_\Delta \tilde{W}_0$ is graded. Moreover, in terms of the matrix algebra $M_k(D)$, this grading is elementary. Thus we obtain the following graded version of Litoff's Theorem [23, IV.15] (cf. Theorem 4 in [12]):

Corollary 4.2. *Let G be a group and let R be a G-graded algebra (or ring). If R is graded simple with minimal graded left ideals, then there exists a graded division algebra D such that R is a direct limit of matrix algebras over D with elementary gradings.* □

4.7 Classification of G-Gradings on the Algebras $\mathfrak{F}_\Pi(U)$

In this work, we are primarily interested in the case $R = \mathfrak{F}_\Pi(U)$ where U is a vector space over $\mathbb{K}$ and Π is a total subspace of U^*. We will assume that $\mathbb{K}$ is *algebraically closed*. Then the algebras of the form $\mathfrak{F}_\Pi(U)$ have the following abstract characterization: they are precisely the locally finite simple algebras with minimal left ideals. Indeed, $\mathfrak{F}_\Pi(U) = U \otimes \Pi$ is a direct limit of matrix algebras over $\mathbb{K}$ and hence is simple and locally finite. Conversely, if R is a locally finite simple algebra with minimal left ideals, then R is isomorphic to $\mathfrak{F}_\Pi(U)$ where U is a right vector space over a division algebra Δ and Π is a total subspace of U^*. But Δ is isomorphic to a subalgebra of R, hence algebraic over $\mathbb{K}$. Since $\mathbb{K}$ is algebraically closed, this implies $\Delta = \mathbb{K}$.

If R is given a G-grading, then R is graded simple with minimal graded left ideals (Lemma 4.1), so we can apply Corollary 4.1. Hence R is isomorphic to some $\mathfrak{F}(G, D, \tilde{V}, \tilde{W})$ as a graded algebra. We claim that, disregarding the grading, D is isomorphic to $M_\ell(\mathbb{K})$ for some ℓ.

Recall from the proof of Theorem 4.4 that we can represent R as $\mathfrak{F}_J^{\mathrm{gr}}(I)$ where $I = R\varepsilon$ is a minimal graded left ideal, ε is a homogeneous idempotent, and $J = \varepsilon R$. Recall also that $D = \mathrm{End}_R^{\mathrm{gr}}(I)$ coincides with $\mathrm{End}_R(I)$ and is isomorphic to $\varepsilon R\varepsilon$. It is known that R is semisimple as a left or right R-module (see e.g., [23, IV.9]). In fact, it is easy to see that if R is represented as $\mathfrak{F}_\Pi(U)$, then the mapping $U_0 \mapsto U_0 \otimes \Pi$ is a one-to-one correspondence between the subspaces of U and the right ideals of R whereas the mapping $\Pi_0 \mapsto U \otimes \Pi_0$ is a one-to-one correspondence between the subspaces of Π and the left ideals of R. Hence we can write $I = R\varepsilon_1 \oplus \cdots \oplus R\varepsilon_\ell$ where ε_i are orthogonal idempotents with $\varepsilon_1 + \cdots + \varepsilon_\ell = \varepsilon$ and $R\varepsilon_i$ are minimal (ungraded) left ideals. Each of the $R\varepsilon_i$ is isomorphic to U as a left R-module. Since $\mathrm{End}_R(U) = \mathbb{K}$, it follows that the algebra $\mathrm{End}_R(I)$ is isomorphic to $M_\ell(\mathbb{K})$, completing the proof of the claim.

If R is represented as $\mathfrak{F}_\Pi(U)$, we can construct this isomorphism explicitly. Namely, write $I = U \otimes \Pi_0$ where $\Pi_0 \subset \Pi$ is an ℓ-dimensional subspace and select $U_0 \subset U$ such that the restriction of the bilinear form $\Pi \times U \to \mathbb{K}$ to $\Pi_0 \times U_0$ is nondegenerate. Let $\{e_1, \ldots, e_\ell\}$ be a basis of U_0 and let $\{e^1, \ldots, e^\ell\}$ be the dual basis of Π_0. Then we can take $\varepsilon_i = e_i \otimes e^i$, so $R\varepsilon_i = U \otimes \mathbb{K}e^i$, and the elements $e_i \otimes e^j$ constitute a basis of matrix units for $\varepsilon R\varepsilon$.

As for the graded division algebra D, this is a matrix algebra which is given a G-grading that makes it a graded division algebra. We have described them completely in Sect. 4.2. In what follows we will use notation introduced therein.

We want to understand the relation between, on the one hand, $\tilde{V}$ and $\tilde{W}$ and, on the other hand, U and Π. We will use I as V and J as W. The mapping $u \mapsto u \otimes e^i$ is an isomorphism of left R-modules $U \to R\varepsilon_i$. Also, the mapping $f \mapsto e_i \otimes f$ is an isomorphism of right R-modules $\Pi \to \varepsilon_i R$. This allows us to identify I with U^ℓ and J with Π^ℓ. Recall that the D-bilinear form $J \times I \to D$ is just the multiplication of R. Hence, under the above identifications, this D-bilinear form maps $(f_1, \ldots, f_\ell) \in \Pi^\ell$ and $(u_1, \ldots, u_\ell) \in U^\ell$ to the matrix $[(f_i, u_j)]_{i,j}$ in D.

Let M be the unique simple right D-module and let N be the unique simple left D-module, i.e., M is $\mathbb{K}^\ell$ written as rows and N is $\mathbb{K}^\ell$ written as columns. Then, disregarding the G-gradings on I and J, we can identify I with $U \otimes M$ as an (R, D)-bimodule and also identify J with $N \otimes \Pi$ as a (D, R)-bimodule. Under these identifications, the D-bilinear form $J \times I \to D$ coincides with the extension of the $\mathbb{K}$-linear form $\Pi \times U \to \mathbb{K}$. Now, we have $U \cong I \otimes_D N$ and $\Pi \cong M \otimes_D J$ as R-modules. If we identify U with $I \otimes_D N$ and Π with $M \otimes_D J$, then the $\mathbb{K}$-bilinear form $\Pi \otimes U \to \mathbb{K}$ is related to the D-bilinear form $J \times I \to D$ by the following formula:

$$(m \otimes y, x \otimes n) = m(y, x)n \quad \text{for all} \quad m \in M, n \in N, x \in I, y \in J, \tag{10}$$

where the right-hand side is the scalar in $\mathbb{K}$ obtained by multiplying a row, a matrix and a column. Recall that $\tilde{V}$ and $\tilde{W}$ associated to V and W are defined in such a way that $V = \tilde{V} \otimes D$ and $W = D \otimes \tilde{W}$ as ungraded D-modules, and the D-bilinear form $W \times V \to D$ is the extension of the $\mathbb{K}$-bilinear form $\tilde{W} \times \tilde{V} \to \mathbb{K}$ (recall that $D_e = \mathbb{K}$). Hence $U = V \otimes_D N = (\tilde{V} \otimes D) \otimes_D N \cong \tilde{V} \otimes N$ and $\Pi = M \otimes_D W = M \otimes_D (D \otimes \tilde{W}) \cong M \otimes \tilde{W}$ as vector spaces over $\mathbb{K}$, where the isomorphism $(\tilde{V} \otimes D) \otimes_D N \cong \tilde{V} \otimes N$ is given by $\tilde{v} \otimes d \otimes n \mapsto \tilde{v} \otimes dn$ and the isomorphism $M \otimes_D (D \otimes \tilde{W}) \to M \otimes \tilde{W}$ is given by $m \otimes d \otimes \tilde{w} \mapsto md \otimes \tilde{w}$. Substituting $x = \tilde{v} \otimes a$ and $y = b \otimes \tilde{w}$, for any $\tilde{v} \in \tilde{V}$, $\tilde{w} \in \tilde{W}$, $a, b \in D$, into (10), we obtain

$$(m \otimes b \otimes \tilde{w}, \tilde{v} \otimes a \otimes n) = m(b \otimes \tilde{w}, \tilde{v} \otimes a)n = mb(\tilde{w}, \tilde{v})an.$$

Hence, if we identify U with $\tilde{V} \otimes N$ and Π with $M \otimes \tilde{W}$, then the $\mathbb{K}$-bilinear forms $\Pi \times U \to \mathbb{K}$ and $\tilde{W} \times \tilde{V} \to \mathbb{K}$ are related by the following formula:

$$(m \otimes \tilde{w}, \tilde{v} \otimes n) = (\tilde{w}, \tilde{v})mn \quad \text{for all} \quad m \in M, n \in N, \tilde{v} \in \tilde{V}, \tilde{w} \in \tilde{W}. \tag{11}$$

In other words, we can identify U with $\tilde{V}^\ell$ and Π with $\tilde{W}^\ell$ so that the above $\mathbb{K}$-bilinear forms are related as follows:

$$((\tilde{w}_1, \dots, \tilde{w}_\ell), (\tilde{v}_1, \dots, \tilde{v}_\ell)) = \sum_{i=1}^{\ell} (\tilde{w}_i, \tilde{v}_i) \quad \text{for all} \quad \tilde{v}_i \in \tilde{V} \text{ and } \tilde{w}_i \in \tilde{W}. \tag{12}$$

Finally, we observe that Definition 4.3 simplifies because $D_e = D'_e = \mathbb{K}$. For brevity, an isomorphism of triples from $(\mathbb{K}, U, \Pi)$ to $(\mathbb{K}, U', \Pi')$ will be called an *isomorphism of pairs* from (U, Π) to (U', Π'). In other words, pairs (U, Π) and (U', Π') are isomorphic if and only if there exists an isomorphism $U \to U'$ of vector spaces (over $\mathbb{K}$) whose adjoint $(U')^* \to U^*$ maps Π' onto Π.

Definition 4.4. We will write $(D, \tilde{V}, \tilde{W}) \sim (D', \tilde{V}', \tilde{W}')$ if there is an element $g \in G$ such that ${}^{[g^{-1}]}D^{[g]} \cong D'$ as graded algebras and, for any $A \in G/T$, we have $(\tilde{V}_A, \tilde{W}_{A^{-1}}) \cong (\tilde{V}'_{Ag}, \tilde{W}'_{g^{-1}A^{-1}})$.

To summarize:

Theorem 4.6. *Let R be a locally finite simple algebra with minimal left ideals over an algebraically closed field $\mathbb{K}$. If R is given a grading by a group G, then R is isomorphic to some $\mathfrak{F}(G, D, \tilde{V}, \tilde{W})$ where D is a matrix algebra over $\mathbb{K}$ equipped with a division grading. Conversely, if $D = M_\ell(\mathbb{K})$ with a division grading, then $\mathfrak{F}(G, D, \tilde{V}, \tilde{W})$ is a locally finite simple algebra with minimal left ideals, which can be represented as $\mathfrak{F}_\Pi(U)$ where $U = \tilde{V}^\ell$, $\Pi = \tilde{W}^\ell$, and the nondegenerate bilinear form $\Pi \times U \to \mathbb{K}$ is given by* (12). *Moreover, two such graded algebras $\mathfrak{F}(G, D, \tilde{V}, \tilde{W})$ and $\mathfrak{F}(G, D', \tilde{V}', \tilde{W}')$ are isomorphic if and only if $(D, \tilde{V}, \tilde{W}) \sim (D', \tilde{V}', \tilde{W}')$ in the sense of Definition 4.4.* □

If G is abelian, then ${}^{[g^{-1}]}D^{[g]} = D$ and the isomorphism class of D is determined by (T, β). Hence we may write $\mathfrak{F}(G, T, \beta, \tilde{V}, \tilde{W})$ for $\mathfrak{F}(G, D, \tilde{V}, \tilde{W})$.

Definition 4.5. We will write $(\tilde{V}, \tilde{W}) \sim (\tilde{V}', \tilde{W}')$ if there is an element $g \in G$ such that for any $A \in G/T$, we have $(\tilde{V}_A, \tilde{W}_{A^{-1}}) \cong (\tilde{V}'_{Ag}, \tilde{W}'_{g^{-1}A^{-1}})$.

Corollary 4.3. *If in Theorem 4.6 the group G is abelian, then R is isomorphic to some $\mathfrak{F}(G, T, \beta, \tilde{V}, \tilde{W})$. Two such graded algebras $\mathfrak{F}(G, T, \beta, \tilde{V}, \tilde{W})$ and $\mathfrak{F}(G, T', \beta', \tilde{V}', \tilde{W}')$ are isomorphic if and only if $T = T'$, $\beta = \beta'$ and $(\tilde{V}, \tilde{W}) \sim (\tilde{V}', \tilde{W}')$ in the sense of Definition 4.5.* □

In the case where R has countable dimension, we can classify G-gradings on R in combinatorial terms. Clearly, $\mathfrak{F}_\Pi(U)$ has countable dimension if and only if both Π and U have countable dimension. It is known that then there exist dual bases in U and Π, hence all such pairs (U, Π) are isomorphic and there is only one such algebra R, which is denoted by $M_\infty(\mathbb{K})$. We will state the classification of G-gradings in a form that also applies to $M_n(\mathbb{K})$, for which this result is known under the assumption that G is abelian [4]. Up to isomorphism, the pairs $(\tilde{V}_A, \tilde{W}_{A^{-1}})$ can be encoded by the function $\kappa\colon G/T \to \{0, 1, 2, \ldots, \infty\}$ that sends A to $\dim \tilde{V}_A$. Note that the support of the function κ is finite or countable, so $|\kappa| = \sum_{A \in G/T} \kappa(A)$ is defined as an element of $\{0, 1, 2, \ldots, \infty\}$. We will denote the associated graded algebra $\mathfrak{F}(G, D, \tilde{V}, \tilde{W})$ by $\mathfrak{F}(G, D, \kappa)$. Finally, for any $g \in G$, define $\kappa^g\colon G/(g^{-1}Tg) \to \{0, 1, 2, \ldots, \infty\}$ by setting $\kappa^g(Ag) = \kappa(A)$ for all $A \in G/T$.

Corollary 4.4. *Let $\mathbb{K}$ be an algebraically closed field and let $R = M_n(\mathbb{K})$ where $n \in \mathbb{N} \cup \{\infty\}$. If R is given a grading by a group G, then R is isomorphic to some $\mathfrak{F}(G, D, \kappa)$ where $D = M_\ell(\mathbb{K})$, with $\ell \in \mathbb{N}$ and $n = |\kappa|\ell$, is equipped with a division grading. Moreover, two such graded algebras $\mathfrak{F}(G, D, \kappa)$ and $\mathfrak{F}(G, D', \kappa')$ are isomorphic if and only if there exists $g \in G$ such that ${}^{[g^{-1}]}D^{[g]} \cong D'$ as graded algebras and $\kappa^g = \kappa'$.* □

The graded algebra $\mathfrak{F}(G, D, \kappa)$ can be constructed explicitly as follows. Select a left transversal S for T and let $\gamma(A)$ be the unique element of $A \cap S$. For each $A \in G/T$, select dual bases $\{v_i(A)\}$ and $\{v^j(A)\}$ for $\tilde{V}_A$ and $\tilde{W}_A$, respectively, consisting each of $\kappa(A)$ vectors. Then the algebra $\mathfrak{F}(G, D, \kappa)$ has a basis

$$\{E_{i,j}^{A,B}(t) \mid A, B \in G/T,\ i, j \in \mathbb{N}, i \leq \kappa(A), j \leq \kappa(B),\ t \in T\},$$

where $E_{i,j}^{A,B}(t) = v_i(A)X_t \otimes v^j(B) = v_i(A) \otimes X_t v^j(B)$. Equations (8) and (9) imply that the multiplication is given by the formula (using Kronecker delta):

$$E_{i,j}^{A,B}(t)E_{i',j'}^{A',B'}(t') = \delta_{B,A'}\delta_{j,i'}\sigma(t,t')E_{i,j'}^{A,B'}(tt')$$

and the G-grading is given by

$$\deg E_{i,j}^{A,B}(t) = \gamma(A)t\gamma(B)^{-1}.$$

Thus we recover Theorem 5 in [12], which asserts the existence of such a basis under the assumption that G is a finite abelian group and $\mathbb{K}$ is algebraically closed of characteristic zero. Theorem 6 in the same paper gives our condition for isomorphism of two G-gradings in the special case $D = \mathbb{K}$.

4.8 Antiautomorphisms and Sesquilinear Forms

We want to investigate under what conditions a graded algebra described by Theorem 4.4 admits an antiautomorphism. So, we temporarily return to the general setting: R is a G-graded primitive algebra (or ring) with minimal graded left ideals.

We may assume $\mathfrak{F}_W^{\mathrm{gr}}(V) \subset R \subset \mathfrak{L}_W^{\mathrm{gr}}(V)$ where V is a right vector space over a graded division algebra D, W is a left vector space over D, and W is identified with a total graded subspace of $V^{\mathrm{gr}\,*}$ by virtue of a D-bilinear form $(\ ,\ )$. Thus, we have

(13)
$$(dw, v) = d(w, v) \quad \text{and} \quad (w, vd) = (w, v)d \quad \text{for all} \quad v \in V,\ w \in W,\ d \in D.$$

Note that, since the adjoint of any operator $r \in R$ leaves W invariant, W becomes a graded right R-module such that

$$(wr, v) = (w, rv) \quad \text{for all} \quad v \in V,\ w \in W,\ r \in R. \tag{14}$$

It follows that $\mathfrak{F}_V^{\mathrm{gr}}(W) \subset R^{\mathrm{op}} \subset \mathfrak{L}_V^{\mathrm{gr}}(W)$ where W is regarded as a right vector space over D^{op}, V as a left vector space over D^{op}, and V is identified with a total graded subspace of $W^{\mathrm{gr}\,*}$ by virtue of $(v, w)^{\mathrm{op}} = (w, v)$. (The gradings on all these objects are by the group G^{op}.)

Now suppose that we have an antiautomorphism φ of the graded algebra R. Since $\mathscr{S} = \mathfrak{F}_W^{\mathrm{gr}}(V)$ is the unique minimal graded two-sided ideal of R, we have $\varphi(\mathscr{S}) = \mathscr{S}$. Thus φ restricts to an antiautomorphism of the graded simple algebra $\mathscr{S}$. It is known (see [9]) that if a graded simple algebra admits an antiautomorphism, then the support of the grading generates an abelian group. Now observe that any element of the support of $\mathrm{End}_D^{\mathrm{gr}}(V)$ has the form gh^{-1} where g and h are in the support of V, and all elements of this form already occur in the support of $\mathfrak{F}_W^{\mathrm{gr}}(V)$. Hence, the support of R equals the support of $\mathscr{S}$ and generates an abelian group. For this reason, we will assume from now on that G is *abelian*.

Applying Theorem 4.5 to the isomorphism $\varphi\colon R \to R^{\mathrm{op}}$ and taking into account that G is abelian, we see that there exist $g_0 \in G$ and an isomorphism $(\varphi_0, \varphi_1, \varphi_2)$ from $(D, V^{[g_0]}, W^{[g_0^{-1}]})$ to (D^{op}, W, V) such that $\varphi(r) = \varphi_1 \circ r \circ \varphi_1^{-1}$. In particular, φ_0 is an antiautomorphism of the graded algebra D and φ_1 is φ_0-semilinear:

(15) $$\varphi_1(vd) = \varphi_0(d)\varphi(v) \quad \text{for all} \quad v \in V \text{ and } d \in D.$$

Now define a nondegenerate $\mathbb{K}$-bilinear form $B\colon V \times V \to D$ as follows:

$$B(u, v) = (\varphi_1(u), v) \quad \text{for all} \quad u, v \in V.$$

Then B has degree g_0 when regarded as a map $V \otimes V \to D$. Combining (13) and (15), we see that, over D, the form B is linear in the second argument and φ_0-semilinear in the first argument, i.e.,

(16) $$B(ud, v) = \varphi_0(d)B(u, v) \quad \text{and} \quad B(u, vd) = B(u, v)d \quad \text{for all} \quad u, v \in V,\ d \in D.$$

For brevity, we will say that B is φ_0-*sesquilinear.*

Applying (14), we obtain for all $u, v \in V$ and $r \in R$:

$$\begin{aligned} B(ru, v) &= (\varphi_1(ru), v) = (\varphi_1(u)\varphi(r), v) \\ &= (\varphi_1(u), \varphi(r)v) = B(u, \varphi(r)v), \end{aligned}$$

which means that $\varphi(r)$ is *adjoint to r with respect to B*. In particular, φ can be recovered from B.

We will need one further property of B. Consider

$$\bar{B}(u, v) = \varphi_0^{-1}(B(v, u)).$$

Then $\bar{B}$ is a nondegenerate φ_0^{-1}-sesquilinear form of the same degree as B. Clearly, we have $\bar{B}(ru, v) = \bar{B}(u, \varphi^{-1}(r)v)$, so $\bar{B}$ is related to φ^{-1} in the same way as B is related to φ. We claim that there exists a φ_0^{-2}-semilinear isomorphism of graded vector spaces $Q\colon V \to V$ such that

(17) $$\bar{B}(u, v) = B(Qu, v) \quad \text{for all} \quad u, v \in V.$$

Indeed, $Q = \varphi_1^{-1} \circ \varphi_2^{-1}$ satisfies the requirements: it is clearly an invertible φ_0^{-2}-semilinear map, homogeneous of degree e, and we have

$$B(Qu, v) = (\varphi_2^{-1}(u), v) = (v, \varphi_2^{-1}(u))^{\mathrm{op}} = \varphi_0^{-1}((\varphi_1(v), u)) = \bar{B}(u, v).$$

It is important to note that the adjoint $Q^* = (\varphi_2^{-1})^* \circ (\varphi_1^{-1})^* = \varphi_1 \circ \varphi_2$ maps W onto W, i.e., Q is a homeomorphism.

Definition 4.6. We will say that a nondegenerate homogeneous φ_0-sesquilinear form $B\colon V \times V \to D$ is *weakly Hermitian* if there exists a φ_0^{-2}-semilinear isomorphism $Q\colon V \to V$ of graded vector spaces such that (17) holds. (Note that, since B is nondegenerate, (17) uniquely determines Q.)

The following is a graded version of the main result in [23, IV.12].

Theorem 4.7. *Let G be an abelian group. Let D be a G-graded division algebra (or ring), let V be a graded right vector space over D, and let W be a total graded subspace of $V^{\mathrm{gr}\,*}$. Let R be a G-graded algebra (or ring) such that*

$$\mathfrak{F}_W^{\mathrm{gr}}(V) \subset R \subset \mathfrak{L}_W^{\mathrm{gr}}(V).$$

If φ is an antiautomorphism of the graded algebra R, then there exist an antiautomorphism φ_0 of the graded algebra D and a weakly Hermitian nondegenerate homogeneous φ_0-sesquilinear form $B : V \times V \to D$, such that the following conditions hold:

(a) *The mapping $V \to V^{\mathrm{gr}\,*}\colon u \mapsto f_u$, where $f_u(v) = B(u, v)$ for all $v \in V$, sends V onto W;*
(b) *For any $r \in R$, $\varphi(r)$ is the adjoint to r with respect to B, i.e., $B(ru, v) = B(u, \varphi(r)v)$, for all $u, v \in V$.*

If φ_0' is an antiautomorphism of D and B' is a φ_0'-sesquilinear form $V \times V \to D$ that define W and φ as in (a) and (b), then there exists a nonzero homogeneous $d \in D$ such that $B' = dB$ and $\varphi_0'(x) = d\varphi_0(x)d^{-1}$ for all $x \in D$.

As a partial converse, if φ_0 is an antiautomorphism of the graded algebra D and $B : V \times V \to D$ is a weakly Hermitian nondegenerate homogeneous φ_0-sesquilinear form, then the adjoint with respect to B defines an antiautomorphism φ of the G-graded algebra $\mathfrak{L}_W^{\mathrm{gr}}(V)$, with $W = \{f_u \mid u \in V\}$, such that $\varphi(\mathfrak{F}_W^{\mathrm{gr}}(V)) = \mathfrak{F}_W^{\mathrm{gr}}(V)$.

Proof. Given an antiautomorphism φ, the existence of the pair (φ_0, B) is already proved. If (φ_0', B') is another such pair, then the corresponding mapping $u \mapsto f_u'$ is an isomorphism of graded R-modules $V^{[g_0']} \to W^{\varphi}$. Hence $\varphi_1' \circ \varphi_1^{-1}$ is a nonzero homogeneous element of $\mathrm{End}_R^{\mathrm{gr}}(W)$, so there exists a nonzero homogeneous $d \in D$ such that $\varphi_1'(v) = d\varphi_1(v)$ for all $v \in V$, which implies $B' = dB$. Now the equation $\varphi_0'(x) = d\varphi_0(x)d^{-1}$ follows easily from (16).

Conversely, for a given antiautomorphism φ_0 and a form B of degree g_0, define $\varphi_1\colon V \to W$ by setting $\varphi_1(u) = f_u$. This is a homogeneous φ_0-semilinear isomorphism of degree g_0. Take $\varphi_2 = Q^{-1} \circ \varphi_1^{-1}$. Then one checks that $(\varphi_0, \varphi_1, \varphi_2)$ is an isomorphism from $(D, V^{[g_0]}, W^{[g_0^{-1}]})$ to (D^{op}, W, V), so Theorem 4.5 tells us that $\varphi(r) = \varphi_1 \circ r \circ \varphi_1^{-1}$ defines an isomorphism of graded algebras $\mathfrak{L}_W^{\text{gr}}(V) \to \mathfrak{L}_V^{\text{gr}}(W) = \mathfrak{L}_W^{\text{gr}}(V)^{\text{op}}$ that restricts to an isomorphism $\mathfrak{F}_W^{\text{gr}}(V) \to \mathfrak{F}_V^{\text{gr}}(W) = \mathfrak{F}_W^{\text{gr}}(V)^{\text{op}}$. It remains to observe that the definition of φ_1 implies that $\varphi(r)$ is the adjoint to r with respect to B, for any $r \in \mathfrak{L}_W^{\text{gr}}(V)$. □

Note that it follows that any antiautomorphism φ of the graded algebra R extends to an antiautomorphism of $\mathfrak{L}_W^{\text{gr}}(V)$ and restricts to an antiautomorphism of $\mathfrak{F}_W^{\text{gr}}(V)$.

Remark 4.1. It is easy to compute φ on $\mathfrak{F}_W^{\text{gr}}(V)$ explicitly. We have $\varphi(v \otimes f_u) = \varphi_2(f_u) \otimes \varphi_1(v)$, so, taking into account $\varphi_1(v) = f_v$ and $\varphi_2 = Q^{-1} \circ \varphi_1^{-1}$, we obtain

$$\varphi(v \otimes f_u) = Q^{-1}u \otimes f_v \quad \text{for all} \quad u, v \in V. \tag{18}$$

4.9 Antiautomorphisms that Are Involutive on the Identity Component

We restrict ourselves to the case where R is a locally finite simple algebra with minimal left ideals over an *algebraically closed* field $\mathbb{K}$. If R is given a grading by an abelian group G, then, by Theorem 4.6, R is isomorphic to some $\mathfrak{F}(G, D, \tilde{V}, \tilde{W})$ where D is a matrix algebra with a division grading. Suppose that the graded algebra R admits an antiautomorphism φ. Then, by Theorem 4.7, we obtain an antiautomorphism φ_0 for D. It is known [11] that this forces the support T of D to be an elementary 2-group and hence char $\mathbb{K} \neq 2$ or $T = \{e\}$. From now on, we assume char $\mathbb{K} \neq 2$.

Since G is abelian, any division grading on a matrix algebra can be realized using generalized Pauli matrices. If the support T is an elementary 2-group, then the matrix transpose preserves this grading. Choose a nonzero element X_t in each component D_t. Then the transpose of X_t equals $\beta(t)X_t$ where $\beta(t) \in \{\pm 1\}$. It is easy to check (see [4]) that $\beta\colon T \to \{\pm 1\}$ is a quadratic form on T if we regard it as a vector space over the field of two elements, with the nondegenerate alternating bicharacter $\beta\colon T \times T \to \mathbb{K}^\times$ being the associated bilinear form: $\beta(tt') = \beta(t)\beta(t')\beta(t, t')$ for all $t, t' \in T$. It is easy to see that any automorphism of the graded algebra D is a conjugation by some X_t. Hence φ_0 is given by $\varphi_0(X_{t'}) = \beta(t')X_t^{-1}X_{t'}X_t$ for some $t \in T$. In particular, φ_0 is an *involution*. Hence, the isomorphism Q associated to the φ_0-sesquilinear form B in Theorem 4.7 is linear over D and thus Q is an invertible element of the identity component of $\mathfrak{L}_W^{\text{gr}}(V)$. Adjusting B, we may assume without loss of generality that

$$\varphi_0(X_t) = \beta(t)X_t \quad \text{for all} \quad t \in T.$$

This convention makes the choice of B unique up to a scalar in $\mathbb{K}^\times$.

Assume that φ restricts to an involution on R_e. Then B has certain symmetry properties, which we are going to investigate now. In particular, B is *balanced*, i.e., for any pair of homogeneous $u, v \in V$, we have

$$B(u, v) = 0 \Leftrightarrow B(v, u) = 0.$$

Recall that $\mathfrak{F}(G, D, \tilde{V}, \tilde{W}) = \mathfrak{F}_W^{\mathrm{gr}}(V)$ where V and W are constructed from $\tilde{V}$ and $\tilde{W}$ as follows. Select a transversal S for T and, for each $A \in G/T$, set $V_A = \tilde{V}_A \otimes D$ and $W_A = D \otimes \tilde{W}_A$, with the degree of the elements of $\tilde{V}_A \otimes 1$ and $1 \otimes \tilde{W}_A$ set to be the unique element of $A \cap S$, which we denote by $\gamma(A)$. It will be convenient to identify $\tilde{V}_A$ with $\tilde{V}_A \otimes 1$ and $\tilde{W}_A$ with $1 \otimes \tilde{W}_A$.

Using the definition of $\bar{B}$ and Eq. (17), we compute, for all $u, v \in V$ and $r \in \mathfrak{L}_W^{\mathrm{gr}}(V)$:

$$\begin{aligned} B(u, \varphi^2(r)v) &= B(\varphi(r)u, v) = \varphi_0(\bar{B}(v, \varphi(r)u)) = \varphi_0(B(Qv, \varphi(r)u)) \\ &= \varphi_0(B(rQv, u)) = \bar{B}(u, rQv) = B(Qu, rQv). \end{aligned} \tag{19}$$

Substituting $r = 1$, we obtain $B(u, v) = B(Qu, Qv)$ for all $u, v \in V$ and hence $B(Qu, v) = B(u, Q^{-1}v)$. Continuing with (19) we obtain, for all $u, v \in V$, $B(u, \varphi^2(r)v) = B(u, Q^{-1}rQv)$. Therefore,

$$\varphi^2(r) = Q^{-1}rQ \quad \text{for all} \quad r \in R. \tag{20}$$

Observe that the identity component R_e is the direct sum of subalgebras R^A, $A \in G/T$, where R^A consists of all operators in R_e that map the isotypic component V_A into itself and other isotypic components to zero. Clearly, R^A is spanned by the operators of the form $w \otimes v$ where $v \in \tilde{V}_A$ and $w \in \tilde{W}_{A^{-1}}$. Being homogeneous of degree e, Q maps V_A onto V_A. The restriction of Q to V_A is linear over D and, by (20), commutes with all elements of R^A. It follows that Q acts on V_A as a scalar $\lambda_A \in \mathbb{K}^\times$. Now (17) implies that B is balanced, as claimed.

The fact that B is balanced allows us to define the concept of orthogonality for homogeneous elements and for graded subspaces of V. Since B is homogeneous of degree, say, g_0, we have, for all $u \in V_{g_1}$ and $v \in V_{g_2}$, that $B(u, v) = 0$ unless $g_0 g_1 g_2 \in T$. It follows that V_A is orthogonal to all isotypic components except $V_{g_0^{-1}A^{-1}}$, and hence the restriction of B to $V_A \times V_{g_0^{-1}A^{-1}}$ is nondegenerate. It will be important to distinguish whether or not A equals $g_0^{-1}A^{-1}$.

If $g_0 A^2 = T$, then the element $g_0\gamma(A)^2 \in T$ does not depend on the choice of the transversal and will be denoted by $\tau(A)$. The restriction of B to $V_A \times V_A$ is a nondegenerate φ_0-sesquilinear form over D. It is uniquely determined by its restriction to $\tilde{V}_A \times \tilde{V}_A$, which is a bilinear form over $\mathbb{K}$ with values in $D_{\tau(A)}$. Set

$$B(u, v) = \tilde{B}_A(u, v)X_{\tau(A)} \quad \text{for all} \quad u, v \in \tilde{V}_A \text{ where } g_0 A^2 = T. \tag{21}$$

Then $\tilde{B}_A$ is a nondegenerate bilinear form on $\tilde{V}_A$ with values in $\mathbb{K}$. Setting $t = \tau(A)$ for brevity, we compute:

$$\tilde{B}_A(v,u)X_t = B(v,u) = \varphi_0(B(Qu,v)) = \varphi_0(B(\lambda_A u,v))$$
$$= \varphi_0(\lambda_A \tilde{B}_A(u,v)X_t) = \lambda_A \tilde{B}_A(u,v)\varphi_0(X_t) = \lambda_A \beta(t)\tilde{B}_A(u,v)X_t,$$

so $\tilde{B}_A(v,u) = \lambda_A\beta(t)\tilde{B}_A(u,v)$.

If $g_0A^2 \neq T$, then we may assume without loss of generality that the transversal is chosen so that

(22) $$g_0\gamma(A)\gamma(g_0^{-1}A^{-1}) = e.$$

Then the restrictions of B to $\tilde{V}_A \times \tilde{V}_{g_0^{-1}A^{-1}}$ and to $\tilde{V}_{g_0^{-1}A^{-1}} \times \tilde{V}_A$ are nondegenerate bilinear forms with values in $D_e = \mathbb{K}$. Denote them by $\tilde{B}_A$ and $\tilde{B}_{g_0^{-1}A^{-1}}$, respectively, i.e., set

(23)
$$B(u,v) = \tilde{B}_A(u,v)1 \quad \text{for all} \quad u \in \tilde{V}_A \text{ and } v \in \tilde{V}_{g_0^{-1}A^{-1}} \text{ where } g_0A^2 \neq T.$$

It is easy to see how $\tilde{B}_A$ and $\tilde{B}_{g_0^{-1}A^{-1}}$ are related: $\tilde{B}_{g_0^{-1}A^{-1}}(v,u) = \lambda_A\tilde{B}_A(u,v)$.

Putting all pieces together, we set $\tilde{V}_{\{A,g_0^{-1}A^{-1}\}} = \tilde{V}_A \oplus \tilde{V}_{g_0^{-1}A^{-1}}$ and

(24) $$\tilde{V} = \bigoplus_{A\in G/T,\, g_0A^2=T} \tilde{V}_A \oplus \bigoplus_{\{A,g_0^{-1}A^{-1}\}\subset G/T,\, g_0A^2\neq T} \tilde{V}_{\{A,g_0^{-1}A^{-1}\}},$$

and define a nondegenerate bilinear form $\tilde{B}\colon \tilde{V} \times \tilde{V} \to \mathbb{K}$ so that all summands in (24) are orthogonal to each other, the restriction of $\tilde{B}$ to $\tilde{V}_A \times \tilde{V}_A$ is $\tilde{B}_A$ if $g_0A^2 = T$ while the restriction of $\tilde{B}$ to $\tilde{V}_A$ is zero and the restriction to $\tilde{V}_A \times \tilde{V}_{g_0^{-1}A^{-1}}$ is $\tilde{B}_A$ if $g_0A^2 \neq T$.

Conversely, let $\tilde{V}$ be a vector space over $\mathbb{K}$ that is given a grading by G/T and let $\tilde{B}$ be a nondegenerate bilinear form on $\tilde{V}$ that is compatible with the grading in the sense that

$$\tilde{B}(\tilde{V}_{g_1T}, \tilde{V}_{g_2T}) = 0 \quad \text{for all} \quad g_1, g_2 \in G \text{ with } g_0g_1g_2T \neq T$$

and, for all $A \in G/T$, satisfies the following symmetry condition:

(25) $$\tilde{B}(v,u) = \mu_A\tilde{B}(u,v) \text{ for all } u \in \tilde{V}_A \text{ and } v \in \tilde{V}_{g_0^{-1}A^{-1}} \text{ where } \mu_A \in \mathbb{K}^\times.$$

It follows that $\mu_A\mu_{g_0^{-1}A^{-1}} = 1$. Hence, if $g_0A^2 = T$, then $\tilde{V}$ restricts to a symmetric or a skew-symmetric form on $\tilde{V}_A$. For any $A \in G/T$, let $\tilde{B}_A$ be the restriction of

$\tilde{B}$ to $\tilde{V}_A \times \tilde{V}_{g_0^{-1}A^{-1}}$. It follows that $\tilde{B}_A$ is nondegenerate. Choose a transversal S for T so that (22) holds for all A with $g_0A^2 \neq T$. Set $V_A = \tilde{V}_A \otimes {}^{[\gamma(A)]}D$. Then V_A is a graded right D-module whose isomorphism class does not depend on the choice of S. Set $V = \bigoplus_{A\in G/T} V_A$. Define $B\colon V \times V \to D$ using (21) and (23), setting B equal to zero in all other cases and then extending by φ_0-sesquilinearity. Clearly, B is nondegenerate. Set $W = \{f_u \mid u \in V\}$ where $f_u(v) = B(u, v)$. We will denote the corresponding G-graded algebra $\mathfrak{F}^{\mathrm{gr}}_W(V)$ by $\mathfrak{F}(G, D, \tilde{V}, \tilde{B}, g_0)$ or $\mathfrak{F}(G, T, \beta, \tilde{V}, \tilde{B}, g_0)$, since D is determined by the support T and bicharacter β. The graded algebra $\mathfrak{F}^{\mathrm{gr}}_W(V)$ has an antiautomorphism φ defined by the adjoint with respect to B. Indeed, let $Q\colon V \to V$ act on V_A as the scalar λ_A where

$$\lambda_A = \begin{cases} \mu_A\beta(\tau(A)) & \text{if } \; g_0A^2 = T; \\ \mu_A & \text{if } \; g_0A^2 \neq T. \end{cases} \tag{26}$$

Then Q satisfies $\bar{B}(u, v) = B(Qu, v)$ for all $u, v \in V$ and hence B is weakly Hermitian. Since Q commutes with the elements of R_e, Eq. (20) tells us that φ^2 is the identity on R_e.

Definition 4.7. With fixed D and φ_0, we will write $(\tilde{V}, \tilde{B}, g_0) \sim (\tilde{V}', \tilde{B}', g_0')$ if there is an element $g \in G$ such that $g_0' = g_0g^{-2}$ and, for any $A \in G/T$ with $g_0A^2 = T$, we have $\tilde{V}_A \cong \tilde{V}'_{gA}$ as inner product spaces while, for any $A \in G/T$ with $g_0A^2 \neq T$, we have $(\tilde{V}_A, \tilde{V}_{g_0^{-1}A^{-1}}) \cong (\tilde{V}'_{gA}, \tilde{V}'_{(g_0')^{-1}g^{-1}A^{-1}})$ and $\mu_A = \mu'_{gA}$ (where μ_A is defined by (25) and μ'_A by the same equation with $\tilde{B}$ replaced by $\tilde{B}'$).

Recall that disregarding the grading, R can be represented as $\mathfrak{F}_\Pi(U)$ with $U = \tilde{V} \otimes N$ and $\Pi = M \otimes \tilde{W}$, where M and N are the natural right and left modules for $D = M_\ell(\mathbb{K})$, respectively (see the analysis preceding Theorem 4.6). In our case, $W = \{f_u \mid u \in V\}$ and $\tilde{W} = \{f_{\tilde{u}} \mid \tilde{u} \in \tilde{V}\}$. Note that $f_{\tilde{u}}(\tilde{v}) = B(\tilde{u}, \tilde{v})$ does not necessarily belong to $D_e = \mathbb{K}$ for all $\tilde{u}, \tilde{v} \in \tilde{V}$, so Eq. (11) for the $\mathbb{K}$-bilinear form $\Pi \times U \to \mathbb{K}$ should be modified as follows:

$$(m \otimes \tilde{u}, \tilde{v} \otimes n) = mB(\tilde{u}, \tilde{v})n \quad \text{for all} \quad m \in M, n \in N, \; \tilde{u}, \tilde{v} \in \tilde{V}. \tag{27}$$

If we identify U and Π with $\tilde{V}^\ell$, then the above $\mathbb{K}$-bilinear form is given by

(28)

$$((\tilde{u}_1, \dots, \tilde{u}_\ell), (\tilde{v}_1, \dots, \tilde{v}_\ell)) = \sum_{i,j=1}^{\ell} x_{ij}\tilde{B}(\tilde{u}_i, \tilde{v}_j) \;\text{ for all }\; \tilde{u}_i \in \tilde{V}_{A_i} \;\text{ and }\; \tilde{v}_j \in \tilde{V}_{A'_j},$$

where x_{ij} is the (i, j)-entry of the matrix $X_{\tau(A)}$ if $A_i = A'_j = A$ with $g_0A^2 = T$ and $x_{ij} = \delta_{i,j}$ otherwise.

Now we are ready to state the result:

Theorem 4.8. *Let R be a locally finite simple algebra with minimal left ideals over an algebraically closed field $\mathbb{K}$, $\operatorname{char}\mathbb{K} \neq 2$. If R is given a grading by an abelian group G and an antiautomorphism φ that preserves the grading and restricts to an involution on R_e, then (R, φ) is isomorphic to some $\mathfrak{F}(G, D, \tilde{V}, \tilde{B}, g_0)$ where D is a matrix algebra over $\mathbb{K}$ equipped with a division grading and an involution φ_0 given by matrix transpose. Conversely, if $D = M_\ell(\mathbb{K})$ with a division grading and φ_0 is the matrix transpose, then $\mathfrak{F}(G, D, \tilde{V}, \tilde{W})$ is a locally finite simple algebra with minimal left ideals, which can be represented as $\mathfrak{F}_U(U)$ where $U = \tilde{V}^\ell$ and the nondegenerate bilinear form $U \times U \to \mathbb{K}$ is given by* (28). *Two such graded algebras $\mathfrak{F}(G, D, \tilde{V}, \tilde{B}, g_0)$ and $\mathfrak{F}(G, D', \tilde{V}', \tilde{B}', g_0')$ are not isomorphic unless $D \cong D'$ as graded algebras, whereas for fixed D and φ_0, $\mathfrak{F}(G, D, \tilde{V}, \tilde{B}, g_0)$ and $\mathfrak{F}(G, D, \tilde{V}', \tilde{B}', g_0')$ are isomorphic as graded algebras with antiautomorphism if and only if $(\tilde{V}, \tilde{B}, g_0) \sim (\tilde{V}', \tilde{B}', g_0')$ in the sense of Definition 4.7.*

In the case where R has finite or countable dimension, i.e., $R = M_n(\mathbb{K})$ with $n \in \mathbb{N} \cup \{\infty\}$, we can express the classification in combinatorial terms. Since $\mathbb{K}$ is algebraically closed, two vector spaces with symmetric inner products are isomorphic if they have the same finite or countable dimension. The same is true for vector spaces with skew-symmetric inner product. Therefore, for $A \in G/T$ with $g_0 A^2 = T$, the isomorphism class of $\tilde{V}_A$ is encoded by μ_A and $\dim \tilde{V}_A$. For $A \in G/T$ with $g_0 A^2 \neq T$, the isomorphism class of $(\tilde{V}_A, \tilde{V}_{g_0^{-1}A^{-1}})$ is encoded by $\dim \tilde{V}_A = \dim \tilde{V}_{g_0^{-1}A^{-1}}$. We introduce functions $\mu\colon G/T \to \mathbb{K}^\times$ sending A to μ_A (where we set $\mu_A = 1$ if $\tilde{V}_A = 0$) and, as before, $\kappa\colon G/T \to \{0, 1, 2, \ldots, \infty\}$ sending A to $\dim \tilde{V}_A$. Recall that μ satisfies $\mu_A \mu_{g_0^{-1}A^{-1}} = 1$ for all $A \in G/T$ and κ has a finite or countable support. We will denote the associated graded algebra with antiautomorphism $\mathfrak{F}(G, D, \tilde{V}, \tilde{B}, g_0)$ by $\mathfrak{F}(G, D, \kappa, \mu, g_0)$ or by $\mathfrak{F}(G, T, \beta, \kappa, \mu, g_0)$. For a given elementary 2-subgroup $T \subset G$ and a bicharacter β, we fix a realization of D using Pauli matrices and thus fix an involution φ_0 on D. Finally, for any $g \in G$, define μ^g and κ^g by setting $\mu^g(gA) = \mu(A)$ and $\kappa^g(gA) = \kappa(A)$ for all $A \in G/T$.

Corollary 4.5. *Let $\mathbb{K}$ be an algebraically closed field, $\operatorname{char}\mathbb{K} \neq 2$, and let $R = M_n(\mathbb{K})$ where $n \in \mathbb{N} \cup \{\infty\}$. If R is given a grading by an abelian group G and an antiautomorphism φ that preserves the grading and restricts to an involution on R_e, then (R, φ) is isomorphic to some $\mathfrak{F}(G, D, \kappa, \mu, g_0)$ where $D = M_\ell(\mathbb{K})$, with $\ell \in \mathbb{N}$ and $n = |\kappa|\ell$, is equipped with a division grading and an involution φ_0 given by matrix transpose. Moreover, $\mathfrak{F}(G, T, \beta, \kappa, \mu, g_0)$ and $\mathfrak{F}(G, T', \beta', \kappa', \mu', g_0')$ are isomorphic as graded algebras with antiautomorphism if and only if $T' = T$, $\beta' = \beta$, and there exists $g \in G$ such that $g_0' = g_0 g^{-2}$, $\kappa' = \kappa^g$ and $\mu' = \mu^g$.* □

4.10 Graded Involutions

We can specialize Theorem 4.8 to obtain a classification of involutions on the graded algebra R. By (20), φ is an involution if and only if $Q\colon V \to V$ is a scalar operator, i.e., $\bar{B} = \lambda B$ for some $\lambda \in \mathbb{K}^\times$, which implies $\lambda \in \{\pm 1\}$. Disregarding the grading, $R = \mathfrak{F}_U(U)$ where U is an inner product space. Since φ_0 is matrix transpose, Eq. (27) implies that $(v, u) = \lambda(u, v)$ for all $u, v \in U$. Hence in the case $\bar{B} = B$, we obtain an *orthogonal involution* on R and write $\mathrm{sgn}(\varphi) = 1$, whereas in the case $\bar{B} = -B$, we obtain a *symplectic involution* and write $\mathrm{sgn}(\varphi) = -1$. Since all λ_A must be equal to $\lambda = \mathrm{sgn}(\varphi)$, Eq. (26) yields

$$\mu_A = \begin{cases} \mathrm{sgn}(\varphi)\beta(\tau(A)) & \text{if} \quad g_0 A^2 = T; \\ \mathrm{sgn}(\varphi) & \text{if} \quad g_0 A^2 \neq T. \end{cases} \tag{29}$$

We note that for $g_0 A^2 \neq T$, although the space $\tilde{V}_{\{A, g_0^{-1}A^{-1}\}}$ now has a symmetric or skew-symmetric inner product, the equivalence relation in Definition 4.7 requires more than just an isomorphism of inner product spaces: the isomorphism must respect the direct sum decomposition $\tilde{V}_{\{A, g_0^{-1}A^{-1}\}} = \tilde{V}_A \oplus \tilde{V}_{g_0^{-1}A^{-1}}$. To summarize:

Proposition 4.1. *Under the conditions of Theorem 4.8, φ is an involution if and only if $\tilde{B}$ satisfies the symmetry condition* (25) *where μ_A is given by* (29). □

In the case of finite or countable dimension, we again can reduce everything to combinatorial terms (for the finite case, this result appeared in [4]). Once $\delta = \mathrm{sgn}(\varphi)$ is specified, the function $\mu\colon G/T \to \mathbb{K}^\times$ is determined by (29), so we will denote the corresponding graded algebra with involution by $\mathfrak{F}(G, D, \kappa, \delta, g_0)$ or $\mathfrak{F}(G, T, \beta, \kappa, \delta, g_0)$.

Corollary 4.6. *Let $R = \mathfrak{F}(G, T, \beta, \kappa)$ where the ground field $\mathbb{K}$ is algebraically closed of characteristic different from 2, and G is an abelian group. The graded algebra R admits an involution φ with* $\mathrm{sgn}(\varphi) = \delta$ *if and only if T is an elementary 2-group and, for some $g_0 \in G$, we have $\kappa(A) = \kappa(g_0^{-1}A^{-1})$ for all $A \in G/T$ and we also have $\beta(g_0 a^2) = \delta$ for all $A = aT \in G/T$ such that $g_0 A^2 = T$ and $\kappa(A)$ is finite and odd. If φ is an involution on R with* $\mathrm{sgn}(\varphi) = \delta$, *then the pair (R, φ) is isomorphic to some $\mathfrak{F}(G, T, \beta, \kappa, \delta, g_0)$. Moreover, $\mathfrak{F}(G, T, \beta, \kappa, \delta, g_0)$ and $\mathfrak{F}(G, T', \beta', \kappa', \delta', g_0')$ are isomorphic as graded algebras with involution if and only if $T' = T$, $\beta' = \beta$, $\delta' = \delta$, and there exists $g \in G$ such that $g_0' = g_0 g^{-2}$ and $\kappa' = \kappa^g$.* □

5 Gradings on Lie Algebras of Finitary Linear Transformations

Throughout this section $\mathbb{K}$ is an algebraically closed field of characteristic different from 2, and G is an abelian group. Our goal is to classify G-gradings on the infinite-dimensional simple Lie algebras $\mathfrak{fsl}(U, \Pi)$, $\mathfrak{fso}(U, \Phi)$ and $\mathfrak{fsp}(U, \Phi)$ of finitary linear transformation over $\mathbb{K}$. We are going to transfer the classification results for the associative algebras $\mathfrak{F}_\Pi(U)$ in Sect. 4.3 to the above Lie algebras using the transfer Theorems 3.4 and 3.5.

We denote by $\mathscr{H}$ the group algebra $\mathbb{K}G$, which is a commutative and cocommutative Hopf algebra. A G-grading on an algebra $\mathscr{U}$ is equivalent to a (right) $\mathscr{H}$-comodule structure $\rho\colon \mathscr{U} \to \mathscr{U} \otimes \mathscr{H}$ that is also a homomorphism of algebras. Indeed, given a G-grading on $\mathscr{U}$, we set $\rho(x) = x \otimes g$ for all $x \in \mathscr{L}_g$ and extend by linearity. Conversely, given ρ, we obtain a G-grading on $\mathscr{U}$ by setting $\mathscr{U}_g = \{x \in \mathscr{U} \mid \rho(x) = x \otimes g\}$. We can extend ρ to a homomorphism $\mathscr{U} \otimes \mathscr{H} \to \mathscr{U} \otimes \mathscr{H}$ by setting $\rho(x \otimes h) = \rho(x)(1 \otimes h)$. It is easy to see that this extended homomorphism is surjective.

5.1 Special Linear Lie Algebras

Let $\mathscr{L} = \mathfrak{fsl}(U, \Pi)$ where U is infinite-dimensional and let $R = \mathfrak{F}_\Pi(U)$. Suppose $\mathscr{L}$ is given a G-grading and let $\rho\colon \mathscr{L} \otimes \mathscr{H} \to \mathscr{L} \otimes \mathscr{H}$ be the Lie homomorphism obtained by extending the comodule structure map. Then $\mathscr{L} = [R, R]$ is a noncentral Lie ideal of R and $\mathscr{M} = \mathscr{L} \otimes \mathscr{H}$ is a Lie ideal of $R \otimes \mathscr{H}$. Moreover, $\langle \mathscr{L} \rangle = R$ implies $\langle \mathscr{M} \rangle = R \otimes \mathscr{H}$, and $[\mathscr{L}, \mathscr{L}] = \mathscr{L}$ implies $[\mathscr{M}, \mathscr{M}] = \mathscr{M}$. Applying Theorem 3.4 and observing that $\tau = 0$, we conclude that ρ extends to a map $\rho'\colon R \otimes \mathscr{H} \to R \otimes \mathscr{H}$ which is a sum of a homomorphism and the negative of an antihomomorphism. Thus, there are central idempotents e_1 and e_2 in $\mathscr{H}$ (which can be identified with the center of $R^\sharp \otimes \mathscr{H}$) with $e_1 + e_2 = 1$ such that the composition of ρ' and the projection $R \otimes \mathscr{H} \to R \otimes e_1\mathscr{H}$ is a homomorphism, while the composition of ρ' and the projection $R \otimes \mathscr{H} \to R \otimes e_2\mathscr{H}$ is the negative of an antihomomorphism.

If ψ is any automorphism of $\mathscr{L}$, it can be extended to a map $\psi'\colon R \to R$ that is a homomorphism or the negative of an antihomomorphism (use [18, Theorem 6.19] or our Theorem 3.4 with $\mathscr{H} = \mathbb{K}$). Clearly, ψ' is surjective. Since R is simple, we conclude that ψ' is an automorphism or the negative of an antiautomorphism. We claim that ψ cannot admit extensions of both types. Indeed, assume that ψ' is an automorphism of R and ψ'' is the negative of an antiautomorphism of R such that both restrict to ψ. Let $\sigma = (\psi')^{-1}\psi''$. Then σ is the negative of an antiautomorphism of R that restricts to identity on $\mathscr{L}$. Hence, for any $x \in \mathscr{L}$ and $r \in R$, we have $[x, r] = \sigma([x, r]) = [\sigma(x), \sigma(r)] = [x, \sigma(r)]$. It follows that $r - \sigma(r)$ belongs to the center of R, which is zero, so σ is the identity map—a contradiction. We have shown that for any automorphism ψ of $\mathscr{L}$, there is a unique extension $\psi'\colon R \to R$ that is either an antiautomorphism or the negative of an antiautomorphism.

Now any character $\chi \in \hat{G}$ acts as an automorphism ψ of $\mathscr{L}$ defined by $\psi(x) = \chi(g)x$ for all $x \in \mathscr{L}_g$. Denote this ψ by $\eta(\chi)$, i.e., $\eta(\chi) = (\mathrm{id}_R \otimes \chi)\rho$. Clearly, η is a homomorphism from $\hat{G}$ to $\mathrm{Aut}\,(\mathscr{L})$. Define $\eta'(\chi) = \eta(\chi)'$. It follows from the uniqueness of extension that η' is a homomorphism from $\hat{G}$ to the group $\overline{\mathrm{Aut}}\,(R)$

consisting of all automorphisms and the negatives of antiautomorphisms of R. We can regard χ as a homomorphism of algebras $\mathscr{H} \to \mathbb{K}$ and define $\eta''(\chi) = (\mathrm{id}_R \otimes \chi)\rho'$. Then $\eta''(\chi)$ is a map $R \to R$ that restricts to $\eta(\chi)$ on $\mathscr{L}$. Clearly, $\chi(e_1)$ is either 1 or 0, and then $\chi(e_2)$ is either 0 or 1, respectively. An easy calculation shows that if $\chi(e_1) = 1$, then $\eta''(\chi)$ is a homomorphism, and if $\chi(e_1) = 0$, then $\eta''(\chi)$ is the negative of an antihomomorphism. We conclude that $\eta''(\chi) = \eta'(\chi)$. Setting $h = e_1 - e_2$, we see that, for any $\chi \in \hat{G}$,

$$\chi(h) = \begin{cases} 1 & \text{if } \eta'(\chi) \in \mathrm{Aut}\,(R); \\ -1 & \text{if } \eta'(\chi) \notin \mathrm{Aut}\,(R). \end{cases} \tag{30}$$

It is well known that any idempotent of $\mathscr{H} = \mathbb{K}G$ is contained in $\mathbb{K}K$ for some finite subgroup $K \subset G$, such that char $\mathbb{K}$ does not divide the order of K. Pick K such that $e_1 \in \mathbb{K}G_0$, then also $h \in \mathbb{K}K$. Since any character of K extends to a character of G, Eq. (30) implies that $(\chi_1\chi_2)(h) = \chi_1(h)\chi_2(h)$ for all $\chi_1, \chi_2 \in \hat{K}$, so h is a group-like element of $(\mathbb{K}\hat{K})^* = \mathbb{K}K$. It follows that $h \in K$. Clearly, the order of h is at most 2. Since char $\mathbb{K} \neq 2$, the characterization given by (30) shows that the element h is uniquely determined by the given G-grading on $\mathscr{L}$. It follows that e_1, e_2 and ρ' are also uniquely determined.

If h has order 1, then $e_1 = 1$ and $e_2 = 0$, so the map $\rho'\colon R \to R \otimes \mathscr{H}$ is a homomorphism of associative algebras. Since $\mathscr{L}$ generates R, it immediately follows that ρ' is a comodule structure. This makes R a G-graded algebra such that the given grading on $\mathscr{L}$ is just the restriction of the grading on R, i.e., $\mathscr{L}_g = R_g \cap \mathscr{L}$ for all $g \in G$. Such gradings on $\mathscr{L}$ will be referred to as *grading of Type I*. We can use Corollary 4.3 to obtain all such gradings.

If h has order 2, then both e_1 and e_2 are nontrivial, so ρ' is not a homomorphism of associative algebras, which means that the algebra R does not admit a G-grading that would restrict to the given grading on $\mathscr{L}$. Such gradings on $\mathscr{L}$ will be referred to as *grading of Type II (with the distinguished element $h = e_1 - e_2$)*. Let $\overline{G} = G/\langle h \rangle$ and $\overline{\mathscr{H}} = \mathbb{K}\overline{G}$. Denote the quotient map $G \to \overline{G}$ by π and extend it to a homomorphism of Hopf algebras $\mathscr{H} \to \overline{\mathscr{H}}$. Since $\pi(e_2) = 0$, the map $\overline{\rho} = (\mathrm{id}_R \otimes \pi)\rho$ is a homomorphism of associative algebras, so $\overline{\rho}\colon R \to R \otimes \overline{\mathscr{H}}$ is a comodule structure, which makes R a $\overline{G}$-graded algebra. The restriction of this grading to $\mathscr{L}$ is the coarsening of the given G-grading induced by $\pi : G \to \overline{G}$, i.e., $R_{\overline{g}} \cap \mathscr{L} = \mathscr{L}_g \oplus \mathscr{L}_{gh}$ for all $g \in G$, where $\overline{g} = \pi(g)$. The original grading can be recovered as follows.

Fix a character $\chi \in \hat{G}$ satisfying $\chi(h) = -1$ and let $\psi = \eta(\chi)$. Then we get

$$\mathscr{L}_g = \{x \in \mathscr{L}_{\overline{g}} \mid \psi(x) = \chi(g)x\}.$$

Indeed, by definition of ψ, $\mathscr{L}_g$ is contained in the right-hand side. Conversely, suppose $x \in \mathscr{L}_{\overline{g}}$ satisfies $\psi(x) = \chi(g)x$. Write $x = y + z$ where $y \in \mathscr{L}_g$ and $y \in \mathscr{L}_{gh}$. Then we have $\chi(g)(y + z) = \psi(x) = \chi(g)(y - z)$. It follows that $z = 0$ and so $x \in \mathscr{L}_g$.

By (30), the extension ψ' of ψ is not an automorphism, so $\psi' = -\varphi$ where φ is an antiautomorphism of R. Since ψ leaves the components $\mathscr{L}_{\overline{g}}$ invariant and $\mathscr{L}$ generates R, it follows that φ leaves the components $R_{\overline{g}}$ invariant. Since $\varphi^2 = (\psi')^2 = \eta'(\chi^2)$ and χ^2 can be regarded as a character of $\overline{G}$, we obtain $\varphi^2(r) = \chi^2(\overline{g})r$ for all $r \in R_{\overline{g}}$. Set

(31) $$R_g = \{r \in R_{\overline{g}} \mid \varphi(r) = -\chi(g)r\}.$$

Then $R_{\overline{g}} = R_g \oplus R_{gh}$ and hence $R = \bigoplus_{g\in G} R_g$. Since ψ' is not an automorphism of R, this is not a G-grading on the associative algebra R. It is, however, a G-grading on the Lie algebra $R^{(-)}$ (corresponding to the comodule structure ρ'). The $\overline{G}$-grading on R and the antiautomorphism φ completely determine the G-grading on $\mathscr{L}$:

(32) $$\mathscr{L}_g = \{x \in R_{\overline{g}} \cap \mathscr{L} \mid \varphi(x) = -\chi(g)x\}.$$

Conversely, suppose we have a $\overline{G}$-grading on R and an antiautomorphism φ of the $\overline{G}$-graded algebra R satisfying the following *compatibility condition*:

(33) $$\varphi^2(r) = \chi^2(\overline{g})r \quad \text{for all} \quad r \in R_{\overline{g}},\ \overline{g} \in \overline{G}.$$

Since $-\varphi$ is an automorphism of $R^{(-)}$, Eq. (31) gives a G-grading on $R^{(-)}$ that refines the given $\overline{G}$-grading. Since $\mathscr{L} = [R, R]$ is a G-graded subalgebra of $R^{(-)}$, we see that (32) defines a Type II grading on $\mathscr{L}$ with distinguished element h. Note that the compatibility condition implies that φ acts as involution on the identity component of the $\overline{G}$-grading of R, so Theorem 4.8 tells us that (R, φ) is isomorphic to some $\mathfrak{F}(\overline{G}, D, \tilde{V}, \tilde{B}, \overline{g}_0)$.

Proposition 5.1. *The graded algebra* $R = \mathfrak{F}(\overline{G}, D, \tilde{V}, \tilde{B}, \overline{g}_0)$ *with its antiautomorphism* φ *satisfies the compatibility condition* (33) *if and only if* $\pi : G \to \overline{G}$ *splits over the support* $\overline{T}$ *of* D *and there exists* $\mu_0 \in \mathbb{K}^\times$ *such that* $\tilde{B}$ *satisfies the symmetry condition* (25) *where* μ_A, *for all* $A \in \overline{G}/\overline{T}$, *is given by*

(34) $$\mu_A = \begin{cases} \mu_0\chi^{-2}(A)\beta(\tau(A)) & \text{if } \overline{g}_0 A^2 = \overline{T}; \\ \mu_0\chi^{-2}(A) & \text{if } \overline{g}_0 A^2 \neq \overline{T}; \end{cases}$$

where we regard χ^2 *as a character of* $\overline{G}/\overline{T}$ *(since* χ^2 *is trivial on* $\overline{T}$*).*

We note that since $\mu_A \mu_{\overline{g}_0^{-1}A^{-1}} = 1$ for all A, the scalar μ_0 satisfies

$$\mu_0^2 = \chi^{-2}(\overline{g}_0)$$

and hence can take only two values.

To state the classification of G-gradings on the Lie algebras $\mathfrak{fsl}(U, \Pi)$, we introduce the model G-graded algebras $\mathfrak{A}^{(\mathrm{I})}(G, T, \beta, \tilde{V}, \tilde{W})$ and $\mathfrak{A}^{(\mathrm{II})}(G, H, h, \beta, \tilde{V}, \tilde{B}, \overline{g}_0)$.

Let $T \subset G$ be a finite subgroup with a nondegenerate alternating bicharacter β. The Lie algebra $\mathfrak{A}^{(\mathrm{I})}(G, T, \beta, \tilde{V}, \tilde{W})$ is just the commutator subalgebra of the G-graded associative algebra $R = \mathfrak{F}(G, T, \beta, \tilde{V}, \tilde{W})$ (introduced before Corollary 4.3). By Theorem 4.6, it is isomorphic to $\mathfrak{fsl}(U, \Pi)$ where $U = \tilde{V}^{\ell}$, $\Pi = \tilde{W}^{\ell}$, $\ell^2 = |T|$, and the bilinear form $\Pi \times U \to \mathbb{K}$ is given by (12).

Let $H \subset G$ be a finite elementary 2-subgroup, $h \neq e$ an element of H, and β a nondegenerate alternating bicharacter on $H/\langle h \rangle$. Fix a character $\chi \in \hat{G}$ with $\chi(h) = -1$. Let $\overline{G} = G/\langle h \rangle$ and $\overline{T} = H/\langle h \rangle$. The Lie algebra $\mathfrak{A}^{(\mathrm{II})}(G, H, h, \beta, \tilde{V}, \tilde{B}, \overline{g}_0)$ is the commutator subalgebra of the Lie algebra $R^{(-)}$ with a G-grading defined by refining the $\overline{G}$-grading as in (31), where R is the $\overline{G}$-graded associative algebra $\mathfrak{F}(\overline{G}, \overline{T}, \beta, \tilde{V}, \tilde{B}, \overline{g}_0)$ with antiautomorphism φ (introduced before Theorem 4.8) and the bilinear form $\tilde{B}$ on $\tilde{V}$ satisfies the symmetry condition (25) with μ_A given by (34) for some $\mu_0 \in \mathbb{K}^{\times}$. By Theorem 4.8, $\mathfrak{A}^{(\mathrm{II})}(G, H, h, \beta, \tilde{V}, \tilde{B}, \overline{g}_0)$ is isomorphic to $\mathfrak{fsl}_U(U)$ where $U = \tilde{V}^{\ell}$, $\ell^2 = |H|/2$, and the bilinear form $U \times U \to \mathbb{K}$ is given by (28).

The definition of $\mathfrak{A}^{(\mathrm{II})}(G, H, h, \beta, \tilde{V}, \tilde{B}, \overline{g}_0)$ depends on the choice of χ. However, regardless of this choice, we obtain the same collection of graded algebras as $(\tilde{V}, \tilde{B})$ ranges over all possibilities allowed by the chosen χ. We assume that a choice of χ is fixed for any element $h \in G$ of order 2.

Now we can state our main result about gradings on special Lie algebras of finitary linear operators on an infinite-dimensional vector space.

Theorem 5.1. *Let G be an abelian group and let $\mathbb{K}$ be an algebraically closed field,* char $\mathbb{K} \neq 2$. *If a special Lie algebra $\mathscr{L}$ of finitary linear operators on an infinite-dimensional vector space over $\mathbb{K}$ is given a G-grading, then $\mathscr{L}$ is isomorphic as a graded algebra to some $\mathfrak{A}^{(\mathrm{I})}(G, T, \beta, \tilde{V}, \tilde{W})$ or $\mathfrak{A}^{(\mathrm{II})}(G, H, h, \beta, \tilde{V}, \tilde{B}, \overline{g}_0)$. No G-graded Lie algebra with superscript* (I) *is isomorphic to one with superscript* (II). *Moreover,*

- *$\mathfrak{A}^{(\mathrm{I})}(G, T, \beta, \tilde{V}, \tilde{W})$ and $\mathfrak{A}^{(\mathrm{I})}(G, T', \beta', \tilde{V}', \tilde{W}')$ are isomorphic if and only if $T' = T$ and either $\beta' = \beta$ and $(\tilde{V}', \tilde{W}') \sim (\tilde{V}, \tilde{W})$, or $\beta' = \beta^{-1}$ and $(\tilde{V}', \tilde{W}') \sim (\tilde{W}, \tilde{V})$ as in Definition 4.5.*
- *$\mathfrak{A}^{(\mathrm{II})}(G, H, h, \beta, \tilde{V}, \tilde{B}, \overline{g}_0)$ and $\mathfrak{A}^{(\mathrm{II})}(G, H', h', \beta', \tilde{V}', \tilde{B}', \overline{g}_0')$ are isomorphic if and only if $H' = H$, $h' = h$, $\beta' = \beta$, and $(\tilde{V}', \tilde{B}', \overline{g}_0') \sim (\tilde{V}, \tilde{B}, \overline{g}_0)$ as in Definition 4.7.*

5.1.1 Graded Bases

The Lie algebra $\mathscr{L} = \mathfrak{A}^{(\mathrm{I})}(G, T, \beta, \tilde{V}, \tilde{W})$ is the commutator subalgebra of $R = \mathfrak{F}(G, T, \beta, \tilde{V}, \tilde{W})$. As a vector space over $\mathbb{K}$, this latter can be written as $\tilde{V} \otimes D \otimes \tilde{W}$—see Sect. 4.6. Recall that, for $A, A' \in G/T$, we have $(\tilde{W}_A, \tilde{V}_{A'}) = 0$ unless $A' = A^{-1}$, and $\tilde{W}_{A^{-1}}$ is a total subspace of $\tilde{V}_A^*$. The action of the tensor $v \otimes d \otimes w \in$

$\tilde{V}_A \otimes D \otimes \tilde{W}_{(A')^{-1}}$ on $u \otimes d' \in \tilde{V}_{A''} \otimes D$ is given by $(v \otimes d \otimes w)(u \otimes d') = v \otimes d(w,u)d' = (w,u)(v \otimes (dd'))$, which is zero unless $A'' = A'$. Also, the degree of the element $v \otimes d \otimes w$ in the G-grading equals $\gamma(A)(\deg d)\gamma(A')^{-1}$. Computing the commutators of such elements and using our standard notation X_t for the basis elements of D, we find that $\mathscr{L}$ is spanned by the elements of the form $v \otimes X_t \otimes w$ where $(w,v) = 0$ or $t = e$. Note that since $\tilde{W}$ is a total subspace of $\tilde{V}^*$, there exist $A_0 \in G/T$, $v_0 \in \tilde{V}_{A_0}$ and $w_0 \in \tilde{W}_{A_0^{-1}}$ such that $(w_0, v_0) = 1$. Then $\mathbb{K}(v_0 \otimes 1 \otimes w_0) \oplus \mathscr{L} = R$. Given bases for the vector spaces $\tilde{V}_A$ and $\tilde{W}_A$ for all $A \in G/T$ such that the basis for $\tilde{V}_{A_0}$ includes v_0 and the basis for $\tilde{W}_{A_0^{-1}}$ includes w_0, we obtain a graded basis for $\mathscr{L} = \mathfrak{A}^{(\mathrm{I})}(G,T,\beta,\tilde{V},\tilde{W})$ consisting of the following two sets: first, the elements of the form $v \otimes 1 \otimes w - (w,v)(v_0 \otimes 1 \otimes w_0)$ with $v \neq v_0$ or $w \neq w_0$, and, second the elements of the form $v \otimes X_t \otimes w$ with $t \neq e$, where v and w range over the given bases of $\tilde{V}$ and $\tilde{W}$, respectively. Hence, for $g \neq e$, a basis for $\mathscr{L}_g$ consists of the elements

$$E^{(\mathrm{I})}_{g,v,w} = v \otimes X_t \otimes w,$$

where v ranges over a basis of $\tilde{V}_A$, w ranges over a basis of $\tilde{W}_{gA^{-1}}$, and A ranges over G/T, while $t = g\gamma(A)^{-1}\gamma(g^{-1}A)$. For $g = e$, we take the elements

$$E^{(\mathrm{I})}_{g,v,w} - (w,v)E^{(\mathrm{I})}_{g,v_0,w_0}$$

and discard the zero obtained when $v = v_0$ and $w = w_0$.

In the case of $\mathscr{L} = \mathfrak{A}^{(\mathrm{II})}(G,H,h,\beta,\tilde{V},\tilde{B},\overline{g}_0)$, we have to start with the $\overline{G}$-grading on $R = \mathfrak{F}(\overline{G},\overline{T},\beta,\tilde{V},\tilde{B},\overline{g}_0)$ where $\overline{G} = G/\langle h\rangle$ and $\overline{T} = H/\langle h\rangle$. Now R can be written as $\tilde{V} \otimes D \otimes \tilde{V}$ where the action of the tensor $v \otimes d \otimes w \in \tilde{V}_A \otimes D \otimes \tilde{V}_{\overline{g}_0^{-1}(A')^{-1}}$ on $u \otimes d' \in \tilde{V}_{A''} \otimes D$ is given by $(v \otimes d \otimes w)(u \otimes d') = v \otimes dB(w,u)d'$, which is zero unless $A'' = A'$. Here we are using the notation of Sect. 4.9—in particular, the φ_0-sesquilinear form B given by (21) and (23)—and we have identified $w \in \tilde{V}$ with $f_w \in W$ given by $f_w(u) = B(w,u)$ for all $u \in V$. The degree of the element $v \otimes d \otimes w$ in the $\overline{G}$-grading equals $\gamma(A)(\deg d)\overline{g}_0\gamma(\overline{g}_0^{-1}(A')^{-1}) = \gamma(A)(\deg d)\tau(A')\gamma(A')^{-1}$ where, as before, $\tau(A') = \overline{g}_0\gamma(A')^2$ if $\overline{g}_0(A')^2 = \overline{T}$, and we have set $\tau(A') = \overline{e}$ if $\overline{g}_0(A')^2 \neq \overline{T}$. Also, (18) yields $\varphi(v \otimes d \otimes w) = Q^{-1}w \otimes \varphi_0(d)v$. Taking $d = X_t$ and recalling that $\varphi_0(X_t) = \beta(t)X_t$, we obtain

$$\varphi(v \otimes X_t \otimes w) = \beta(t)(Q^{-1}(w) \otimes X_t \otimes v).$$

With our fixed character $\chi : G \to \mathbb{K}^\times$ satisfying $\chi(h) = -1$, the G-grading on the vector space R is given by (31). Taking into account that $\varphi^2(r) = \chi(g)^2 r$ for any $r \in R_{\overline{g}}$, we can write

$$r = \frac{1}{2}\left(r - \frac{1}{\chi(g)}\varphi(r)\right) + \frac{1}{2}\left(r + \frac{1}{\chi(g)}\varphi(r)\right)$$

where

$$\frac{1}{2}\left(r - \frac{1}{\chi(g)}\varphi(r)\right) \in R_g \quad \text{and} \quad \frac{1}{2}\left(r + \frac{1}{\chi(g)}\varphi(r)\right) \in R_{gh}.$$

Since $\chi(gh) = -\chi(g)$, the second expression is identical to the first one if g is replaced by gh. Substituting $r = v \otimes X_t \otimes w$ and the above expression for $\varphi(v \otimes X_t \otimes w)$, we find that R_g is spanned by the elements

$$v \otimes X_t \otimes w - \frac{\beta(t)}{\chi(g)} Q^{-1} w \otimes X_t \otimes v,$$

where $v \in \tilde{V}_A$, $w \in \tilde{V}_{\overline{g}_0^{-1}(A')^{-1}}$, and $\gamma(A)t\tau(A')\gamma(A')^{-1} = \overline{g}$. The eigenvalues of Q are given by (26), and $\mu_{\overline{g}_0^{-1}(A')^{-1}} = \mu_{A'}^{-1}$, so we can rewrite the above spanning elements as follows:

$$v \otimes X_t \otimes w - \mu_{A'}\beta(t)\beta(\tau(A'))\chi^{-1}(g)(w \otimes X_t \otimes v).$$

Finally, recalling that $\mu_{A'}$ is determined in Proposition 5.1 and $A' = \overline{g}^{-1}A$, we conclude that R_g is spanned by the elements

$$E^{(\mathrm{II})}_{g,v,w} = v \otimes X_t \otimes w - \frac{\mu_0\beta(t)\chi(g)}{\chi^2(A)}(w \otimes X_t \otimes v), \tag{35}$$

where v ranges over a basis of $\tilde{V}_A$, w ranges over a basis of $\tilde{V}_{\overline{g}_0^{-1}\overline{g}A^{-1}}$, and A ranges over $\overline{G}/\overline{T} = G/H$, while $t = \overline{g}\gamma(A)^{-1}\gamma(\overline{g}^{-1}A)\tau(\overline{g}^{-1}A)$. If we discard zeros among the elements $E^{(\mathrm{II})}_{g,v,w}$, the remaining ones form a basis for R_g. This is also a basis for $\mathscr{L}_g$ unless $g = e$ or $g = h$. For these two cases, bases can be obtained using the same idea as for Type I, namely, subtracting a suitable scalar multiple of $E^{(\mathrm{II})}_{g,v_0,w_0}$, where $v_0 \in \tilde{V}_{A_0}$ and $w_0 \in \tilde{V}_{\overline{g}_0^{-1}A_0^{-1}}$ are such that $\tilde{B}(w_0, v_0) = 1$. Specifically, for $g = e$ and $g = h$, we replace the elements (35) by

$$E^{(\mathrm{II})}_{g,v,w} - \beta(t_0)\beta(t)\tilde{B}(w, v)E^{(\mathrm{II})}_{g,v_0,w_0}$$

where $t = \tau(A)$ and $t_0 = \tau(A_0)$.

5.1.2 Countable Case

In the case $\mathscr{L} = \mathfrak{sl}(\infty)$, we can express the classification of G-gradings in combinatorial terms. Here $R = M_\infty(\mathbb{K})$, whose G-gradings are classified in Corollary 4.4. Namely, R is isomorphic to $\mathfrak{F}(G, D, \kappa)$, which we can write as $\mathfrak{F}(G, T, \beta, \kappa)$ because G is abelian. Recall that the function $\kappa\colon G/T \to \{0, 1, 2, \ldots, \infty\}$ has a finite or countable support; here $|\kappa| = \sum_{A \in G/T} \kappa(A)$ must be infinite. The

G-grading on $\mathfrak{F}(G, T, \beta, \kappa)$ restricts to $\mathscr{L}$, and we denote the resulting G-graded Lie algebra by $\mathfrak{A}^{(\mathrm{I})}(G, T, \beta, \kappa)$. In this way we obtain all Type I gradings on $\mathscr{L}$.

For Type II, we can use Corollary 4.5, with G replaced by $\overline{G} = G/\langle h\rangle$ and with the function $\mu\colon A \mapsto \mu_A$ ($A \in \overline{G}/\overline{T}$) satisfying (34). With the other parameters fixed, there are at most two such functions. Indeed, if there is $A_0 \in \overline{G}/\overline{T}$ such that $\overline{g}_0 A_0^2 = \overline{T}$ and $\kappa(A_0)$ is finite and odd (forcing $\mu_{A_0} = 1$), then there is at most one admissible function μ, defined by (34) with $\mu_0 = \chi^2(A_0)\beta(\tau(A_0))$. Such function exists if and only if all A_0 of this kind produce the same value μ_0. If there is no A_0 of this kind, then there are exactly two admissible functions μ, defined by (34) where μ_0 satisfies $\mu_0^2 = \chi^{-2}(\overline{g}_0)$. Denote by $\mathfrak{A}^{(\mathrm{II})}(G, H, h, \beta, \kappa, \mu_0, \overline{g}_0)$ the G-graded Lie algebra obtained by restricting to $\mathscr{L}$ the refinement (31) of the $\overline{G}$-grading on $\mathfrak{F}(G, \overline{T}, \beta, \kappa, \mu, \overline{g}_0)$.

Recall that κ^g is defined by $\kappa^g(gA) = \kappa(A)$ for all $A \in G/T$. Set $\tilde{\kappa}(A) = \kappa(A^{-1})$.

Corollary 5.1. *Let $\mathbb{K}$ be an algebraically closed field,* char $\mathbb{K} \neq 2$, *G an abelian group. Suppose $\mathscr{L} = \mathfrak{sl}(\infty)$ over $\mathbb{K}$ is given a G-grading. Then, as a graded algebra, $\mathscr{L}$ is isomorphic to one of $\mathfrak{A}^{(\mathrm{I})}(G, T, \beta, \kappa)$ or $\mathfrak{A}^{(\mathrm{II})}(G, H, h, \beta, \kappa, \mu_0, \overline{g}_0)$. No G-graded Lie algebra with superscript* (I) *is isomorphic to one with superscript* (II). *Moreover,*

- $\mathfrak{A}^{(\mathrm{I})}(G, T, \beta, \kappa) \cong \mathfrak{A}^{(\mathrm{I})}(G, T', \beta', \kappa')$ *if and only if $T' = T$ and either $\beta' = \beta$ and $\kappa' = \kappa^g$ for some $g \in G$, or $\beta' = \beta^{-1}$ and $\kappa' = \tilde{\kappa}^g$ for some $g \in G$.*
- $\mathfrak{A}^{(\mathrm{II})}(G, H, h, \beta, \kappa, \mu_0, \overline{g}_0) \cong \mathfrak{A}^{(\mathrm{II})}(G, H', h', \beta', \kappa', \mu_0', \overline{g}_0')$ *if and only if $H' = H$, $h' = h$, $\beta' = \beta$, and there exists $g \in G$ such that $\kappa' = \kappa^{\overline{g}}$, $\mu_0' = \mu_0\chi^2(g)$ and $\overline{g}_0' = \overline{g}_0\overline{g}^{-2}$.* □

5.2 Orthogonal and Symplectic Lie Algebras

In the case $\mathscr{L} = \mathfrak{fso}(U, \Phi)$ or $\mathscr{L} = \mathfrak{fsp}(U, \Phi)$, we deal with simple Lie algebras of skew-symmetric elements in the associative algebra $R = \mathfrak{F}_\Pi(U)$ with respect to the involution φ determined by the nondegenerate form Φ, which is either orthogonal or symplectic. Here Π is identified with U by virtue of Φ. We continue to assume that U is infinite-dimensional. Suppose that $\mathscr{L}$ is given a G-grading. Applying Theorem 3.5 to the corresponding Lie homomorphism $\rho\colon \mathscr{L} \otimes \mathscr{H} \to \mathscr{L} \otimes \mathscr{H}$ and observing that $\tau = 0$ (because $\mathscr{L} = [\mathscr{L}, \mathscr{L}]$), we conclude that ρ extends to a homomorphism of associative algebras $\rho'\colon R \otimes \mathscr{H} \to R \otimes \mathscr{H}$. Since $\mathscr{L}$ generates R, it follows that ρ' is a comodule structure. This gives R a G-grading that restricts to the given grading on $\mathscr{L}$, i.e., $\mathscr{L}_g = R_g \cap \mathscr{L}$ for all $g \in G$. Moreover, since φ restricts to the negative identity on $\mathscr{L}$, the restriction of the map $\varphi \otimes \mathrm{id}_{\mathscr{H}}$ to $\mathscr{L} \otimes \mathscr{H}$ commutes with ρ, which implies that $\varphi \otimes \mathrm{id}_{\mathscr{H}}$ commutes with ρ'. This means that φ is an involution of R as a G-graded algebra. Theorem 4.8 and Proposition 4.1 classify all the pairs (R, φ) up to isomorphism.

To state the classification of G-gradings on the Lie algebras $\mathfrak{fso}(U,\Phi)$ and $\mathfrak{fsp}(U,\Phi)$, we introduce the model algebras $\mathfrak{B}(G,T,\beta,\tilde{V},\tilde{B},g_0)$ and $\mathfrak{C}(G,T,\beta,\tilde{V},\tilde{B},g_0)$.

Let $T\subset G$ be a finite elementary 2-subgroup with a nondegenerate alternating bicharacter β. Let $\mathscr{L}$ be the Lie algebra of skew-symmetric elements in the G-graded associative algebra $R=\mathfrak{F}(G,T,\beta,\tilde{V},\tilde{B},g_0)$ with involution, where the bilinear form $\tilde{B}$ on $\tilde{V}$ satisfies the symmetry condition (25) with μ_A given by (29). If $\mathrm{sgn}(\varphi)=1$, we will denote $\mathscr{L}$ by $\mathfrak{B}(G,T,\beta,\tilde{V},\tilde{B},g_0)$. By Theorem 4.8, it is isomorphic to $\mathfrak{fso}(U,\Phi)$ where $U=\tilde{V}^\ell$, $\ell^2=|T|$, and the bilinear form $\Phi\colon U\times U\to\mathbb{K}$ is given by (28). If $\mathrm{sgn}(\varphi)=-1$, we will denote $\mathscr{L}$ by $\mathfrak{C}(G,T,\beta,\tilde{V},\tilde{B},g_0)$. By Theorem 4.8, it is isomorphic to $\mathfrak{fsp}(U,\Phi)$ where U and Φ are as above.

Theorem 5.2. *Let G be an abelian group and let $\mathbb{K}$ be an algebraically closed field,* char $\mathbb{K}\neq 2$. *If an orthogonal or symplectic Lie algebra $\mathscr{L}$ of finitary linear operators on an infinite-dimensional vector space over $\mathbb{K}$ is given a G-grading, then $\mathscr{L}$ is isomorphic as a graded algebra to some $\mathfrak{B}(G,T,\beta,\tilde{V},\tilde{B},g_0)$ in the orthogonal case or $\mathfrak{C}(G,T,\beta,\tilde{V},\tilde{B},g_0)$ in the symplectic case. The G-graded algebras with parameters $(T,\beta,\tilde{V},\tilde{B},g_0)$ and $(T',\beta',\tilde{V}',\tilde{B}',g_0')$ are isomorphic if an only if $T'=T$, $\beta'=\beta$, and $(\tilde{V}',\tilde{B}',g_0')\sim(\tilde{V},\tilde{B},g_0)$ as in Definition 4.7.*

5.2.1 Graded Bases

The calculations which we used to find graded bases in the case of Type II gradings on special linear Lie algebras apply also for orthogonal and symplectic Lie algebras, with the following simplifications: we always deal with the group G (hence we omit bars and χ) and Q is id in the orthogonal case and $-$id in the symplectic case. Thus, $\mathfrak{B}(G,T,\beta,\tilde{V},\tilde{B},g_0)$ is the set of skew-symmetric elements in $\mathfrak{F}(G,D,\tilde{V},\tilde{B},g_0)$, where $\tilde{B}$ is an "orthosymplectic" form on $\tilde{V}$ that determines a "Hermitian" form on V, as described in Sect. 4.10. The skew-symmetric elements

$$v\otimes X_t\otimes w-\beta(t)(w\otimes X_t\otimes v) \tag{36}$$

span $\mathscr{L}=\mathfrak{B}(G,T,\beta,\tilde{V},\tilde{B},g_0)$, so we obtain a basis for $\mathscr{L}_g$ by letting A range over G/T, v over a basis of $\tilde{V}_A$, and w over a basis of $\tilde{V}_{g_0^{-1}gA^{-1}}$, while $t=g\gamma(A)^{-1}\gamma(g^{-1}A)\tau(g^{-1}A)$, and taking the nonzero elements given by (36).

In a similar way, $\mathfrak{C}(G,T,\beta,\tilde{V},\tilde{B},g_0)$ is the set of skew-symmetric elements in the algebra $\mathfrak{F}(G,D,\tilde{V},\tilde{B},g_0)$, where $\tilde{B}$ is an "orthosymplectic" form on $\tilde{V}$ that determines a "skew-Hermitian" form on V. The skew-symmetric elements

$$v\otimes X_t\otimes w+\beta(t)(w\otimes X_t\otimes v) \tag{37}$$

span $\mathscr{L}=\mathfrak{C}(G,T,\beta,\tilde{V},\tilde{B},g_0)$, and we obtain a basis for $\mathscr{L}_g$ as above, but using (37) instead of (36).

5.2.2 Countable Case

In the cases $\mathscr{L} = \mathfrak{so}(\infty)$ and $\mathscr{L} = \text{Span}\{()\}\infty)$, we can express the classification of G-gradings in combinatorial terms. Here we have to deal with the pairs (R, φ) where $R = M_\infty(\mathbb{K})$ is endowed with a G-grading and an involution φ that respects this grading. Such pairs are classified in Corollary 4.6. Namely, (R, φ) is isomorphic to some $\mathfrak{F}(G, T, \beta, \kappa, \delta, g_0)$ where $|\kappa|$ must be infinite and $\delta = \text{sgn}(\varphi)$.

We denote by $\mathfrak{B}(G, T, \beta, \kappa, g_0)$ and $\mathfrak{C}(G, T, \beta, \kappa, g_0)$ the G-graded Lie algebras of skew-symmetric elements (with respect to φ) in $\mathfrak{F}(G, T, \beta, \kappa, \delta, g_0)$ where $\delta = 1$ and $\delta = -1$, respectively.

Corollary 5.2. *Let* $\mathbb{K}$ *be an algebraically closed field,* char $\mathbb{K} \neq 2$, G *an abelian group. Let* $\mathscr{L}$ *be* $\mathfrak{so}(\infty)$, *respectively* $\text{Span}\{()\}\infty)$, *over* $\mathbb{K}$. *Suppose* $\mathscr{L}$ *is given a* G*-grading. Then, as a graded algebra,* $\mathscr{L}$ *is isomorphic to some* $\mathfrak{B}(G, T, \beta, \kappa, g_0)$, *respectively* $\mathfrak{C}(G, T, \beta, \kappa, g_0)$. *No* G*-graded Lie algebra* $\mathfrak{B}(G, T, \beta, \kappa, g_0)$ *is isomorphic to* $\mathfrak{C}(G, T', \beta', \kappa', g_0')$. *Moreover,* $\mathfrak{B}(G, T, \beta, \kappa, g_0)$ *and* $\mathfrak{B}(G, T', \beta', \kappa', g_0')$ *are isomorphic if and only if* $T' = T$, $\beta' = \beta$, *and there exists* $g \in G$ *such that* $g_0' = g_0 g^{-2}$ *and* $\kappa' = \kappa^g$. *The same holds for* $\mathfrak{C}(G, T, \beta, \kappa, g_0)$ *and* $\mathfrak{C}(G, T', \beta', \kappa', g_0')$. □

6 Group Gradings on Nilpotent Lie Algebras

Historically one of the deepest result on the grading of Lie algebras was Cartan's theorem describing the symmetric decompositions (that is, $\mathbb{Z}_2$-graded) of simple Lie algebras. A notable consequence of this theorem is the classification of compact riemannian symmetric spaces. In the previous sections we summarized recent results on general gradings on simple Lie algebras. This will permit us to define in the second part new homogeneous reductive non-symmetric spaces associated to these gradings. The gradings on nilpotent Lie algebras have not been so much explored. It is then interesting on the one hand to classify the symmetric decompositions of a nilpotent Lie algebra and on the other hand to consider the most general gradings by finite abelian groups. The geometrical interest is double: on the one hand, to construct numerous examples of symmetric nilpotent riemannian and non-riemannian spaces and on the other hand to obtain a new approach of the affine geometry on nilpotent Lie groups. In this paper we restrict ourselves to the study of gradings on filiform Lie algebras of positive rank, in the case where the grading group is finitely generated abelian without elements of order p in case char $\mathbb{K} = p > 0$.

6.1 *An Example of an* Ad Hoc *Argument*

In some cases all gradings, up to isomorphism, can be determined based on the properties of algebras themselves rather than on the general theory of gradings.

As an example, let us consider *free nilpotent algebras*. By definition, each such algebra $L = L(X)$ is generated by a set of elements X such that any map of X to L extends to a homomorphism of L. Additionally, L has a finite $\mathbb{Z}$-grading $L = \bigoplus_{k\in\mathbb{Z}} L_k$, where L_k is the linear span of all monomials of degree k in X. Any set of generators with the same property as X is call a *free base* of L. A well-known example of such an algebra is the 3-dimensional *Heisenberg Lie algebra* $H = \langle x, y, z \mid [x, y] = z\rangle$. Here $X = \{x, y\}$ but actually any set Y of two elements linearly independent modulo the derived subalgebra $[H, H] = \langle z\rangle$ is also a free base of H.

Now let us assume the base field $\mathbb{K}$ infinite, the grading group G abelian and the base $X = \{x_1, \ldots, x_n\}$ finite (one says that $L(X)$ has *finite rank* n). In this case, it is easy to show, by a Vandermonde's type argument [1], that the structural $\mathbb{Z}$-grading on L can be refined to a $\mathbb{Z}^n$-grading, where given $\alpha = (i_1, \ldots, i_n) \in \mathbb{Z}^n$, the component L_α is defined as the linear span of monomials in $x_1, \ldots, x_n$ whose degree in x_k equals i_k, for any $k = 1, \ldots, n$. Now consider $\gamma = (g_1, \ldots, g_n) \in G^n$. Given $\alpha \in \mathbb{Z}^n$, define $\gamma^\alpha = g_1^{i_1} \cdots g_n^{i_n} \in G$ and set $L_\gamma^{(g)} = \bigoplus_{\gamma^\alpha = g} L_\alpha$. Then $\Gamma_\gamma : \bigoplus_{g\in G} L_\gamma^{(g)}$ is a G-grading of L.

Proposition 6.1. *Let* $\Gamma : L = \bigoplus_{g\in G} L_g$ *be a* G*-grading of a free nilpotent algebra* $L = L(X)$ *of finite rank over an infinite field* $\mathbb{K}$. *If* G *is an abelian group, then there is* $\gamma \in G^n$ *such that* Γ *is isomorphic to* Γ_γ. *If also* Γ *is isomorphic to* Γ_δ, *for some* $\delta \in G^n$, *then* δ *can be obtained from* γ *by permuting its components.*

Proof. Notice that the subalgebra $L^2 = \bigoplus_{k\geq 2} L_k$ is G-graded. Hence we have graded vector space bases B and C in L and L^2. We have $\#B = \#C + n$. This shows that L has a basis $\{y_1, \ldots, y_n\} \cup C$, where $\{y_1, \ldots, y_n\} \subset B$. The elements of $Y = \{y_1, \ldots, y_n\}$ are graded of degrees $g_1, \ldots, g_n$. By a Nakayama type argument (see, for example, [1, 1.7.2]), since L is nilpotent, a set generating L modulo L^2 is a generating set of L. Therefore, the map $\varphi : X \to L$ given by $x_i \mapsto y_i, i = 1, \ldots, n$, extends to a homomorphism of L onto itself, hence to an isomorphism of L. If $\gamma = (g_1, \ldots, g_n)$, this map is an isomorphism of Γ_γ and our original map Γ.

Since any permutation of the elements of X leads to an isomorphism of L, it is obvious that Γ_γ is isomorphic to Γ_δ if δ is a permutation of γ. Conversely, such an isomorphism leads to the isomorphism of the graded vector space L_1 onto itself. Now, for each $g \in G$, $\dim L_\gamma^{(g)} = \dim L_\delta^{(g)}$. Clearly then the number of entries of g to both γ and δ is the same, hence δ is a permutation of δ. □

6.2 Filiform Lie Algebras

Let $\mathbb{K}$ be a field and L be a Lie algebra over $\mathbb{K}$. We denote by $\{L^k \mid k = 1, 2, \ldots\}$ the lower central series of L defined by $L^1 = L$ and $L^k = [L^{k-1}, L]$, for $k = 2, 3, \ldots$. One calls L *nilpotent* if there is a natural n such that $L^{n+1} = \{0\}$. If $L^n \neq \{0\}$, then n is called the *nilpotent index* of L. As just above, a set of elements $\{x_1, \ldots, x_m\} \subset L$

generates L if and only if the $\{x_1 + L^2, \dots, x_m + L^2\}$ is the spanning set in the vector space L/L^2. As a result, a nilpotent Lie algebra with $\dim L/L^2 \leq 1$ is at most 1-dimensional. If $n > 1$, then the nilpotent index of an n-dimensional nilpotent Lie algebra never exceeds $n - 1$.

Definition 6.1. Given a natural number n, an n-dimensional Lie algebra L is called *n-dimensional filiform* if the nilpotent index of L is maximal possible, that is, $n-1$. In this case we must have $\dim L/L^2 = 2$, $\dim L^k/L^{k-1} = 1$ for $k = 2, 3, \dots, n-1$.

In the situation described in Definition 6.1, the lower central series of L is "thready", whence the French name "filiform". If we choose a basis $\{e_1, e_2, \dots, e_{n-2}, e_{n-1}, e_n\}$ of L so that $\{e_n\}$ is a basis of L^{n-1}, $\{e_{n-1}, e_n\}$ is a basis of L^{n-2}, $\{e_{n-2}, e_{n-1}, e_n\}$ is a basis of L^{n-3}, etc., it is easy to observe that the center of L is 1-dimensional and equals L^{n-1}.
Thus, the lower central sequence of L takes the form

$$L = L^1 \supset L^2 \supset \dots \supset L^{n-1} = Z(L) \supset \{0\},$$

where $Z(L)$ is the center of L and all containments are proper. In any Lie algebra the lower central series is a *filtration* in the sense that $[L^i, L^j] \subset L^{i+j}$.

Theorem 6.1 ([31]). *Any n-dimensional filiform $\mathbb{K}$-Lie algebra L admits an* adapted basis $\{X_1, \dots, X_n\}$, *that is, a basis satisfying:*

$$[X_1, X_i] = X_{i+1},\ i = 2, \dots, n-1,$$

$$[X_2, X_3] = \sum_{k \geq 5} \gamma_k X_k,$$

$$[X_i, X_{n-i+1}] = (-1)^{i+1} \alpha X_n, \text{ where } \alpha = 0 \text{ when } n = 2m+1,$$

$$L^i = \mathbb{K}\{X_{i+1}, \dots, X_n\} \text{ for all } i \geq 2.$$

Now let us consider a collection of vector spaces $W_i = L^i/L^{i+1}$, $i = 1, 2, \dots, n-1$. The vector space direct sum $\mathrm{gr}\, L = \bigoplus_{i=1}^{n-1} W_i$ becomes a Lie algebra if one defines the bracket of the elements by setting $[X + L^{i+1}, Y + L^{j+1}] = [X, Y] + L^{i+j+1}$, for $X \in L^i$, $Y \in L^j$, $1 \leq i, j \leq n-1$. It follows from the above theorem that all the associated graded algebras for filiform Lie algebras are again filiform. They belong to one of the two types as follows. Note for the future that when we define a Lie algebra by writing a list of commutators $[X_i, X_j]$ of the elements of the basis, we always mean $[X_k, X_l] = 0$ for (k, l) not on the list, except, naturally, that $[X_j, X_i] = -[X_i, X_j]$.

L_n: Each of these algebras has an adapted basis $\{X_1, X_2, \dots, X_n\}$ such that $[X_1, X_i] = X_{i+1}$, for $i = 2, 3, \dots, n-1$.

Q_n: Here $n = 2m$. Each of these algebras has an adapted basis $\{X_1, X_2, \ldots, X_n\}$ such that $[X_1, X_i] = X_{i+1}$, if $i = 2, 3, \ldots, n-1$, and $[X_j, X_{2m-j+1}] = (-1)^{j+1} X_{2m}$, if $2 \leq j \leq m$.

If gr $L \cong L$, then we call L *naturally graded* (by the group $\mathbb{Z}$). So there are only two types of naturally graded algebras: L_n and Q_n.

Corollary 6.1 ([31]). *Any naturally graded filiform Lie algebra is isomorphic to*

- L_n *if* n *is odd,*
- L_n *or* Q_n *if* n *is even.*

We deduce that L_n and Q_n admits a $\mathbb{Z}$-grading with support $\{1, 2, \ldots, n\}$. From what follows, we will see the existence of other non-isomorphic $\mathbb{Z}$-gradings and we will determine the filiform Lie algebras admitting such gradings.

6.3 The Automorphisms Group of a Filiform Lie Algebra

A special feature of filiform Lie algebras is the following.

Theorem 6.2. *Let* L *be an* n*-dimensional filiform* $\mathbb{K}$*-Lie algebra with* $n \geq 4$. *Then the group* Aut L *is a solvable algebraic group, of toral rank at most 2.*

Proof. Assume $\sigma \in \operatorname{Aut} L$ and let $\{X_1, \ldots, X_n\}$ be an adapted basis of L. We know that in any algebra L, $\sigma(L^i) = L^i$ and so for any $i \geq 3$, $\sigma(X_i) = \lambda_i X_i + U_i$, where $\lambda_i \neq 0$ and $U_i \in L^i$. We also know that $[X_1, X_i] = X_{i+1}$ and $[X_2, X_3] = \nu_3 X_5 + V_i$ where $V_i \in L^5$. Assume that $\sigma(X_2) = \mu_1 X_1 + \mu_2 X_2 + Y$, where $Y \in L^2$. Then

$$\begin{aligned}
&[\sigma(X_2), \sigma(X_3)] = [\mu_1 X_1 + \mu_2 X_2 + Y, \lambda_3 X_3 + U_3] \\
&= \mu_1 \lambda_3 X_4 + \mu_1 [X_1, U_3] + \mu_2 \lambda_3 (\nu_3 X_5 + V_i) + \mu_2 [X_1, U_3] + \lambda_3 [Y, X_3] + [Y, U_3].
\end{aligned}$$

Because $[L^i, L^j] \subset L^{i+j}$, all $[X_1, U_3]$, $[X_1, U_3]$, $[Y, X_3]$,$[Y, U_3]$ are in L^4. So $[\sigma(X_2), \sigma(X_3)] = \mu_1 \lambda_3 X_4 + Z$, where $Z \in L^4$. On the other hand,

$$\sigma([X_2, X_3]) = \sigma(\nu_3 X_5 + V_i) \in L^4.$$

Thus $\mu_1 \lambda_3 = 0$. Since $\dim L \geq 4$, we have $\lambda_3 \neq 0$, and then $\mu_1 = 0$. If we set, for each $i = 1, \ldots, n$, $F_i = \operatorname{Span}\{X_i, \ldots, X_n\}$, then

$$\mathscr{F} : L = F_1 \supset F_2 \supset \cdots \supset F_n \supset \{0\}$$

is a flag of subspaces in L. It is well-known that the set of all automorphisms of a linear space respecting a flag is a solvable subgroup of GL (L). Now we have just proved that Aut L respects $\mathscr{F}$; hence Aut L is a solvable group, as claimed.

Finally, the matrices of the elements of Aut L with respect to an adapted basis are triangular. Let T_n is the group of all triangular matrices, U_n the subgroup of unitriangular matrices, D_n is the subgroup of the diagonal matrices and $C_n = \{\text{diag}\,(t, s, ts, \ldots, t^{n-2}s) \mid s, t \in \mathbb{K} \setminus \{0\}\}$. Then $G/G \cap U_n \cong GU_n/U_n = C_nU_n/U_n \cong C_n$. Since C_n is a 2-dimensional torus, the (toral) rank of G does not exceed 2. □

Remark. If $\dim L = 2$ or 3, then Aut L contains a subgroup isomorphic to $GL(2, \mathbb{K})$, hence not solvable.

6.4 Filiform Lie Algebras of Rank 1 or 2

In this section we list filiform Lie algebras of nonzero rank. It is proven in [21] that over $\mathbb{K} = \mathbb{C}$ every filiform algebra of nonzero rank is isomorphic to one of the algebras on the list.

Let L be a n-dimensional filiform $\mathbb{K}$-Lie algebra whose rank $r(L)$ is not 0. Thus $r(L) = 2$ or 1 and

(1) If $r(L) = 2$, L is isomorphic to

(a) L_n, $n \geq 3$

$$[X_1, X_i] = X_{i+1}, \quad 1 \leq i \leq n-1$$

(b) Q_n, $n = 2m$, $m \geq 3$

$$\begin{aligned} &[Y_1, Y_i] = Y_{i+1}, && 1 \leq i \leq n-2, \\ &[Y_i, Y_{n-i+1}] = (-1)^{i+1} Y_n, && 2 \leq 1 \leq m \end{aligned}$$

(2) If $r(L) = 1$, L is isomorphic to

(a) $A_n^p(\alpha_1, \ldots, \alpha_t)$, $n \geq 4$, $t = [\frac{n-p}{2}]$, $1 \leq p \leq n-4$

$$\begin{aligned} &[X_1, X_i] = X_{i+1}, \quad 1 \leq i \leq n-1, \\ &[X_i, X_{i+1}] = \alpha_{i-1} X_{2i+p}, && 2 \leq i \leq t, \\ &[X_i, X_j] = a_{i-1,j-1} X_{i+j+p-1}, && 2 \leq i < j, \quad i + j \leq n - p + 1 \end{aligned}$$

(b) $B_n^p(\alpha_1, \cdots, \alpha_t)$, $n = 2m$, $m \geq 3$, $1 \leq p \leq n-5$, $t = [\frac{n-p-3}{2}]$

$$\begin{aligned} &[Y_1, Y_i] = Y_{i+1}, && 1 \leq i \leq n-2, \\ &[Y_i, Y_{n-i+1}] = (-1)^{i+1} Y_n, && 2 \leq i \leq m \\ &[Y_i, Y_{i+1}] = \alpha_{i-1} Y_{2i+p}, && 2 \leq i \leq t+1 \\ &[Y_i, Y_j] = a_{i-1,j-1} Y_{i+j+p-1}, && 2 \leq i < j,\ i + j \leq n - p \end{aligned}$$

In both cases

$$\begin{cases} a_{i,i} = 0, \\ a_{i,i+1} = \alpha_i, \\ a_{i,j} = a_{i+1,j} + a_{i,j+1}, \end{cases}$$

where $\{X_1, \dots, X_n\}$ is an adapted basis and $\{Y_1, \dots, Y_n\}$ a quasi-adapted basis.

Note that if L is of the type A_n^p then gr L is of the type L_n. If L is of the type B_n^p then gr L is of the type Q_n.

Remark.

- In this proposition, the basis used to define the brackets is not always an adapted basis. More precisely, it is adapted for Lie algebras L_n and A_n^k, and it is not adapted for Lie algebras Q_n and B_n^k. But if $\{Y_1, \dots, Y_n\}$ is a quasi-adapted basis, that is a basis which diagonalizes the semisimple derivations, then the basis $\{X_1 = Y_1 - Y_2, X_2 = Y_2, \dots, X_n = Y_n\}$ is adapted.
- In [21], the proof of the main result is given for $\mathbb{K} = \mathbb{C}$. Without any further restrictions, we can extend this result to arbitrary fields $\mathbb{K}$ which are algebraically closed and of characteristic 0. Since the proof is based on a simultaneous reduction of a semisimple endomorphism and a nilpotent adjoint operator, this result can be extended to any field $\mathbb{K}$ which contains the eigenvalues of the semisimple endomorphism. Thus the result is true over any algebraically closed field.
- In the statement of the above result in [21], there is a third family denoted C_n. But all the Lie algebras of this family are isomorphic to Q_n. This error was noticed after the publication of the paper (see also [20]).

6.5 Standard Gradings of Filiform Lie Algebras of Nonzero Rank

In each of the four types of filiform algebras introduced in the previous section there are standard gradings, as follows.

(1) If $L = L_n$, then $L = \bigoplus_{(a,b)\in\mathbb{Z}^2} L_{(a,b)}$ where $L_{(a,b)} = \{0\}$ except $L_{(1,0)} = \langle X_1\rangle$, $L_{(s-2,1)} = \langle X_s\rangle$, for all $s = 2, \dots, n$.
(2) If $L = Q_n$, then $L = \bigoplus_{(a,b)\in\mathbb{Z}^2} L_{(a,b)}$ where $L_{(a,b)} = \{0\}$ except $L_{(1,0)} = \langle(X_1 + X_2)\rangle$, $L_{(s-2,1)} = \langle X_s\rangle$, for $s = 2, \dots, n-1$, $L_{(n-3,2)} = \langle X_n\rangle$.
(3) If $L = A_n^p$, then $L = \bigoplus_{a\in\mathbb{Z}} L_a$ where $L_a = \{0\}$ except $L_1 = \langle X_1\rangle$, $L_{s+p-1} = \langle X_s\rangle$, for $s = 2, \dots n$.
(4) If $L = B_n^p$, then $L = \bigoplus_{a\in\mathbb{Z}} L_a$ where $L_a = \{0\}$ except $L_1 = \langle(X_1 + X_2)\rangle$, $L_{s+p-1} = \langle X_s\rangle$, for $s = 2, \dots, n-1$, $L_{n+2p-1} = \langle X_n\rangle$,

where $\{X_1, \cdots, X_n\}$ is an adapted basis. Our main result says the following.

Theorem 6.3. *Let L be a finite-dimensional filiform $\mathbb{K}$-algebra of nonzero rank r over an algebraically closed field $\mathbb{K}$ of characteristic 0. If G is an abelian finitely* *generated group, then any G-grading of L is a coarsening of a standard grading by $\mathbb{Z}^r$. The same result is true for the filiform Lie algebras of the types L, Q, A, B over an algebraically closed field of characteristic $p >$, provided G has no elements of order p.*

Proof. Each of the four cases above gives rise to a maximal torus D in Aut L. To describe D we only need to indicate the action of an element of D on the generators of L, which will be X_1 and X_2 in the cases of L_n and A_n or $X_1 + X_2$ and X_2 in the cases of Q_n and B_n

(1) If $L = L_n$, then $D = \{\varphi_{u,t} \mid u, t \in \mathbb{K}^\times\}$ where $\varphi_{u,t}(X_1) = uX_1$, $\varphi_{u,t}(X_2) = tX_2$.
(2) If $L = Q_n$, then $D = \{\varphi_{u,t} \mid u, t \in \mathbb{K}^\times\}$ where $\varphi_{u,t}(X_1 + X_2) = u(X_1 + X_2)$, $\varphi_{u,t}(X_2) = tX_2$.
(3) If $L = A_n^p$, then $D = \{\varphi_u \mid u \in \mathbb{K}^\times\}$ where $\varphi_u(X_1) = uX_1$, $\varphi_u(X_2) = u^{p+1}X_2$.
(4) If $L = B_n^p$, then $D = \{\varphi_u \mid u \in \mathbb{K}^\times\}$ where $\varphi_u(X_1 + X_2) = u(X_1 + X_2)$, $\varphi_u(X_2) = u^{p+1}X_2$. □

Lemma 6.1. *In each of the four cases in Theorem 6.3, the centralizer of D is equal to D.*

Proof. If L is of the type L_n, then we have to determine all $\varphi \in$ Aut L such that $\varphi_{u,t}\varphi = \varphi\varphi_{v,s}$, for all $u, t, v, s \in \mathbb{K}^\times$. Notice that $\varphi_{u,t}(X_1) = uX_1$ and $\varphi_{u,t}(X_i) = u^{i-2}tX_i$, for all $i \geq 2$. Now let $\varphi(X_i) = \sum_{j\geq i} a_{ji}X_j$, for $i = 1, 2, \ldots, n$. Then $\varphi_{u,t}\varphi(X_1) = a_{11}uX_1 + \sum_{j\geq 2} a_{ij}u^{j-2}tX_j$ whereas $\varphi\varphi_{v,s}(X_1) = \sum_{j\geq 1} a_{ij}vX_j$. Also, $\varphi_{u,t}\varphi(X_i) = \sum_{j\geq i} a_{ij}u^{j-2}tX_j$ whereas $\varphi\varphi_{v,s}(X_i) = \sum_{j\geq i} a_{ij}v^{i-2}sX_j$, if $i \geq 2$. Thus we have $a_{11}u = a_{11}v$ and since φ is a triangular automorphism, $a_{11} \neq 0$, hence $v = u$. Similarly, comparing $\varphi_{u,t}\varphi(X_2)$ and $\varphi\varphi_{v,s}(X_2)$, we obtain $s = t$. Thus the normalizer of D is equal to its centralizer. Now we have $a_{j1}(u^{j-2}t - u) = 0$, for all $j \neq 1$ and $a_{ji}(u^{j-2}t - u^{i-2}t) = 0$, for all $j \neq i$. Here u, t are arbitrary elements of an infinite set $\mathbb{K}^\times$. It follows that all a_{ji} are zero as soon as $j \neq i$. Notice that in any case it follows from $[X_1, X_i] = X_{i+1}$ for $i \geq 2$, that $a_{ii} = a_{11}^{i-2}a_{22}$. As a result, $\varphi = \varphi_{a_{11},a_{22}}$, proving that the normalizer of D is indeed, D itself.

Now assume L is of the type Q_n. Then, instead of comparing the values of the sides of $\varphi_{u,t}\varphi = \varphi\varphi_{s,v}$ at $X_1, X_2, \ldots, X_n$ we can compare them on the quasi-adapted basis $X_1 + X_2, X_2, \ldots, X_n$. Then we obtain $a_{ji} = 0$, for $j \neq i, i \geq 2$. At the same time,

$$\varphi(X_1 + X_2) = a_{11}X_1 + \sum_{j\geq 2}(a_{j1} + a_{j2})X_j = a_{11}(X_1 + X_2) + (a_{21} + a_{22}$$
$$+ a_{11})X_2 + \sum_{j\geq 3}(a_{j1} - a_{j2})X_j.$$

Applying the same argument, as before, we obtain $a_{j1} = a_{j2} = 0$, for $j \geq 3$ and $a_{21} = -a_{11} - a_{22}$. Hence $\varphi(X_1 + X_2) = a_{11}(X_1 + X_2)$. Thus, $\varphi = \varphi_{a_{11},a_{22}}$, as previously.

Next, assume L is of one of the types A_n^p, where $p \geq 1$. In this case we can repeat the argument of the case L_n, bearing in mind that $t = u^{p+1}$. Then we will have equations $a_{j1}(u^{j-2}u^{p+1} - u) = 0$, or $a_{j1}(u^{j+p-2} - 1) = 0$, for all $j \neq 1$ and $a_{ji}(u^{j-2}u^{p+1} - u^{i-2}u^{p+1}) = 0$, or $a_{ji}(u^j - u^i) = 0$ for all $j \neq i$ and all $u \in \mathbb{K}^{\times}$. Since $j + p - 2 \neq 0$, we again can make the same conclusion $a_{ji} = 0$, for all $j \neq i$.

The case B_n^p, where $k \geq 1$, is reduced to the case Q_n in the same manner as A_n^p to L_n.

Thus the proof of our lemma is complete. □

To complete the proof of Theorem 6.3, we need only to refer to Theorem 2.3 from the Introduction.

References

1. Bahturin, Yuri; Identical Relations in Lie Algebras. *VNU Science Press*, Utrecht, 1987, x+309pp.
2. Bahturin, Yuri; Brešar, Matej, *Lie gradings on associative algebras*, J. Algebra, **321** (2009), 264–283.
3. Bahturin, Yuri; Brešar, Matej; Kochetov, Mikhail, *Group gradings on finitary simple Lie algebras*, Int.J. Algebra Comp, **22** (2012), 125–146.
4. Bahturin, Y.; Kochetov, M. *Classification of group gradings on simple Lie algebras of types A, B, C and D*. J. Algebra **324** (2010), 2971–2989.
5. Bahturin, Yuri; Kochetov, Mikhail. Group gradings on the Lie algebra $\mathfrak{psl}_n$ in positive characteristic. *J. Pure Appl. Algebra* 213 (2009), no. 9, 1739–1749.
6. Bahturin, Yuri; Kochetov, Mikhail. Classification of group gradings on simple Lie algebras of types A,B,C and D. *J. Algebra* 324 (2010), no. 11, 2971–2989.
7. Bahturin, Yuri; Kochetov, Mikhail, *Group gradings on restricted Cartan-type Lie algebras*, Pacific J. Math. **253** (2011), no. 2, 289–319.
8. Bahturin, Y.; Kochetov, M.; Montgomery, S. *Group gradings on simple Lie algebras in positive characteristic*. Proc. Amer. Math. Soc., **137** (2009), no. 4, 1245–1254.
9. Bahturin, Y.; Shestakov, I.; Zaicev, M. *Gradings on simple Jordan and Lie algebras*. J. Algebra, **283** (2005), no. 2, 849–868.
10. Bahturin, Y. A.; Sehgal, S. K.; Zaicev, M. V. *Group gradings on associative algebras*, J. Algebra **241** (2001), no. 2, 677–698.
11. Bahturin, Yuri; Zaicev, Mikhail, *Group gradings on simple Lie algebras of type "A"*, J. Lie Theory **16**(2006), 719–742.
12. Bahturin, Y.; Zaicev, M. *Gradings on simple algebras of finitary matrices*, J. Algebra **324** (2010), no. 6, 1279–1289.
13. Baranov, A. A. *Finitary simple Lie algebras*. J. Algebra, **219** (1999), no. 1, 299–329.
14. Baranov, A. A.; Strade, H. *Finitary Lie algebras*. J. Algebra, **254** (2002), no. 1, 173–211.
15. Beidar, K. I.; Brešar, M.; Chebotar, M. A.; Martindale, 3rd,W. S. *On Herstein's Lie map conjectures. I*, Trans. Amer. Math. Soc. **353** (2001), no. 10, 4235–4260 (electronic).
16. ———, *On Herstein's Lie map conjectures. II*, J. Algebra **238** (2001), no. 1, 239–264.
17. ———, *On Herstein's Lie map conjectures. III*, J. Algebra **249** (2002), no. 1, 59–94.
18. Brešar, M.; Chebotar, M. A.; Martindale, 3rd, W. S. *Functional identities*, Frontiers in Mathematics, Birkhäuser Verlag, Basel, 2007.
19. Elduque, Alberto; Kochetov, Mikhail. *Gradings on Simple Lie Algebras* AMS Mathematical Surveys and Monographs, 189 (2013), 336 pp.

20. Goze, Michel; Khakimdjanov, Yusupdjan. Some nilpotent Lie algebras and its applications. *Algebra and operator theory* (Tashkent, 1997), 4964, Kluwer Acad. Publ., Dordrecht, 1998.
21. Goze, Michel; Hakimjanov, Yusupdjan. Sur les algèbres de Lie nilpotentes admettant un tore de dérivations. *Manuscripta Math.* 84 (1994), no. 2, 115–124.
22. Havliček, Miloslav; Patera, Jiři; Pelantova, Edita, *On Lie gradings. II,* Linear Algebra Appl. 277 (1998), no. 1–3, 97–125.
23. Jacobson, N. *Structure of Rings*. Colloquium Publications, **37**, American Math. Society, Providence, RI, 1964.
24. Montgomery, S., *Hopf Algebras and Their Actions on Rings*, CBMS Regional Conference Series in Mathematics, **82**. American Mathematical Society, Providence, RI, 1993.
25. Năstăsescu, C.; Van Oystaeyen, F.; *Methods of graded rings*, Lecture Notes in Mathematics, Vol. 1836, Springer-Verlag, Berlin, 2004.
26. Patera, Jiri; Zassenhaus, Hans, On Lie gradings. I, *Linear Algebra. Appl.*, **112** (1989), 87–159
27. Platonov, Vladimir P. Subgroups of algebraic groups over a local or global field containing a maximal torus. *C. R. Acad. Sci. Paris* Sr. I Math. 318 (1994), no. 10, 899–903.
28. Springer, T.A. and Steinberg, R. *Conjugacy classes*. Seminar on Algebraic Groups and Related Finite Groups (The Institute for Advanced Study, Princeton, N.J., 1968/69), Lecture Notes in Mathematics, **131**, Springer, Berlin, 1970, pp. 167–266.
29. Strade, H.; *Simple Lie algebras over fields of positive characteristic. I*, de Gruyter Expositions in Mathematics, Vol. 38, Walter de Gruyter & Co., Berlin, 2004, Structure theory.
30. ______, *Simple Lie algebras over fields of positive characteristic. II*, de Gruyter Expositions in Mathematics, Vol. 42, Walter de Gruyter & Co., Berlin, 2009, Classifying the absolute toral rank two case.
31. Vergne, Michèle. Cohomologie des algèbres de Lie nilpotentes. Application l'étude de la variété des algèbres de Lie nilpotentes. *C. R. Acad. Sci. Paris* , Sr. A-B 267 1968, A867A870.
32. Waterhouse, W.C., *Introduction to affine group schemes*, GTM, Vol 66, Springer Verlag, New York, 1979.

Bounding the Dimensions of Rational Cohomology Groups

Christopher P. Bendel, Brian D. Boe, Christopher M. Drupieski, Daniel K. Nakano, Brian J. Parshall, Cornelius Pillen, and Caroline B. Wright

Abstract Let k be an algebraically closed field of characteristic $p > 0$, and let G be a simple simply-connected algebraic group over k. In this paper we investigate situations where the dimension of a rational cohomology group for G can be bounded by a constant times the dimension of the coefficient module. As an application, effective bounds on the first cohomology of the symmetric group are obtained. We also show how, for finite Chevalley groups, our methods permit significant improvements over previous estimates for the dimensions of second cohomology groups.

Research of Parshall was partially supported by NSF grant DMS 1001900.

C.P. Bendel
Department of Mathematics, Statistics and Computer Science, University of Wisconsin–Stout, Menomonie, WI 54751, USA
e-mail: bendelc@uwstout.edu

B.D. Boe • D.K. Nakano (✉)
Department of Mathematics, University of Georgia, Athens, GA 30602, USA
e-mail: brian@math.uga.edu; nakano@math.uga.edu

C.M. Drupieski
Department of Mathematical Sciences, DePaul University, Chicago, IL 60614, USA
e-mail: cdrupies@depaul.edu

B.J. Parshall
Department of Mathematics, University of Virginia, Charlottesville, VA 22903, USA
e-mail: bjp8w@virginia.edu

C. Pillen
Department of Mathematics and Statistics, University of South Alabama, Mobile, AL 36688, USA
e-mail: pillen@southalabma.edu

C.B. Wright
Louise S. McGehee School, 2343 Prytania Street, New Orleans, LA 70130, USA
e-mail: carriew@mcgeheeschool.com

G. Mason et al. (eds.), *Developments and Retrospectives in Lie Theory: Algebraic Methods*, Developments in Mathematics 38,
DOI 10.1007/978-3-319-09804-3_2

Key words Cohomology • Representation theory • Algebraic Groups

Mathematics Subject Classification (2010): 20G10.

1 Introduction

1.1 Cohomology Bounds

Let k be a field, let S be a finite group, and let V be an absolutely irreducible kS-module on which S acts faithfully. In 1986, Guralnick conjectured the existence of a universal upper bound, independent of k, S, or V, for the dimension of the first cohomology group $\mathrm{H}^1(S, V)$ [11]. Based on the evidence available at the time, Guralnick suggested that a suitable upper bound might be 2, though later work by Scott and his student McDowell showed that if $S = PSL_6(\mathbb{F}_p)$ with p sufficiently large, then there exists an absolutely irreducible kS-module V on which S acts faithfully with $\dim \mathrm{H}^1(S, V) = 3$ [24]. Still, the existence of *some* universal upper bound remained a plausible idea, and the best guess for a particular bound remained 3 until the recent American Institute of Mathematics (AIM) workshop "Cohomology bounds and growth rates" in June 2012.

As reported by AIM and the workshop organizers [12], on day 3 of the workshop, Scott reported on calculations conducted by his student Sprowl [25], from which they could deduce the existence of 4- and 5-dimensional examples for $\mathrm{H}^1(S, V)$ when $S = PSL_7(\mathbb{F}_p)$ with p sufficiently large. These calculations were independently confirmed by Lübeck, who subsequently showed that large-dimensional examples also arise when $S = G(\mathbb{F}_p)$ is a finite group of Lie type with underlying root system of type E_6 or F_4. Among all the dimensions computed during and in the weeks after the workshop, the largest was $\dim \mathrm{H}^1(S, V) = 469$ for $S = PSL_8(\mathbb{F}_p)$ with p some sufficiently large prime number. These particular large-dimensional examples do not disprove Guralnick's conjecture, but they do make it seem less likely that any universal upper bound exists.

Even if no universal upper bound exists for the dimensions of the cohomology groups $\mathrm{H}^1(S, V)$, the computer calculations of Scott, Sprowl, and Lübeck demonstrate how, by exploiting connections between the cohomology of semisimple algebraic groups and the combinatorics of Kazhdan–Lusztig polynomials, it is possible to obtain much information about the size of $\mathrm{H}^1(S, V)$ when S is a finite group of Lie type.

Indeed, a thread of research leading up to the 2012 AIM workshop, and since, has been the desire to obtain, or show the existence of, bounds on the dimensions of the cohomology groups $\mathrm{H}^m(G(\mathbb{F}_q), V)$ that depend only on the (rank of the) underlying root system. Specifically, let G be a simple simply-connected algebraic group over an algebraically closed field k of characteristic $p > 0$. Assume G is defined and split over $\mathbb{F}_p$. Given $q = p^r$ with $r \geq 1$, let $G(\mathbb{F}_q)$ be the finite subgroup of $\mathbb{F}_q$-rational points in G. Cline, Parshall, and Scott [7] proved that there exists a constant $C(\Phi)$, depending only on the underlying root system Φ, such that for each irreducible $kG(\mathbb{F}_q)$-module V,

$$\dim \mathrm{H}^1(G(\mathbb{F}_q), V) \leq C(\Phi). \tag{1}$$

Arguing via different methods, Parker and Stewart [19] determined an explicit (large) constant that can be used for $C(\Phi)$ in (1), and which is given by a formula depending on the rank of Φ. On the other hand, if K is an algebraically closed field of characteristic relatively prime to q, and if G' is a finite simple group of Lie type (defined over $\mathbb{F}_q$), then Guralnick and Tiep showed for each irreducible KG'-module V that $\dim \mathrm{H}^1(G', V) \leq |W| + e$ [15, Theorem 1.3]. Here W is the Weyl group of G', and e is the twisted Lie rank of G'. Except for certain small values of q depending on the Lie type of G, the finite group $G(\mathbb{F}_q)$ is a central extension of a nonabelian simple group of Lie type; see [10, 2.2.6–2.2.7].

Parshall and Scott [21] later extended the Cline–Parshall–Scott result (1) to show for each irreducible rational G-module that

$$\dim \mathrm{H}^m(G, V) \leq c(\Phi, m) \tag{2}$$

for some constant $c(\Phi, m)$ depending only on Φ and the degree m. Then Bendel et al. [3] succeeded in finding a simultaneous generalization of (1) and (2), showing for each irreducible $kG(\mathbb{F}_q)$-module V that

$$\dim \mathrm{H}^m(G(\mathbb{F}_q), V) \leq C(\Phi, m) \tag{3}$$

for some constant $C(\Phi, m)$ depending only on Φ and m. Similar results were also obtained in [21] bounding the higher extension groups $\mathrm{Ext}^m_G(V_1, V_2)$, and in [3] bounding the groups $\mathrm{Ext}^m_{G(\mathbb{F}_q)}(V_1, V_2)$, assuming that V_1 and V_2 are irreducible rational G-modules (resp. $kG(\mathbb{F}_q)$-modules), though some additional restrictions on V_1 are necessary when $m > 1$.

1.2 Bounds Based on Dimension

In a different direction from the results described above, in this paper we explore bounds on the dimensions of cohomology groups that depend not on the rank of an underlying root system, but on the dimension of the coefficient module. This is in the spirit of a number of earlier general results providing bounds on the dimensions of $\mathrm{H}^1(S, V)$ and $\mathrm{H}^2(S, V)$ for S a finite group. Specifically, for $m = 1$, Guralnick and Hoffman proved:

Theorem 1 ([13, Theorem 1]). *Let S be a finite group, and let V be an irreducible kS-module on which S acts faithfully. Then*

$$\dim \mathrm{H}^1(S, V) \leq \tfrac{1}{2} \dim V.$$

In the case $m = 2$, the cohomology group $\mathrm{H}^2(S, V)$ parametrizes non-equivalent group extensions of V by S, and has connections with the study of profinite presentations. Guralnick, Kantor, Kassabov, and Lubotsky verified an earlier conjecture of Holt by proving the following theorem:

Theorem 2 ([14, Theorem B]). *Let S be a finite quasi-simple group, and let V be a kS-module. Then*

$$\dim \mathrm{H}^2(S, V) \leq (17.5) \dim V.$$

Guralnick et al. also showed that if S is an arbitrary finite group and if V is an irreducible kS-module that is faithful for S, then $\dim \mathrm{H}^2(S, V) \leq (18.5) \dim V$ [14, Theorem C], but that in general no analogue of this result can hold for $\mathrm{H}^m(S, V)$ when $m \geq 3$ [14, Theorem G]. Still, their work leaves open the possibility of finding constants $C(m)$ for each $m \geq 3$, such that $\dim \mathrm{H}^m(S, V) \leq C(m) \cdot \dim V$ when S is restricted to a suitable collection of finite groups. If such constants exist, we say that the cohomology groups $\mathrm{H}^m(S, V)$, for S in the specified collection of finite groups and V in the specified collection of kS-modules, are *linearly bounded*. We call the existence of such constants the *linear boundedness question* for the given groups and modules. More generally, we can consider the linear boundedness question for collections of algebraic groups and for accompanying collections of rational modules.

1.3 Overview

This paper investigates the linear boundedness question for the rational cohomology of a simple simply-connected algebraic group G over k that is defined and split over $\mathbb{F}_p$. In other words, we investigate, for $m \leq 3$, upper bounds on the dimension of $\mathrm{H}^m(G, M)$ for M a rational G-module (which the reader may assume to always be finite-dimensional, though we do not always make this assumption explicit, nor is it necessary for the validity of every result in this paper).

In this context we are able to exploit the existence of a Borel subgroup B in G (i.e., a maximal closed connected solvable subgroup in G) and a maximal torus T in B. In Sect. 2 we apply intricate calculations of Bendel, Nakano, and Pillen [5], Wright [27], and Andersen and Rian [2], summarized in Theorem 7, to prove for a finite-dimensional rational B-module M that

$$\dim \mathrm{H}^m(B, M) \leq \begin{cases} \dim M & \text{if } m = 1 \text{ or } 2, \\ 2 \dim M & \text{if } m = 3 \text{ and } p > h. \end{cases} \tag{4}$$

Here h is the Coxeter number for Φ. For a rational G-module M, it is well known that $\mathrm{H}^m(G, M) \cong \mathrm{H}^m(B, M)$, so the above inequalities yield general bounds on the dimensions of rational cohomology groups for G when m equals 1, 2, or 3. Further refinements are given in the case $m = 1$ through the explicit computation of $\mathrm{H}^1(B, \mu)$ for μ a one-dimensional B-module.

If M is a rational T-module, then M admits a weight space decomposition $M = \bigoplus_{\lambda \in X(T)} M_\lambda$. Here $X(T)$ is the character group of T. In particular, let M be a rational G-module. Using the weight space decomposition of M, in Sect. 3 we establish bounds on the dimension of $\mathrm{H}^1(G, M)$ in terms of the dimensions of the weight spaces of M.

This new idea gives much finer information than previous bounds depending only on the dimension of M, since it enables us to produce formulas in terms of the differences of dimensions of weight spaces. For example, given knowledge about the weight space decomposition of M, one can use these formulas in various situations to prove the vanishing of cohomology groups. More generally, when the dimension of M is relatively small, our bounds provide much more effective estimates on the dimension of $\mathrm{H}^1(G, M)$ than the estimates that arise through the methods of Parshall and Scott [21] or of Parker and Stewart [19].

As an application of our techniques, in Sect. 4 we show how to obtain effective bounds on the dimension of first cohomology groups for $S = \Sigma_d$, the symmetric group on d letters. We also demonstrate for $S = G(\mathbb{F}_q)$, with $q = p^r$ and r sufficiently large, that

$$\dim \mathrm{H}^m(G(\mathbb{F}_q), V) \leq \begin{cases} \frac{1}{h} \dim V & \text{if } m = 1, \\ \dim V & \text{if } m = 2, \\ 2 \dim V & \text{if } m = 3 \text{ and } p > h \end{cases} \tag{5}$$

for each $kG(\mathbb{F}_q)$-module V. Our results for $m = 2$ indicate that the bound given in Theorem 2 can be significantly improved when S is a finite Chevalley group. We do not treat the twisted finite groups of Lie type in this paper, but invite the reader to consider how our results could be extended to those cases.

1.4 Notation

We generally follow the notation and terminology of [18]. Let k be an algebraically closed field of characteristic $p > 0$. Let G be a simple simply-connected algebraic group scheme over k, defined and split over $\mathbb{F}_p$, and let $F : G \to G$ be the standard Frobenius morphism on G. For $r \geq 1$ and $q = p^r$, denote by G_r the r-th Frobenius kernel of G, and by $G(\mathbb{F}_q)$ the finite subgroup of $\mathbb{F}_q$-rational points in G, consisting of the fixed-points in $G(k)$ of F^r. Then $G(\mathbb{F}_q)$ is the universal version of an untwisted finite group of Lie type, as defined in [10, 2.2].

Let $T \subset G$ be a maximal torus, which we assume to be defined and split over $\mathbb{F}_p$. Let Φ be the set of roots of T in G, and let h be the Coxeter number of Φ. Then Φ is an indecomposable root system. Fix a set of simple roots $\Delta \subset \Phi$, and denote the corresponding sets of positive and negative roots in Φ by Φ^+ and Φ^-, respectively. Write $W = N_G(T)/T$ for the Weyl group of Φ. Let $B = T \cdot U \subset G$ be a Borel subgroup containing T, with unipotent radical U corresponding to Φ^-. Set $B(\mathbb{F}_q) = B \cap G(\mathbb{F}_q)$, $U(\mathbb{F}_q) = U \cap G(\mathbb{F}_q)$, and $T(\mathbb{F}_q) = T \cap G(\mathbb{F}_q)$. Similarly, set $B_r = B \cap G_r$, $U_r = U \cap G_r$, and $T_r = T \cap G_r$. We write $\mathfrak{g} = \mathrm{Lie}(G)$ and $\mathfrak{u} = \mathrm{Lie}(U)$ for the Lie algebras of G and U, respectively. Then $\mathfrak{g}$ and $\mathfrak{u}$ are naturally restricted Lie algebras. We write $u(\mathfrak{g})$ for the restricted enveloping algebra of $\mathfrak{g}$, and $U(\mathfrak{g})$ for the ordinary universal enveloping algebra of $\mathfrak{g}$.

Let $X(T)$ be the character group of T. Write

$$X(T)_+ = \{\lambda \in X(T) : (\lambda, \alpha^\vee) \geq 0 \text{ for all } \alpha \in \Delta\}$$

for the set of dominant weights in $X(T)$, and for $r \geq 1$, write

$$X_r(T) = \{\lambda \in X(T)_+ : (\lambda, \alpha^\vee) < p^r \text{ for all } \alpha \in \Delta\}$$

for the set of p^r-restricted dominant weights in $X(T)$. For each $\lambda \in X(T)_+$, let $H^0(\lambda) = \mathrm{ind}_B^G(\lambda)$ be the corresponding induced module, which has irreducible socle $\mathrm{soc}_G H^0(\lambda) = L(\lambda)$. Each irreducible rational G-module is isomorphic to $L(\lambda)$ for some $\lambda \in X(T)_+$. Since G is assumed to be simply-connected, the $L(\lambda)$ for $\lambda \in X_r(T)$ form a complete set of pairwise nonisomorphic irreducible $kG(\mathbb{F}_q)$-modules.

2 Bounds on Rational Cohomology Groups

2.1 *Weight Spaces and B-Cohomology*

The irreducible B-modules are one-dimensional and are identified with elements of $X(T)$ via inflation from T to B. For a finite-dimensional rational B-module M, the B-module composition factors of M can be read off with multiplicities from its weight space decomposition. By considering the long exact sequence in cohomology, it follows for each $m \geq 0$ that one has the inequality

$$\dim \mathrm{H}^m(B, M) \leq \textstyle\sum_{\mu \in X(T)} \dim M_\mu \cdot \dim \mathrm{H}^m(B, \mu). \qquad (6)$$

Thus, if one can determine a bound on the dimension of $\mathrm{H}^m(B, \mu)$ that depends only on m and not on μ, then one can obtain a similar bound on the dimension of $\mathrm{H}^m(B, M)$ that depends only on m and the dimension of M. In particular, if M is a rational G-module considered also as a rational B-module by restriction, then one has $\mathrm{H}^\bullet(G, M) \cong \mathrm{H}^\bullet(B, M)$ by [18, II.4.7], so a bound on the dimension of $\mathrm{H}^m(B, M)$ automatically yields a bound on the dimension of $\mathrm{H}^m(G, M)$.

For $m = 1$, we can use Andersen's explicit computation of $\mathrm{H}^1(B, \mu)$ for each $\mu \in X(T)$, together with the formula (6), to give a general upper bound on $\dim \mathrm{H}^1(B, M)$.

Theorem 3 ([1, Corollary 2.4]). *Let $\mu \in X(T)$. Then*

$$\mathrm{H}^1(B, \mu) \cong \begin{cases} k & \text{if } \mu = -p^t\alpha \text{ for some } \alpha \in \Delta \text{ and } t \geq 0, \\ 0 & \text{otherwise.} \end{cases}$$

Corollary 4. *Let M be a finite-dimensional rational B-module. Then*

$$\dim \mathrm{H}^1(B, M) \leq \sum_{\alpha \in \Delta, t \geq 0} \dim M_{-p^t\alpha} \leq \dim M.$$

In particular, if M is a finite-dimensional rational G-module, then

$$\dim \mathrm{H}^1(G, M) \leq \sum_{\alpha \in \Delta,\, t \geq 0} \dim M_{-p^t\alpha}.$$

2.2 Applications to G-Cohomology

We can now apply the results of the preceding section to give bounds on the dimension of $\mathrm{H}^1(G, M)$ for M a rational G-module, by considering the action of the Weyl group on the set of weights of M.

Theorem 5. *Let M be a finite-dimensional rational G-module. Then*

$$\dim \mathrm{H}^1(G, M) \leq \tfrac{1}{h} \dim M.$$

Proof. Recall that restriction from G to B induces an isomorphism $\mathrm{H}^\bullet(G, M) \cong \mathrm{H}^\bullet(B, M)$. Then

$$\dim \mathrm{H}^1(G, M) \leq \textstyle\sum_{\alpha \in \Delta,\, t \geq 0} \dim M_{-p^t\alpha}$$

by Corollary 4. Next, the set of weights of a rational G-module is invariant under the action of the ambient Weyl group W, and all roots in Φ of a given root length lie in a single W-orbit. In particular, if $t \geq 0$, and if $\alpha, \beta \in \Phi$ are of the same length, then $\dim M_{-p^t\alpha} = \dim M_{-p^t\beta}$. Let Φ_s (resp. Φ_l) denote the set of short (resp. long) roots in Φ, and set $\Delta_s = \Delta \cap \Phi_s$ (resp. $\Delta_l = \Delta \cap \Phi_l$). Then one can check that $|\Phi_s| = h \cdot |\Delta_s|$ and $|\Phi_l| = h \cdot |\Delta_l|$; cf. [16, Proposition 3.18] for the case of one root length. Together these equalities imply that $\sum_{\alpha \in \Delta,\, t \geq 0} \dim M_{-p^t\alpha} \leq \frac{1}{h} \dim M$. □

Corollary 6. *Let M be a finite-dimensional rational G-module. Then*

$$\dim \mathrm{H}^1(G, M) \leq \tfrac{1}{2} \dim M.$$

Proof. This follows from Theorem 5 since the Coxeter number is always at least 2. □

If $M = L(\lambda)$ is an irreducible rational G-module, then we can use Theorem 1 to give an alternate proof of Corollary 6, as follows. To begin, we may assume by the Linkage Principle that $\lambda \in \mathbb{Z}\Phi$, and also that $\lambda \neq 0$, since $L(0) = k$ and $\mathrm{H}^1(G, k) = 0$ [18, II.4.11]. Next, choose $r > 1$ such that $\lambda \in X_r(T)$, and set $q = p^r$. Then $L(\lambda)$ is an irreducible $kG(\mathbb{F}_q)$-module, and the restriction map $\mathrm{H}^1(G, L(\lambda)) \to \mathrm{H}^1(G(\mathbb{F}_q), L(\lambda))$ is injective by [8, 7.4]. Now write $Z(G(\mathbb{F}_q))$ for the center of $G(\mathbb{F}_q)$. Since G is simply-connected (i.e., is of universal type), $G'(\mathbb{F}_q) := G(\mathbb{F}_q)/Z(G(\mathbb{F}_q))$ is a nonabelian finite simple group by [10, 2.2.6–2.2.7]; this uses the fact that $r > 1$. Also, $Z(G(\mathbb{F}_q))$ is a subgroup of $Z(G)$ by [10, 2.5.9]. Since $Z(G) = \bigcap_{\alpha \in \Phi} \ker(\alpha) \subset T$ acts trivially on $L(\lambda)$ whenever $\lambda \in \mathbb{Z}\Phi$, it follows that $L(\lambda)$ is naturally a nontrivial irreducible module for $G'(\mathbb{F}_q)$. In particular, $G'(\mathbb{F}_q)$ must act faithfully on $L(\lambda)$. The group $Z(G(\mathbb{F}_q))$ has order prime to p, so it follows from considering the Lyndon–Hochschild–Serre spectral sequence for the group extension $1 \to Z(G(\mathbb{F}_q)) \to G(\mathbb{F}_q) \to G'(\mathbb{F}_q) \to 1$ that $\mathrm{H}^1(G(\mathbb{F}_q), L(\lambda)) \cong \mathrm{H}^1(G'(\mathbb{F}_q), L(\lambda))$. Then $\dim \mathrm{H}^1(G(\mathbb{F}_q), L(\lambda)) \leq \frac{1}{2} \dim L(\lambda)$ by Theorem 1.

2.3 Bounds for Second and Third Cohomology Groups

Now we consider $\mathrm{H}^m(B, \mu)$, $\mathrm{H}^m(B, M)$, and $\mathrm{H}^m(G, M)$ for m equal to 2 or 3. First recall the following results:

Theorem 7. *Let $\mu \in X(T)$. Then*

(a) $\dim \mathrm{H}^2(B, \mu) \leq 1$.
(b) *If $p > h$, then* $\dim \mathrm{H}^3(B, \mu) \leq 2$.

Proof. For (a), see [5, Theorem 5.8] and [27, Theorem 4.1.1]. For (b), see [2, Theorem 5.2]. □

Applying (6) to the preceding theorem, and using the fact that $\mathrm{H}^\bullet(G, M) \cong \mathrm{H}^\bullet(B, M)$ for each rational G-module M, one obtains:

Corollary 8. *Let M be a finite-dimensional rational B-module. Then*

(a) $\dim \mathrm{H}^2(B, M) \leq \dim M$.
(b) *If $p > h$, then* $\dim \mathrm{H}^3(B, M) \leq 2 \cdot \dim M$.

In particular, if M is a finite-dimensional rational G-module, then these inequalities also hold with B replaced by G.

As in Corollary 4, a stronger version of the above result can be obtained by considering precisely which weights μ of M satisfy $\mathrm{H}^m(B, \mu) \neq 0$. The results in the corollary motivate posing the following question, an affirmative answer to which would yield, for each $m \geq 1$, upper bounds on the dimensions of $\mathrm{H}^m(B, M)$ and $\mathrm{H}^m(G, M)$ for each finite-dimensional rational G-module M.

Question 9 (Linear boundedness for Borel subgroups). *For each $m \geq 1$, does there exist a constant $C(m)$, depending on m but independent of the rank of G or of the weight $\mu \in X(T)$, such that* $\dim \mathrm{H}^m(B, \mu) \leq C(m)$*?*

2.4 Bounds for Finite Chevalley Groups

In this section we discuss some rough analogues for the finite subgroup $G(\mathbb{F}_q)$ of G of the results in Sects. 2.2 and 2.3. While the bounds presented here are often significantly worse than those given by Theorem 1, we point out that they can be obtained using only purely elementary methods. Recall that p is *nonsingular* for G if $p > 2$ when Φ is of type B, C, or F, and if $p > 3$ when Φ is of type G_2.

Theorem 10. *Let M be a finite-dimensional $kG(\mathbb{F}_q)$-module, and suppose p is nonsingular for G. Then*

$$\dim \mathrm{H}^1(G(\mathbb{F}_q), M) \leq r \cdot |\Delta| \cdot \dim M.$$

Proof. Recall that $U(\mathbb{F}_q)$ is a Sylow p-subgroup of $G(\mathbb{F}_q)$ [10, 2.3.4]. In particular, the index of $U(\mathbb{F}_q)$ in $G(\mathbb{F}_q)$ is prime to p, so restriction from $G(\mathbb{F}_q)$ to $U(\mathbb{F}_q)$ defines an injection

$$\mathrm{H}^\bullet(G(\mathbb{F}_q), M) \hookrightarrow \mathrm{H}^\bullet(U(\mathbb{F}_q), M).$$

Since $U(\mathbb{F}_q)$ is a p-group, each irreducible $kU(\mathbb{F}_q)$-module is isomorphic to k. Then considering a $U(\mathbb{F}_q)$-composition series for M, and using the long exact sequence in cohomology, it follows by induction on the dimension of M that

$$\dim \mathrm{H}^1(U(\mathbb{F}_q), M) \leq \dim \mathrm{H}^1(U(\mathbb{F}_q), k) \cdot \dim M.$$

Now $\mathrm{H}^1(U(\mathbb{F}_q), k)$ identifies with the space of k-linear maps $kU(\mathbb{F}_q)_{ab} \to k$. Here $U(\mathbb{F}_q)_{ab}$ is the abelianization of $U(\mathbb{F}_q)$. Since p is nonsingular, it follows from [10, 3.3.1] that $U(\mathbb{F}_q)_{ab} \cong (\mathbb{F}_q)^{|\Delta|}$ as an abelian group. Specifically, $U(\mathbb{F}_q)_{ab}$ identifies with the direct product of the root subgroups in $U(\mathbb{F}_q)$ corresponding to simple roots. Since $q = p^r$, $k\mathbb{F}_q = k \otimes_{\mathbb{F}_p} \mathbb{F}_q$ has k-dimension r. Then $\dim \mathrm{H}^1(U(\mathbb{F}_q), k) = r \cdot |\Delta|$. □

More generally, one can argue as in the proof of the theorem to show for each $m \geq 0$ that

$$\dim \mathrm{H}^m(G(\mathbb{F}_q), M) \leq \dim \mathrm{H}^m(U(\mathbb{F}_q), k) \cdot \dim M.$$

In turn, the dimension of $\mathrm{H}^m(U(\mathbb{F}_q), k)$ is bounded above by the dimension of the cohomology group $\mathrm{H}^m(u(\mathfrak{u}^{\oplus r}), k)$ for the restricted enveloping algebra $u(\mathfrak{u}^{\oplus r})$; see [26, 2.4].

3 Bounds Depending on Weight Space Multiplicities

3.1 Bounds for Simple Modules

Next we explore some bounds on dim $\mathrm{H}^1(G, M)$ that depend on weight space multiplicities.

Lemma 11. *Let* $\lambda \in X(T)_+$. *Then* dim $\mathrm{H}^1(G, L(\lambda)) \leq \dim H^0(\lambda)_0$.

Proof. If $\lambda = 0$, then the result follows because $H^0(0) = L(0) = k$ and $\mathrm{H}^1(G, k) = 0$. So assume that $\lambda \neq 0$. There exists a short exact sequence of G-modules

$$0 \to L(\lambda) \to H^0(\lambda) \to Q \to 0.$$

Since $\mathrm{H}^1(G, H^0(\lambda)) = 0$ by [18, II.4.13], and since $\mathrm{Hom}_G(k, L(\lambda)) = 0$ by the assumption $\lambda \neq 0$, the corresponding long exact sequence in cohomology yields $\mathrm{H}^1(G, L(\lambda)) \cong \mathrm{Hom}_G(k, Q)$. Now the multiplicity of the trivial module in $\mathrm{soc}_G\, Q$ is bounded above by the composition multiplicity of the trivial module in $H^0(\lambda)$, which is bounded above by the weight space multiplicity dim $H^0(\lambda)_0$. □

In some cases we can improve the conclusion of the lemma to a strict inequality. Suppose that the set of weights of T in $L(\lambda)$ is equal to that of $H^0(\lambda)$. By [23], this condition is known to hold if $\lambda \in X_1(T)$ and p is good for Φ. Recall that p is good for Φ provided $p > 2$ when Φ is of type B_n, C_n, or D_n; $p > 3$ when Φ is of type F_4, G_2, E_6, or E_7; and provided $p > 5$ when Φ is of type E_8. By the Linkage Principle, $\mathrm{H}^1(G, L(\lambda)) = 0$ unless $\lambda \in \mathbb{Z}\Phi$. It is well known that the set of weights of $H^0(\lambda)$ is saturated. Thus, if $\lambda \in \mathbb{Z}\Phi \cap X(T)_+$, then 0 is a weight of $H^0(\lambda)$, and hence also of $L(\lambda)$. In particular, dim $Q_0 <$ dim $H^0(\lambda)_0$. Since dim $\mathrm{Hom}_G(k, Q)$ is bounded above by dim Q_0, we obtain in this case that dim $\mathrm{H}^1(G, L(\lambda)) <$ dim $H^0(\lambda)_0$.

3.2 Relating G-Cohomology to $\mathfrak{g}$-Cohomology

We now present a preliminary result that relates $\mathrm{H}^1(G, M)$ to cohomology for $\mathfrak{g} = \mathrm{Lie}(G)$, by way of cohomology for the Frobenius kernel G_1. Write $\mathrm{H}^\bullet(\mathfrak{g}, M) = \mathrm{H}^\bullet(U(\mathfrak{g}), M)$ for the ordinary Lie algebra cohomology of $\mathfrak{g}$ with coefficients in the $\mathfrak{g}$-module M. If M is a rational G-module, then the adjoint action of G on $\mathfrak{g}$, together with the given action of G on M, induce a rational G-module structure on $\mathrm{H}^\bullet(\mathfrak{g}, M)$.

Lemma 12. *Let M be a rational G-module, and suppose that $M^{G_1} = 0$. Then*

(a) *Restriction from G to G_1 induces an isomorphism* $\mathrm{H}^1(G, M) \cong \mathrm{H}^1(G_1, M)^{G/G_1}$.
(b) *There exists a G-equivariant isomorphism* $\mathrm{H}^1(G_1, M) \cong \mathrm{H}^1(\mathfrak{g}, M)$.
(c) $\mathrm{H}^1(G, M) \cong \mathrm{H}^1(\mathfrak{g}, M)^G$.

Proof. Consider the Lyndon–Hochschild–Serre spectral sequence

$$E_2^{i,j} = \mathrm{H}^i(G/G_1, \mathrm{H}^j(G_1, M)) \Rightarrow \mathrm{H}^{i+j}(G, M), \tag{7}$$

and its associated five-term exact sequence

$$0 \to E_2^{1,0} \to \mathrm{H}^1(G, M) \to E_2^{0,1} \to E_2^{2,0} \to \mathrm{H}^2(G, M). \tag{8}$$

Since $\mathrm{Hom}_{G_1}(k, M) = M^{G_1} = 0$, one has $E_2^{i,0} = 0$ for all $i \geq 0$. Then (8) yields that restriction from G to G_1 defines an isomorphism $\mathrm{H}^1(G, M) \cong \mathrm{H}^1(G_1, M)^{G/G_1}$. This proves (a).

For (b), recall that the representation theory of the restricted enveloping algebra $u(\mathfrak{g})$ is naturally equivalent to that of the Frobenius kernel G_1 [18, I.9.6]. We thus identify $\mathrm{H}^1(G_1, M)$ and $\mathrm{H}^1(u(\mathfrak{g}), M)$ via this equivalence. Since $u(\mathfrak{g})$ is a homomorphic image of $U(\mathfrak{g})$, and since the quotient map $U(\mathfrak{g}) \to u(\mathfrak{g})$ is compatible with the adjoint action of G, there exists a corresponding G-module homomorphism $\mathrm{H}^1(G_1, M) \to \mathrm{H}^1(\mathfrak{g}, M)$, which by [18, I.9.19(1)] fits into an exact sequence

$$0 \to \mathrm{H}^1(G_1, M) \to \mathrm{H}^1(\mathfrak{g}, M) \to \mathrm{Hom}^s(\mathfrak{g}, M^{\mathfrak{g}}). \tag{9}$$

Here, given a vector space V, $\mathrm{Hom}^s(\mathfrak{g}, V)$ denotes the set of additive functions $\varphi : \mathfrak{g} \to V$ satisfying the property $\varphi(ax) = a^p\varphi(x)$ for all $a \in k$ and $x \in \mathfrak{g}$. Since M is a rational G-module, it is in particular a restricted $\mathfrak{g}$-module, i.e., a $u(\mathfrak{g})$-module. Then $M^{\mathfrak{g}} = M^{U(\mathfrak{g})} = M^{u(\mathfrak{g})} = M^{G_1}$. This space is zero by assumption, so we conclude that $\mathrm{H}^1(G_1, M) \cong \mathrm{H}^1(\mathfrak{g}, M)$. This proves (b). Now (c) follows immediately from (a) and (b). □

Replacing G by B in the previous proof, one obtains the following lemma:

Lemma 13. *Let M be a rational B-module, and suppose that $M^{B_1} = 0$. Then restriction from B to B_1 induces an isomorphism $\mathrm{H}^1(B, M) \cong \mathrm{H}^1(B_1, M)^{B/B_1}$. In particular, restriction from B to U_1 defines an injection $\mathrm{H}^1(B, M) \hookrightarrow \mathrm{H}^1(U_1, M)^T$.*

Proof. It remains to explain the last statement in the lemma. Since B_1 is the semidirect product of U_1 and the diagonalizable group scheme T_1, it follows that restriction from B_1 to U_1 defines an isomorphism $\mathrm{H}^\bullet(B_1, M) \cong \mathrm{H}^\bullet(U_1, M)^{T_1}$ [18, I.6.9]. Then restriction from B to U_1 defines an isomorphism $\mathrm{H}^1(B, M) \cong \mathrm{H}^1(U_1, M)^{B/U_1}$, and the latter space is a subspace of $\mathrm{H}^1(U_1, M)^T$. □

3.3 Bounds Based on Weight Spaces

The results in the preceding section can be employed to establish upper bounds for $\mathrm{H}^1(G, M)$ in terms of specific weight space multiplicities.

Proposition 14. *Let M be a rational B-module, and suppose that $M^{B_1} = 0$. Let $\Delta' \subseteq \Phi^+$ be a set of roots such that the root spaces $\mathfrak{u}_{-\alpha}$ for $\alpha \in \Delta'$ generate $\mathfrak{u}$ as a Lie algebra. Then*

$$\dim \mathrm{H}^1(B, M) \leq \textstyle\sum_{\alpha \in \Delta'} \dim M_{-\alpha} - \dim M_0.$$

Proof. First, $\mathrm{H}^1(B, M)$ injects into $\mathrm{H}^1(U_1, M)^T$ by Lemma 13. Next, replacing G by U in (9), there exists a B-equivariant injection $\mathrm{H}^1(U_1, M) \hookrightarrow \mathrm{H}^1(\mathfrak{u}, M)$. Now recall that $\mathrm{H}^1(\mathfrak{u}, M)$ fits into an exact sequence

$$0 \to \mathrm{Inn}(\mathfrak{u}, M) \to \mathrm{Der}(\mathfrak{u}, M) \to \mathrm{H}^1(\mathfrak{u}, M) \to 0. \tag{10}$$

Here $\mathrm{Der}(\mathfrak{u}, M)$ is the space of all Lie algebra derivations of $\mathfrak{u}$ into M, and $\mathrm{Inn}(\mathfrak{u}, M)$ is the space of all inner derivations of $\mathfrak{u}$ into M. Since M is a rational B-module, the conjugation action of B on $\mathfrak{u}$ makes (10) into an exact sequence of rational B-modules. Then applying the exact functor $(-)^T$ to (10), one obtains

$$\dim \mathrm{H}^1(B, M) \leq \dim \mathrm{H}^1(\mathfrak{u}, M)^T = \dim \mathrm{Der}(\mathfrak{u}, M)^T - \dim \mathrm{Inn}(\mathfrak{u}, M)^T.$$

As rational B-modules, $\mathrm{Inn}(\mathfrak{u}, M) \cong M/M^{\mathfrak{u}}$. Observe that $(M^{\mathfrak{u}})^T = (M^{U_1})^T \subseteq M^{B_1} = 0$. Then it follows that $\mathrm{Inn}(\mathfrak{u}, M)^T \cong (M/M^{\mathfrak{u}})^T \cong M_0$. Finally, a Lie algebra derivation $\mathfrak{u} \to M$ is completely determined by its action on a set of Lie algebra generators for $\mathfrak{u}$, say, the root subspaces $\mathfrak{u}_{-\alpha}$ for $\alpha \in \Delta'$. Moreover, a T-invariant derivation $\mathfrak{u} \to M$ must map $\mathfrak{u}_{-\beta}$ into $M_{-\beta}$ for each $\beta \in \Phi^+$. Then $\dim \mathrm{Der}(\mathfrak{u}, M)^T \leq \sum_{\alpha \in \Delta'} \dim M_{-\alpha}$. Combining this with the previous observations, we obtain the inequality in the statement of the proposition. □

Remark 15. If p is nonsingular for G, then one can take $\Delta' = \Delta$ in Proposition 14.

The next result is an analogue for algebraic groups of [6, Corollary 2.9]. Let α_0 be the highest short root in Φ, and let $\tilde{\alpha}$ be the highest long root in Φ.

Corollary 16. *Let M be a rational G-module, and suppose that $M^{B_1} = 0$. Let $\Delta' \subseteq \Phi^+$ be a set of roots such that the root spaces $\mathfrak{u}_{-\alpha}$ for $\alpha \in \Delta'$ generate $\mathfrak{u}$ as a Lie algebra. Then*

$$\dim \mathrm{H}^1(G, M) \leq \textstyle\sum_{\alpha\in\Delta'} \dim M_\alpha - \dim M_0.$$

In particular, suppose p is nonsingular for G. Then

$$\dim \mathrm{H}^1(G, M) \leq |\Delta_s| \cdot \dim M_{\alpha_0} + |\Delta_l| \cdot \dim M_{\tilde{\alpha}} - \dim M_0,$$

where by convention we consider all roots in Φ as long, and set $\Delta_s = \emptyset$, whenever Φ has roots of only a single root length.

Proof. Observe that if the root spaces $\mathfrak{u}_{-\alpha}$ for $\alpha \in \Delta'$ generate $\mathfrak{u}$ as a Lie algebra, then so do the root spaces $\mathfrak{u}_{w_0\alpha}$ for $\alpha \in \Delta'$. Here $w_0 \in W$ is the longest element of W. One has $\mathrm{H}^\bullet(G, M) \cong \mathrm{H}^\bullet(B, M)$, so $\dim \mathrm{H}^1(G, M) \leq \sum_{\alpha\in\Delta'} \dim M_{w_0\alpha} - \dim M_0$. But $\dim M_\lambda = \dim M_{w\lambda}$ whenever $w \in W$, so we conclude that $\dim \mathrm{H}^1(G, M) \leq \sum_{\alpha\in\Delta'} \dim M_\alpha - \dim M_0$. In particular, if p is nonsingular for G, then we can take $\Delta' = \Delta$. Since all roots of a given root length in Φ are conjugate under W, we then have $\dim M_\alpha = \dim M_{\alpha_0}$ whenever $\alpha \in \Delta_s$, and $\dim M_\alpha = \dim M_{\tilde{\alpha}}$ whenever $\alpha \in \Delta_l$, so that $\sum_{\alpha\in\Delta} \dim M_\alpha = |\Delta_s| \cdot \dim M_{\alpha_0} + |\Delta_l| \cdot \dim M_{\tilde{\alpha}}$. □

The bounds established in Corollary 16 can be improved if we assume that the Weyl group has order prime to p. For the next theorem, recall that if M is a rational G-module, then the 0-weight space M_0 of M is naturally a module for $W = N_G(T)/T$.

Theorem 17. *Let M be a rational G-module. Assume that $M^{G_1} = 0$, and that $p \nmid |W|$. Then*

$$\dim \mathrm{H}^1(G, M) \leq \dim M_{\alpha_0} + \dim M_{\tilde{\alpha}} - \dim M_0^W,$$

where by convention we say $M_{\alpha_0} = 0$ if Φ has roots of only a single root length. In particular,

$$\dim \mathrm{H}^1(G, M) \leq \begin{cases} \frac{1}{|\Phi|} \cdot \dim M & \text{if } \Phi \text{ has one root length,} \\ \frac{2}{|\Phi|} \cdot \dim M & \text{if } \Phi \text{ has two root lengths.} \end{cases}$$

Proof. The second statement follows from the first by applying the argument given in the proof of Theorem 5, so we proceed to prove the first statement. Set $N = N_G(T)$, and observe that the fixed-point functor $(-)^N$ factors as the composition of the exact functor $(-)^T$ with the functor $(-)^{N/T} = (-)^W$, which is also exact by the assumption that the finite group W has order prime to p. Then $(-)^N$ is exact. One has $\mathrm{H}^1(G,M) \cong \mathrm{H}^1(\mathfrak{g},M)^G$ by Lemma 12(c), so in particular $\dim \mathrm{H}^1(G,M) \leq \dim \mathrm{H}^1(\mathfrak{g},M)^N$. Now applying the exact functor $(-)^N$ to the exact sequence of rational G-modules

$$0 \to \mathrm{Inn}(\mathfrak{g},M) \to \mathrm{Der}(\mathfrak{g},M) \to \mathrm{H}^1(\mathfrak{g},M) \to 0,$$

one obtains

$$\dim \mathrm{H}^1(\mathfrak{g},M)^N = \dim \mathrm{Der}(\mathfrak{g},M)^N - \dim \mathrm{Inn}(\mathfrak{g},M)^N.$$

As a rational G-module, $\mathrm{Inn}(\mathfrak{g},M) \cong M/M^{\mathfrak{g}}$. But $M^{\mathfrak{g}} = M^{G_1} = 0$ by assumption, so

$$\dim \mathrm{H}^1(\mathfrak{g},M)^N = \dim \mathrm{Der}(\mathfrak{g},M)^N - \dim M^N.$$

Observe that $M^N = (M^T)^W = M_0^W$. Also, an N-invariant derivation $\delta : \mathfrak{g} \to M$ is in particular T-invariant, and so must map $\mathfrak{g}_\beta$ into M_β for each $\beta \in \Phi$. All roots of a given length in Φ are conjugate under W, so it follows that an N-invariant derivation δ is uniquely determined by its values on any single root space in $\mathfrak{g}$ if Φ has only one root length, and by its values on any pair of root spaces in $\mathfrak{g}$ corresponding to a long root and a short root if Φ has two root lengths. In particular, $\dim \mathrm{Der}(\mathfrak{g},M)^N \leq \dim M_{\alpha_0} + \dim M_{\tilde{\alpha}}$, where by convention we say that $M_{\alpha_0} = 0$ if Φ only has roots of a single root length. Combining this and the preceding observations, one obtains the first statement of the theorem. □

Remark 18. The assumption $M^{B_1} = 0$, and hence also $M^{G_1} = 0$, is satisfied if $M = L(\lambda)$ for some $\lambda \in X(T)_+$ with $\lambda \notin pX(T)$. Indeed, in this case $\lambda = \lambda_0 + p\lambda_1$ for some $0 \neq \lambda_0 \in X_1(T)$ and some $\lambda_1 \in X(T)_+$. Then $L(\lambda) \cong L(\lambda_0) \otimes L(\lambda_1)^{(1)}$ by the Steinberg tensor product theorem, so that $L(\lambda)^{B_1} \cong L(\lambda_0)^{B_1} \otimes L(\lambda_1)^{(1)}$. Now $L(\lambda_0)^{U_1} = L(\lambda_0)_{w_0\lambda_0}$ (cf. [18, II.3.12]), so it follows that $L(\lambda_0)^{B_1} = (L(\lambda_0)^{U_1})^{T_1} = 0$ because $w_0\lambda_0 \notin pX(T)$ by the condition $0 \neq \lambda_0 \in X_1(T)$.

If $\lambda = 0$, then $L(\lambda) = k$, and one has $\mathrm{H}^1(G,k) = 0$ and $\mathrm{H}^1(\mathfrak{g},k) \cong (\mathfrak{g}/[\mathfrak{g},\mathfrak{g}])^* = 0$, so that Lemma 12(c) holds in this case. If also $p \neq 2$ when Φ is of type C_n, then $\mathrm{H}^1(G_1,k) = 0$ [18, II.12.2], which recovers all parts of Lemma 12 when $M = k$. Now let $\lambda \in X(T)_+ \cap pX(T)$ be nonzero. Then $\lambda = p^s\mu$ for some $\mu \in X(T)_+$ with $\mu \notin pX(T)$, and $L(\lambda) \cong L(\mu)^{(s)}$ is trivial as a G_s-module. Suppose that $p \neq 2$ if Φ is of type C_n. Then $\mathrm{H}^1(G_s, L(\lambda)) \cong \mathrm{H}^1(G_s,k) \otimes L(\lambda) = 0$ by [18, II.12.2], and $\mathrm{Hom}_{G_s}(k, L(\lambda)) \cong L(\lambda)$, so replacing G_1 by G_s in (7), the corresponding five-term exact sequence shows that the inflation map induces an isomorphism

$$\mathrm{H}^1(G/G_s, L(\mu)^{(s)}) \cong \mathrm{H}^1(G, L(\lambda)).$$

Identifying G/G_s with the Frobenius twist $G^{(s)}$ of G, we can make the identification

$$\mathrm{H}^1(G/G_s, L(\mu)^{(s)}) \cong \mathrm{H}^1(G, L(\mu)).$$

Then there exists a vector space isomorphism $\mathrm{H}^1(G, L(\lambda)) \cong \mathrm{H}^1(G, L(\mu))$. This shows that, assuming that $p \neq 2$ when Φ is of type C_n, Corollary 16 and Theorem 17 can still be applied to provide bounds on dim $\mathrm{H}^1(G, L(\lambda))$ when $\lambda \in pX(T)$.

Remark 19. The results in Sects. 3.2 and 3.3 remain true, with exactly the same proofs, under the weaker assumption that G is a connected reductive algebraic group over k that is defined and split over $\mathbb{F}_p$, that $T \subset G$ is a maximal split torus in G, that the root system Φ of T in G is indecomposable, that $B \subset G$ is a Borel subgroup of G, etc. In particular, the results hold for $G = GL_n(k)$ when $n \geq 2$.

4 Applications

4.1 Bounds for Σ_d

In this section only, let $S = \Sigma_d$ be the symmetric group on d letters ($d \geq 2$), and let $G = GL_d(k)$ be the general linear group. It is well known, from considering commuting actions on tensor space, that there are close connections between the representation theories of S and G.

Let $T \subset G$ be the subgroup of diagonal matrices, and write $X(T) = \bigoplus_{i=1}^{d} \mathbb{Z}\varepsilon_i$ for the character group of T. Here $\varepsilon_i : T \to k$ is the i-th diagonal coordinate function on T. Recall that the set $X(T)_+$ of dominant weights on T consists of the weights $\lambda = \sum_{i=1}^{d} a_i\varepsilon_i \in X(T)$ with $a_i - a_{i+1} \geq 0$ for each $1 \leq i < d$. Then identifying a partition $\lambda = (\lambda_1, \lambda_2, \ldots)$ of d with the weight $\lambda = \sum_{i=1}^{d} \lambda_i\varepsilon_i$, the set of partitions of d is naturally a subset of $X(T)_+$. Moreover, this subset parametrizes the irreducible degree-d polynomial representations of G. Recall that a partition $\lambda = (\lambda_1, \lambda_2, \ldots)$ is p-restricted if $\lambda_i - \lambda_{i+1} < p$ for each $i \geq 1$, and is p-regular if no nonzero part λ_i of the partition is repeated p or more times. Then the irreducible $k\Sigma_d$-modules are indexed by the set Λ_{res} of p-restricted partitions of d. Given $\lambda \in \Lambda_{\mathrm{res}}$, let D_λ be the corresponding irreducible $k\Sigma_d$-module, and write sgn for the sign representation of Σ_d. Given a partition λ, write λ' for the transpose partition. Then the irreducible $k\Sigma_d$-modules can be indexed by p-regular partitions by setting $D^{\lambda'} = D_\lambda \otimes \mathrm{sgn}$ for each $\lambda \in \Lambda_{\mathrm{res}}$.

Doty, Erdmann, and Nakano constructed a spectral sequence relating the cohomology theories for GL_d and Σ_d, showing for $p \geq 3$ and $\lambda \in \Lambda_{\mathrm{res}}$ that

$$\mathrm{H}^1(\Sigma_d, D^{\lambda'}) = \mathrm{H}^1(\Sigma_d, D_\lambda \otimes \mathrm{sgn}) \cong \mathrm{Ext}^1_G(\delta, L(\lambda)), \tag{11}$$

where $\delta = (1^d)$ is the one-dimensional determinant representation of GL_d [9, Theorem 5.4(a)]; cf. also [20, Theorem 4.6(b)]. One can now apply Corollary 16 to obtain the following bound for the first cohomology of symmetric groups.

Theorem 20. *Suppose $p \geq 3$, and let $\lambda \in \Lambda_{\text{res}}$ with $\lambda \neq (1^d)$. Then*

$$\dim \mathrm{H}^1(\Sigma_d, D^{\lambda'}) \leq \textstyle\sum_{\alpha\in\Delta} \dim L(\lambda)_{\delta+\alpha} - \dim L(\lambda)_\delta.$$

Proof. One has $\mathrm{H}^1(\Sigma_d, D^{\lambda'}) \cong \mathrm{Ext}^1_G(\delta, L(\lambda))$ by (11). Then by Corollary 16 and Remark 19,

$$\begin{aligned}\dim \mathrm{Ext}^1_G(\delta, L(\lambda)) &= \dim \mathrm{H}^1(G, L(\lambda) \otimes -\delta)\\ &\leq \textstyle\sum_{\alpha\in\Delta} \dim(L(\lambda) \otimes -\delta)_\alpha - \dim(L(\lambda) \otimes -\delta)_0\\ &= \textstyle\sum_{\alpha\in\Delta} \dim L(\lambda)_{\delta+\alpha} - \dim L(\lambda)_\delta.\end{aligned}$$

□

We present the following example to illustrate how Theorem 20 can provide a more effective upper bound on cohomology than earlier established bounds involving $\dim D^{\lambda'}$.

Example 21. Let $\mathrm{char}(k) = p \geq 3$, and consider $S = \Sigma_d$ with $p \mid d$. In this case

$$\Delta = \{\alpha_1, \alpha_2, \ldots, \alpha_{d-1}\} = \{\varepsilon_1 - \varepsilon_2, \varepsilon_2 - \varepsilon_3, \ldots, \varepsilon_{d-1} - \varepsilon_d\},$$

and $\delta = (1, 1, \ldots, 1)$. Set $\lambda = (2, 1, 1, \ldots, 1, 0)$. Then $\lambda' = (d-1, 1, 0, \ldots, 0)$. Moreover, $L(\lambda)$ identifies with the $(d^2 - 2)$-dimensional irreducible $G = GL_d(k)$-module that can be realized as a quotient of the adjoint representation of G tensored by δ. Also, observe that $\dim D_\lambda = \dim D^{\lambda'} = \dim L(2, 1, \ldots, 1, 0)_\delta = d - 2$. Now by Theorem 20,

$$\dim \mathrm{H}^1(\Sigma_d, D^{(d-1,1,0,\ldots,0)}) \leq \textstyle\sum_{\alpha\in\Delta} \dim L(\lambda)_{\delta+\alpha} - \dim L(\lambda)_\delta = |\Delta| - (d-2) = 1.$$

This bound is an equality, because the $(d-1)$-dimensional Specht module $S^{\lambda'}$ is a nonsplit extension of $D^{\lambda'}$ by the trivial module k [17, Theorem 24.1]. The equality $\dim \mathrm{Ext}^1_G(\delta, L(\lambda)) = 1$ can also be seen from observing that the Weyl module $\Delta(\lambda)$ for G is a nonsplit extension of $L(\lambda)$ by δ. On the other hand, for $p \geq 5$, we claim that Theorem 1 yields the (weaker) estimate

$$\dim \mathrm{H}^1(\Sigma_d, D^{(d-1,1,0,\ldots,0)}) \leq \tfrac{1}{2} \dim D^{(d-1,1,0,\ldots,0)} = \tfrac{1}{2}(d-2).$$

In order to apply Theorem 1, we must explain for $p \geq 5$ why the action of Σ_d on $D^{\lambda'}$ is faithful. Write $\rho : \Sigma_d \to GL(D^{\lambda'})$ for the map defining the representation of Σ_d on $D^{\lambda'}$, and write A_d for the alternating group on d letters. Observe that $\ker(\rho) \cap A_d$ is a normal subgroup of the nonabelian simple group A_d, so either

$\ker(\rho) \cap A_d = \{1\}$, or $\ker(\rho) \cap A_d = A_d$. The latter equality is false, because $A_d \not\subset \ker(\rho)$, so we have $\ker(\rho) \cap A_d = \{1\}$. This implies that $\ker(\rho)$ contains only odd permutations and the identity. Since the product of any two odd permutations is an element of A_d, and since $A_d \cap \ker(\rho) = \{1\}$, we conclude that in fact $\ker(\rho)$ can contain at most two elements, namely, the identity and an odd permutation that is equal to its own inverse. Subgroups of this type are not normal in Σ_d, whereas $\ker(\rho)$ is normal in Σ_d, so we conclude that $\ker(\rho) = \{1\}$. Thus, $D^{\lambda'}$ is an irreducible faithful Σ_d-module.

4.2 Bounds for $G(\mathbb{F}_q)$

Assume once again that G is as defined in Sect. 1.4. One can apply Cline, Parshall, Scott, and van der Kallen's [8] rational and generic cohomology results to identify certain cohomology groups for $G(\mathbb{F}_q)$ with cohomology groups for G. Then applying our results on the dimensions of rational cohomology groups, one can obtain corresponding bounds for $G(\mathbb{F}_q)$. For sufficiently large q, this approach can be used to recover, and in general, improve upon, the bounds in Theorem 1. The following theorem demonstrates this approach.

Theorem 22. *Let $r \geq 2$, and set $s = \left[\frac{r}{2}\right]$. Assume that $p^{s-1}(p-1) > h$. Then for each finite-dimensional $kG(\mathbb{F}_q)$-module V, one has*

$$\dim \mathrm{H}^1(G(\mathbb{F}_q), V) \leq \tfrac{1}{h} \dim V.$$

Proof. Arguing by induction on the composition length, and using the long exact sequence in cohomology, it suffices to assume that V is an irreducible $kG(\mathbb{F}_q)$-module. Then $V = L(\lambda)$ for some $\lambda \in X_r(T)$. Write λ in the form $\lambda = \lambda_0 + p^s\lambda_1$ with $\lambda_0 \in X_s(T)$ and $\lambda_1 \in X(T)_+$, and set $\tilde{\lambda} = \lambda_1 + p^{r-s}\lambda_0$. Then $L(\tilde{\lambda}) \cong L(\lambda)^{(r-s)}$ as $kG(\mathbb{F}_q)$-modules. In particular, $\dim L(\lambda) = \dim L(\tilde{\lambda})$. Now by [4, Theorem 5.5], the stated hypotheses imply that either $\mathrm{H}^1(G(\mathbb{F}_q), L(\lambda)) \cong \mathrm{H}^1(G, L(\lambda))$, or $\mathrm{H}^1(G(\mathbb{F}_q), L(\lambda)) \cong \mathrm{H}^1(G, L(\tilde{\lambda}))$. In either case, the inequality $\mathrm{H}^1(G(\mathbb{F}_q), L(\lambda)) \leq \frac{1}{h}\dim L(\lambda)$ then follows from Theorem 5. □

If V is an irreducible $kG(\mathbb{F}_q)$-module and if the Weyl group W is of order prime to p, then the inequality in Theorem 22 can be improved by applying Theorem 17.

4.3 Bounds for Degrees 2 and 3

For higher degrees, one can make use of recent work of Parshall, Scott, and Stewart [22] to apply the approach of the previous section. Given a positive integer n, they show that there exists an integer r_0, depending on n and on the underlying root system Φ, such that if $r \geq r_0$, $q = p^r$, and $\lambda \in X_r(T)$, then $\mathrm{H}^n(G(\mathbb{F}_q), L(\lambda)) \cong \mathrm{H}^n(G, L(\lambda'))$; see [22, Theorem 5.8]. Here λ' is a certain "q-shift" of λ, similar to the weight $\tilde{\lambda}$ used in the previous proof. Of importance for our purposes is that $\dim L(\lambda') = \dim L(\lambda)$. With this, one can recover Theorem 22 for arbitrary primes, but at the expense of requiring a potentially larger r. For degrees 2 and 3, one can use this idea along with Corollary 8 to improve, for sufficiently large r, upon the bound in Theorem 2. As above, the proofs of the following two theorems reduce to the case where V is an irreducible $kG(\mathbb{F}_q)$-module.

Theorem 23. *There exists a constant $D(\Phi, 2)$, depending on Φ, such that if $r \geq D(\Phi, 2)$ and if $q = p^r$, then, for each finite-dimensional $kG(\mathbb{F}_q)$-module V, one has*

$$\dim \mathrm{H}^2(G(\mathbb{F}_q), V) \leq \dim V.$$

Theorem 24. *Suppose $p > h$. Then there exists a constant $D(\Phi, 3)$, depending on Φ, such that if $r \geq D(\Phi, 3)$ and if $q = p^r$, then, for each finite-dimensional $kG(\mathbb{F}_q)$-module V, one has*

$$\dim \mathrm{H}^3(G(\mathbb{F}_q), V) \leq 2 \cdot \dim V.$$

The constants $D(\Phi, 2)$ and $D(\Phi, 3)$ in the previous two theorems can be determined recursively. This is done in the proofs of Theorems 5.2 and 5.8 in [22].

Acknowledgements The authors thank the American Institute of Mathematics for hosting the workshops "Cohomology and representation theory for finite groups of Lie type" in June 2007, and "Cohomology bounds and growth rates" in June 2012. Many of the results described in Sect. 1.1 were motivated by the ideas exchanged at the 2007 workshop, and provided impetus for the organization of the 2012 meeting. The results of this paper were obtained in the AIM working group format at the 2012 workshop, which promoted a productive exchange of ideas between the authors at the meeting.

The fourth author (Nakano) presented talks at the workshop "Lie Groups, Lie Algebras and their Representations" at U.C. Santa Cruz in 1999 and at Louisiana State University (LSU) in 2011. At the LSU meeting, his lecture was devoted to explaining the various connections between the cohomology theories for reductive algebraic groups and their associated Frobenius kernels and finite Chevalley groups. The results in this paper are a natural extension of the results discussed in his presentation.

References

1. H. H. Andersen, *Extensions of modules for algebraic groups*, Amer. J. Math. **106** (1984), no. 2, 489–504.
2. H. H. Andersen and T. Rian, *B-cohomology*, J. Pure Appl. Algebra **209** (2007), no. 2, 537–549.
3. C. P. Bendel, D. K. Nakano, B. J. Parshall, C. Pillen, L. L. Scott, and D. I. Stewart, *Bounding extensions for finite groups and Frobenus kernels*, preprint, 2012, `arXiv:1208.6333`.
4. C. P. Bendel, D. K. Nakano, and C. Pillen, *Extensions for finite groups of Lie type. II. Filtering the truncated induction functor*, Representations of algebraic groups, quantum groups, and Lie algebras, Contemp. Math., Vol. 413, Amer. Math. Soc., Providence, RI, 2006, pp. 1–23.

5. C. P. Bendel, D. K. Nakano, and C. Pillen, *Second cohomology groups for Frobenius kernels and related structures*, Adv. Math. **209** (2007), no. 1, 162–197.
6. E. Cline, B. Parshall, and L. Scott, *Cohomology of finite groups of Lie type. I*, Inst. Hautes Études Sci. Publ. Math. (1975), no. 45, 169–191.
7. E. Cline, B. Parshall, and L. Scott, *Reduced standard modules and cohomology*, Trans. Amer. Math. Soc. **361** (2009), no. 10, 5223–5261.
8. E. Cline, B. Parshall, L. Scott, and W. van der Kallen, *Rational and generic cohomology*, Invent. Math. **39** (1977), no. 2, 143–163.
9. S. R. Doty, K. Erdmann, and D. K. Nakano, *Extensions of modules over Schur algebras, symmetric groups and Hecke algebras*, Algebr. Represent. Theory **7** (2004), no. 1, 67–100.
10. D. Gorenstein, R. Lyons, and R. Solomon, *The classification of the finite simple groups. Number 3. Part I. Chapter A*, Vol. 40, Mathematical Surveys and Monographs, no. 3, American Mathematical Society, Providence, RI, 1998.
11. R. M. Guralnick, *The dimension of the first cohomology group*, Representation theory, II (Ottawa, Ont., 1984), Lecture Notes in Math., Vol. 1178, Springer, Berlin, 1986, pp. 94–97.
12. R. M. Guralnick, T. L. Hodge, B. J. Parshall, and L. L. Scott, *Counterexample to Wall's conjecture* [online], http://www.aimath.org/news/wallsconjecture/ [cited September 20, 2012].
13. R. M. Guralnick and C. Hoffman, *The first cohomology group and generation of simple groups*, Groups and geometries (Siena, 1996), Trends Math., Birkhäuser, Basel, 1998, pp. 81–89.
14. R. M. Guralnick, W. M. Kantor, M. Kassabov, and A. Lubotzky, *Presentations of finite simple groups: profinite and cohomological approaches*, Groups Geom. Dyn. **1** (2007), no. 4, 469–523.
15. R. M. Guralnick and P. H. Tiep, *First cohomology groups of Chevalley groups in cross characteristic*, Ann. of Math. (2) **174** (2011), no. 1, 543–559.
16. J. E. Humphreys, *Reflection groups and Coxeter groups*, Cambridge Studies in Advanced Mathematics, Vol. 29, Cambridge University Press, Cambridge, 1990.
17. G. D. James, *The Representation Theory of the Symmetric Groups*, Lecture Notes in Mathematics, Vol. 682, Springer, Berlin, 1978.
18. J. C. Jantzen, *Representations of Algebraic Groups*, second ed., Mathematical Surveys and Monographs, Vol. 107, American Mathematical Society, Providence, RI, 2003.
19. A. E. Parker and D. I. Stewart, *First cohomology groups for finite groups of Lie type in defining characteristic*, Bull. London Math. Soc. **46** (2014), 227–238.
20. B. J. Parshall and L. L. Scott, *Quantum Weyl reciprocity for cohomology*, Proc. London Math. Soc. (3) **90** (2005), no. 3, 655–688.
21. B. J. Parshall and L. L. Scott, *Bounding Ext for modules for algebraic groups, finite groups and quantum groups*, Adv. Math. **226** (2011), no. 3, 2065–2088.
22. B. J. Parshall, L. L. Scott, and D. I. Stewart, *Shifted generic cohomology*, Compositio Math. **149** (2013), 1765–1788.
23. A. A. Premet, *Weights of infinitesimally irreducible representations of Chevalley groups over a field of prime characteristic*, Mat. Sb. (N.S.) **133(175)** (1987), no. 2, 167–183, 271.
24. L. L. Scott, *Some new examples in 1-cohomology*, J. Algebra **260** (2003), no. 1, 416–425, Special issue celebrating the 80th birthday of Robert Steinberg.
25. L. L. Scott and T. Sprowl, *Computing individual Kazhdan-Lusztig basis elements*, preprint 2013, `arXiv:1309.7265`.
26. University of Georgia VIGRE Algebra Group, *First cohomology for finite groups of Lie type: simple modules with small dominant weights*, Trans. Amer. Math. Soc. **365** (2013), no. 2, 1025–1050.
27. C. B. Wright, *Second cohomology groups for algebraic groups and their Frobenius kernels*, J. Algebra **330** (2011), 60–75.

Representations of the General Linear Lie Superalgebra in the BGG Category $\mathcal{O}$

Jonathan Brundan

Abstract This is a survey of some recent developments in the highest weight repesentation theory of the general linear Lie superalgebra $\mathfrak{gl}_{n|m}(\mathbb{C})$. The main focus is on the analog of the Kazhdan–Lusztig conjecture as formulated by the author in 2002, which was finally proved in 2011 by Cheng, Lam and Wang. Recently another proof has been obtained by the author joint with Losev and Webster, by a method which leads moreover to the construction of a Koszul-graded lift of category $\mathcal{O}$ for this Lie superalgebra.

Key words General linear Lie superalgebra • Category $\mathcal{O}$

Mathematics Subject Classification (2010): 17B10, 17B37.

1 Introduction

The representation theory of the general linear Lie superalgebra (as well as the other classical families) was first investigated seriously by Victor Kac [30, 31] around 1976. Kac classified the finite-dimensional irreducible representations and proved character formulae for the typical ones. Then in the 1980s work of Sergeev [44] and Berele–Regev [5] exploited the superalgebra analog of Schur–Weyl duality to work out character formulae for the irreducible polynomial representations. It took another decade before Serganova [43] explained how the characters of arbitrary finite-dimensional irreducible representations could be approached. Subsequent

Research supported in part by NSF grant no. DMS-1161094.

J. Brundan (✉)
Department of Mathematics, University of Oregon, Eugene, OR 07403, USA
e-mail: brundan@uoregon.edu

G. Mason et al. (eds.), *Developments and Retrospectives in Lie Theory: Algebraic Methods*, Developments in Mathematics 38,
DOI 10.1007/978-3-319-09804-3_3

work of the author and others [10, 16, 17, 47] means that by now the category of finite-dimensional representations is well understood (although there remain interesting questions regarding the tensor structure).

One can also ask about the representation theory of the general linear Lie superalgebra in the analog of the Bernstein–Gelfand–Gelfand category $\mathcal{O}$ from [7]. This is the natural home for the irreducible highest weight representations. The classical theory of category $\mathcal{O}$ for a semisimple Lie algebra, as in for example Humphreys' book [27] which inspired this article, sits at the heart of modern geometric representation theory. Its combinatorics is controlled by the underlying Weyl group, and many beautiful results are deduced from the geometry of the associated flag variety via the Beilinson–Bernstein localization theorem [3]. There still seems to be no satisfactory substitute for this geometric part of the story for $\mathfrak{gl}_{n|m}(\mathbb{C})$ but at least the combinatorics has now been worked out: in [10] it was proposed that the combinatorics of the Weyl group (specifically the Kazhdan–Lusztig polynomials arising from the associated Iwahori–Hecke algebra) should simply be replaced by the combinatorics of a canonical basis in a certain $U_q\mathfrak{sl}_\infty$-module $V^{\otimes n} \otimes W^{\otimes m}$. This idea led in particular to the formulation of a superalgebra analog of the Kazhdan–Lusztig conjecture.

The super Kazhdan–Lusztig conjecture is now a theorem. In fact there are two proofs, first by Cheng, Lam and Wang [18], then more recently in joint work of the author with Losev and Webster [15]. In some sense both proofs involve a reduction to the ordinary Kazhdan–Lusztig conjecture for the general linear Lie algebra. Cheng, Lam and Wang go via some infinite dimensional limiting versions of the underlying Lie (super)algebras using the technique of "super duality," which originated in [17, 22]. On the other hand the proof in [15] involves passing from category $\mathcal{O}$ for $\mathfrak{gl}_{n|m}(\mathbb{C})$ to some subquotients which, thanks to results of Losev and Webster from [36], are equivalent to sums of blocks of parabolic category $\mathcal{O}$ for some other general linear Lie algebra. The approach of [15] allows also for the construction of a graded lift of $\mathcal{O}$ which is Koszul, in the spirit of the famous results of Beilinson, Ginzburg and Soergel [4] in the classical setting. The theory of categorification developed by Rouquier [42] and others, and the idea of Schur–Weyl duality for higher levels from [14], both play a role in this work.

This article is an attempt to give a brief overview of these results. It might serve as a useful starting point for someone trying to learn about the combinatorics of category $\mathcal{O}$ for the general linear Lie superalgebra for the first time. We begin with the definition of $\mathcal{O}$ and the basic properties of Verma supermodules and their projective covers. Then we formulate the super Kazhdan–Lusztig conjecture precisely and give some examples, before fitting it into the general framework of tensor product categorifications. Finally we highlight one of the main ideas from [15] involving a double centralizer property (an analog of Soergel's Struktursatz from [45]), and suggest a related question which we believe should be investigated further. In an attempt to maximize the readability of the article, precise references to the literature have been deferred to notes at the end of each section.

We point out in conclusion that there is also an attractive Kazhdan–Lusztig conjecture for the Lie superalgebra $\mathfrak{q}_n(\mathbb{C})$ formulated in [11], which remains quite untouched. One can also ponder Kazhdan–Lusztig combinatorics for the other classical families of Lie superalgebra. Dramatic progress in the case of $\mathfrak{osp}_{n|2m}(\mathbb{C})$ has been made recently in [2]; see also [25].

2 Super Category $\mathcal{O}$ and Its Blocks

Fix $n, m \geq 0$ and let $\mathfrak{g}$ denote the *general linear Lie superalgebra* $\mathfrak{gl}_{n|m}(\mathbb{C})$. As a vector superspace this consists of $(n+m) \times (n+m)$ complex matrices $\left(\begin{array}{c|c} * & * \\ \hline * & * \end{array}\right)$ with $\mathbb{Z}/2$-grading defined so that the ij-matrix unit $e_{i,j}$ is even for $1 \leq i, j \leq n$ or $n+1 \leq i, j \leq n+m$, and $e_{i,j}$ is odd otherwise. It is a Lie superalgebra via the *supercommutator*

$$[x, y] := xy - (-1)^{\bar{x}\bar{y}} yx$$

for homogeneous $x, y \in \mathfrak{g}$ of parities $\bar{x}, \bar{y} \in \mathbb{Z}/2$, respectively.

By a $\mathfrak{g}$-*supermodule* we mean a vector superspace $M = M_{\bar{0}} \oplus M_{\bar{1}}$ equipped with a graded linear left action of $\mathfrak{g}$, such that $[x, y]v = x(yv) - (-1)^{\bar{x}\bar{y}} y(xv)$ for all homogeneous $x, y \in \mathfrak{g}$ and $v \in M$. For example, we have the *natural representation* U of $\mathfrak{g}$, which is just the superspace of column vectors on standard basis $u_1, \ldots, u_{n+m}$, where $\bar{u}_i = \bar{0}$ for $1 \leq i \leq n$ and $\bar{u}_i = \bar{1}$ for $n+1 \leq i \leq n+m$. We write $\mathfrak{g}$-smod for the category of all $\mathfrak{g}$-supermodules. A morphism $f : M \to N$ in this category means a linear map such that $f(M_i) \subseteq N_i$ for $i \in \mathbb{Z}/2$ and $f(xv) = xf(v)$ for $x \in \mathfrak{g}$, $v \in M$. This is obviously a $\mathbb{C}$-linear abelian category. It is also a *supercategory*, that is, it is equipped with the additional data of an endofunctor $\Pi : \mathfrak{g}\text{-smod} \to \mathfrak{g}\text{-smod}$ with $\Pi^2 \cong \mathrm{id}$. The functor Π here is the *parity switching functor*, which is defined on a supermodule M by declaring that ΠM is the same underlying vector space as M but with the opposite $\mathbb{Z}/2$-grading, viewed as a $\mathfrak{g}$-supermodule with the new action $x \cdot v := (-1)^{\bar{x}} xv$. On a morphism $f : M \to N$ we take $\Pi f : \Pi M \to \Pi N$ to be the same underlying linear map as f. Clearly $\Pi^2 = \mathrm{id}$.

Remark 2.1. Given any $\mathbb{C}$-linear supercategory $\mathscr{C}$, one can form the *enriched category* $\hat{\mathscr{C}}$. This is a category enriched in the monoidal category of vector superspaces. It has the same objects as in $\mathscr{C}$, and its morphisms are defined from $\mathrm{Hom}_{\hat{\mathscr{C}}}(M, N) := \mathrm{Hom}_{\mathscr{C}}(M, N)_{\bar{0}} \oplus \mathrm{Hom}_{\mathscr{C}}(M, N)_{\bar{1}}$ where

$$\mathrm{Hom}_{\mathscr{C}}(M, N)_{\bar{0}} := \mathrm{Hom}_{\mathscr{C}}(M, N), \qquad \mathrm{Hom}_{\mathscr{C}}(M, N)_{\bar{1}} := \mathrm{Hom}_{\mathscr{C}}(M, \Pi N).$$

The composition law is obvious (but involves the isomorphism $\Pi^2 \cong \mathrm{id}$ which is given as part of the data of $\mathscr{C}$). This means one can talk about *even* and *odd* morphisms between objects of $\mathscr{C}$. In the case of $\mathfrak{g}$-smod, an odd homomorphism $f : M \to N$ is a linear map such that $f(M_i) \subseteq N_{i+\bar{1}}$ for $i \in \mathbb{Z}/2$ and $f(xv) = (-1)^{|x|}xf(v)$ for homogeneous $x \in \mathfrak{g}$, $v \in M$.

Let $\mathfrak{b}$ be the *standard Borel subalgebra* consisting of all upper triangular matrices in $\mathfrak{g}$. It is the stabilizer of the *standard flag* $\langle u_1 \rangle < \langle u_1, u_2 \rangle < \cdots < \langle u_1, \ldots, u_{n+m} \rangle$ in the natural representation V. More generally a *Borel subalgebra* of $\mathfrak{g}$ is the stabilizer of an arbitrary homogeneous flag in V. Unlike in the purely even setting, it is not true that all Borel subalgebras are conjugate under the appropriate action of the general linear supergroup $G = \mathrm{GL}_{n|m}$. This leads to some combinatorially interesting variants of the theory which are also well understood, but our focus in this article will just be on the standard choice of Borel.

Let $\mathfrak{t}$ be the Cartan subalgebra of $\mathfrak{g}$ consisting of diagonal matrices. Let $\delta_1, \ldots, \delta_{n+m}$ be the basis for $\mathfrak{t}^*$ such that δ_i picks out the ith diagonal entry of a diagonal matrix. Define a non-degenerate symmetric bilinear form $(?, ?)$ on $\mathfrak{t}^*$ by setting $(\delta_i, \delta_j) := (-1)^{\bar{u}_i}\delta_{i,j}$. The *root system* of $\mathfrak{g}$ is

$$R := \{\delta_i - \delta_j \mid 1 \leq i, j \leq n+m, i \neq j\},$$

which decomposes into even and odd roots $R = R_{\bar{0}} \sqcup R_{\bar{1}}$ so that $\delta_i - \delta_j$ is of parity $\bar{u}_i + \bar{u}_j$. Let $R^+ = R^+_{\bar{0}} \sqcup R^+_{\bar{1}}$ denote the positive roots associated to the Borel subalgebra $\mathfrak{b}$, i.e., $\delta_i - \delta_j$ is positive if and only if $i < j$. The *dominance order* $\trianglerighteq$ on $\mathfrak{t}^*$ is defined so that $\lambda \trianglerighteq \mu$ if $\lambda - \mu$ is a sum of positive roots. Let

$$\rho := -\delta_2 - 2\delta_3 - \cdots - (n-1)\delta_n + (n-1)\delta_{n+1} + (n-2)\delta_{n+2} + \cdots + (n-m)\delta_{n+m}.$$

One can check that 2ρ is congruent to the sum of the positive even roots minus the sum of the positive odd roots modulo $\delta := \delta_1 + \cdots + \delta_n - \delta_{n+1} - \cdots - \delta_{n+m}$.

Let $s\mathscr{O}$ be the full subcategory of $\mathfrak{g}$-smod consisting of all finitely generated $\mathfrak{g}$-supermodules which are locally finite dimensional over $\mathfrak{b}$ and satisfy

$$M = \bigoplus_{\lambda \in \mathfrak{t}^*} M_\lambda,$$

where for $\lambda \in \mathfrak{t}^*$ we write $M_\lambda = M_{\lambda,\bar{0}} \oplus M_{\lambda,\bar{1}}$ for the λ-weight space of M with respect to $\mathfrak{t}$ defined in the standard way. This is an abelian subcategory of $\mathfrak{g}$-smod closed under Π. It is the analog for $\mathfrak{gl}_{n|m}(\mathbb{C})$ of the Bernstein–Gelfand–Gelfand category $\mathscr{O}$ for a semisimple Lie algebra. All of the familiar basic properties from the purely even setting generalize rather easily to the super case. For example all supermodules in $s\mathscr{O}$ have finite length, there are enough projectives, and so on. An easy way to prove these statements is to compare $s\mathscr{O}$ to the classical BGG category $\mathscr{O}_{ev}$ for the even part $\mathfrak{g}_{\bar{0}} \cong \mathfrak{gl}_n(\mathbb{C}) \oplus \mathfrak{gl}_m(\mathbb{C})$ of $\mathfrak{g}$. One can restrict any supermodule in $s\mathscr{O}$ to $\mathfrak{g}_{\bar{0}}$ to get a module in $\mathscr{O}_{ev}$; conversely for any $M \in \mathscr{O}_{ev}$ we can view it as a supermodule concentrated in a single parity then induce to get $U(\mathfrak{g}) \otimes_{U(\mathfrak{g}_{\bar{0}})} M \in \mathscr{O}$. This relies on the fact that $U(\mathfrak{g})$ is free of finite rank as a $U(\mathfrak{g}_{\bar{0}})$-module, thanks to the PBW theorem for Lie superalgebras. Then the fact that $s\mathcal{O}$ has enough projectives follows because $\mathcal{O}_{ev}$ does, and induction sends projectives to projectives as it is left adjoint to an exact functor.

In fact it is possible to eliminate the "super" in the supercategory $s\mathcal{O}$ entirely by passing to a certain subcategory $\mathcal{O}$. To explain this let $\hat{\mathbb{C}}$ be some set of representatives for the cosets of $\mathbb{C}$ modulo $\mathbb{Z}$ such that $0 \in \hat{\mathbb{C}}$. Then define $p_{z+n} := \bar{n} \in \mathbb{Z}/2$ for each $z \in \hat{\mathbb{C}}$ and $n \in \mathbb{Z}$. Finally for $\lambda \in \mathfrak{t}^*$ let $p(\lambda) := p_{(\lambda,\delta_{n+1}+\cdots+\delta_{n+m})}$. This defines a *parity function* $p : \mathfrak{t}^* \to \mathbb{Z}/2$ with the key property that $p(\lambda + \delta_i) = p(\lambda) + \bar{u}_i$. If $M \in s\mathcal{O}$ then M decomposes as a direct sum of $\mathfrak{g}$-supermodules as

$$M = M_+ \oplus M_- \qquad \text{where} \qquad M_+ := \bigoplus_{\lambda \in \mathfrak{t}^*} M_{\lambda, p(\lambda)}, \quad M_- := \bigoplus_{\lambda \in \mathfrak{t}^*} M_{\lambda, p(\lambda)+\bar{1}}.$$

Let $\mathcal{O}$ (resp. $\Pi\mathcal{O}$) be the full subcategory of $s\mathcal{O}$ consisting of all supermodules M such that $M = M_+$ (resp. $M = M_-$). Both are Serre subcategories of $s\mathcal{O}$, hence they are abelian, and the functor Π defines an equivalence between $\mathcal{O}$ and $\Pi\mathcal{O}$. Moreover there are no nonzero odd homomorphisms between objects of $\mathcal{O}$; equivalently there are no nonzero even homomorphisms between an object of $\mathcal{O}$ and an object of $\Pi\mathcal{O}$. Hence:

Lemma 2.2. $s\mathcal{O} = \mathcal{O} \oplus \Pi\mathcal{O}$.

Remark 2.3. Let $s\hat{\mathcal{O}}$ be the enriched category arising from the supercategory $s\mathcal{O}$ as in Remark 2.1. Lemma 2.2 implies that the natural inclusion functor $\mathcal{O} \to s\hat{\mathcal{O}}$ is fully faithful and essentially surjective, hence it defines an equivalence between $\mathcal{O}$ and $s\hat{\mathcal{O}}$. In particular $s\hat{\mathcal{O}}$ is itself abelian, although the explicit construction of kernels and cokernels of inhomogeneous morphisms in $s\hat{\mathcal{O}}$ is a bit awkward.

Henceforth we will work just with the category $\mathcal{O}$ rather than the supercategory $s\mathcal{O}$. Note in particular that $\mathcal{O}$ contains the natural supermodule U and its dual $U^\vee$, and it is closed under tensoring with these objects. For each $\lambda \in \mathfrak{t}^*$ we have the *Verma supermodule*

$$M(\lambda) := U(\mathfrak{g}) \otimes_{U(\mathfrak{b})} \mathbb{C}_{\lambda, p(\lambda)} \in \mathcal{O},$$

where $\mathbb{C}_{\lambda,p(\lambda)}$ is a one-dimensional $\mathfrak{b}$-supermodule of weight λ concentrated in parity $p(\lambda)$. The usual argument shows that $M(\lambda)$ has a unique irreducible quotient, which we denote by $L(\lambda)$. The supermodules $\{L(\lambda) \mid \lambda \in \mathfrak{t}^*\}$ give a complete set of pairwise nonisomorphic irreducibles in $\mathcal{O}$. We say that $\lambda \in \mathfrak{t}^*$ is *dominant* if

$$\begin{cases} (\lambda, \delta_i - \delta_{i+1}) \in \mathbb{Z}_{\geq 0} \text{ for } i = 1, \ldots, n-1, \\ (\lambda, \delta_i - \delta_{i+1}) \in \mathbb{Z}_{\leq 0} \text{ for } i = n+1, \ldots, n+m-1. \end{cases}$$

Then the supermodules $\{L(\lambda) \mid \text{for all dominant } \lambda \in \mathfrak{t}^*\}$ give a complete set of pairwise non-isomorphic finite-dimensional irreducible $\mathfrak{g}$-supermodules (up to parity switch). This is an immediate consequence of the following elementary but important result.

Theorem 2.4 (Kac). *For $\lambda \in \mathfrak{t}^*$ the irreducible supermodule $L(\lambda)$ is finite dimensional if and only if λ is dominant.*

Proof. Let $L_{ev}(\lambda)$ be the irreducible highest weight module for $\mathfrak{g}_{\bar{0}}$ of highest weight λ. Classical theory tells us that $L_{ev}(\lambda)$ is finite dimensional if and only if λ is dominant. Since $L(\lambda)$ contains a highest weight vector of weight λ, its restriction to $\mathfrak{g}_{\bar{0}}$ has $L_{ev}(\lambda)$ as a composition factor, hence if $L(\lambda)$ is finite dimensional then λ is dominant. Conversely, let $\mathfrak{p}$ be the maximal parabolic subalgebra of $\mathfrak{g}$ consisting of block upper triangular matrices of the form $\left(\begin{array}{c|c} * & * \\ \hline 0 & * \end{array}\right)$. There is an obvious projection $\mathfrak{p} \twoheadrightarrow \mathfrak{g}_{\bar{0}}$, allowing us to view $L_{ev}(\lambda)$ as a $\mathfrak{p}$-supermodule concentrated in parity $p(\lambda)$. Then for any $\lambda \in \mathfrak{t}^*$ we can form the *Kac supermodule*

$$K(\lambda) := U(\mathfrak{g}) \otimes_{U(\mathfrak{p})} L_{ev}(\lambda) \in \mathcal{O}.$$

Since $K(\lambda)$ is a quotient of $M(\lambda)$, it has irreducible head $L(\lambda)$. Moreover the PBW theorem implies that $K(\lambda)$ is finite dimensional if and only if $L_{ev}(\lambda)$ is finite dimensional. Hence if λ is dominant we deduce that $L(\lambda)$ is finite dimensional. □

The *degree of atypicality* of $\lambda \in \mathfrak{t}^*$ is defined to be the maximal number of mutually orthogonal odd roots $\beta \in R^+_{\bar{1}}$ such that $(\lambda + \rho, \beta) = 0$. In particular λ is *typical* if $(\lambda + \rho, \beta) \neq 0$ for all $\beta \in R^+_{\bar{1}}$. For typical $\lambda \in \mathfrak{t}^*$, Kac showed further that the Kac supermodules $K(\lambda)$, defined in the proof of Theorem 2.4, are actually irreducible. Thus most questions about typical irreducible supermodules in $\mathcal{O}$ reduce to the purely even case. For example, using the Weyl character formula one can deduce in this way a simple formula for the character of an arbitrary typical finite-dimensional irreducible $\mathfrak{g}$-supermodule. It is not so easy to compute the characters of *atypical* finite-dimensional irreducible supermodules, but this has turned out still to be combinatorially quite tractable. We will say more about the much harder problem of finding characters of arbitrary (not necessarily typical or finite-dimensional) irreducible supermodules in $\mathcal{O}$ in the next section; inevitably this involves some Kazhdan–Lusztig polynomials.

Let $P(\mu)$ be a projective cover of $L(\mu)$ in $\mathcal{O}$. We have the usual statement of *BGG reciprocity*: each $P(\mu)$ has a *Verma flag*, i.e., a finite filtration whose sections are Verma supermodules, and the multiplicity $(P(\mu) : M(\lambda))$ of $M(\lambda)$ as a section of a Verma flag of $P(\mu)$ is given by

$$(P(\mu) : M(\lambda)) = [M(\lambda) : L(\mu)],$$

where the right-hand side denotes composition multiplicity. Of course $[M(\lambda) : L(\mu)]$ is zero unless $\mu \trianglelefteq \lambda$ in the dominance ordering, while $[M(\lambda) : L(\lambda)] = 1$. Thus $\mathcal{O}$ is a *highest weight category* in the formal sense of Cline, Parshall and Scott, with weight poset $(\mathfrak{t}^*, \trianglelefteq)$.

The partial order $\trianglelefteq$ on $\mathfrak{t}^*$ being used here is rather crude. It can be replaced with a more intelligent order $\leq$, called the *Bruhat order*. To define this, given $\lambda \in \mathfrak{t}^*$, let

$$A(\lambda) := \{\alpha \in R_{\bar{0}}^+ \mid (\lambda + \rho, \alpha^\vee) \in \mathbb{Z}_{>0}\}, \qquad B(\lambda) := \{\beta \in R_{\bar{1}}^+ \mid (\lambda + \rho, \beta) = 0\},$$

where $\alpha^\vee$ denotes $2\alpha/(\alpha, \alpha)$. Then introduce a relation $\uparrow$ on $\mathfrak{t}^*$ by declaring that $\mu \uparrow \lambda$ if we either have that $\mu = s_\alpha \cdot \lambda$ for some $\alpha \in A(\lambda)$ or we have that $\mu = \lambda - \beta$ for some $\beta \in B(\lambda)$; here, for $\alpha = \delta_i - \delta_j \in R_{\bar{0}}^+$ and $\lambda \in \mathfrak{t}^*$, we write $s_\alpha \cdot \lambda$ for $s_\alpha(\lambda + \rho) - \rho$, where $s_\alpha : \mathfrak{t}^* \to \mathfrak{t}^*$ is the reflection transposing δ_i and δ_j and fixing all other δ_k. Finally define $\leq$ to be the transitive closure of the relation $\uparrow$, i.e., we have that $\mu \leq \lambda$ if there exists $r \geq 0$ and weights $\nu_0, \ldots, \nu_r \in \mathfrak{t}^*$ with $\mu = \nu_0 \uparrow \nu_1 \uparrow \cdots \uparrow \mu_r = \lambda$.

Lemma 2.5. *If* $[M(\lambda) : L(\mu)] \neq 0$ *then* $\mu \leq \lambda$ *in the Bruhat order.*

Proof. This is a consequence of the super analog of the Jantzen sum formula from [39, §10.3]; see also [26]. In more detail, the Jantzen filtration on $M(\lambda)$ is a certain exhaustive descending filtration $M(\lambda) = M(\lambda)_0 \supset M(\lambda)_1 \supseteq M(\lambda)_2 \supseteq \cdots$ such that $M(\lambda)_0/M(\lambda)_1 \cong L(\lambda)$, and the sum formula shows that

$$\sum_{k\geq 1} \mathrm{ch}\, M(\lambda)_k = \sum_{\alpha \in A(\lambda)} \mathrm{ch}\, M(s_\alpha \cdot \lambda) + \sum_{\beta \in B(\lambda)} \sum_{k \geq 1} (-1)^{k-1} \mathrm{ch}\, M(\lambda - k\beta).$$

To deduce the lemma from this, suppose that $[M(\lambda) : L(\mu)] \neq 0$. Then $\mu \trianglelefteq \lambda$, so that $\lambda - \mu$ is a sum of N simple roots $\delta_i - \delta_{i+1}$ for some $N \geq 0$. We proceed by induction on N, the case $N = 0$ being vacuous. If $N > 0$, then $L(\mu)$ is a composition factor of $M(\lambda)_1$ and the sum formula implies that $L(\mu)$ is a composition factor either of $M(s_\alpha \cdot \lambda)$ for some $\alpha \in A(\lambda)$ or that $L(\mu)$ is a composition factor of $M(\lambda - k\beta)$ for some odd $k \geq 1$ and $\beta \in B(\lambda)$. It remains to apply the induction hypothesis and the definition of $\uparrow$. □

Let $\approx$ be the equivalence relation on $\mathfrak{t}^*$ generated by the Bruhat order $\leq$. We refer to the $\approx$-equivalence classes as *linkage classes*. For a linkage class $\xi \in \mathfrak{t}^*/\approx$, let $\mathcal{O}_\xi$ be the Serre subcategory of $\mathcal{O}$ generated by the irreducible supermodules $\{L(\lambda) \mid \lambda \in \xi\}$. Then, as a purely formal consequence of Lemma 2.5, we get that the category $\mathcal{O}$ decomposes as

$$\mathcal{O} = \bigoplus_{\xi \in \mathfrak{t}^*/\approx} \mathcal{O}_\xi.$$

In fact this is the finest possible such direct sum decomposition, i.e., each $\mathcal{O}_\xi$ is an indecomposable subcategory of $\mathcal{O}$. In other words, this is precisely the decomposition of $\mathcal{O}$ into *blocks*. An interesting open problem here is to classify the blocks $\mathcal{O}_\xi$ up to equivalence.

Let us describe the linkage class ξ of $\lambda \in \mathfrak{t}^*$ more explicitly. Let k be the degree of atypicality of λ and $\beta_1, \ldots, \beta_k \in R_{\bar{1}}^+$ be distinct mutually orthogonal odd roots such that $(\lambda+\rho,\beta_i)=0$ for each $i=1,\dots,k$. Also let W_λ be the *integral Weyl group* corresponding to λ, that is, the subgroup of $GL(\mathfrak{t}^*)$ generated by the reflections s_α for $\alpha\in R_{\bar 0}^+$ such that $(\lambda+\rho,\alpha)\in\mathbb{Z}$. Then

$$\xi=\left\{w\cdot(\lambda+n_1\beta_1+\cdots+n_k\beta_k)\;\middle|\;n_1,\dots,n_k\in\mathbb{Z},\,w\in W_\lambda\right\},$$

where $w\cdot\nu=w(\nu+\rho)-\rho$ as before. Note in particular that all $\mu\approx\lambda$ have the same degree of atypicality k as λ.

The following useful result reduces many questions about $\mathcal{O}$ to the case of *integral blocks*, that is, blocks corresponding to linkage classes of *integral weights* belonging to the set

$$\mathfrak{t}^*_{\mathbb{Z}}:=\mathbb{Z}\delta_1\oplus\cdots\oplus\mathbb{Z}\delta_{n+m}.$$

Theorem 2.6 (Cheng, Mazorchuk, Wang). *Every block $\mathcal{O}_\xi$ of $\mathcal{O}$ is equivalent to a tensor product of integral blocks of general linear Lie superalgebras of the same total rank as* $\mathfrak{g}$.

If λ is atypical, then the linkage class ξ containing λ is infinite. This is a key difference between the representation theory of Lie superalgebras and the classical representation theory of a semisimple Lie algebra, in which all blocks are finite (bounded by the order of the Weyl group). It means that the highest weight category $\mathcal{O}_\xi$ cannot be viewed as a category of modules over a finite-dimensional quasi-hereditary algebra. Nevertheless one can still consider the underlying basic algebra

$$A_\xi:=\bigoplus_{\lambda,\mu\in\xi}\operatorname{Hom}_{\mathfrak{g}}(P(\lambda),P(\mu))$$

with multiplication coming from composition. This is a *locally unital algebra*, meaning that it is equipped with the system of mutually orthogonal idempotents $\{1_\lambda\mid\lambda\in\xi\}$ such that

$$A_\xi=\bigoplus_{\lambda,\mu\in\xi}1_\mu A_\xi 1_\lambda,$$

where 1_λ denotes the identity endomorphism of $P(\lambda)$. Writing mof-A_ξ for the category of finite dimensional locally unital right A_ξ-modules, i.e., modules M with $M=\bigoplus_{\lambda\in\xi}M1_\lambda$, the functor

$$\mathcal{O}_\xi\to\text{mof-}A_\xi,\qquad M\mapsto\bigoplus_{\lambda\in\xi}\operatorname{Hom}_{\mathfrak{g}}(P(\lambda),-)$$

is an equivalence of categories. Note moreover that each right ideal $1_\lambda A_\xi$ and each left ideal $A_\xi 1_\lambda$ is finite dimensional; these are the indecomposable projectives and the linear duals of the indecomposable injectives in mof-A_ξ, respectively.

Remark 2.7. It is also natural to view A_ξ as a superalgebra concentrated in parity $\bar{0}$. Then the block $s\mathcal{O}_\xi = \mathcal{O}_\xi \oplus \Pi\mathcal{O}_\xi$ of the supercategory $s\mathcal{O}$ associated to the linkage class ξ is equivalent to the category of finite-dimensional locally unital right A_ξ*-supermodules*. This gives another point of view on Lemma 2.2.

Example 2.8. Let us work out in detail the example of $\mathfrak{gl}_{1|1}(\mathbb{C})$. This is easy but nevertheless very important: often $\mathfrak{gl}_{1|1}(\mathbb{C})$ plays a role parallel to that of $\mathfrak{sl}_2(\mathbb{C})$ in the classical theory. So now $\rho = 0$ and the only postive root is $\alpha = \delta_1 - \delta_2 \in R^+_{\bar{1}}$. The Verma supermodules $M(\lambda)$ are the same as the Kac supermodules $K(\lambda)$ from the proof of Theorem 2.4; they are two-dimensional with weights λ and $\lambda - \alpha$. Moreover $M(\lambda)$ is irreducible for typical λ. If λ is atypical, then $\lambda = c\alpha$ for some $c \in \mathbb{C}$, and the irreducible supermodule $L(\lambda)$ comes from the one-dimensional representation $\mathfrak{g} \to \mathbb{C}, x \mapsto c\,\mathrm{str}\,x$ where str denotes *supertrace*. Finally let us restrict attention just to the *principal block* $\mathcal{O}_0$ containing the irreducible supermodules $L(i) := L(i\alpha)$ for each $i \in \mathbb{Z}$. We have shown that $M(i) := M(i\alpha)$ has length two with composition factors $L(i)$ and $L(i-1)$; hence by BGG reciprocity the projective indecomposable supermodule $P(i) := P(i\alpha)$ has a two-step Verma flag with sections $M(i)$ and $M(i+1)$. We deduce that the Loewy series of $P(i)$ looks like $P(i) = P^0(i) > P^1(i) > P^2(i) > 0$ with

$$P^0(i)/P^1(i) \cong L(i), \qquad P^1(i)/P^2(i) \cong L(i-1) \oplus L(i+1), \qquad P^2(i) \cong L(i).$$

From this one obtains the following presentation for the underlying basic algebra A_0: it is the path algebra of the quiver

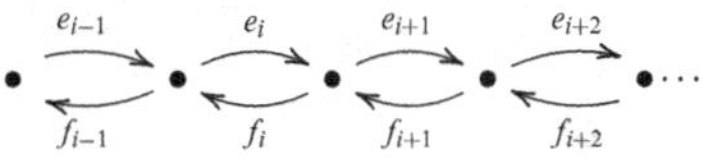

with vertex set $\mathbb{Z}$, modulo the relations $e_i f_i + f_{i+1} e_{i+1} = 0$, $e_{i+1} e_i = f_i f_{i+1} = 0$ for all $i \in \mathbb{Z}$. We stress the similarity between these and the relations $ef + fe = c$, $e^2 = f^2 = 0$ in $U(\mathfrak{g})$ itself (where $c = e_{1,1} + e_{2,2} \in \mathfrak{z}(\mathfrak{g})$, $e = e_{1,2}$ and $f = e_{2,1}$). One should also observe at this point that these relations are homogeneous, so that A_0 can be viewed as a positively graded algebra, with grading coming from path length. In fact this grading makes A_0 into a (locally unital) Koszul algebra.

To conclude the section, we offer one piece of justification for focussing so much attention on category $\mathcal{O}$. The study of primitive ideals of universal enveloping algebras of Lie algebras, especially semisimple ones, has classically proved to be very rich and inspired many important discoveries. So it is natural to ask about the space of all primitive ideals $\mathrm{Prim}\,U(\mathfrak{g})$ in our setting too. It turns out for $\mathfrak{gl}_{n|m}(\mathbb{C})$ that all primitive ideals are automatically homogeneous. In fact one just needs to consider annihilators of irreducible supermodules in $\mathcal{O}$:

Theorem 2.9 (Musson). $\mathrm{Prim}\,U(\mathfrak{g}) = \{\mathrm{Ann}_{U(\mathfrak{g})} L(\lambda) \mid \lambda \in \mathfrak{t}^*\}$.

This is the analog of a famous theorem of Duflo in the context of semisimple Lie algebras. Letzter showed subsequently that there is a bijection

$$\operatorname{Prim} U(\mathfrak{g}_{\bar{0}}) \xrightarrow{\sim} \operatorname{Prim} U(\mathfrak{g}), \qquad \operatorname{Ann}_{U(\mathfrak{g}_{\bar{0}})} L_{ev}(\lambda) \mapsto \operatorname{Ann}_{U(\mathfrak{g})} L(\lambda).$$

Combined with classical results of Joseph, this means that the fibers of the map

$$\mathfrak{t}^* \to \operatorname{Prim} U(\mathfrak{g}), \qquad \lambda \mapsto \operatorname{Ann}_{U(\mathfrak{g})} L(\lambda)$$

can be described in terms of the Robinson–Schensted algorithm. Hence we get an explicit description of the set $\operatorname{Prim} U(\mathfrak{g})$.

Notes. For the basic facts about super category $\mathcal{O}$ for basic classical Lie superalgebras, see §8.2 of Musson's book [39]. Lemma 2.2 was pointed out originally in [10, §4-e]. The observation that $s\hat{\mathcal{O}}$ is abelian from Remark 2.3 is due to Cheng and Lam [17]; in fact these authors work entirely with the equivalent category $s\hat{\mathcal{O}}$ in place of our $\mathcal{O}$.

The classification of finite-dimensional irreducible supermodules from Theorem 2.4 is due to Kac [30]. The irreducibilty of the typical Kac supermodules was established soon after in [31]. Kac only considered finite-dimensional representations at the time but the same argument works in general. The characters of the atypical finite-dimensional irreducible representations were first described algorithmically by Serganova in [43]. Then an easier approach was developed in [10], confirming some conjectures from [29]; see also [41] for a direct combinatorial proof of the equivalence of the formulae in [43] and [10]. Yet another approach via "super duality" was developed in [17, 22], based on the important observation that the Kazhdan–Lusztig polynomials appearing in [10, 43] are the same as certain Kazhdan–Lusztig polynomials for Grassmannians as computed originally by Lascoux and Schützenberger [33]. Subsequently Su and Zhang [47] were able to use the explicit formulae for these Kazhdan–Lusztig polynomials to extract some more explicit (but cumbersome) closed character and dimension formulae. There is also an elegant diagrammatic description of the basic algebra that is Morita equivalent to the subcategory $\mathcal{F}$ of $\mathcal{O}$ consisting of all its finite-dimensional supermodules in terms of Khovanov's arc algebra; see [16].

The analog of BGG reciprocity for $\mathfrak{gl}_{n|m}(\mathbb{C})$ as stated here was first established by Zou [48]; see also [12]. For the classification of blocks of $\mathcal{O}$ and proof of Theorem 2.6, see [19, Theorems 3.10–3.12]. A related problem is to determine when two irreducible highest weight supermodules have the same central character. This is solved via the explicit description of the center $Z(\mathfrak{g})$ of $U(\mathfrak{g})$ in terms the Harish-Chandra homomorphism and supersymmetric polynomials, which is due to Kac, Sergeev and Gorelik; see [39, §13.1] or [21, §2.2] for recent expositions. Lemma 2.5 is slightly more subtle and cannot be deduced just from central character considerations. Musson has recently proved a refinement of the sum formula recorded in the proof of Lemma 2.5, in which the right-hand side is rewritten as a finite sum of the characters of highest weight modules; details will appear in [40].

The results of Musson, Letzter and Joseph classifying primitive ideals of $U(\mathfrak{g})$ are in [28,34,38]; see also [39, Ch. 15]. The recent preprint [24] makes some further progress towards determining all inclusions between primitive ideals.

3 Kazhdan–Lusztig Combinatorics and Categorification

In this section we restrict attention just to the highest weight subcategory $\mathcal{O}_{\mathbb{Z}}$ of $\mathcal{O}$ consisting of supermodules M such that $M = \bigoplus_{\lambda \in \mathfrak{t}^*_{\mathbb{Z}}} M_{\lambda, p(\lambda)}$. In other words we only consider *integral blocks*. This is justified by Theorem 2.6. The goal is to understand the composition multiplicities

$$[M(\lambda) : L(\mu)]$$

of the Verma supermodules in $\mathcal{O}_{\mathbb{Z}}$. It will be convenient as we explain this to represent $\lambda \in \mathfrak{t}^*_{\mathbb{Z}}$ instead by the $n|m$-tuple $(\lambda_1, \dots, \lambda_n|\lambda_{n+1}, \dots, \lambda_{n+m})$ of integers defined from $\lambda_i := (\lambda + \rho, \delta_i)$.

Let P denote the free abelian group $\bigoplus_{i\in\mathbb{Z}} \mathbb{Z}\varepsilon_i$ and let $Q \subset P$ be the subgroup generated by the *simple roots* $\alpha_i := \varepsilon_i - \varepsilon_{i+1}$. Thus Q is the root lattice of the Lie algebra $\mathfrak{sl}_\infty$. Let $\leq$ be the usual dominance ordering on P defined by $\xi \leq \varpi$ if $\varpi - \xi$ is a sum of simple roots. For $\lambda = (\lambda_1, \dots, \lambda_n|\lambda_{n+1}, \dots, \lambda_{n+m}) \in \mathfrak{t}^*_{\mathbb{Z}}$ we let

$$|\lambda| := \varepsilon_{\lambda_1} + \cdots + \varepsilon_{\lambda_n} - \varepsilon_{\lambda_{n+1}} - \cdots - \varepsilon_{\lambda_{n+m}} \in P.$$

Then it is clear that two weights $\lambda, \mu \in \mathfrak{t}^*_{\mathbb{Z}}$ are linked if and only if $|\lambda| = |\mu|$, i.e., the fibers of the map $\mathfrak{t}^*_{\mathbb{Z}} \to P, \lambda \mapsto |\lambda|$ are exactly the linkage classes. The Bruhat order $\leq$ on $\mathfrak{t}^*_{\mathbb{Z}}$ can also be interpreted in these terms: let

$$|\lambda|_i := \begin{cases} \varepsilon_{\lambda_i} & \text{for } 1 \leq i \leq n, \\ -\varepsilon_{\lambda_i} & \text{for } n+1 \leq i \leq n+m, \end{cases}$$

so that $|\lambda| = |\lambda|_1 + \cdots + |\lambda|_{n+m}$. Then one can show that $\lambda \leq \mu$ in the Bruhat order if and only if $|\lambda|_1 + \cdots + |\lambda|_i \geq |\mu|_1 + \cdots + |\mu|_i$ in the dominance ordering on P for all $i = 1, \dots, n+m$, with equality when $i = n+m$.

Let V be the natural $\mathfrak{sl}_\infty$-module on basis $\{v_i \mid i \in \mathbb{Z}\}$ and W be its dual on basis $\{w_i \mid i \in \mathbb{Z}\}$. The Chevalley generators $\{f_i, e_i \mid i \in \mathbb{Z}\}$ of $\mathfrak{sl}_\infty$ act by

$$f_i v_j = \delta_{i,j} v_{i+1}, \quad e_i v_j = \delta_{i+1,j} v_i, \quad f_i w_j = \delta_{i+1,j} w_i, \quad e_i w_j = \delta_{i,j} w_{i+1}.$$

The tensor space $V^{\otimes n} \otimes W^{\otimes m}$ has the obvious basis of monomials

$$v_\lambda := v_{\lambda_1} \otimes \cdots \otimes v_{\lambda_n} \otimes w_{\lambda_{n+1}} \otimes \cdots \otimes w_{\lambda_{n+m}}$$

indexed by $n|m$-tuples $\lambda = (\lambda_1, \dots, \lambda_n | \lambda_{n+1}, \dots, \lambda_{n+m})$ of integers. In other words the monomial basis of $V^{\otimes n} \otimes W^{\otimes m}$ is parametrized by the set $\mathfrak{t}_{\mathbb{Z}}^*$ of integral weights for $\mathfrak{g} = \mathfrak{gl}_{n|m}(\mathbb{C})$.

This prompts us to bring category $\mathcal{O}$ back into the picture. Let $\mathcal{O}_{\mathbb{Z}}^{\Delta}$ be the exact subcategory of $\mathcal{O}_{\mathbb{Z}}$ consisting of all supermodules with a Verma flag, and denote its complexified Grothendieck group by $K(\mathcal{O}_{\mathbb{Z}}^{\Delta})$. Thus $K(\mathcal{O}_{\mathbb{Z}}^{\Delta})$ is the complex vector space on basis $\{[M(\lambda)] \mid \lambda \in \mathfrak{t}_{\mathbb{Z}}^*\}$. Henceforth we *identify*

$$K(\mathcal{O}_{\mathbb{Z}}^{\Delta}) \leftrightarrow V^{\otimes n} \otimes W^{\otimes m}, \qquad [M(\lambda)] \leftrightarrow v_\lambda.$$

Since projectives have Verma flags we have that $P(\mu) \in \mathcal{O}_{\mathbb{Z}}^{\Delta}$; let $b_\mu \in V^{\otimes n} \otimes W^{\otimes m}$ be the corresponding tensor under the above identification, i.e.,

$$[P(\mu)] \leftrightarrow b_\mu.$$

By BGG reciprocity we have that

$$b_\mu = \sum_{\lambda \in \mathfrak{t}_{\mathbb{Z}}^*} [M(\lambda) : L(\mu)] v_\lambda.$$

Now the punchline is that the vectors $\{b_\mu | \mu \in \mathfrak{t}_{\mathbb{Z}}^*\}$ turn out to coincide with Lusztig's *canonical basis* for the tensor space $V^{\otimes n} \otimes W^{\otimes m}$. The definition of the latter goes via some quantum algebra introduced in the next few paragraphs.

Let $U_q\mathfrak{sl}_\infty$ be the quantized enveloping algebra associated to $\mathfrak{sl}_\infty$. This is the $\mathbb{Q}(q)$-algebra on generators $\{\dot{f}_i, \dot{e}_i, \dot{k}_i, \dot{k}_i^{-1} | i \in \mathbb{Z}\}$[1] subject to well-known relations. We view $U_q\mathfrak{sl}_\infty$ as a Hopf algebra with comultiplication

$$\Delta(\dot{f}_i) = 1 \otimes \dot{f}_i + \dot{f}_i \otimes \dot{k}_i, \quad \Delta(\dot{e}_i) = \dot{k}_i^{-1} \otimes \dot{e}_i + \dot{e}_i \otimes 1, \quad \Delta(\dot{k}_i) = \dot{k}_i \otimes \dot{k}_i.$$

We have the natural $U_q\mathfrak{sl}_\infty$-module $\dot{V}$ on basis $\{\dot{v}_i \mid i \in \mathbb{Z}\}$ and its dual $\dot{W}$ on basis $\{\dot{w}_i \mid i \in \mathbb{Z}\}$. The Chevalley generators $\dot{f}_i$ and $\dot{e}_i$ of $U_q\mathfrak{sl}_\infty$ act on these basis vectors by exactly the same formulae as at $q = 1$, and also $\dot{k}_i \dot{v}_j = q^{\delta_{i,j} - \delta_{i+1,j}} \dot{v}_j$ and $\dot{k}_i \dot{w}_j = q^{\delta_{i+1,j} - \delta_{i,j}} \dot{w}_j$. There is also an R-matrix giving some distinguished intertwiners $\dot{V} \otimes \dot{V} \xrightarrow{\sim} \dot{V} \otimes \dot{V}$ and $\dot{W} \otimes \dot{W} \xrightarrow{\sim} \dot{W} \otimes \dot{W}$, from which we produce the following $U_q\mathfrak{sl}_\infty$-module homomorphisms:

$$\dot{c} : \dot{V} \otimes \dot{V} \to \dot{V} \otimes \dot{V}, \qquad \dot{v}_i \otimes \dot{v}_j \mapsto \begin{cases} \dot{v}_j \otimes \dot{v}_i + q^{-1} \dot{v}_i \otimes \dot{v}_j & \text{if } i < j, \\ (q + q^{-1}) \dot{v}_j \otimes \dot{v}_i & \text{if } i = j, \\ \dot{v}_j \otimes \dot{v}_i + q \dot{v}_i \otimes \dot{v}_j & \text{if } i > j; \end{cases}$$

[1]We follow the convention of adding a dot to all q-analogs to distinguish them from their classical counterparts.

$$\dot{c} : \dot{W} \otimes \dot{W} \to \dot{W} \otimes \dot{W}, \qquad \dot{w}_i \otimes \dot{w}_j \mapsto \begin{cases} \dot{w}_j \otimes \dot{w}_i + q\dot{w}_i \otimes \dot{w}_j & \text{if } i < j, \\ (q+q^{-1})\dot{w}_j \otimes \dot{w}_i & \text{if } i = j, \\ \dot{w}_j \otimes \dot{w}_i + q^{-1}\dot{w}_i \otimes \dot{w}_j & \text{if } i > j. \end{cases}$$

Then we form the tensor space $\dot{V}^{\otimes n} \otimes \dot{W}^{\otimes m}$, which is a $U_q\mathfrak{sl}_\infty$-module with its monomial basis $\{\dot{v}_\lambda \mid \lambda \in \mathfrak{t}^*_{\mathbb{Z}}\}$ defined just like above. Let $\dot{c}_k := 1^{\otimes(k-1)} \otimes \dot{c} \otimes 1^{n+m-1-k}$ for $k \neq n$, which is a $U_q\mathfrak{sl}_\infty$-module endomorphism of $\dot{V}^{\otimes n} \otimes \dot{W}^{\otimes m}$.

Next we must pass to a formal completion $\dot{V}^{\otimes n}\hat{\otimes}\dot{W}^{\otimes m}$ of our q-tensor space. Let $I \subset \mathbb{Z}$ be a finite subinterval and $I_+ := I \cup (I+1)$. Let $\dot{V}_I$ and $\dot{W}_I$ be the subspaces of $\dot{V}$ and $\dot{W}$ spanned by the basis vectors $\{\dot{v}_i \mid i \in I_+\}$ and $\{\dot{w}_i \mid i \in I_+\}$, respectively. Then $\dot{V}_I^{\otimes n} \otimes \dot{W}_J^{\otimes m}$ is a subspace of $\dot{V}^{\otimes n} \otimes \dot{W}^{\otimes m}$. For $J \subseteq I$ there is an obvious projection $\pi_J : \dot{V}_I^{\otimes n} \otimes \dot{W}_I^{\otimes m} \twoheadrightarrow \dot{V}_J^{\otimes n} \otimes \dot{W}_J^{\otimes m}$ mapping $\dot{v}_\lambda$ to $\dot{v}_\lambda$ if all the entries of the tuple λ lie in J_+, or to zero otherwise. Then we set

$$\dot{V}^{\otimes n}\hat{\otimes}\dot{W}^{\otimes m} := \varprojlim \dot{V}_I^{\otimes n} \otimes \dot{W}_I^{\otimes m},$$

taking the inverse limit over all finite subintervals $I \subset \mathbb{Z}$ with respect to the projections π_J just defined. The action of $U_q\mathfrak{sl}_\infty$ and of each $\dot{c}_k$ extend naturally to the completion.

Lemma 3.1. *There is a unique continuous antilinear involution*

$$\psi : \dot{V}^{\otimes n}\hat{\otimes}\dot{W}^{\otimes m} \to \dot{V}^{\otimes n}\hat{\otimes}\dot{W}^{\otimes m}$$

such that

- *ψ commutes with the actions of $\dot{f}_i$ and $\dot{e}_i$ for all $i \in \mathbb{Z}$ and with the endomorphisms $\dot{c}_k$ for all $k \neq n$;*
- *Each $\psi(\dot{v}_\lambda)$ is equal to $\dot{v}_\lambda$ plus a (possibly infinite) $\mathbb{Z}[q,q^{-1}]$-linear combination of $\dot{v}_\mu$ for $\mu > \lambda$ in the Bruhat order.*

Proof. For each finite subinterval $I \subset \mathbb{Z}$, let $U_q\mathfrak{sl}_I$ be the subalgebra of $U_q\mathfrak{sl}_\infty$ generated by $\{\dot{f}_i, \dot{e}_i, \dot{k}_i^{\pm1} \mid i \in I\}$. A construction of Lusztig [37, §27.3] involving the quasi-R-matrix Θ_I for $U_q\mathfrak{sl}_I$ gives an antilinear involution $\psi_I : \dot{V}_I^{\otimes n} \otimes \dot{W}_I^{\otimes m} \to \dot{V}_I^{\otimes n} \otimes \dot{W}_I^{\otimes m}$ commuting with the actions of $\dot{f}_i$ and $\dot{e}_i$ for $i \in I$. Moreover for $J \subset I$ the involutions ψ_I and ψ_J are intertwined by the projection $\pi_J : \dot{V}_I^{\otimes n} \otimes \dot{W}_I^{\otimes m} \twoheadrightarrow \dot{V}_J^{\otimes n} \otimes \dot{W}_J^{\otimes m}$, as follows easily from the explicit form of the quasi-R-matrix. Hence the involutions ψ_I for all I induce a well-defined involution ψ on the inverse limit. The fact that the resulting involution commutes with each $\dot{c}_k$ can be deduced from the formal definition of the latter in terms of the R-matrix. Finally the uniqueness is a consequence of the existence of an algorithm to uniquely compute the canonical basis using just the given two properties (as sketched below). □

This puts us in position to apply Lusztig's lemma to deduce for each $\mu \in \mathfrak{t}_{\mathbb{Z}}^*$ that there is a unique vector $\dot{b}_\mu \in \dot{V}^{\otimes n} \hat{\otimes} \dot{W}^{\otimes m}$ such that

- $\psi(\dot{b}_\mu) = \dot{b}_\mu$;
- $\dot{b}_\mu$ is equal to $\dot{v}_\mu$ plus a (possibly infinite) $q\mathbb{Z}[q]$-linear combination of $\dot{v}_\lambda$ for $\lambda > \mu$.

We refer to the resulting topological basis $\{\dot{b}_\mu \mid \mu \in \mathfrak{t}_{\mathbb{Z}}^*\}$ for $\dot{V}^{\otimes n} \hat{\otimes} \dot{W}^{\otimes m}$ as the *canonical basis*. In fact, but this is in no way obvious from the above definition, each $\dot{b}_\mu$ is always a *finite* sum of $\dot{v}_\lambda$'s, i.e., $\dot{b}_\mu \in \dot{V}^{\otimes n} \otimes \dot{W}^{\otimes m}$ before completion. Moreover the polynomials $d_{\lambda,\mu}(q)$ arising from the expansion

$$\dot{b}_\mu = \sum_{\lambda \in \mathfrak{t}_{\mathbb{Z}}^*} d_{\lambda,\mu}(q) \dot{v}_\lambda$$

are known always to be some finite type A parabolic Kazhdan–Lusztig polynomials (suitably normalized). In particular $d_{\lambda,\mu}(q) \in \mathbb{N}[q]$.

Now we can state the following fundamental theorem, formerly known as the "super Kazhdan–Lusztig conjecture."

Theorem 3.2 (Cheng, Lam, Wang). *For any $\lambda, \mu \in \mathfrak{t}_{\mathbb{Z}}^*$ we have that*

$$[M(\lambda) : L(\mu)] = d_{\lambda,\mu}(1).$$

In other words, the vectors $\{b_\mu \mid \mu \in \mathfrak{t}_{\mathbb{Z}}^*\}$ arising from the projective indecomposable supermodules in $\mathcal{O}_{\mathbb{Z}}$ via the identification $K(\mathcal{O}_{\mathbb{Z}}^\Delta) \leftrightarrow V^{\otimes n} \otimes W^{\otimes m}$ coincide with the specialization of Lusztig's canonical basis $\{\dot{b}_\mu \mid \mu \in \mathfrak{t}_{\mathbb{Z}}^*\}$ at $q = 1$.

We are going to do two more things in this section. First, we sketch briefly how one can compute the canonical basis algorithmically. Then we will explain how Theorem 3.2 should really be understood in terms of a certain graded lift $\dot{\mathcal{O}}_{\mathbb{Z}}$ of $\mathcal{O}_{\mathbb{Z}}$, using the language of categorification.

The algorithm to compute the canonical basis goes by induction on the degree of atypicality. Recall that a weight $\mu \in \mathfrak{t}_{\mathbb{Z}}^*$ is *typical* if $\{\mu_1, \dots, \mu_n\} \cap \{\mu_{n+1}, \dots, \mu_{n+m}\} = \varnothing$. We also say it is *weakly dominant* if $\mu_1 \geq \cdots \geq \mu_n$ and $\mu_{n+1} \leq \cdots \leq \mu_{n+m}$ (equivalently $\mu + \rho$ is dominant in the earlier sense). The weights that are both typical and weakly dominant are maximal in the Bruhat ordering, so that $\dot{b}_\mu = \dot{v}_\mu$. Then to compute $\dot{b}_\mu$ for an arbitrary typical but not weakly dominant μ we just have to follow the usual algorithm to compute Kazhdan–Lusztig polynomials. Pick $k \neq n$ such that *either* $k < n$ and $\mu_k < \mu_{k+1}$ *or* $k > n$ and $\mu_k > \mu_{k+1}$. Let λ be the weight obtained from μ by interchanging μ_k and μ_{k+1}. By induction on the Bruhat ordering we may assume that $\dot{b}_\lambda$ is already computed. Then $\dot{c}_k \dot{b}_\lambda$ is ψ-invariant and has $\dot{v}_\mu$ as its leading term with coefficient 1, i.e., it equals $\dot{v}_\mu$ plus a $\mathbb{Z}[q, q^{-1}]$-linear combination of $\dot{v}_\nu$ for $\nu > \mu$. It just remains to adjust this vector by subtracting bar-invariant multiples of inductively computed canonical basis vectors $\dot{b}_\nu$ for $\nu > \mu$ to obtain a vector that is both ψ-invariant and lies in $\dot{v}_\mu + \sum_{\lambda > \mu} q\mathbb{Z}[q] \dot{v}_\lambda$. This must equal $\dot{b}_\mu$ by the uniqueness.

Now suppose that $\mu \in \mathfrak{t}_{\mathbb{Z}}^*$ is not typical. The idea to compute $\dot{b}_\mu$ then is to apply a certain *bumping procedure* to produce from μ another weight λ of strictly smaller atypicality, together with a monomial $\dot{x}$ of quantum divided powers of Chevalley generators of $U_q\mathfrak{sl}_\infty$, such that $\dot{x}\dot{b}_\lambda$ has $\dot{v}_\mu$ as its leading term with coefficient 1. Then we can adjust this ψ-invariant vector by subtracting bar-invariant multiples of recursively computed canonical basis vectors $\dot{b}_\nu$ for $\nu > \mu$, to obtain $\dot{b}_\mu$ as before. The catch is that (unlike the situation in the previous paragraph) there are infinitely many weights $\nu > \mu$ so that it is not clear that the recursion always terminates in finitely many steps. Examples computed using a GAP implementation of the algorithm suggest that it always does; our source code is available at [13]. (In any case one can always find a finite interval I such that $\dot{x}\dot{b}_\lambda \in \dot{V}_I^{\otimes n} \otimes \dot{W}_I^{\otimes m}$; then by some nontrivial but known positivity of structure constants we get that $\dot{b}_\mu \in \dot{V}_I^{\otimes n} \otimes \dot{W}_I^{\otimes m}$ too; hence one can apply π_I prior to making any subsequent adjustments to guarantee that the algorithm terminates in finitely many steps.)

Example 3.3. With this example we outline the bumping procedure. Given an atypical μ, we let i be the largest integer that appears both to the left and to the right of the separator | in the tuple μ. Pick one of the two sides of the separator and let $j \geq i$ be maximal such that all of $i, i+1, \ldots, j$ appear on this side of μ. Add 1 to all occurrences of $i, i+1, \ldots, j$, on the chosen side. Then if $j+1$ also appears on the other side of μ, we repeat the bumping procedure on that side with i replaced by $j+1$. We continue in this way until $j+1$ is not repeated on the other side. This produces the desired output weight λ of strictly smaller atypicality. For example, if $\mu = (0, 5, 2, 2|0, 1, 3, 4)$ of atypicality one we bump as follows:

$$(1,6,3,3|0,2,4,5) \xleftarrow{\dot{e}_5} (1,5,3,3|0,2,4,5) \xleftarrow{\dot{f}_4\dot{f}_3} (1,5,3,3|0,2,3,4)$$
$$\xleftarrow{\dot{e}_2^{(2)}} (1,5,2,2|0,2,3,4) \xleftarrow{\dot{f}_1} (1,5,2,2|0,1,3,4) \xleftarrow{\dot{e}_0} (0,5,2,2|0,1,3,4).$$

The labels on the edges here are the appropriate monomials that reverse the bumping procedure; then the final monomial $\dot{x}$ output by the bumping procedure is the product $\dot{e}_5\dot{f}_4\dot{f}_3\dot{e}_2^{(2)}\dot{f}_1\dot{e}_0$ of all of these labels. Thus we should compute $\dot{e}_5\dot{f}_4\dot{f}_3\dot{e}_2^{(2)}\dot{f}_1\dot{e}_0\dot{b}_{(1,6,3,3|0,2,4,5)}$, where $\dot{b}_{(1,6,3,3|0,2,4,5)}$ can be worked out using the typical algorithm. The result is a ψ-invariant vector equal to $\dot{b}_{(0,5,2,2|0,1,3,4)}$ plus some higher terms which can be computed recursively (specifically one finds that $(q+q^{-1})\dot{b}_{(2,5,2,2|1,2,3,4)}$ needs to be subtracted).

Example 3.4. Here we work out the combinatorics in the principal block for $\mathfrak{gl}_{2|1}(\mathbb{C})$. The weights are $\{(0, i|i), (i, 0|i) \mid i \in \mathbb{Z}\}$. The corresponding canonical basis vectors $\dot{b}_\mu$ are represented in the following diagram which is arranged according to the Bruhat graph; we show just enough vertices for the generic pattern to be apparent.

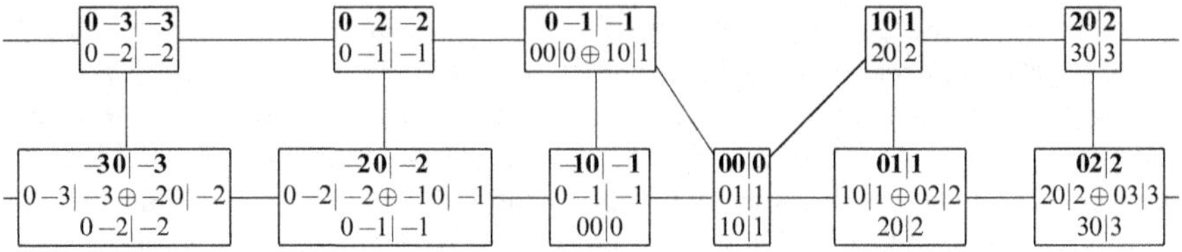

For example, the center node of this diagram encodes $\dot{b}_{(0,0|0)} = \dot{v}_{(0,0|0)} + q\dot{v}_{(0,1|1)} + q^2\dot{v}_{(1,0|1)}$; the node to the right of that encodes $\dot{b}_{(0,1|1)} = \dot{v}_{(0,1|1)} + q\dot{v}_{(1,0|1)} + q\dot{v}_{(0,2|2)} + q^2\dot{v}_{(2,0|2)}$. Let us explain in more detail how we computed $\dot{b}_{(-1,0|-1)}$ here. The bumping procedure tells us to look at $\dot{e}_0\dot{e}_{-1}\dot{b}_{(0,1|-1)}$. As $(0,1|-1)$ is typical we get easily from the typical algorithm that $\dot{b}_{(0,1|-1)} = \dot{c}_1\dot{v}_{(1,0|-1)} = \dot{v}_{(0,1|-1)} + q\dot{v}_{(1,0|-1)}$. Hence

$$\dot{e}_0\dot{e}_{-1}\dot{b}_{(0,1|-1)} = \dot{v}_{(-1,0|-1)} + (1+q^2)\dot{v}_{(0,0|0)} + q\dot{v}_{(0,-1|-1)} + q\dot{v}_{(0,1|1)} + q^2\dot{v}_{(1,0|1)}.$$

This vector is ψ-invariant with the right leading term $\dot{v}_{(-1,0|-1)}$, but we must make one correction to remove a term $\dot{v}_{(0,0|0)}$, i.e., we must subtract $\dot{b}_{(0,0|0)}$ as already computed, to obtain that $\dot{b}_{(-1,0|-1)} = \dot{v}_{(-1,0|-1)} + q\dot{v}_{(0,-1|-1)} + q^2\dot{v}_{(0,0|0)}$.

Returning to more theoretical considerations, the key point is that the category $\mathcal{O}_{\mathbb{Z}}$ is an example of an $\mathfrak{sl}_\infty$-tensor product categorification of $V^{\otimes n} \otimes W^{\otimes m}$. This means in particular that there exist some exact endofunctors F_i and E_i of $\mathcal{O}_{\mathbb{Z}}^{\Delta}$ whose induced actions on $K(\mathcal{O}_{\mathbb{Z}}^{\Delta})$ match the actions of the Chevalley generators f_i and e_i on $V^{\otimes n} \otimes W^{\otimes m}$ under our identification. To define these functors, recall that U denotes the natural $\mathfrak{g}$-module of column vectors. Let $U^\vee$ be its dual. Introduce the biadjoint projective functors

$$F := - \otimes U : \mathcal{O}_{\mathbb{Z}} \to \mathcal{O}_{\mathbb{Z}}, \qquad E := - \otimes U^\vee : \mathcal{O}_{\mathbb{Z}} \to \mathcal{O}_{\mathbb{Z}}.$$

The action of the Casimir tensor

$$\Omega := \sum_{i,j=1}^{n+m} (-1)^{\bar{u}_j} e_{i,j} \otimes e_{j,i} \in \mathfrak{g} \otimes \mathfrak{g}$$

defines an endomorphism of $FM = M \otimes U$ for each $M \in \mathcal{O}_{\mathbb{Z}}$. Let F_i be the summand of the functor F defined so that $F_i M$ is the generalized i-eigenspace of Ω for each $i \in \mathbb{Z}$. We then have that $F = \bigoplus_{i \in \mathbb{Z}} F_i$. Similarly the functor E decomposes as $E = \bigoplus_{i \in \mathbb{Z}} E_i$ where each E_i is biadjoint to F_i; explicitly one can check that $E_i M$ is the generalized $(m-n-i)$-eigenspace of Ω on $EM = M \otimes U^\vee$. Now it is an instructive exercise to prove:

Lemma 3.5. *The exact functors F_i and E_i send supermodules with Verma flags to supermodules with Verma flags. Moreover the induced endomorphisms $[F_i]$ and*

$[E_i]$ *of* $K(\mathcal{O}_{\mathbb{Z}}^{\Delta})$ *agree under the above identification with the endomorphisms* f_i *and* e_i *of* $V^{\otimes n} \otimes W^{\otimes m}$ *defined by the action of the Chevalley generators of* $\mathfrak{sl}_\infty$.

In fact much more is true here. The action of Ω on each FM defines a natural transformation $x \in \mathrm{End}(F)$. Also let $t \in \mathrm{End}(F^2)$ be such that $t_M : F^2M \to F^2M$ is the endomorphism

$$t_M : M \otimes U \otimes U \to M \otimes U \otimes U, \qquad v \otimes u_i \otimes u_j \mapsto (-1)^{\bar{u}_i \bar{u}_j} v \otimes u_j \otimes u_i.$$

From x and t one obtains $x_i := F^{d-i} x F^{i-1} \in \mathrm{End}(F^d)$ and $t_j := F^{d-j-1} t F^{j-1} \in \mathrm{End}(F^d)$ for each $d \geq 0, 1 \leq i \leq d$ and $1 \leq j \leq d-1$. It is straightforward to check that these natural transformations satisfy the defining relations of the degenerate affine Hecke algebra H_d. This shows that the category $\mathcal{O}_{\mathbb{Z}}$ equipped with the biadjoint pair of endofunctors F and E, plus the endomorphisms $x \in \mathrm{End}(F)$ and $t \in \mathrm{End}(F^2)$, is an $\mathfrak{sl}_\infty$*-categorification* in the sense of Chuang and Rouquier. In addition $\mathcal{O}_{\mathbb{Z}}$ is a highest weight category, and Lemma 3.5 checks some appropriate compatibility of the categorical action with this highest weight structure. The conclusion is that $\mathcal{O}_{\mathbb{Z}}$ is actually an $\mathfrak{sl}_\infty$*-tensor product categorification* of $V^{\otimes n} \otimes W^{\otimes m}$ in a formal sense introduced by Losev and Webster.

We are ready to state the following extension of the super Kazhdan–Lusztig conjecture, which incorporates a $\mathbb{Z}$-grading in the spirit of the classic work of Beilinson, Ginzburg and Soergel on Koszulity of category $\mathcal{O}$ in the purely even setting.

Theorem 3.6 (Brundan, Losev, Webster). *There exists a unique (up to equivalence) graded lift* $\dot{\mathcal{O}}_{\mathbb{Z}}$ *of* $\mathcal{O}_{\mathbb{Z}}$ *that is a* $U_q\mathfrak{sl}_\infty$*-tensor product categorification of* $\dot{V}^{\otimes n} \otimes \dot{W}^{\otimes m}$. *Moreover the category* $\dot{\mathcal{O}}_{\mathbb{Z}}$ *is a standard Koszul highest weight category, and its graded decomposition numbers* $[\dot{M}(\lambda) : \dot{L}(\mu)]_q$ *are given by the parabolic Kazhdan–Lusztig polynomials* $d_{\lambda,\mu}(q)$ *as defined above.*

A few more explanations are in order. To start with we should clarify what it means to say that $\dot{\mathcal{O}}_{\mathbb{Z}}$ is a graded lift of $\mathcal{O}_{\mathbb{Z}}$. The easiest way to understand this is to remember as discussed in the previous section that $\mathcal{O}_{\mathbb{Z}}$ is equivalent to the category mof-A of finite-dimensional locally unital right A-modules, where A is the locally unital algebra

$$A := \bigoplus_{\lambda,\mu \in \mathfrak{t}_{\mathbb{Z}}^*} \mathrm{Hom}_{\mathfrak{g}}(P(\lambda), P(\mu)).$$

To give a graded lift $\dot{\mathcal{O}}_{\mathbb{Z}}$ of $\mathcal{O}_{\mathbb{Z}}$ amounts to exhibiting some $\mathbb{Z}$-grading on the algebra A with respect to which each of its distinguished idempotents 1_λ are homogeneous; then the category grmof-A of *graded* finite-dimensional locally unital right A-modules gives a graded lift of $\mathcal{O}_{\mathbb{Z}}$. Of course there can be many ways to do this, including the trivial way that puts all of A in degree zero! Theorem 3.6 asserts in particular that the algebra A admits a positive grading making it into a (locally unital) Koszul algebra; as is well known such a grading (if it exists) is unique up to automorphism.

For this choice of grading, the category $\dot{\mathcal{O}}_{\mathbb{Z}} :=$ grmof-A is a graded highest weight category with distinguished irreducible objects $\{\dot{L}(\lambda) \mid \lambda \in \mathfrak{t}_{\mathbb{Z}}^*\}$, standard objects $\{\dot{M}(\lambda) \mid \lambda \in \mathfrak{t}_{\mathbb{Z}}^*\}$ and indecomposable projective objects $\{\dot{P}(\lambda) := 1_\lambda A \mid \lambda \in \mathfrak{t}_{\mathbb{Z}}^*\}$; these are graded lifts of the modules $L(\lambda)$, $M(\lambda)$ and $P(\lambda)$, respectively, such that the canonical maps $\dot{P}(\lambda) \twoheadrightarrow \dot{M}(\lambda) \twoheadrightarrow \dot{L}(\lambda)$ are all homogeneous of degree zero. Then the assertion from Theorem 3.6 that $\dot{\mathcal{O}}_{\mathbb{Z}}$ is standard Koszul means that each $\dot{M}(\lambda)$ possesses a *linear projective resolution*, that is, there is an exact sequence

$$\cdots \to \dot{P}^2(\lambda) \to \dot{P}^1(\lambda) \to \dot{P}(\lambda) \to \dot{M}(\lambda) \to 0,$$

such that for each $i \geq 1$ the module $\dot{P}^i(\lambda)$ is a direct sum of graded modules $q^i \dot{P}(\mu)$ for $\mu > \lambda$. Here q denotes the degree shift functor defined on a graded module M by letting qM be the same underlying module with new grading defined from $(qM)_i := M_{i-1}$.

Let $\dot{\mathcal{O}}_{\mathbb{Z}}^{\Delta}$ be the exact subcategory of $\dot{\mathcal{O}}_{\mathbb{Z}}$ consisting of modules with a graded Δ-flag. Its Grothendieck group is a $\mathbb{Z}[q, q^{-1}]$-module with q acting by degree shift. Let $K_q(\dot{\mathcal{O}}_{\mathbb{Z}}^{\Delta})$ be the $\mathbb{Q}(q)$-vector space obtained by extending scalars, i.e., it is the $\mathbb{Q}(q)$-vector space on basis $\{[\dot{M}(\lambda)] \mid \lambda \in \mathfrak{t}_{\mathbb{Z}}^*\}$. Then again we *identify*

$$K_q(\dot{\mathcal{O}}_{\mathbb{Z}}^{\Delta}) \leftrightarrow \dot{V}^{\otimes n} \otimes \dot{W}^{\otimes m}, \qquad [\dot{M}(\lambda)] \leftrightarrow \dot{v}_\lambda.$$

The assertion about graded decomposition numbers in Theorem 3.6 means under this identification that

$$[\dot{P}(\mu)] \leftrightarrow \dot{b}_\mu.$$

The assertion that $\dot{\mathcal{O}}_{\mathbb{Z}}$ is a $U_q\mathfrak{sl}_\infty$-tensor product categorification means in particular that the biadjoint endofunctors F_i and E_i of $\mathcal{O}_{\mathbb{Z}}$ admit graded lifts $\dot{F}_i$ and $\dot{E}_i$, which are also biadjoint up to approriate degree shifts. Moreover these graded functors preserve modules with a graded Verma flag, and their induced actions on $K_q(\dot{\mathcal{O}}_{\mathbb{Z}}^{\Delta})$ agree with the actions of $\dot{f}_i, \dot{e}_i \in U_q\mathfrak{sl}_\infty$ under our identification.

We will say more about the proof of Theorem 3.6 in the next section.

Notes. The identification of the Bruhat order on $\mathfrak{t}_{\mathbb{Z}}^*$ with the "reverse dominance ordering" is justified in [10, Lemma 2.5]. Our Lemma 3.1 is a variation on [10, Theorem 2.14]; the latter theorem was used in [10] to define a twisted version of the canonical basis which corresponds to the indecomposable tilting supermodules rather than the indecomposable projectives in $\mathcal{O}_{\mathbb{Z}}$. The super Kazhdan–Lusztig conjecture as formulated here is equivalent to [10, Conjecture 4.32]; again the latter was expressed in terms of tilting supermodules. The equivalence of the two versions of the conjecture can be deduced from the Ringel duality established in [12, (7.4)]; see also [15, Remark 5.30]. The algorithm for computing the canonical basis sketched here is a variation on an algorithm described in detail in [10, §2-h];

the latter algorithm computes the twisted canonical basis rather than the canonical basis. Example 3.4 was worked out already in [20, §9.5].

Theorems 3.2 and 3.6 are proved in [18] and [15], respectively. In fact both of these articles also prove a more general form of the super Kazhdan–Lusztig conjecture which is adapted to arbitrary Borel subalgebras $\mathfrak{b}$ of $\mathfrak{g}$; at the level of combinatorics this amounts to shuffling the tensor factors in the tensor product $\dot{V}^{\otimes n} \otimes \dot{W}^{\otimes m}$ into more general orders. The article [15] also considers parabolic analogs. The idea that blocks of category $\mathcal{O}$ should possess Koszul graded lifts goes back to the seminal work of Beilinson, Ginzburg and Soergel [4] in the context of semisimple Lie algebras. The notion of $\mathfrak{sl}$-categorification was introduced by Chuang and Rouquier following their joint work [23]. The definition was recorded for the first time in the literature in [42, Definition 5.29]. For the definition of tensor product categorification, see [36, Definition 3.2] and also [15, Definition 2.9]. A full proof of Lemma 3.5 (and its generalization to the parabolic setting) can be found in [15, Theorem 3.9].

4 Principal W-Algebras and the Double Centralizer Property

By a *prinjective object* we mean an object that is both projective and injective. To set the scene for this section we recall a couple of classical results. Let $\mathcal{O}_0$ be the principal block of category $\mathcal{O}$ for a semisimple Lie algebra $\mathfrak{g}$, and recall that the irreducible modules in $\mathcal{O}_0$ are the modules $\{L(w \cdot 0) \mid w \in W\}$ parametrized by the Weyl group W. There is a unique indecomposable prinjective module in $\mathcal{O}_0$ up to isomorphism, namely, the projective cover $P(w_0 \cdot 0)$ of the "antidominant" Verma module $L(w_0 \cdot 0)$; here w_0 is the longest element of the Weyl group.

Theorem 4.1 (Soergel's Endomorphismensatz). *The endomorphism algebra*

$$C_0 := \mathrm{End}_{\mathfrak{g}}(P(w_0 \cdot 0))$$

is generated by the center $Z(\mathfrak{g})$ of the universal enveloping algebra of $\mathfrak{g}$. Moreover C_0 is canonically isomorphic to the coinvariant algebra, i.e., the cohomology algebra $H^*(G/B, \mathbb{C})$ of the flag variety associated to $\mathfrak{g}$.

Theorem 4.2 (Soergel's Struktursatz). *The functor*

$$\mathbb{V}_0 := \mathrm{Hom}_{\mathfrak{g}}(P(w_0 \cdot 0), -) : \mathcal{O}_0 \to \text{mof-}C_0$$

is fully faithful on projectives.

With these two theorems in hand, we can explain Soergel's approach to the construction of the Koszul graded lift of the category $\mathcal{O}_0$. Introduce the *Soergel modules*

$$Q(w) := \mathbb{V}_0 P(w \cdot 0) \in \text{mof-}C_0$$

for each $w \in W$. The Struktursatz implies that the finite-dimensional algebra

$$A_0 := \bigoplus_{x,y \in W} \mathrm{Hom}_{C_0}(Q(x), Q(y))$$

is isomorphic to the endomorphism algebra of a minimal projective generator for $\mathcal{O}_0$. The algebra C_0 is naturally graded as it is a cohomology algebra. It turns out that each Soergel module $Q(w)$ also admits a unique graded lift $\dot{Q}(w)$ that is a self-dual graded C_0-module. Hence we get induced a grading on the algebra A_0. This is the grading making A_0 into a Koszul algebra. The resulting category grmof-A_0 is the appropriate graded lift $\dot{\mathcal{O}}_0$ of $\mathcal{O}_0$.

Now we return to the situation of the previous section, so $\mathcal{O}_{\mathbb{Z}}$ is the integral part of category $\mathcal{O}$ for $\mathfrak{g} = \mathfrak{gl}_{n|m}(\mathbb{C})$ and we represent integral weights $\lambda \in \mathfrak{t}^*_{\mathbb{Z}}$ as $n|m$-tuples of integers. The proof of Theorem 3.6 stated above follows a similar strategy to Soergel's construction in the classical case but there are several complications. To start with, in any atypical block, there turn out to be infinitely many isomorphism classes of indecomposable prinjective supermodules:

Lemma 4.3. *For $\lambda \in \mathfrak{t}^*_{\mathbb{Z}}$, the projective supermodule $P(\lambda) \in \mathcal{O}_{\mathbb{Z}}$ is injective if and only if λ is* antidominant, *i.e., $\lambda_1 \leq \cdots \leq \lambda_n$ and $\lambda_{n+1} \geq \cdots \geq \lambda_{n+m}$. (Recall λ_i denotes $(\lambda + \rho, \delta_i) \in \mathbb{Z}$).*

Proof. This follows by a special case of [15, Theorem 2.22]. More precisely, there is an $\mathfrak{sl}_\infty$-crystal with vertex set $\mathfrak{t}^*_{\mathbb{Z}}$, namely, Kashiwara's crystal associated to the $\mathfrak{sl}_\infty$-module $V^{\otimes n} \otimes W^{\otimes m}$. Then [15, Theorem 2.22] shows that the set of $\lambda \in \mathfrak{t}^*_{\mathbb{Z}}$ such that $P(\lambda)$ is injective is the vertex set of the connected component of this crystal containing any weight $(i, \ldots, i|j, \ldots, j)$ for $i < j$. Now it is a simple combinatorial exercise to see that the vertices in this connected component are exactly the antidominant $\lambda \in \mathfrak{t}^*_{\mathbb{Z}}$. □

Remark 4.4. More generally, for $\lambda \in \mathfrak{t}^*$, the projective $P(\lambda) \in \mathcal{O}$ is injective if and only if λ is antidominant in the sense that $(\lambda, \delta_i - \delta_j) \notin \mathbb{Z}_{\geq 0}$ for $1 \leq i < j \leq n$ and $(\lambda, \delta_i - \delta_j) \notin \mathbb{Z}_{\leq 0}$ for $n+1 \leq i < j \leq n+m$. This follows from Lemma 4.3 and Theorem 2.6. In other words, the projective $P(\lambda)$ is injective if and only if the irreducible supermodule $L(\lambda)$ is of maximal Gelfand–Kirillov dimension amongst all supermodules in $\mathcal{O}$.

Then, fixing $\xi \in \mathfrak{t}^*_{\mathbb{Z}}/\approx$, the appropriate analog of the coinvariant algebra for the block $\mathcal{O}_\xi$ is the locally unital algebra

$$C_\xi := \bigoplus_{\text{Antidominant } \lambda, \mu \in \xi} \mathrm{Hom}_{\mathfrak{g}}(P(\lambda), P(\mu)).$$

For atypical blocks this algebra is infinite dimensional and no longer commutative. Still there is an analog of the Struktursatz:

Theorem 4.5 (Brundan, Losev, Webster). *The functor* $\mathbb{V}_\xi : \mathcal{O}_\xi \to \text{mof-}C_\xi$ *sending* $M \in \mathcal{O}_\xi$ *to*

$$\mathbb{V}_\xi M := \bigoplus_{\text{Antidominant } \lambda \in \xi} \operatorname{Hom}_{\mathfrak{g}}(P(\lambda), M)$$

is fully faithful on projectives.

However we do not at present know of any explicit description of the algebra C_ξ. Instead the proof of Theorem 3.6 involves another abelian category mod-H_ξ. This notation is strange because actually there is no single algebra H_ξ here; rather, there is an infinite *tower of cyclotomic quiver Hecke algebras* $H_\xi^1 \subset H_\xi^2 \subset H_\xi^3 \subset \cdots$, which arise as the endomorphism algebras of larger and larger finite direct sums of indecomposable prinjective supermodules (with multiplicities). Then the category mod-H_ξ consists of sequences of finite-dimensional modules over this tower of Hecke algebras subject to some stability condition. Moreover there is an explicitly constructed exact functor $\mathbb{U}_\xi : \mathcal{O}_\xi \to \text{mod-}H_\xi$. The connection between this and the functor $\mathbb{V}_\xi$ comes from the following lemma.

Lemma 4.6. *There is a unique (up to isomorphism) equivalence of categories*

$$\mathbb{I}_\xi : \text{mod-}H_\xi \xrightarrow{\sim} \text{mof-}C_\xi$$

such that $\mathbb{V}_\xi \cong \mathbb{I}_\xi \circ \mathbb{U}_\xi$.

Proof. This follows because both of the functors $\mathbb{U}_\xi$ and $\mathbb{V}_\xi$ are quotient functors, i.e., they satisfy the universal property of the Serre quotient of $\mathcal{O}_\xi$ by the subcategory generated by $\{L(\lambda) \mid \lambda \in \xi$ such that λ is *not* antidominant$\}$. For $\mathbb{U}_\xi$ this universal property is established in [15, Theorem 4.9]. It is automatic for $\mathbb{V}_\xi$. □

Each of the algebras H_ξ^r in the tower of Hecke algebras is naturally graded, so that we are able to define a corresponding graded category grmod-H_ξ. Then we prove that the modules $Y(\lambda) := \mathbb{U}_\xi P(\lambda) \in \text{mod-}H_\xi$ admit unique graded lifts $\dot{Y}(\lambda) \in \text{grmod-}H_\xi$ which are self-dual in an appropriate sense. Since the functor $\mathbb{U}_\xi$ is also fully faithful on projectives (e.g., by Theorem 4.5 and Lemma 4.6), we thus obtain a $\mathbb{Z}$-grading on the basic algebra

$$A_\xi := \bigoplus_{\lambda,\mu \in \xi} \operatorname{Hom}_{\mathfrak{g}}(P(\lambda), P(\mu)) \cong \bigoplus_{\lambda,\mu \in \xi} \operatorname{Hom}_{H_\xi}(\dot{Y}(\lambda), \dot{Y}(\mu)),$$

that is Morita equivalent to $\mathcal{O}_\xi$. This grading turns out to be Koszul, and grmof-A_ξ gives the desired graded lift $\dot{\mathcal{O}}_\xi$ of the block $\mathcal{O}_\xi$ from Theorem 3.6.

The results just described provide a substitute for Soergel's Endomorphismensatz for $\mathfrak{gl}_{n|m}(\mathbb{C})$, with the tower of cyclotomic quiver Hecke algebras replacing the coinvariant algebra. However we still do not find this completely satisfactory, and actually believe that it should be possible to give an explicit (graded!) description of the basic algebra C_ξ itself. This seems like a tractable problem whose solution could suggest some more satisfactory geometric picture underpinning the rich structure of super category $\mathcal{O}$.

Example 4.7. Here we give explicit generators and relations for the algebra C_0 for the principal block of $\mathcal{O}$ for $\mathfrak{gl}_{2|1}(\mathbb{C})$. The prinjectives are indexed by $\mathbb{Z}$ and their Verma flags are as displayed on the bottom row of the diagram in Example 3.4. The algebra C_0 is isomorphic to the path algebra of the same infinite linear quiver as in Example 2.8 modulo the relations

$$\begin{aligned} e_{i+1}e_i &= f_i f_{i+1} = 0 && \text{for all } i \in \mathbb{Z}, \\ f_{i+1}e_{i+1}f_{i+1}e_{i+1} + e_i f_i e_i f_i &= 0 && \text{for } i \leq -2 \text{ or } i \geq 1, \\ f_0 e_0 + e_{-1}f_{-1}e_{-1}f_{-1} &= 0, \\ f_1 e_1 f_1 e_1 + e_0 f_0 &= 0. \end{aligned}$$

Moreover the appropriate grading on C_0 is defined by setting $\deg(e_i) = \deg(f_i) = 1 + \delta_{i,0}$. Here is a brief sketch of how one can see this. The main point is to exploit Theorem 3.6: the grading on $\dot{\mathcal{O}}_0$ induces a positive grading on C_0 with degree zero component $\bigoplus_{i\in\mathbb{Z}} \mathbb{C}1_i$. Let $D(i)$ be the one-dimensional irreducible C_0-module corresponding to $i \in \mathbb{Z}$ and let $Q(i)$ be its projective cover (equivalently, injective hull). The proof of Theorem 3.6 implies further that these modules possess self-dual graded lifts $\dot{D}(i)$ and $\dot{Q}(i)$. A straightforward calculation using the graded version of BGG reciprocity and the information in Example 3.4 gives the graded composition multiplicities of each $\dot{Q}(i)$. From this one deduces for each $i \in \mathbb{Z}$ that there are unique (up to scalars) nonzero homomorphisms $e_i : \dot{Q}(i-1) \to \dot{Q}(i)$ and $f_i : \dot{Q}(i) \to \dot{Q}(i-1)$ that are homogeneous of degree $1 + \delta_{i,0}$. By considering images and kernels of these homomorphisms and using self-duality, it follows that each $\dot{Q}(i)$ has irreducible head $q^{-2}\dot{D}(i)$, irreducible socle $q^2\dot{D}(i)$, and heart $\operatorname{rad}\dot{Q}(i)/\operatorname{soc}\dot{Q}(i) \cong \dot{Q}_-(i) \oplus \dot{Q}_+(i)$, where $\dot{Q}_-(0) := \dot{D}(-1)$, $\dot{Q}_+(-1) := \dot{D}(0)$ and all other $\dot{Q}_\pm(i)$ are uniserial with layers $q^{-1}\dot{D}(i \pm 1)$, $\dot{D}(i)$, $q\dot{D}(i \pm 1)$ in order from top to bottom. Hence $(e_i f_i)^{2-\delta_{i,0}} \neq 0 \neq (f_{i+1}e_{i+1})^{2-\delta_{i+1,0}}$ for each $i \in \mathbb{Z}$. Since each $\operatorname{End}_{C_0}(\dot{Q}(i))$ is one-dimensional in degree 4, it is then elementary to see that e_i and f_i can be scaled to ensure that the given relations hold, and the result follows.

Remark 4.8. With a similar analysis, one can show for any $n \geq 1$ that the algebra C_ξ associated to the block ξ of $\mathfrak{gl}_{n|1}(\mathbb{C})$ containing the weight $-\rho = (0, \dots, 0|0)$ is described by the same quiver as in Examples 2.8 and 4.7 subject instead to the relations

$$e_{i+1}e_i = f_i f_{i+1} = (f_{i+1}e_{i+1})^{n-\delta_{i+1,0}(n-1)} + (e_i f_i)^{n-\delta_{i,0}(n-1)} = 0$$

for all $i \in \mathbb{Z}$. This time $\deg(e_i) = \deg(f_i) = 1 + \delta_{i,0}(n-1)$.

To finish the article we draw attention to one more piece of this puzzle. First, we need to introduce the *principal W-superalgebra* $W_{n|m}$ associated to $\mathfrak{g} = \mathfrak{gl}_{n|m}(\mathbb{C})$. Let π be a two-rowed array of boxes with a connected strip of $\min(n, m)$ boxes in its first (top) row and a connected strip of $\max(n, m)$ boxes in its second (bottom) row; each box in the first row should be immediately above a box in the second row but the boxes in the rows need not be left-justified. We write the numbers $1, \dots, n$ in order into the boxes on a row of length n and the numbers $n+1, \dots, n+m$ in order into the boxes on the other row. Also let s_- (resp. s_+) be the number of boxes overhanging on the left hand side (resp. the right-hand side) of this diagram. For example, here is a choice of the diagram π for $\mathfrak{gl}_{5,2}(\mathbb{C})$:

	6	7		
1	2	3	4	5

For this $s_- = 1$ and $s_+ = 2$. Numbering the columns of π by $1, 2, \dots$ from left to right, we let $\mathrm{col}(i)$ be the column number of the box containing the entry i. Then define a $\mathbb{Z}$-grading $\mathfrak{g} = \bigoplus_{d \in \mathbb{Z}} \mathfrak{g}(d)$ by declaring that the matrix unit $e_{i,j}$ is of degree $\mathrm{col}(j) - \mathrm{col}(i)$, and let

$$\mathfrak{p} := \bigoplus_{d \geq 0} \mathfrak{g}(d), \qquad \mathfrak{h} := \mathfrak{g}(0), \qquad \mathfrak{m} := \bigoplus_{d<0} \mathfrak{g}(d).$$

Let $e := e_{1,2} + \cdots + e_{n-1,n} + e_{n+1,n+2} + \cdots + e_{n+m-1,n+m} \in \mathfrak{g}(1)$. This is a representative for the principal nilpotent orbit in $\mathfrak{g}$. Let $\chi : \mathfrak{m} \to \mathbb{C}$ be the one-dimensional representation with $\chi(x) := \mathrm{str}(xe)$. Finally set

$$W_{n|m} := \{u \in U(\mathfrak{p}) \mid u\mathfrak{m}_\chi \subseteq \mathfrak{m}_\chi U(\mathfrak{g})\}$$

where $\mathfrak{m}_\chi := \{x - \chi(x) \mid x \in \mathfrak{m}\} \subset U(\mathfrak{m})$. It is easy to check that $W_{n|m}$ is a subalgebra of $U(\mathfrak{p})$.

Theorem 4.9 (Brown, Brundan, Goodwin). *The superalgebra $W_{n|m}$ contains some explicit even elements $\{d_i^{(r)} \mid i = 1, 2, r > 0\}$ and odd elements $\{f^{(r)} \mid r > s_-\} \cup \{e^{(r)} \mid r > s_+\}$. These elements generate $W_{n|m}$ subject only to the following relations:*

$$\begin{aligned}
d_1^{(r)} &= 0 \qquad \text{if } r > \min(m, n),\\
[d_i^{(r)}, d_j^{(s)}] &= 0,\\
[e^{(r)}, e^{(s)}] &= 0,\\
[f^{(r)}, f^{(s)}] &= 0,\\
[d_i^{(r)}, e^{(s)}] &= (-1)^p \sum_{a=0}^{r-1} d_i^{(a)} e^{(r+s-1-a)},
\end{aligned}$$

$$[d_i^{(r)}, f^{(s)}] = -(-1)^p \sum_{a=0}^{r-1} f^{(r+s-1-a)} d_i^{(a)},$$

$$[e^{(r)}, f^{(s)}] = (-1)^p \sum_{a=0}^{r+s-1} \tilde{d}_1^{(a)} d_2^{(r+s-1-a)}.$$

Here $d_i^{(0)} = 1$ *and* $\tilde{d}_i^{(r)}$ *is defined recursively from* $\sum_{a=0}^{r} \tilde{d}_i^{(a)} d_i^{(r-a)} = \delta_{r,0}$. *Also* $p := \bar{0}$ *if the numbers* $1, \dots, n$ *appear on the first row of* π, *and* $p := \bar{1}$ *otherwise.*

The relations in Theorem 4.9 arise from the defining relations for the Yangian $Y(\mathfrak{gl}_{1|1})$. In fact the structure of the superalgebra $W_{n|m}$ is quite interesting. To start with there is an explicit description of its center $Z(W_{n|m})$, which is canonically isomorphic to the center $Z(\mathfrak{g})$ of the universal enveloping superalgebra of $\mathfrak{g}$ itself. All of the supercommutators $[e^{(r)}, f^{(s)}]$ are central. Then $W_{n|m}$ possesses a triangular decomposition $W_{n|m} = W_{n|m}^- W_{n|m}^0 W_{n|m}^+$, i.e., the multiplication map $W_{n|m}^- \otimes W_{n|m}^0 \otimes W_{n|m}^+ \to W_{n|m}$ is a vector space isomorphism where

- $W_{n|m}^-$ is a Grassmann algebra generated freely by $\{f^{(r)} \mid s_- < r \leq s_- + \min(m, n)\}$;
- $W_{n|m}^+$ is a Grassmann algebra generated freely by $\{e^{(r)} \mid s_+ < r \leq s_+ + \min(m, n)\}$;
- $W_{n|m}^0$ is the polynomial algebra on $\{d_1^{(r)}, d_2^{(s)} \mid 0 < r \leq \min(m, n), 0 < s \leq \max(m, n)\}$.

Using this one can label its irreducible representations by mimicking the usual arguments of highest weight theory. They are all finite dimensional, in fact of dimension $2^{\min(m,n)-k}$ where k is the degree of atypicality of the corresponding central character. Moreover the irreducible representations of integral central character are parametrized by the antidominant weights in $\mathfrak{t}_{\mathbb{Z}}^*$, i.e., the same weights that index the prinjective supermodules in $\mathcal{O}_{\mathbb{Z}}$.

The principal W-superalgebra is relevant to our earlier discussion because of the existence of the *Whittaker coinvariants functor*

$$\mathbb{W} := H_0(\mathfrak{m}_\chi, -) : \mathcal{O} \to W_{n|m}\text{-smod}.$$

This is an exact functor sending $M \in \mathcal{O}$ to the vector superspace $H_0(\mathfrak{m}_\chi, M) := M/\mathfrak{m}_\chi M$ of $\mathfrak{m}_\chi$-coinvariants in M. The definition of $W_{n|m}$ ensures that this is a (finite-dimensional) left $W_{n|m}$-supermodule in the natural way. Then it turns out that the functor $\mathbb{W}$ sends the irreducible $L(\lambda) \in \mathcal{O}$ to an irreducible $W_{n|m}$-supermodule if λ is antidominant or to zero otherwise, and every irreducible $W_{n|m}$-supermodule arises in this way (up to isomorphism and parity switch).

Theorem 4.10 (Brown, Brundan, Goodwin). *For $\xi \in \mathfrak{t}^*_{\mathbb{Z}}/\approx$, let $\mathbb{W}_\xi : \mathcal{O}_\xi \to W_{n|m}$-smod be the restriction of the Whittaker coinvariants functor. As $\mathbb{V}_\xi$ is a quotient functor, there exists a unique (up to isomorphism) exact functor*

$$\mathbb{J}_\xi : \text{mof-}C_\xi \to W_{n|m}\text{-smod}$$

such that $\mathbb{W}_\xi \cong \mathbb{J}_\xi \circ \mathbb{V}_\xi$. This defines an equivalence of categories between mof-C_ξ *and a certain full subcategory $\mathcal{R}_\xi$ of $W_{n|m}$*-smod *which is closed under taking submodules, quotients and finite direct sums.*

Thus $\mathbb{W}_\xi : \mathcal{O}_\xi \to \mathcal{R}_\xi$ is another quotient functor which is fully faithful on projectives. One intriguing consequence is that for blocks ξ of maximal atypicality (in which all the irreducible $W_{n|m}$-supermodules are one-dimensional), the algebra C_ξ can be realized also as an "idempotented quotient" of $W_{n|m}$. We already saw a very special case of this in Example 2.8, and describe the next easiest case in Example 4.11 below; hopefully this gives a rough idea of what we mean by "idempotented quotient." Unfortunately in general we still have no idea of the precise form that the relations realizing C_ξ as a quotient of $W_{n|m}$ should take.

Example 4.11. The generators and relations for the principal W-superalgebra $W_{2|1}$ from Theorem 4.9 collapse to just requiring generators $c := d_2^{(1)} - d_1^{(1)}, d := -d_1^{(1)}, e := e^{(1+s_+)}$ and $f := f^{(1+s_-)}$, subject to the relations $[c,d] = [c,e] = [c,f] = 0$ (i.e., c is central), $[d,e] = e, [d,f] = -f$ and $e^2 = f^2 = 0$. Let C_0 be the algebra described explicitly in Example 4.7 and let $\hat{C}_0$ be its completion consisting of (possibly infinite) formal sums $\{\sum_{i,j\in\mathbb{Z}} a_{i,j} \mid a_{i,j} \in 1_i C_0 1_j\}$. The relations imply that there is a homomorphism

$$\phi : W_{2|1} \to \hat{C}_0, \qquad c \mapsto 0, \quad d \mapsto \sum_{i\in\mathbb{Z}} i 1_i, \quad e \mapsto \sum_{i\in\mathbb{Z}} e_i, \quad f \mapsto \sum_{i\in\mathbb{Z}} f_i.$$

Then we have that $C_0 = \bigoplus_{i,j\in\mathbb{Z}} 1_i \phi(W_{2|1}) 1_j$.

Notes. Soergel's Theorems 4.1–4.2 were proved originally in [45]. Soergel's proof of the Endomorphismensatz goes via deformed category $\mathcal{O}$; Bernstein subsequently gave a more elementary proof in [6]. We also mention [46] which contains a generalization of the Struktursatz to parabolic category $\mathcal{O}$; our Theorem 4.5 is a close relative of that. For the formal definition of the category mod-H_ξ of stable modules over the tower of Hecke algebras mentioned briefly here we refer to [15, §4]. Theorem 4.5 follows from [15, Theorem 4.10] and Lemma 4.6. Example 4.7 was computed with help from Catharina Stroppel.

The definition of principal W-algebras for semisimple Lie algebras is due to Kostant [32], although of course the language is much more recent. Kostant showed in the classical case that the principal W-algebra is canonically isomorphic to the center $Z(\mathfrak{g})$ of the universal enveloping algebra of $\mathfrak{g}$. The explicit presentation for the principal W-superalgebra for $\mathfrak{gl}_{n|m}(\mathbb{C})$ from Theorem 4.9 is proved in [8, Theorem 4.5]. For the classification of irreducible $W_{n|m}$-supermodules we refer to [8, Theorems 7.2–7.3].

The Whittaker coinvariants functor $\mathbb{W}$ for the principal nilpotent orbit of $\mathfrak{gl}_{n|m}(\mathbb{C})$ is studied in detail in [9], culminating in the proof of Theorem 4.10. The identification of C_ξ as an idempotented quotient of $W_{n|m}$ is also explained more fully there. The idea that Soergel's functor $\mathbb{V}$ is related to $\mathbb{W}$ goes back to the work of Backelin [1] in the classical case. In [35, Theorem 4.7], Losev has developed a remarkably general theory of Whittaker coinvariant functors associated to arbitrary nilpotent orbits in semisimple Lie algebras; see also the brief discussion of Lie superalgebras in [35, §6.3.2].

Acknowledgements Special thanks go to Catharina Stroppel for several discussions which influenced this exposition. I also thank Geoff Mason, Ivan Penkov and Joe Wolf for providing me the opportunity to write a survey article of this nature. In fact I gave a talk on exactly this topic at the Seminar "Lie Groups, Lie Algebras and their Representations" at Riverside in November 2002, when the super Kazhdan–Lusztig conjecture was newborn.

References

1. E. Backelin, Representation theory of the category $\mathcal{O}$ in Whittaker categories, *Internat. Math. Res. Notices* **4** (1997), 153–172.
2. H. Bao and W. Wang, A new approach to Kazhdan-Lusztig theory of type B via quantum symmetric pairs; `arxiv:1310.0103`.
3. A. Beilinson and J. Bernstein, Localisation de $\mathfrak{g}$-modules, *C. R. Acad. Sci. Paris Ser. I Math.* **292** (1981), 15–18.
4. A. Beilinson, V. Ginzburg and W. Soergel, Koszul duality patterns in representation theory, *J. Amer. Math. Soc.* **9** (1996), 473–527.
5. A. Berele and A. Regev, Hook Young diagrams with applications to combinatorics and to representations of Lie superalgebras, *Advances Math.* **64** (1987), 118–175.
6. J. Bernstein, Trace in categories, in: *Operator Algebras, Unitary Representations and Invariant Theory*, eds. A. Connes, M. Duflo, A. Joseph and R. Rentschler, pp. 417–432, Birkhäuser, 1990.
7. J. Bernstein, I. M. Gelfand and S. I. Gelfand, Structure of representations generated by vectors of highest weight, *Func. Anal. Appl.* **5** (1971), 1–9.
8. J. Brown, J. Brundan and S. Goodwin, Principal W-algebras for $GL(m|n)$, *Alg. Numb. Theory* **7** (2013), 1849–1882.
9. J. Brown, J. Brundan and S. Goodwin, Whittaker coinvariants for $GL(m|n)$, in preparation.
10. J. Brundan, Kazhdan-Lusztig polynomials and character formulae for the Lie superalgebra $\mathfrak{gl}(m|n)$, *J. Amer. Math. Soc.* **16** (2003), 185–231.
11. J. Brundan, Kazhdan-Lusztig polynomials and character formulae for the Lie superalgebra $\mathfrak{q}(n)$, *Advances Math.* **182** (2004), 28–77.
12. J. Brundan, Tilting modules for Lie superalgebras, *Comm. Algebra* **32** (2004), 2251–2268.
13. J. Brundan, `http://pages.uoregon.edu/brundan/papers/supero.gap`.
14. J. Brundan and A. Kleshchev, Schur-Weyl duality for higher levels, *Selecta Math.* **14** (2008), 1–57.
15. J. Brundan, I. Losev and B. Webster, Tensor product categorifications and the super Kazhdan-Lusztig conjecture; `arxiv:1310.0349`.

16. J. Brundan and C. Stroppel, Highest weight categories arising from Khovanov's diagram algebra IV: the general linear supergroup, *J. Eur. Math. Soc.* **14** (2012), 373–419.
17. S.-J. Cheng and N. Lam, Irreducible characters of the general linear superalgebra and super duality, *Commun. Math. Phys.* **298** (2010), 645–672.
18. S.-J. Cheng, N. Lam and W. Wang, Brundan-Kazhdan-Lusztig conjecture for general linear Lie superalgebras; `arxiv:1203.0092`.
19. S.-J. Cheng, V. Mazorchuk and W. Wang, Equivalence of blocks for the general linear Lie superalgebra, *Lett. Math. Phys.* **103** (2013), 1313–1327.
20. S.-J. Cheng and W. Wang, Brundan-Kazhdan-Lusztig and super duality conjectures, *Publ. RIMS* **44** (2008), 1219–1272.
21. S.-J. Cheng and W. Wang, *Dualities and Representations of Lie Superalgebras*, Graduate Studies in Mathematics, Vol. 144, AMS, 2012.
22. S.-J. Cheng, W. Wang and R.B. Zhang, Super duality and Kazhdan-Lusztig polynomials, *Trans. Amer. Math. Soc.* **360** (2008), 5883–5924.
23. J. Chuang and R. Rouquier, Derived equivalences for symmetric groups and $\mathfrak{sl}_2$-categorification, *Ann. of Math.* **167** (2008), 245–298.
24. K. Coulembier and V. Mazorchuk, Primitive ideals, twisting functors and star actions for classical Lie superalgebras; `arXiv:1401.3231`.
25. M. Ehrig and C. Stroppel, Nazarov-Wenzl algebras, coideal subalgebras and categorified skew Howe duality; `arXiv:1310.1972`.
26. M. Gorelik, The Kac construction of the center of $U(\mathfrak{g})$ for Lie superalgebras, *J. Nonlinear Math. Phys.* **11** (2004), 325–349.
27. J. Humphreys, *Representations of Semisimple Lie Algebras in the BGG Category $\mathcal{O}$*, Graduate Studies in Mathematics, Vol. 94, AMS, 2008.
28. A. Joseph, Sur la classification des idéaux primitifs dans l'algèbre enveloppante de $\mathfrak{sl}(n+1,\mathbb{C})$, *C. R. Acad. Sci. Paris* **287** (1978), A303–A306.
29. J. van der Jeugt and R.-B. Zhang, Characters and composition factor multiplicities for the Lie superalgebra $\mathfrak{gl}(m|n)$, *Lett. Math. Phys.* **47** (1999), 49–61.
30. V. Kac, Characters of typical representations of classical Lie superalgebras, *Commun. in Algebra* **5** (1977), 889–897.
31. V. Kac, Representations of classical Lie superalgebras, in: "*Differential geometrical methods in mathematical physics II*", Lecture Notes in Math. no. 676, pp. 597–626, Springer-Verlag, Berlin, 1978.
32. B. Kostant, On Whittaker modules and representation theory, *Invent. Math.* **48** (1978), 101–184.
33. A. Lascoux and M.-P. Schützenberger, Polynômes de Kazhdan et Lusztig pour les Grassmanniennes, *Astérisque* **87–88** (1981), 249–266.
34. E. Letzter, A bijection of primitive spectra for classical Lie superalgebras of type I, *J. London Math. Soc.* **53** (1996), 39–49.
35. I. Losev, Dimensions of irreducible modules over W-algebras and Goldie ranks; `arxiv:1209.1083`. To appear in Invent. Math.
36. I. Losev and B. Webster, On uniqueness of tensor products of irreducible categorifications; `arxiv:1303.4617`.
37. G. Lusztig, *Introduction to Quantum Groups*, Birkhäuser, 1993.
38. I. Musson, A classication of primitive ideals in the enveloping algebra of a classical simple Lie superalgebra, *Adv. Math.* **91** (1992), 252–268.
39. I. Musson, *Lie Superalgebras and Enveloping Algebras*, Graduate Studies in Mathematics, Vol. 131, AMS, 2012.
40. I. Musson, The Jantzen filtration and sum formula for basic classical Lie superalgebras, in preparation.
41. I. Musson and V. Serganova, Combinatorics of character formulas for the Lie superalgebra $\mathfrak{gl}(m,n)$, *Transform. Groups* **16** (2011), 555–578.
42. R. Rouquier, 2-Kac-Moody algebras (2008); `arXiv:0812.5023`.

43. V. Serganova, Kazhdan-Lusztig polynomials for the Lie superalgebra $GL(m|n)$, *Adv. Sov. Math.* **16** (1993), 151–165.
44. A. Sergeev, Tensor algebra of the identity representation as a module over the Lie superalgebras $GL(n, m)$ and $Q(n)$, *Math. USSR Sbornik* **51** (1985), 419–427.
45. W. Soergel, Kategorie $\mathcal{O}$, perverse Garben und Moduln über den Koinvarianten zur Weylgruppe, *J. Amer. Math. Soc.* **3** (1990), 421–445.
46. C. Stroppel, Category $\mathcal{O}$: quivers and endomorphisms of projectives, *Represent. Theory* **7** (2003), 322–345.
47. Y. Su and R.-B. Zhang, Character and dimension formulae for general linear superalgebra, *Advances Math.* **211** (2007), 1–33.
48. Y. M. Zou, Categories of finite dimensional weight modules over type I classical Lie superalgebras, *J. Algebra* **180** (1996), 459–482.

Three Results on Representations of Mackey Lie Algebras

Alexandru Chirvasitu

Abstract I. Penkov and V. Serganova have recently introduced, for any nondegenerate pairing $W \otimes V \to \mathbb{C}$ of vector spaces, the Lie algebra $\mathfrak{gl}^M = \mathfrak{gl}^M(V, W)$ consisting of endomorphisms of V whose duals preserve $W \subseteq V^*$. In their work, the category $\mathbb{T}_{\mathfrak{gl}^M}$ of $\mathfrak{gl}^M$-modules, which are finite length subquotients of the tensor algebra $T(W \otimes V)$, is singled out and studied. Denoting by $\mathbb{T}_{V \otimes W}$ the category with the same objects as $\mathbb{T}_{\mathfrak{gl}^M}$ but regarded as $V \otimes W$-modules, we first show that when W and V are paired by dual bases, the functor $\mathbb{T}_{\mathfrak{gl}^M} \to \mathbb{T}_{V \otimes W}$ taking a module to its largest weight submodule with respect to a sufficiently nice Cartan subalgebra of $V \otimes W$ is a tensor equivalence. Secondly, we prove that when W and V are countable-dimensional, the objects of $\mathbb{T}_{\mathrm{End}(V)}$ have finite-length as $\mathfrak{gl}^M$-modules. Finally, under the same hypotheses, we compute the socle filtration of a simple object in $\mathbb{T}_{\mathrm{End}(V)}$ as a $\mathfrak{gl}^M$-module.

Key words Mackey Lie algebra • Finite length module • Large annihilator • Weight module • Socle filtration

Mathematics Subject Classification (2010): 17B10, 17B20, 17B65.

Introduction

In the recent paper [4], the authors study various categories of representations for Lie algebras associated to pairs of complex vector spaces V, W endowed with a nondegenerate bilinear form $W \otimes V \to \mathbb{C}$. This datum realizes W as a

This work was partially supported by the Danish National Research Foundation through the QGM Center at Aarhus University, and by the Chern-Simons Chair in Mathematical Physics at UC Berkeley.

A. Chirvasitu (✉)
Department of Mathematics, University of Washington, Box 354350 Seattle, WA 98195-4350, USA
e-mail: chirvasitua@gmail.com

G. Mason et al. (eds.), *Developments and Retrospectives in Lie Theory: Algebraic Methods*, Developments in Mathematics 38,
DOI 10.1007/978-3-319-09804-3_4

subspace of the full dual V^*, and vice versa. The associated Mackey Lie algebra $\mathfrak{g}^M = \mathfrak{gl}^M(V, W)$ is then simply the set of all endomorphisms of V whose duals leave $W \subseteq V^*$ invariant. It can be shown that the definition is symmetric in the sense that reversing the roles of V and W produces canonically isomorphic Lie algebras. When $W = V^*$, the resulting Lie algebra is simply $\mathrm{End}(V)$.

Categories $\mathbb{T}_{\mathfrak{g}^M}$ of $\mathfrak{g}^M$-representations are then introduced. They consist of modules for which all elements have appropriately large annihilators; see [4, § 7.3]. One remarkable result is that all these categories, for all possible nondegenerate pairs (V, W), are in fact equivalent as tensor categories (i.e., symmetric monoidal abelian categories). Moreover, they are also equivalent to the categories $\mathbb{T}_{\mathfrak{sl}(V,W)}$ from [4, § 3.5] and $\mathbb{T} = \mathbb{T}_{\mathfrak{sl}(\infty)}$ introduced and studied earlier in [1]; all of this follows from [4, Theorems 5.1 and 7.9].

In view of the abstract equivalence between $\mathbb{T}_{\mathrm{End}(V)} \simeq \mathbb{T}_{\mathfrak{sl}(V,W)}$ noted above, it is a natural problem to try to find as explicit and natural a functor as possible that implements this equivalence. In order to do this, we henceforth specialize to the case when $W = V_*$ is a vector space whose pairing with V is given by a pair of dual bases $v_\gamma \in V$, $v_\gamma^* \in V_*$ for γ ranging over some (possibly uncountable) set I. This assumption ensures the existence of a so-called *local Cartan subalgebra* $\mathfrak{h} \subseteq \mathfrak{sl}(V, V_*)$ [4, 1.4].

Denote $\mathfrak{g} = \mathfrak{sl}(V, V_*)$. In our setting, for a local Cartan subalgebra $\mathfrak{h} \subseteq \mathfrak{g}$, let $\Gamma_{\mathfrak{h}}^{\mathrm{wt}}$ be the functor from $\mathrm{End}(V)$-modules to $\mathfrak{g}$-modules which picks out the $\mathfrak{h}$-weight part of a representation. Similarly, denote by Γ^{wt} the functor $\bigcap_{\mathfrak{h}} \Gamma_{\mathfrak{h}}^{\mathrm{wt}}$, where the intersection ranges over all local Cartan subalgebras of $\mathfrak{g}$. We will abuse notation and denote by these same symbols the restrictions of $\Gamma_{\mathfrak{h}}^{\mathrm{wt}}$ and Γ^{wt} to various categories of $\mathrm{End}(V)$-modules.

With these preparations (and keeping the notations we've been using), the following seems reasonable [4, 8.4].

Conjecture 1. The functor Γ^{wt} implements an equivalence from $\mathbb{T}_{\mathrm{End}(V)}$ onto $\mathbb{T}_{\mathfrak{g}}$.

One of the main results of this paper is a proof of this conjecture. The outline is as follows:

In the next section we prove Conjecture 1, making use of the results in [6] on a certain universality property for the category $\mathbb{T}_{\mathfrak{g}}$.

In Sect. 2 we specialize to a pairing $V_* \otimes V \to \mathbb{C}$ of countable-dimensional vector spaces V, V_*. In this case, noting that V^*/V_* is a simple $\mathfrak{g}^M = \mathfrak{sl}^M(V, V_*)$-module, the authors of [4] ask whether *all* objects of $\mathbb{T}_{\mathrm{End}(V)}$ are finite-length $\mathfrak{g}^M$-modules. We show that this is indeed the case in Theorem 3.

Finally, Theorem 4 in Sect. 3 contains the description of the socle filtration as a $\mathfrak{g}^M$-module of a simple object in $\mathbb{T}_{\mathrm{End}(V)}$. This solves a third problem posed in the cited paper.

1 Explicit Equivalence Between $\mathbb{T}_{\mathrm{End}(V)}$ and $\mathbb{T}_{\mathfrak{g}}$

We will actually prove a slightly strengthened version of Conjecture 1. Before formulating it, recall our setting: We are considering a pairing between V and V_* determined by dual bases $v_\gamma \in V$ and $v_\gamma^* \in V_*$, and $\mathfrak{g}$ stands for $\mathfrak{sl}(V, V_*)$. By $\mathfrak{h} \subseteq \mathfrak{g}$ we denote the local Cartan subalgebra spanned linearly by the elements $v_\gamma \otimes v_\gamma^* - v_{\bar{\gamma}} \otimes v_{\bar{\gamma}}^* \in \mathfrak{g} \subseteq V \otimes V_*$ for $\gamma \neq \bar{\gamma} \in I$.

Throughout, "tensor category" means symmetric monoidal, $\mathbb{C}$-linear and abelian. Similarly, tensor functors are symmetric monoidal and $\mathbb{C}$-linear, and tensor natural transformations are symmetric monoidal.

Our main result in the present section reads as follows.

Theorem 1. *The functor $\Gamma_{\mathfrak{h}}^{\mathrm{wt}}$ implements a tensor equivalence from $\mathbb{T}_{\mathrm{End}(V)}$ onto $\mathbb{T}_{\mathfrak{g}}$.*

Before embarking on the proof, note that as claimed above, the theorem implies the conjecture.

Corollary 1. *Conjecture 1 is true.*

Proof. On the one hand, the functor Γ^{wt} from the statement of the conjecture is a subfunctor of $\Gamma_{\mathfrak{h}}^{\mathrm{wt}}$. On the other hand though, the theorem says that $\Gamma_{\mathfrak{h}}^{\mathrm{wt}}$ already lands inside the category $\mathbb{T}_{\mathfrak{g}}$ which consists of weight modules for any local Cartan subalgebra of $\mathfrak{g}$ (because it consists of modules emendable in finite direct sums of copies of the tensor algebra $T(V \oplus V_*)$; see e.g., [4, 7.9]). In conclusion, we must have $\Gamma^{\mathrm{wt}} = \Gamma_{\mathfrak{h}}^{\mathrm{wt}}$, and we are done. □

We will make use of the following simple observation.

Lemma 1. *Let H be a cocommutative Hopf algebra over an arbitrary field $\mathbb{F}$, and* $\mathrm{FIN} : H - \mathrm{mod} \to H - \mathrm{mod}$ *the functor sending an H-module M to the largest H-submodule of M which is a union of finite-dimensional H-modules. Then,* FIN *is a tensor functor.*

Proof. Let S be the antipode of H. For an H-module V, denote by V^* the algebraic dual of V made into an H-module via $(hf)(v) = f(S(h)v)$ for $h \in H$, $v \in V$ and $f \in V^*$. Then the usual evaluation $V^* \otimes V \to \mathbb{F}$ is an H-module map if $V^* \otimes V$ is a module via the tensor cateory structure of $H - \mathrm{mod}$.

Now let M, N be H-modules, and $V \subseteq \mathrm{FIN}(M \otimes N)$ be a finite-dimensional H-submodule. We need to show that V is in fact a submodule of $\mathrm{FIN}(M) \otimes \mathrm{FIN}(N)$.

Denote by $N_V \subseteq N$ the image of the H-module morphism $M^* \otimes V \subseteq M^* \otimes M \otimes N \to N$, where the last arrow is evaluation on the first two tensorands. Similarly, denote by $M_V \subseteq M$ the image of $V \otimes N^* \subseteq M \otimes N \otimes N^* \to M$. Then M_V and N_V are H-submodules of M and N respectively, being images of module maps. It is now easily seen from their definition that M_V and N_V are finite-dimensional, and that the inclusion $V \subseteq M \otimes N$ factors through $M_V \otimes N_V \subseteq M \otimes N$. □

Remark 1. The cocommutativity of H is used in the proof to conclude that the category $H-\text{mod}$ is symmetric monoidal, and hence $N \otimes N^* \to \mathbb{F}$ is an H-module map because its domain is isomorphic to $N^* \otimes N$.

Although we do not need this in the sequel, as Lemma 1 will only be applied to universal envelopes of Lie algebras, the above proof can be generalized to show that the functor FIN, defined in the obvious fashion, is monoidal for any Hopf algebra with bijective antipode. In the definition of V_N one would need to use the evaluation map $N \otimes {}^*N \to \mathbb{F}$ instead, where ${}^*N$ is the full dual of N made into an H-module using the inverse of the antipode instead of the antipode.

We now need a characterization of the category $\mathbb{T}_{\text{End}(V)} \simeq \mathbb{T}_{\mathfrak{g}}$ in terms of a universality property which defines it uniquely up to tensor equivalence. The following result is Theorem 3.4.2 from [6], where the category $\mathbb{T}_{\mathfrak{g}}$ is denoted by Rep(**GL**).

Theorem 2. *For any tensor category $\mathcal{C}$ with monoidal unit* **1** *and any morphism $b : x \otimes y \to \mathbf{1}$ in $\mathcal{C}$, there is a left exact tensor functor $F : \mathbb{T}_{\text{End}(V)} \to \mathcal{C}$ sending the pairing $V^* \otimes V \to \mathbb{C}$ in $\mathbb{T}_{\text{End}(V)}$ to b. Moreover, F is unique up to tensor natural isomorphism.* □

As an immediate consequence we have:

Corollary 2. *A left exact tensor functor $\mathbb{T}_{\text{End}(V)} \to \mathbb{T}_{\mathfrak{g}}$ turning the pairing $V^* \otimes V \to \mathbb{C}$ into the pairing $V_* \otimes V \to \mathbb{C}$, is a tensor equivalence.*

Proof. The abstract tensor equivalence $\mathbb{T}_{\text{End}(V)} \simeq \mathbb{T}_{\mathfrak{g}}$ established in [4, 5.1,7.9] identifies the two bilinear pairings in the statement. The conclusion then follows from Theorem 2 in the usual manner (a universality property implies uniqueness up to equivalence). □

The proof of Theorem 1 makes use of the following auxiliary result.

Lemma 2. *Let $\mathfrak{h}$ be a complex abelian Lie algebra. For any functional $\varphi \in \mathfrak{h}^*$, let $M^\varphi \in \mathfrak{h}-\text{mod}$ be an $\mathfrak{h}$-module all of whose elements are vectors of weight φ. Then, using the notation* FIN *from Lemma 1, we have*

$$\text{FIN}\left(\prod_{\varphi\in\mathfrak{h}^*} M^\varphi\right) = \bigoplus_{\varphi\in\mathfrak{h}^*} M^\varphi.$$

Proof. We denote the direct product $\prod_{\varphi} M^\varphi$ by M. Let $x \in M$ be an element contained in some d-dimensional $\mathfrak{h}$-submodule N of M.

Assume there are $d+1$ distinct functionals φ_0 up to φ_d such that the components x_i, $0 \le i \le d$ of x in M^{φ_i} are all nonzero. Because $\varphi_i \in \mathfrak{h}^*$ are distinct, we can find some element $h \in \mathfrak{h}$ such that the scalars $t_i = \varphi_i(h)$ are distinct (as $\mathfrak{h}$ cannot be the union of the kernels of $\varphi_i - \varphi_j$, $0 \le i \ne j \le d$; here we use the fact that we are working over $\mathbb{C}$, or more generally, over an infinite field). The claim now is that $x, hx, \ldots, h^d x$ are linearly independent, contradicting the assumption $\dim N = d$.

To prove the claim, consider the images of the vectors $h^i x$, $0 \leq i \leq d$, through the projection $M \to \prod_{i=0}^{d} M^{\varphi_i}$. They are linear combinations of the x_i's, and their coefficients form the columns of the $(d+1) \times (d+1)$ non-singular Vandermonde matrix

$$\begin{pmatrix} 1 & t_0 & \cdots & t_0^d \\ 1 & t_1 & \cdots & t_1^d \\ \vdots & \vdots & \ddots & \vdots \\ 1 & t_d & \cdots & t_d^d \end{pmatrix}$$

This finishes the proof. □

We are now ready to prove our first main result.

Proof of Theorem 1. First, recall from [1, §4] that $\mathbb{T}_{\mathrm{End}(V)}$ has enough injectives, that the tensor products $(V^*)^{\otimes m} \otimes V^{\otimes n}$ contain all indecomposable injectives as summands, and also [5] that all morphisms between such tensor products are built out of the pairing $V^* \otimes V \to \mathbb{C}$ by taking tensor products, permutations, and linear combinations. Therefore, if we show that $\Gamma = \Gamma_{\mathfrak{h}}^{\mathrm{wt}} : \mathbb{T}_{\mathrm{End}(V)} \to \mathfrak{g} - \mathrm{mod}$ sends $V^* \otimes V \to \mathbb{C}$ to $V_* \otimes V \to \mathbb{C}$ and is a left exact tensor functor, its image will automatically lie in $\mathbb{T}_{\mathfrak{g}}$. We can then apply Corollary 2 to conclude that the resulting functor from $\mathbb{T}_{\mathrm{End}(V)} \to \mathbb{T}_{\mathfrak{g}}$ is a tensor equivalence.

The functor Γ, regarded as a functor from $\mathfrak{g} - \mathrm{mod}$ to $\mathfrak{h}$-weight $\mathfrak{g}$-modules, is the right adjoint of the exact inclusion functor going in the opposite direction; it's thus clear that it is left exact.

Since the pairing $V_* \otimes V \to \mathbb{C}$ is simply the restriction of the full pairing $V^* \otimes V \to \mathbb{C}$ and Γ is compatible with inclusions, we will be done as soon as we prove that it is a tensor functor and it sends V^* to V_*.

We prove tensoriality first. In fact, since compatibility with the symmetry is clear, it is enough to prove monoidality. That is, that the inclusion $\Gamma(M) \otimes \Gamma(N) \subseteq \Gamma(M \otimes N)$ is actually an isomorphism for any $M, N \in \mathbb{T}_{\mathrm{End}(V)}$. To see that this is indeed the case, note that every object of $\mathbb{T}_{\mathrm{End}(V)}$, being embedded in some finite direct sum of tensor products $(V^*)^{\otimes m} \otimes V^{\otimes n}$, is certainly a submodule of a direct product of $\mathfrak{h}$-weight spaces. Lemma 2 now shows that every finite-dimensional $\mathfrak{h}$-submodule of an object in $\mathbb{T}_{\mathrm{End}(V)}$ is automatically an $\mathfrak{h}$-weight module. Conversely, $\mathfrak{h}$-weight modules are unions of finite-dimensional $\mathfrak{h}$-modules. It follows that Γ coincides with the functor FIN considered in Lemma 1 for the Hopf algebra $H = U(\mathfrak{h})$ (i.e., the universal enveloping algebra of $\mathfrak{h}$); the lemma finishes the job of proving monoidality.

Finally, it is almost immediate that $\Gamma(V^*) = V_*$: simply note that the $\mathfrak{h}$-weight subspaces of V^* are the lines spanned by the basis elements v_γ^*. □

2 Restrictions from $\mathbb{T}_{\mathrm{End}(V)}$ to $\mathfrak{g}^M(V, V_*)$ Have Finite Length

In what follows $V_* \otimes V \to \mathbb{C}$ will be a nondegenerate pairing between countable-dimensional vector spaces. In this case, it is shown in [3] that we can find dual bases $v_i, v_i^*, i \in \mathbb{N} = \{0, 1, \ldots\}$ for V and V_* respectively, in the sense that $v_i^*(v_j) = \delta_{ij}$.

Denote $\mathfrak{g} = \mathfrak{sl}(V, V_*)$ and $\mathfrak{g}^M = \mathfrak{sl}^M(V, V_*)$. In general, for a vector space W and a partition $\lambda = (\lambda_1 \geq \lambda_2 \geq \ldots \geq \lambda_m \geq 0)$, denote by W_λ the image of the Schur functor corresponding to λ applied to W.

We think of elements of V^* as row vectors indexed by $\mathbb{N}$, on which the Lie algebra $\mathrm{End}(V)$ of $\mathbb{N} \times \mathbb{N}$ matrices with finite columns acts on the left via − right multiplication. The subspace $V_* \subseteq V^*$ consists of row vectors with only finitely many nonzero entries. We will often think of elements of V^*/V_* as row vectors as well, keeping in mind that changing finitely many entries does not alter the element. For a subset $I \subseteq \mathbb{N}$ and a vector $x \in V^*$, the *restriction* $x|_I$ is the vector obtained by keeping the entries of x indexed by I intact, and turning all other entries to zero. The same terminology applies to $x \in V^*/V_*$.

Let $I \subseteq \mathbb{N}$ be a subset. An element of V^* (respectively V^*/V_*) is *I-concentrated*, or *concentrated in I* if all of its nonzero entries (respectively all but finitely many of its nonzero entries) belong to I. Similarly, a matrix in $\mathrm{End}(V)$ is I-concentrated if all of its nonzero entries are in $I \times I$.

Our result is:

Theorem 3. *The objects of $\mathbb{T}_{\mathrm{End}(V)}$ have finite length when regarded as $\mathfrak{g}^M$-modules.*

Remark 2. In fact, the proof of the theorem could be adapted to the more general case covered in the previous section: V_* and V could be allowed to be uncountable-dimensional, so long as they are still paired by means of dual bases.

We need some preparations. The following result is very likely well known.

Lemma 3. *Let $\mathfrak{G}$ be a Lie algebra over some field $\mathbb{F}$, and $I \subseteq \mathfrak{G}$ an ideal. Let U be a simple $\mathfrak{G}/I$-module, and W an $\mathfrak{G}$-module on which I acts densely and irreducibly, and such that* $\mathrm{End}_I(W) = \mathbb{F}$. *Then, $U \otimes W$ is a simple $\mathfrak{G}$-module.*

Proof. Let $x = \sum_{i=1}^n u_i \otimes w_i$ be a nonzero element of $U \otimes W$, with the tensor product decomposition chosen such that the u_i are linearly independent and all w_i are nonzero. We have to show that x generates $U \otimes W$ as an $\mathfrak{G}$-module.

Because I annihilates U, it acts on $\bigoplus_{i=1}^{n} \mathbb{F}u_i \otimes W \cong W^{\oplus n}$. By the simplicity of W over I, there are vectors $w_i' \in W$, $i \in \{1, \ldots, n\}$, with $w_1' = w_1$ and such that the projection $W^{\oplus n} \to W$ onto the first component maps the I-submodule of $W^{\oplus n}$ generated by $\sum u_i \otimes w_i'$ isomorphically onto W.

Now note that $w_1' \mapsto w_i'$, $i > 1$ extend to I-module automorphisms of W. By the condition $\mathrm{End}_I(W) = \mathbb{F}$, these automorphisms are scalar: $w_i' = t_i w_1' = t_i w_1$ for all $i > 1$; for simplicity, set $t_1 = 1$ so that this identity holds for all i. Now, substituting $\sum_i t_i u_i$ for u_1 and denoting $u_1 = u$, we may assume that a nonzero simple tensor $u \otimes w \in U \otimes W$ belongs to the I-span of x.

Starting with a simple tensor $u \otimes w$ as above, note first that the enveloping algebra $U(I)$ can act so as to obtain any other tensor of the form $u \otimes w'$ (because I annihilates U and acts irreducibly on W).

On the other hand, for $h \in \mathfrak{G}$, we have $h(u \otimes w) = hu \otimes w + u \otimes hw$. Since I acts densely on W, we can find $k \in I$ such that $kw = hw$. In this case we have $(h - k)(u \otimes w) = hu \otimes w$; since U is simple over $\mathfrak{G}$, all simple tensors of the form $u' \otimes w$ are in the $\mathfrak{G}$-span of $u \otimes w$. Combining this with the previous paragraph, we get the desired conclusion. □

We now need the following infinite-dimensional analogue of Schur–Weyl duality.

Proposition 1. *For any partition $\lambda = (\lambda_1 \geq \ldots \geq \lambda_m \geq 0)$, the $\mathfrak{g}^M$-module $(V^*/V_*)_\lambda$ is simple.*

Proof. Let $k = |\lambda| = \sum_{i=1}^{m} \lambda_i$. Choose an arbitrary nonzero $x \in (V^*/V_*)^{\otimes k}$, thought of as a sum $\sum_\ell x^\ell$ for $x^\ell = x_1^\ell \otimes \ldots \otimes x_k^\ell$.

We denote the symmetric group on k letters by S_k. Partition $\mathbb{N}$ into k infinite subsets $I_1, \ldots, I_k$ such that the element of $(V^*/V_*)^{\otimes k}$ defined by

$$x_{\mathrm{RES}} = \sum_\ell \sum_{\sigma \in S_k} (x_1^\ell|_{I_{\sigma(1)}}) \otimes \ldots \otimes (x_k^\ell|_{I_{\sigma(k)}})$$

is nonzero; we leave it to the reader to show that this is possible. Now choose k complex numbers t_j, $j \in \{1, \ldots, k\}$ such the sums $\sum_j m_j t_j$ for nonnegative integers $\sum_j m_j = k$ are distinct for different choices of tuples $m_1, \ldots, m_k$ (e.g., t_j could equal $(k+1)^j$), and let $h \in \mathfrak{g}^M$ be the diagonal matrix whose I_j-indexed entries are equal to t_j. By breaking everything up into h-eigenspaces, we see that the $\mathfrak{g}^M$-module generated by x contains x_{RES}. In order to keep notation simple, we substitute x_{RES} for x and assume that the individual tensorands x_j^ℓ of each summand x^ℓ of x are concentrated in distinct I_j's.

Now consider the subspace W_1 of V^*/V_* generated by all I_1-concentrated x_j^ℓ's, and let $p_1, \ldots, p_s$ be rank one idempotent I_1-concentrated matrices in $\mathfrak{g}^M$ such that $\sum_i p_i$ acts as the identity on W_1. Since $x = \sum_i p_i x$, some $p_i x$ must be nonzero. Substitute it for x, and repeat the process with I_2 in place of I_1, etc. The resulting nonzero element, again denoted by x, will now be a linear combination of simple tensors x^ℓ as before, with tensorands x^ℓ_j concentrated in distinct I_j's for each ℓ, and such that all x^ℓ_i's concentrated in I_j (for all ℓ) are equal. Denoting by x_j this common I_j-concentrated vector, our element x is a linear combination of permutations of $x_1 \otimes \ldots \otimes x_k$.

Note that the entire procedure we have just described is S_k-equivariant: If the vector we started out with was in $(V^*/V_*)_\lambda \subseteq (V^*/V_*)^{\otimes k}$, then so is the output of the process. We assume this to be the case for the rest of the proof.

Because $\mathfrak{g}^M$ acts transitively on V^*/V_* (in the sense that any nonzero element can be transformed into any other element by acting on it with some matrix in $\mathfrak{g}^M$), we can find k^2 elements a_{ij} of $\mathfrak{g}^M$ such that $a_{pq}x_r = \delta_{qr}x_p$. The elements a_{ij} generate a Lie algebra isomorphic to $\mathfrak{gl}(k)$, and by ordinary Schur–Weyl duality we conclude that the $\mathfrak{g}^M$-module generated by x contains $c_\lambda(W^{\otimes k})$, where W is the linear space spanned by the x_j, and c_λ is the Young symmetrizer corresponding to λ.

Since for each j the vector x_j can be transformed into any other I_j-concentrated vector by acting on it with some I_j-concentrated matrix, the conclusion from the previous paragraph applies to any choice of x_j's. The desired result follows from the fact that every element of $c_\lambda(V^*/V_*)^{\otimes k}$ is a sum of elements from $c_\lambda(W^{\otimes k})$ for various W spanned by various tuples $\{x_j\}$. □

As a consequence of Lemma 3 and Proposition 1 we get:

Corollary 3. *Let W be a simple object in $\mathbb{T}_{\mathfrak{g}^M}$, and λ be a partition. Then, the $\mathfrak{g}^M$-module $(V^*/V_*)_\lambda \otimes W$ is simple.*

Proof. We apply Lemma 3 to the Lie algebra $\mathfrak{G} = \mathfrak{g}^M$, the ideal $I = \mathfrak{g}$, and the modules $U = (V^*/V_*)_\lambda$ and W. We already know that W is simple over $\mathfrak{g}$ and is acted upon densely by the latter Lie algebra [4, Corollary 7.6], and the remaining condition $\mathrm{End}_{\mathfrak{g}}(W) = \mathbb{C}$ follows for example from the fact that all simple modules in $\mathbb{T}_{\mathfrak{g}} = \mathbb{T}_{\mathfrak{g}^M}$ are highest weight modules with respect to a certain Borel subalgebra of $\mathfrak{g}$. □

We can now turn to Theorem 3.

Proof of Theorem 3. Since all objects of $\mathbb{T}_{\mathrm{End}(V)}$ are isomorphic to subquotients of finite direct sums of tensor products $(V^*)^{\otimes m} \otimes V^{\otimes n}$, it suffices to prove the conclusion for these tensor products. In turn, when regarded as $\mathfrak{g}^M$-modules, these tensor products have filtrations by finite direct sums of objects of the form $(V^*/V_*)^{\otimes m_1} \otimes V_*^{\otimes m_2} \otimes V^{\otimes n}$. The leftmost tensorand $(V^*/V_*)^{\otimes m_1}$ breaks up as a direct sum of images $(V^*/V_*)_\lambda$ of Schur functors, while $V_*^{\otimes m_2} \otimes V^{\otimes n}$ has a finite filtration by simple modules from the category $\mathbb{T}_{\mathfrak{g}^M}$. The conclusion now follows from Corollary 3 above. □

3 Socle Filtrations of $\mathbb{T}_{\mathrm{End}(V)}$-Objects Over $\mathfrak{g}^M(V, V_*)$

We now tackle the problem of finding the socle filtrations of simples in $\mathbb{T}_{\mathrm{End}(V)}$ as $\mathfrak{g}^M$-modules. We start with a definition.

Definition 1. A filtration $M^0 \subseteq M^1 \subseteq \ldots \subseteq M^n = M$ of an object M in an abelian category is *essential* if for every $p < q < r$, the module M^q/M^p is essential in M^r/M^q, i.e. intersects every nonzero submodule of M^r/M^q non-trivially.

Remark 3. It can be shown by induction on $r - p$ that the condition in Definition 1 is equivalent to M^{p+1}/M^p being essential in M^{p+2}/M^p for all p.

For dealing with tensor products of copies of V^* and V_* we use the following notation: For a binary word $\mathbf{r} = (r_1, \ldots, r_k)$, $r_i \in \{0, 1\}$, let $V^{\mathbf{r}}$ be the tensor product $\bigotimes_{i=1}^{k} V^{r_i}$, where $V^0 = V_*$ and $V^1 = V^*$. We denote $\sum_i r_i$ by $|\mathbf{r}|$.

Now consider the following (ascending) filtration of $W = (V^*)^{\otimes m} \otimes V^{\otimes n}$:

$$W^k = \sum_{|\mathbf{r}| \subseteq k} V^{\mathbf{r}} \otimes V^{\otimes n}, \quad \text{for every } 0 \subseteq k \subseteq m. \tag{1}$$

Proposition 2. *The filtration (1) of* $W = (V^*)^{\otimes m} \otimes V^{\otimes n}$ *is essential in* $\mathfrak{g}^M$ – mod.

Proof. By Remark 3, it suffices to show that for any $k \geq 0$, the $\mathfrak{g}^M$-module generated by any element $x \in W^{k+2} - W^{k+1}$ intersects $W^{k+1} - W^k$ (the minus signs stand for set difference). Moreover, it is enough to assume that x is a sum of simple tensors $y = y_1 \otimes \ldots \otimes y_m \otimes x_1 \otimes \ldots \otimes x_n$ for $y_i \in V^*$ and $x_j \in V$ such that exactly $k + 2$ of the y_i's are in V_*.

Acting on a term y as above with an element g of $\mathfrak{g}$ which annihilates all $y_i \in V_*$ and all x_j will produce an element of W^{k+1}, which belongs to $W^{k+1} - W^k$ provided it is nonzero; this element can be written as a sum of simple tensors, each of which has the tensorands $y_i \in V_*$ and x_j in common with the original term y. Hence, focusing on the action of g on only those tensorands y_i which do not belong to V_*, it is enough to prove the following claim (which we apply to $s = m - (k + 2)$):

The annihilator in $\mathfrak{g}$ of an element $z \in (V^*)^{\otimes s}$ whose image in $(V^*/V_*)^{\otimes s}$ is nonzero does not contain a finite corank subalgebra (as defined in [4, § 3.5]).

Fixing $p \in \mathbb{N}$, we have to prove that some matrix $a \in \mathfrak{g}$ concentrated in $\mathbb{N}_{\geq p} = \{p, p+1, \ldots\}$ does not annihilate z. In fact, it is enough to prove this for $a \in \mathfrak{g}^M$. Indeed, it would then follow that for sufficiently large $q > p$, the vector $(az)_{\leq q}$ obtained by annihilating all coordinates with index larger than q is nonzero. But we can find some large r such that $(az)_{\leq q}$ equals $(a_{\leq r} z)_{\leq r}$, where $a_{\leq r}$ is the $\{p, \ldots, r\}$-concentrated truncation of a. We would then conclude that $a_{\leq r} z$ is nonzero, and the proof would be complete.

Finally, to show that some $\mathbb{N}_{\geq p}$-concentrated $a \in \mathfrak{g}^M$ does not annihilate z, it suffices to pass to the quotient by V_*, and regard z as a nonzero element of $(V^*/V_*)^{\otimes s}$. Since $\mathfrak{g}^M$ acts on V^*/V_* via its quotient $\mathfrak{g}^M/\mathfrak{g}$, being $\mathbb{N}_{\geq p}$-concentrated no longer matters: any element of $\mathfrak{g}^M$ can be brought into

$\mathbb{N}_{\geq p}$-concentrated form by adding an element of $\mathfrak{g}$. In conclusion, the desired result is now simply that no nonzero element of $(V^*/V_*)^{\otimes s}$ is annihilated by $\mathfrak{g}^M$; this follows immediately from Proposition 1, for example. □

The proof is easily applicable to traceless tensors in $(V^*)^{\otimes m} \otimes V^{\otimes n}$, i.e., the intersection of the kernels of all mn evaluation maps

$$(V^*)^{\otimes m} \otimes V^{\otimes n} \to (V^*)^{\otimes (m-1)} \otimes V^{\otimes (n-1)}.$$

In other words:

Corollary 4. *Let $W \subseteq (V^*)^{\otimes m} \otimes V^{\otimes n}$ be the space of traceless tensors, and set*

$$W^k = W \cap \left(\sum_{|\mathbf{r}| \leq k} V^{\mathbf{r}} \otimes V^{\otimes n} \right), \quad \text{for every } 0 \leq k \leq m. \tag{2}$$

The filtration $\{W^k\}$ of W is essential over $\mathfrak{g}^M$. □

We can push this even further, making use of the $S_m \times S_n$-equivariance of the corollary. Recall that the irreducible objects in $\mathbb{T}_{\mathrm{End}(V)}$ are precisely the modules $W_{\lambda,\mu}$ of traceless tensors in $(V^*)^{\otimes|\lambda|} \otimes V^{\otimes|\mu|}$, for partitions λ, μ (see e.g., [4, Theorem 4.1] and discussion preceding it). For any pair of partitions, the intersection of (2) (for $m = |\lambda|$ and $n = |\nu|$) with $W_{\lambda,\mu}$ is a filtration of $W_{\lambda,\mu}$ by $\mathfrak{g}^M$-modules. It turns out that it is precisely what we are looking for:

Theorem 4. *For any two partitions λ, μ, the intersection of (2) with $W_{\lambda,\mu}$ is the socle filtration of this latter module over $\mathfrak{g}^M$.*

Proof. Immediate by the proof of Proposition 2: simply work with $W_{\lambda,\mu}$ instead of $(V^*)^{\otimes|\lambda|} \otimes V^{\otimes n}$. □

We now rephrase the theorem slightly, to give a more concrete description of the quotients in the socle filtration. To this end, recall that the ring SYM of symmetric functions is a Hopf algebra over $\mathbb{Z}$ (e.g., §2 of [2]), with comultiplication Δ. We regard partitions as elements of SYM by identifying them with the corresponding Schur functions, and we always think of $\Delta(\lambda)$ as a $\mathbb{Z}$-linear combination of tensor products $\mu \otimes \nu$ of partitions.

For a partition λ we denote $\Delta(\lambda)$ by $\lambda_{(1)} \otimes \lambda_{(2)}$. Note that this is a slight notational abuse, as $\Delta(\lambda)$ is not a simple tensor but rather a sum of tensors; we are suppressing the summation symbol to streamline the notation. The summation suppression extends to Schur functors: The expression $M_{\lambda_{(1)}} \otimes M_{\lambda_{(2)}}$, for instance, denotes a direct sum over all summands $\mu \otimes \nu$ of $\Delta(\lambda)$.

Finally, one last piece of notation: For an element $\nu \in$ SYM and $k \in \mathbb{N}$, we denote by $\nu^k \in$ SYM the degree-k homogeneous component of ν with respect to the usual grading of SYM.

We can now state the following consequence of Theorem 4, whose proof we leave to the reader (it consists simply of running through the definition of the comultiplication of SYM). Recall that we denote simple modules in $\mathbb{T}_{\mathrm{End}(V)}$ by $W_{\lambda,\mu}$; similarly, simple modules in $\mathbb{T}_{\mathfrak{g}^M}$ are denoted by $V_{\mu,\nu}$.

Corollary 5. *Let λ, μ be two partitions. The semisimple subquotient W^k/W^{k-1}, $k \geq 0$ of the socle filtration $0 = W^{-1} \subseteq W^0 \subseteq W^1 \ldots$ of $W_{\lambda,\nu}$ in $\mathfrak{g}^M$ – mod is isomorphic to*

$$(V^*/V_*)_{\lambda^k_{(1)}} \otimes V_{\lambda^{|\lambda|-k}_{(2)},\mu}.$$

Finally, it seems likely that as a category of $\mathfrak{g}^M$-modules, $\mathbb{T}_{\mathrm{End}(V)}$ has a universal property of its own, reminiscent of Theorem 2. Denoting by $\mathbb{T}_{\mathrm{RES}}$ the full (tensor) subcategory of $\mathfrak{g}^M$ – mod on the objects of $\mathbb{T}_{\mathrm{End}(V)}$, the following seems sensible.

Conjecture 2. For any tensor category $\mathcal{C}$ with monoidal unit $\mathbf{1}$, any morphism $b : x \otimes y \to \mathbf{1}$ in $\mathcal{C}$, and any subobject $x' \subseteq x$, there is a left exact tensor functor $F : \mathbb{T}_{\mathrm{RES}} \to \mathcal{C}$ sending the pairing $V^* \otimes V \to \mathbb{C}$ to b and turning the inclusion $V_* \subseteq V^*$ into $x' \subseteq x$. Moreover, F is unique up to tensor natural isomorphism.

Acknowledgements I would like to thank Ivan Penkov and Vera Serganova for useful discussions on the contents of [1, 4] and for help editing the manuscript.

References

1. E. Dan-Cohen, I. Penkov, and V. Serganova, *A Koszul category of representations of finitary Lie algebras*. ArXiv e-prints, May 2011.
2. M. Hazewinkel, *Symmetric functions, noncommutative symmetric functions, and quasisymmetric functions*. ArXiv Mathematics e-prints, October 2004.
3. G. W. Mackey, On infinite-dimensional linear spaces. *Trans. Amer. Math. Soc.*, **57** (1945), 155–207.
4. I. Penkov and V. Serganova, Representation theory of Mackey Lie algebras and their dense subalgebras. *This volume*.
5. I. Penkov and K. Styrkas, Tensor representations of classical locally finite Lie algebras, in: *Developments and trends in infinite-dimensional Lie theory*, Progress in Mathematics, **288**, Birkhäuser Boston, 2011, pp. 127–150.
6. S. V. Sam and A. Snowden, *Stability patterns in representation theory*. ArXiv e-prints, February 2013.

Free Field Realizations of the Date-Jimbo-Kashiwara-Miwa Algebra

Ben Cox, Vyacheslav Futorny, and Renato Alessandro Martins

Abstract We use the description of the universal central extension of the DJKM algebra $\mathfrak{sl}(2, R)$ where $R = \mathbb{C}[t, t^{-1}, u \mid u^2 = t^4 - 2ct^2 + 1]$ given in [CF11] to construct realizations of the DJKM algebra in terms of sums of partial differential operators.

Key words Wakimoto modules • DJKM algebras • Affine Lie algebras • Fock spaces

Mathematics Subject Classification (2010): 17B65, 17B69.

1 Introduction

Suppose R is a commutative algebra and $\mathfrak{g}$ is a simple Lie algebra, both defined over the complex numbers. From the work of C. Kassel and J.L. Loday (see [KL82], and [Kas84]) it is shown that the universal central extension $\hat{\mathfrak{g}}$ of $\mathfrak{g}\otimes R$ is the vector space $(\mathfrak{g} \otimes R) \oplus \Omega_R^1/dR$ where Ω_R^1/dR is the space of Kähler differentials modulo exact forms (see [Kas84]). Let $\overline{a}$ denote the image of $a \in \Omega_R^1$ in the quotient Ω_R^1/dR and let $(-,-)$ denote the Killing form on $\mathfrak{g}$. Then more precisely the universal central extension $\hat{\mathfrak{g}}$ is the vector space $(\mathfrak{g} \otimes R) \oplus \Omega_R^1/dR$ made into a Lie algebra by defining

$$[x \otimes f, y \otimes g] := [xy] \otimes fg + (x, y)\overline{fdg}, \quad [x \otimes f, \omega] = 0$$

B. Cox (✉)
Department of Mathematics, The College of Charleston, 66 George Street, Charleston SC 29424, USA
e-mail: coxbl@cofc.edu

V. Futorny • R.A. Martins
Departamento de Matemática, Instituto de Matemática e Estatística, São Paulo, Brasil
e-mail: vfutorny@gmail.com; renatoam@ime.usp.br

G. Mason et al. (eds.), *Developments and Retrospectives in Lie Theory: Algebraic Methods*, Developments in Mathematics 38,
DOI 10.1007/978-3-319-09804-3_5

for $x, y \in \mathfrak{g}$, $f, g \in R$, and $\omega \in \Omega_R^1/dR$. A natural and useful question comes to mind as to whether there exists free field or Wakimoto type realizations of these algebras. M. Wakimoto and B. Feigin and E. Frenkel answered this quesiton when R is the ring of Laurent polynomials in one variable (see [Wak86] and [FF90]).

The goal of this paper is to describe such a realization for the universal central extension of $\mathfrak{g} \otimes R = \mathfrak{sl}(2, R)$ where $R = \mathbb{C}[t, t^{-1}, u | u^2 = t^4 - 2ct^2 + 1]$, $c \neq \pm 1$, which we call a DJKM algebra. More precisely, for this particular ring R, we call $\mathfrak{sl}(2, R)$ the current DJKM algebra and its universal central extension $(\mathfrak{g} \otimes R) \oplus \Omega_R^1/dR$ the DJKM algebra. There are many other interesting commutative rings that one can use instead of this one and we discuss some of them below and what is known about their free field realizations.

There are a number of different veins that lead up to algebraic and geo-metric research on free field realizations of Lie algebras of the universal central extension of $\mathfrak{g} \otimes R$. One of them was M. Wakimoto's motivation for the use of a free field realization was to prove a conjecture of V. Kac and D. Kazhdan on the character of certain irreducible representations of affine Kac–Moody algebras at the critical level (see [Wak86] and [Fre05]). Previous related work on highest weight modules of $\mathfrak{sl}(2, R)$ written in terms of infinite sums of partial differential operators is be found in the early paper of H. P. Jakobsen and V. Kac [JK85]. Interestingly free field realizations have been used by V. V. Schechtman and A. N. Varchenko (and others) to provide integral solutions to the KZ-equations (see for example [SV90] and [EFK98] and their references).

Another application is to help in describing the center of a certain completion of the enveloping algebra of an affine Lie algebra at the critical level and this determination of the center is an important ingredient in the formulation of the geometric Langland's correspondence [Fre07]. As a last bit of motivation, the work B. Feigin and E. Frenkel has shown that free field realizations of an affine Lie algebra appear naturally in the context of the generalized AKNS hierarchies [FF99]. There are numerous other authors who have done interesting work on free field realizations for affine Lie algebras, but we will only mention just the few above.

A separate vein appears as follows: In Kazhdan and Luszig's explicit study of the tensor structure of modules for affine Lie algebras (see [KL93] and [KL91]) the ring of functions, regular everywhere except at a finite number of points, appears naturally. This algebra M. Bremner gave the name *n-point algebra*. In particular in the monograph [FBZ01, Ch. 12] algebras of the form $\oplus_{i=1}^n \mathfrak{g}((t - x_i)) \oplus \mathbb{C}c$ appear in the description of the conformal blocks. These contain the n-point algebras $\mathfrak{g} \otimes \mathbb{C}[(t - x_1)^{-1}, \ldots, (t - x_N)^{-1}] \oplus \mathbb{C}c$ modulo part of the center Ω_R/dR. M. Bremner explicitly described the universal central extension of such an algebra in [Bre94a] and called it the n-point algebra. To our knowledge free field realizations of a general n-point algebra have not been constructed with the exception of when $n = 2, 3, 4$ and we describe this work in this following paragraphs. The case $n = 2$ is just the affine case and is roughly described in the previous paragraph.

If these realizations had been constructed in a logically historical fashion one would have then described them for a three-point algebra. This is the case where

R denotes the ring of rational functions with poles only in the set of distinct complex numbers $\{a_1, a_2, a_3\}$. This algebra is isomorphic to $\mathbb{C}[s, s^{-1}, (s-1)^{-1}]$. M. Schlichenmaier gave an isomorphic description of the three-point algebra as $\mathbb{C}[(z^2-a^2)^k, z(z^2-a^2)^k \mid k \in \mathbb{Z}]$ where $a \neq 0$ (see [Sch03a]). E. Jurisich and the first author of the present paper proved that $R \cong \mathbb{C}[t, t^{-1}, u \mid u^2 = t^2 + 4t]$ so that the three-point algebra looks more like S_b above. The main result of [CJ] provides a natural free field realization in terms of a β-γ-system and the oscillator algebra of the three-point affine Lie algebra when $\mathfrak{g} = \mathfrak{sl}(2, \mathbb{C})$. Besides M. Bremner's article mentioned above, other work on the universal central extension of 3-point algebras can be found in [BT07].

Let R denote the ring of rational functions on the Riemann sphere $S^2 = \mathbb{C} \cup \{\infty\}$ with poles only in the set of distinct points $\{a_1, a_2, a_3, \infty\} \subset S^2$. In the literature one can find the fact that the automorphism group $PGL_2(\mathbb{C})$ of $\mathbb{C}(s)$ is simply 3-transitive and R is a subring of $\mathbb{C}(s)$, so that R is isomorphic to the ring of rational functions with poles at $\{\infty, 0, 1, a\}$. This isomorphism motivates one setting $a = a_4$ and then defining the *4-point ring* as $R = R_a = \mathbb{C}[s, s^{-1}, (s-1)^{-1}, (s-a)^{-1}]$ where $a \in \mathbb{C}\backslash\{0, 1\}$. Letting $S := S_b = \mathbb{C}[t, t^{-1}, u]$ where $u^2 = t^2 - 2bt + 1$ with b a complex number not equal to ± 1, M. Bremner showed us that $R_a \cong S_b$. (This in fact lead to the description above for the three-point algebra given in [CJ].) The latter ring is $\mathbb{Z}_2$-graded where t is even and u is odd, and is a cousin to super Lie algebras, so this ring lends itself to the techniques of conformal field theory. M. Bremner gave an explicit description of the universal central extension of $\mathfrak{g} \otimes R$, in terms of ultraspherical (Gegenbauer) polynomials where R is this four-point algebra (see [Bre95]). Motivated by talks with M. Bremner, the first author gave free field realizations for the four-point algebra where the center acts nontrivially (see [Cox08]).

In Bremner's study of the elliptic affine Lie algebras, $\mathfrak{sl}(2, R) \oplus (\Omega_R/dR)$ where $R = \mathbb{C}[x, x^{-1}, y \mid y^2 = 4x^3 - g_2x - g_3]$, he next explicitly described the universal central extension of this algebra in terms of Pollaczek polynomials (see [Bre94b]). Variations of these algebras appear in the work of A. Fialowski and M. Schlichenmaier [FS06] and [FS05]. Together with André Bueno, the first and second authors of this article described free field type realizations of the elliptic Lie algebra where $R = \mathbb{C}[t, t^{-1}, u \mid, u^2 = t^3 - 2bt^2 - t]$, $b \neq \pm 1$ (see [BCF09]).

In [DJKM83] Date, Jimbo, Kashiwara and Miwa described integrable systems arising from the Landau–Lifshitz differential equation. The integrable hierarchy of this equation was shown to be written in terms of free fermions defined on an elliptic curve. These authors introduced an infinite-dimensional Lie algebra which is a one-dimensional central extension of $\mathfrak{g} \otimes \mathbb{C}[t, t^{-1}, u|u^2 = (t^2 - b^2)(t^2 - c^2)]$ where $b \neq \pm c$ are complex constants and $\mathfrak{g}$ is a simple finite-dimensional Lie algebra. This algebra, which we call the DJKM algebra, acts on the space of solutions of the Landau–Lifshitz equation as infinitesimal Bäcklund transformations.

In [CF11] the first and second authors provided the commutation relations of the universal central extension of the DJKM Lie algebra in terms of a basis of the algebra and certain polynomials. More precisely, in order to pin down this central extension, one needed to describe four families of polynomials that appeared as coefficients in the commutator formulae. Two of these families of polynomials are given in terms of elliptic integrals and the other two families are slight variations of ultraspherical polynomials. One of the elliptic families is precisely given by the associated ultraspherical polynomials (see [BI82]). The associated ultraspherical polynomials in turn are, up to factors of ascending factorials, particular associated Jacobi polynomials. The associated Jacobi polynomials are known to satisfy certain fourth-order linear differential equations (see [Ism05] and formula (48) in [Wim87]). In [CFT13] the first and second authors together with Juan Tirao showed that the remaining elliptic family of polynomials are not of classical type and are orthogonal. We call this later elliptic family, DJKM polynomials. It is not clear to us whether a four-point algebra or an elliptic affine Lie algebra or a DJKM algebra can be realized as either a quotient or subalgebra of one of the others. One should however note that the coordinate ring of the four-point algebra is of genus 0, whereas for the elliptic affine Lie algebra is of genus 1.

The Lie algebras above are examples of Krichever–Novikov algebras (see ([KN87b,KN87a,KN89]). A fair amount of interesting and fundamental work has be done by Krichever, Novikov, Schlichenmaier, and Sheinman on the representation theory of these algebras. In particular, Wess–Zumino–Witten–Novikov theory and analogues of the Knizhnik–Zamolodchikov equations are developed for these algebras (see the survey article [She05], and for example [SS99, SS99, She03, Sch03a, Sch03b], and [SS98]).

Here is an outline of the paper. We review in the Sect. 2 the description of the universal central extension of $\mathfrak{g} \otimes R$ where $\mathfrak{g}$ is a simple finite-dimensional Lie algebra defined over $\mathbb{C}$ and R is a commutative algebra defined over $\mathbb{C}$. We describe in particular a basis for the universal central extension in the case $R = \mathbb{C}[t, t^{-1}, u \,|\, u^2 = (t^2 - a^2)(t^2 - b^2)$ with $a \neq \pm b$ and neither a nor b are zero. The associated ultraspherical polynomials and another new family that we call the DJKM polynomials play an important role in the explicit description of the free field realization of the DJKM algebras. These latter two families of orthogonal polynomials are not of classical type.

In Sect. 3 we give two distinct sets of generators and relations for the DJKM algebra in order to make explicit the relationship between the current algebra type description and the Chevalley type basis. This allows us to provide a free field description based on just fields coming from the Chevalley basis. In Sect. 4 we give a weak triangular decomposition of $\mathfrak{g} \otimes R$, review the notion of formal distributions, and Wick's Theorem. In Sect. 5 we recall the $\beta - \gamma$ system, Taylor's theorem in the setting of formal distributions and give the definition of the DJKM Heisenberg algebra and realization of it in terms of partial differential operators. In the last section we give our main result describing two free field realizations of the DJKM algebra. The proof relies heavily on Wick's Theorem together with Taylor's Theorem given in Sect. 4 together with the results in Sect. 4.

We thank the referee for useful suggestions on the exposition of this paper.

2 Description of the Universal Central Extension of DJKM Algebras

As we described in the introduction, C. Kassel showed the universal central extension of the current algebra $\mathfrak{g} \otimes R$, where $\mathfrak{g}$ is a simple finite-dimensional Lie algebra defined over $\mathbb{C}$, is the vector space $\hat{\mathfrak{g}} = (\mathfrak{g} \otimes R) \oplus \Omega_R^1/dR$ with Lie bracket given by

$$[x \otimes f, Y \otimes g] = [xy] \otimes fg + (x, y)\overline{fdg}, [x \otimes f, \omega] = 0, [\omega, \omega'] = 0,$$

where $x, y \in \mathfrak{g}$, and $\omega, \omega' \in \Omega_R^1/dR$ and (x, y) denotes the Killing form on $\mathfrak{g}$.

Consider the polynomial

$$p(t) = t^n + a_{n-1}t^{n-1} + \cdots + a_0$$

where $a_i \in \mathbb{C}$ and $a_n = 1$. Fundamental to the explicit description of the universal central extension for $R = \mathbb{C}[t, t^{-1}, u | u^2 = p(t)]$ is the following:

Theorem 1 ([Bre94b], Theorem 3.4). *Let R be as above. The set*

$$\{\overline{t^{-1}\,dt}, \overline{t^{-1}u\,dt}, \ldots, \overline{t^{-n}u\,dt}\}$$

forms a basis of Ω_R^1/dR (omitting $\overline{t^{-n}u\,dt}$ if $a_0 = 0$).

Lemma 2.1 ([CF11]). *If $u^m = p(t)$ and $R = \mathbb{C}[t, t^{-1}, u | u^m = p(t)]$, then in Ω_R^1/dR, one has*

$$((m+1)n + im)t^{n+i-1}u\,dt \equiv -\sum_{j=0}^{n-1}((m+1)j + mi)a_j t^{i+j-1}u\,dt \mod dR. \quad (1)$$

In the DJKM algebra setting one takes $m = 2$ and $p(t) = (t^2 - a^2)(t^2 - b^2)$ with $a \neq \pm b$ and neither a nor b is zero. The lemma above leads one to introduce the polynomials $P_k := P_k(c)$ in $c = (a^2 + b^2)/2$ to satisfy the recursion relation

$$(6 + 2k)P_k(c) = 4kcP_{k-2}(c) - 2(k-3)P_{k-4}(c)$$

for $k \geq 0$. Setting

$$P(c, z) := \sum_{k \geq -4} P_k(c)z^{k+4} = \sum_{k \geq 0} P_{k-4}(c)z^k$$

one proves in [CF11] that

$$\frac{d}{dz}P(c,z) - \frac{3z^4 - 4cz^2 + 1}{z^5 - 2cz^3 + z}P(c,z) = \frac{2\,(P_{-1} + cP_{-3})\,z^3 + P_{-2}z^2 + (4cz^2 - 1)\,P_{-4}}{z^5 - 2cz^3 + z}. \quad (2)$$

There are four cases to consider when solving this differential equation depending on the parameters P_k, $k = -1, -2, -3, -4$.

2.1 Elliptic Case 1

Taking the initial conditions $P_{-3}(c) = P_{-2}(c) = P_{-1}(c) = 0$ and $P_{-4}(c) = 1$ we then arrive at a generating function

$$P_{-4}(c,z) := \sum_{k\geq -4} P_{-4,k}(c)z^{k+4} = \sum_{k\geq 0} P_{-4,k-4}(c)z^k,$$

expressed in terms of an elliptic integral

$$P_{-4}(c,z) = z\sqrt{1 - 2cz^2 + z^4}\int \frac{4cz^2 - 1}{z^2(z^4 - 2cz^2 + 1)^{3/2}}\,dz.$$

The way that we interpret the right-hand integral is to expand $(z^4 - 2cz^2 + 1)^{-3/2}$ as a Taylor series about $z = 0$ and then formally integrate term-by-term and multiply the result by the Taylor series of $z\sqrt{1 - 2cz^2 + z^4}$. More precisely, one integrates formally with zero constant term

$$\int (4c - z^{-2})\sum_{n=0}^{\infty} Q_n^{(3/2)}(c)z^{2n}\,dz = \sum_{n=0}^{\infty}\frac{4cQ_n^{(3/2)}(c)}{2n+1}z^{2n+1} - \sum_{n=0}^{\infty}\frac{Q_n^{(3/2)}(c)}{2n-1}z^{2n-1}$$

where $Q_n^{(\lambda)}(c)$ is the n-th Gegenbauer polynomial. When multiplying this by

$$z\sqrt{1 - 2cz^2 + z^4} = \sum_{n=0}^{\infty} Q_n^{(-1/2)}(c)z^{2n+1}$$

one obtains the series $P_{-4}(c,z)$.

The polynomials $P_{-4,k}$ are a special case of associated ultraspherical polynomials (see [CFT13] and its references).

2.2 Elliptic Case 2

If we take initial conditions $P_{-4}(c) = P_{-3}(c) = P_{-1}(c) = 0$ and $P_{-2}(c) = 1$, then we arrive at a generating function defined in terms of another elliptic integral:

$$P_{-2}(c,z) = z\sqrt{1-2cz^2+z^4}\int \frac{1}{(z^4-2cz^2+1)^{3/2}}\,dz.$$

The polynomials $P_{-2,k}$ are a new family of polynomials related to associated ultraspherical polynomials and we call them the DJKM polynomials (see [CFT13] and its references).

2.3 Gegenbauer Case 3

If we take $P_{-1}(c) = 1$, and $P_{-2}(c) = P_{-3}(c) = P_{-4}(c) = 0$ and set

$$P_{-1}(c,z) = \sum_{n\geq 0} P_{-1,n-4}z^n,$$

then we get a solution which after solving for the integration constant can be turned into a power series solution

$$P_{-1}(c,z) = \frac{1}{c^2-1}\left(cz - z^3 - cz + c^2z^3 - \sum_{k=2}^{\infty} cQ_n^{(-1/2)}(c)z^{2n+1}\right)$$

where $Q_n^{(-1/2)}(c)$ is the n-th Gegenbauer polynomial. Hence

$$P_{-1,-4}(c) = P_{-1,-3}(c) = P_{-1,-2}(c) = P_{-1,2m}(c) = 0, \quad P_{-1,-1}(c) = 1,$$

$$P_{-1,2n-3}(c) = \frac{-cQ_n(c)}{c^2-1},$$

for $m \geq 0$ and $n \geq 2$.

2.4 Gegenbauer Case 4

Next we consider the initial conditions $P_{-1}(c) = 0 = P_{-2}(c) = P_{-4}(c) = 0$ with $P_{-3}(c) = 1$ and set

$$P_{-3}(c,z) = \sum_{n\geq 0} P_{-3,n-4}(c)z^n.$$

then we get a power series solution

$$P_{-3}(c,z)=\frac{1}{c^2-1}\left(c^2z-cz^3-z+cz^3-\sum_{k=2}^{\infty}Q_n^{(-1/2)}(c)z^{2n+1}\right)$$

where $Q_n^{(-1/2)}(c)$ is the n-th Gegenbauer polynomial. Hence

$$P_{-3,-4}(c)=P_{-3,-2}(c)=P_{-3,-1}(c)=P_{-1,2m}(c)=0,\quad P_{-3,-3}(c)=1,$$
$$P_{-3,2n-3}(c)=\frac{-Q_n(c)}{c^2-1},$$

for $m\geq 0$ and $n\geq 2$.

3 Generators and Relations

Given the universal central extension of the current algebra $\mathfrak{g}\otimes R$ one can define it using generators and relations in more than at least two ways. One is in terms of the generators $x\otimes r$ where $x\in\mathfrak{g}$ and $r\in R$, plus the center and the other is in terms of Chevalley generators. It is the latter which we need in order to describe a free field realization of the universal central extension. In the case of three-point algebras there are at least four different sets of generators and relations, three of which are described in [CJ]. One of the sets of generators and relations for the three-point algebra given in [Sch03a] seems to be missing some relations. Nevertheless, in the previous work cited in the introduction we neglected to explicitly prove the isomorphism of these two sets of generators and relations for respectively the four point and elliptic algebras. In the theorems below we make explicit this isomorphism.

Set $\omega_0=\overline{t^{-1}\,dt}$, and $\omega_{-k}=\overline{t^{-k}u\,dt}$, $k=1,2,3,4$.

Theorem 2 (see [CF11]). *Let $\mathfrak{g}$ be a simple finite-dimensional Lie algebra over the complex numbers with the Killing form $(\,|\,)$ and define $\psi_{ij}(c)\in\Omega^1_R/dR$ by*

$$\psi_{ij}(c)=\begin{cases}\omega_{i+j-2} & \textit{for}\quad i+j=1,0,-1,-2\\ P_{-3,i+j-2}(c)(\omega_{-3}+c\omega_{-1}) & \textit{for}\quad i+j=-1+2n\geq 3,\\ P_{-3,i+j-2}(c)(c\omega_{-3}+\omega_{-1}) & \textit{for}\quad i+j=1-2n\leq -3,\\ P_{-4,|i+j|-2}(c)\omega_{-4}+P_{-2,|i+j|-2}(c)\omega_{-2} & \textit{for}\quad |i+j|=2n\geq 2,\end{cases}\tag{3}$$

with $n \in \mathbb{Z}$. The universal central extension of the Date–Jimbo–Kashiwara–Miwa algebra is the $\mathbb{Z}_2$-graded Lie algebra

$$\hat{\mathfrak{g}} = \hat{\mathfrak{g}}^0 \oplus \hat{\mathfrak{g}}^1,$$

where

$$\hat{\mathfrak{g}}^0 = \left(\mathfrak{g} \otimes \mathbb{C}[t, t^{-1}]\right) \oplus \mathbb{C}\omega_0,$$
$$\hat{\mathfrak{g}}^1 = \left(\mathfrak{g} \otimes \mathbb{C}[t, t^{-1}]u\right) \oplus \mathbb{C}\omega_{-4} \oplus \mathbb{C}\omega_{-3} \oplus \mathbb{C}\omega_{-2} \oplus \mathbb{C}\omega_{-1}$$

with bracket

$$[x \otimes t^i, y \otimes t^j] = [x, y] \otimes t^{i+j} + \delta_{i+j,0}\, j(x, y)\omega_0,$$
$$[x \otimes t^{i-1}u, y \otimes t^{j-1}u] = [x, y] \otimes (t^{i+j+2} - 2ct^{i+j} + t^{i+j-2})$$
$$+ \left(\delta_{i+j,-2}(j+1) - 2cj\delta_{i+j,0} + (j-1)\delta_{i+j,2}\right)(x, y)\omega_0,$$
$$[x \otimes t^{i-1}u, y \otimes t^j] = [x, y]u \otimes t^{i+j-1} + j(x, y)\psi_{ij}(c).$$

The theorem above is similar to the results that M. Bremner obtained for the elliptic and four-point affine Lie algebra cases ([Bre94b, Theorem 4.6] and [Bre95, Theorem 3.6] respectively) and with the isomorphism obtained for the three point algebra given in [CJ].

Theorem 3. *The universal central extension of the algebra $\mathfrak{sl}(2, \mathbb{C}) \otimes \mathcal{R}$ is isomorphic to the Lie algebra with generators e_n, e_n^1, f_n, f_n^1, h_n, h_n^1, $n \in \mathbb{Z}$, ω_0, ω_{-1}, ω_{-2}, ω_{-3}, ω_{-4} and relations given by*

$$[x_m, x_n] := [x_m, x_n^1] = [x_m^1, x_n^1] = 0, \quad \text{for } x = e, f \tag{4}$$

$$[h_m, h_n] := -2m\delta_{m,-n}\omega_0 = (n-m)\delta_{m,-n}\omega_0, \tag{5}$$

$$[h_m^1, h_n^1] := 2\left((n+2)\delta_{m+n,-4} - 2c(n+1)\delta_{m+n,-2} + n\delta_{m+n,0}\right)\omega_0, \tag{6}$$

$$[h_m, h_n^1] := -2m\psi_{mn}(c), \tag{7}$$

$$[\omega_i, x_m] = [\omega_i, \omega_j] = 0, \quad \text{for } x = e, f, h, \quad i, j \in \{0, 1\} \tag{8}$$

$$[e_m, f_n] = h_{m+n} - m\delta_{m,-n}\omega_0, \tag{9}$$

$$[e_m, f_n^1] = h_{m+n}^1 - m\psi_{mn}(c) =: [e_m^1, f_n], \tag{10}$$

$$[e_m^1, f_n^1] := h_{m+n+4} - 2ch_{m+n+2} + h_{m+n} \tag{11}$$
$$+ \left((n+2)\delta_{m+n,-4} - 2c(n+1)\delta_{m+n,-2} + n\delta_{m+n,0}\right)\omega_0,$$

$$[h_m, e_n] := 2e_{m+n}, \tag{12}$$

$$[h_m, e_n^1] := 2e_{m+n}^1 =: [h_m^1, e_m], \tag{13}$$

$$[h_m^1, e_n^1] := 2e_{m+n+4} - 4ce_{m+n+2} + 2e_{m+n}, \tag{14}$$

$$[h_m, f_n] := -2f_{m+n}, \tag{15}$$

$$[h_m, f_n^1] := -2f_{m+n}^1 =: [h_m^1, f_m], \tag{16}$$

$$[h_m^1, f_n^1] := -2f_{m+n+4} + 4cf_{m+n+2} - 2f_{m+n}, \tag{17}$$

for all $m, n \in \mathbb{Z}$.

Proof. Let $\mathfrak{f}$ denote the free Lie algebra with the generators e_n, e_n^1, f_n, f_n^1, h_n, h_n^1, $n \in \mathbb{Z}$, ω_0, ω_{-1}, ω_{-2}, ω_{-3}, ω_{-4} and relations given above (4)– (17). The map $\phi : \mathfrak{f} \to (\mathfrak{sl}(2, \mathbb{C}) \otimes \mathcal{R}) \oplus (\Omega_{\mathcal{R}}/d\mathcal{R})$ given by

$$\phi(e_n) := e \otimes t^n, \quad \phi(e_n^1) = e \otimes ut^n,$$

$$\phi(f_n) := f \otimes t^n, \quad \phi(f_n^1) = f \otimes ut^n,$$

$$\phi(h_n) := h \otimes t^n, \quad \phi(h_n^1) = h \otimes ut^n,$$

$$\phi(\omega_0) := \overline{t^{-1}\,dt}, \quad \phi(\omega_{-1}) = \overline{t^{-1}u\,dt},$$

$$\phi(\omega_{-2}) := \overline{t^{-2}u\,dt}, \quad \phi(\omega_{-3}) = \overline{t^{-3}u\,dt},$$

$$\phi(\omega_{-4}) := \overline{t^{-4}u\,dt},$$

for $n \in \mathbb{Z}$ is a surjective Lie algebra homomorphism.

Consider the subalgebras

$$S_+ = \langle e_n, e_n^1 \mid n \in \mathbb{Z} \rangle$$

$$S_0 = \langle h_n, h_n^1, \omega_0, \omega_{-1}, \omega_{-2}, \omega_{-3}, \omega_{-4} \mid n \in \mathbb{Z} \rangle$$

$$S_- = \langle f_n, f_n^1 \mid n \in \mathbb{Z} \rangle$$

where $\langle X \rangle$ means spanned by the set X and set $S = S_- + S_0 + S_+$. By (4)–(8) we have

$$S_+ = \sum_{n \in \mathbb{Z}} \mathbb{C}e_n + \sum_{n \in \mathbb{Z}} \mathbb{C}e_n^1, \quad S_- = \sum_{n \in \mathbb{Z}} \mathbb{C}f_n + \sum_{n \in \mathbb{Z}} \mathbb{C}f_n^1$$

$$S_0 = \sum_{n \in \mathbb{Z}} \mathbb{C}h_n + \sum_{n \in \mathbb{Z}} \mathbb{C}h_n^1 + \mathbb{C}\omega_0 + \mathbb{C}\omega_{-1} + \mathbb{C}\omega_{-2} + \mathbb{C}\omega_{-3} + \mathbb{C}\omega_{-4}$$

By (9)–(14) we see that

$$[e_n, S_+] = [e_n^1, S_+] = 0, \quad [h_n, S_+] \subseteq S_+, \quad [h_n^1, S_+] \subseteq S_+,$$
$$[f_n, S_+] \subseteq S_0, \quad [f_n^1, S_+] \subseteq S_0,$$

and similarly $[x_n, S_-] = [x_n^1, S_-] \subseteq S$, $[x_n, S_0] = [x_n^1, S_0] \subseteq S$ for $x = e, f, h$. To sum it up, we observe that $[x_n, S] \subseteq S$ and $[x_n^1, S] \subseteq S$ for $n \in \mathbb{Z}$, $x = h, e, f$. Thus $[S, S] \subset S$. Hence S contains the generators of $\mathfrak{f}$ and is a subalgebra. Hence $S = \mathfrak{f}$. One can now see that ϕ is a Lie algebra isomorphism. □

4 A Triangular Decomposition of DJKM Current Algebras $\mathfrak{g} \otimes R$

From now on we identify R_a with $\mathcal{S}$ and set $R = \mathcal{S}$ which has a basis $t^i, t^i u, i \in \mathbb{Z}$. Let $p : R \to R$ be the automorphism given by $p(t) = t$ and $p(u) = -u$. Then one can decompose $R = R^0 \oplus R^1$ where $R^0 = \mathbb{C}[t^{\pm 1}] = \{r \in R \mid p(r) = r\}$ and $R^1 = \mathbb{C}[t^{\pm 1}]u = \{r \in R \mid p(r) = -r\}$ are the eigenspaces of p. From now on $\mathfrak{g}$ will denote a simple Lie algebra over $\mathbb{C}$ with triangular decomposition $\mathfrak{g} = \mathfrak{n}_- \oplus \mathfrak{h} \oplus \mathfrak{n}_+$ and then the *DJKM loop algebra* $L(\mathfrak{g}) := \mathfrak{g} \otimes R$ has a corresponding $\mathbb{Z}/2\mathbb{Z}$-grading: $L(\mathfrak{g})^i := \mathfrak{g} \otimes R^i$ for $i = 0, 1$. However the degree of t does not render $L(\mathfrak{g})$ a $\mathbb{Z}$-graded Lie algebra. This leads one to the following notion.

Suppose I is an additive subgroup of the rational numbers $\mathbb{P}$ and $\mathcal{A}$ is a $\mathbb{C}$-algebra such that $\mathcal{A} = \oplus_{i \in I} \mathcal{A}_i$ and there exists a fixed $l \in \mathbb{N}$, with

$$\mathcal{A}_i \mathcal{A}_j \subset \oplus_{|k-(i+j)| \leq l} \mathcal{A}_k$$

for all $i, j \in \mathbb{Z}$. Then $\mathcal{A}$ is said to be an *l-quasigraded algebra*. For $0 \neq x \in \mathcal{A}_i$ one says that x is *homogeneous of degree i* and one writes $\deg x = i$.

For example, R has the structure of a 1-quasigraded algebra where $I = \frac{1}{2}\mathbb{Z}$ and $\deg t^i = i$, $\deg t^i u = i + \frac{1}{2}$.

A *weak triangular decomposition* of a Lie algebra $\mathfrak{l}$ is a triple $(\mathfrak{H}, \mathfrak{l}_+, \sigma)$ satisfying

1. $\mathfrak{H}$ and $\mathfrak{l}_+$ are subalgebras of $\mathfrak{l}$,
2. $\mathfrak{H}$ is abelian and $[\mathfrak{H}, \mathfrak{l}_+] \subset \mathfrak{l}_+$,
3. σ is an antiautomorphism of $\mathfrak{l}$ of order 2 which is the identity on $\mathfrak{h}$ and
4. $\mathfrak{l} = \mathfrak{l}_+ \oplus \mathfrak{H} \oplus \sigma(\mathfrak{l}_+)$.

We will let $\sigma(\mathfrak{l}_+)$ be denoted by $\mathfrak{l}_-$.

Theorem 4. *The DJKM current algebra $L(\mathfrak{g})$ is 1-quasigraded Lie algebra where $\deg(x \otimes f) = \deg f$ for f homogeneous. Set $R_+ = \mathbb{C}(1 + u) \oplus \mathbb{C}[t, u]t$ and $R_- = p(R_+)$. Then $L(\mathfrak{g})$ has a weak triangular decomposition given by*

$$L(\mathfrak{g})_\pm = \mathfrak{g} \otimes R_\pm, \quad \mathcal{H} := \mathfrak{h} \otimes \mathbb{C}.$$

Proof. This is essentially the same proof as [Bre95], Theorem 2.1 and so will be omitted. □

4.1 Formal Distributions

We need some more notation that will make some of the arguments later more transparent. Our notation follows roughly [Kac98] and [MN99]: The *formal (Dirac) delta function* $\delta(z/w)$ is the formal distribution

$$\delta(z/w) = z^{-1}\sum_{n\in\mathbb{Z}} z^{-n}w^{n} = w^{-1}\sum_{n\in\mathbb{Z}} z^{n}w^{-n}.$$

For any sequence of elements $\{a_m\}_{m\in\mathbb{Z}}$ in the ring $\mathrm{End}(V)$, V a vector space, the *formal distribution*

$$a(z) := \sum_{m\in\mathbb{Z}} a_m z^{-m-1}$$

is called a *field*, if for any $v \in V$, $a_m v = 0$ for $m \gg 0$. If $a(z)$ is a field, then we set

$$a(z)_- := \sum_{m\geq 0} a_m z^{-m-1}, \quad \text{and} \quad a(z)_+ := \sum_{m<0} a_m z^{-m-1}. \tag{18}$$

The *normal ordered product* of two distributions $a(z)$ and $b(w)$ (and their coefficients) is defined by

$$\sum_{m\in\mathbb{Z}}\sum_{n\in\mathbb{Z}} :a_m b_n: z^{-m-1}w^{-n-1} =: a(z)b(w) := a(z)_+b(w) + b(w)a(z)_-. \tag{19}$$

Then one defines recursively

$$:a^1(z_1)\cdots a^k(z_k) :=: a^1(z_1)\left(:a^2(z_2)\left(:\cdots :a^{k-1}(z_{k-1})a^k(z_k):\right)\cdots:\right):,$$

while normal ordered product

$$:a^1(z)\cdots a^k(z) := \lim_{z_1,z_2,\cdots,z_k\to z} :a^1(z_1)\left(:a^2(z_2)\left(:\cdots :a^{k-1}(z_{k-1})a^k(z_k):\right)\cdots\right):$$

will only be defined for certain k-tuples $(a^1,\ldots,a^k)$.

Set

$$\lfloor ab\rfloor = a(z)b(w) - :a(z)b(w) := [a(z)_-, b(w)], \tag{20}$$

which is called the *contraction* of the two formal distributions $a(z)$ and $b(w)$.

Theorem 5 (Wick's Theorem, [BS83, Hua98] or [Kac98]). *Let $a^i(z)$ and $b^j(z)$ be formal distributions with coefficients in the associative algebra* $\mathrm{End}(\mathbb{C}[\mathbf{x}]\otimes\mathbb{C}[\mathbf{y}])$, *satisfying*

1. $[\lfloor a^i(z)b^j(w)\rfloor, c^k(x)_\pm] = [\lfloor a^i b^j\rfloor, c^k(x)_\pm] = 0$, *for all* i, j, k *and* $c^k(x) = a^k(z)$ *or* $c^k(x) = b^k(w)$.
2. $[a^i(z)_\pm, b^j(w)_\pm] = 0$ *for all* i *and* j.
3. *The products*

$$\lfloor a^{i_1}b^{j_1}\rfloor\cdots\lfloor a^{i_s}b^{i_s}\rfloor : a^1(z)\cdots a^M(z)b^1(w)\cdots b^N(w) :_{(i_1,\ldots,i_s;j_1,\ldots,j_s)}$$

have coefficients in $\mathrm{End}(\mathbb{C}[\mathbf{x}]\otimes\mathbb{C}[\mathbf{y}])$ *for all subsets* $\{i_1,\ldots,i_s\}\subset\{1,\ldots,M\}$, $\{j_1,\ldots,j_s\}\subset\{1,\cdots N\}$. *Here the subscript* $(i_1,\ldots,i_s;\, j_1,\ldots,j_s)$ *means that those factors* $a^i(z)$, $b^j(w)$ *with indices* $i\in\{i_1,\ldots,i_s\}$, $j\in\{j_1,\ldots,j_s\}$ *are to be omitted from the product* $: a^1\cdots a^M b^1\cdots b^N :$ *and when* $s=0$ *we do not omit any factors.*

Then

$$: a^1(z)\cdots a^M(z) :: b^1(w)\cdots b^N(w) :=$$

$$\sum_{\substack{0\le s\le\min(M,N),\\ i_1<\cdots<i_s,\ j_1\neq\cdots\neq j_s}} \lfloor a^{i_1}b^{j_1}\rfloor\cdots\lfloor a^{i_s}b^{j_s}\rfloor : a^1(z)\cdots a^M(z)b^1(w)\cdots b^N(w) :_{(i_1,\ldots,i_s;j_1,\ldots,j_s)}.$$

Setting $m = i-\frac{1}{2}$, $i\in\mathbb{Z}+\frac{1}{2}$ and $x\in\mathfrak{g}$, define $x_{m+\frac{1}{2}} = x\otimes t^{i-\frac{1}{2}}u = x^1_m$ and $x_m := x\otimes t^m$. Define

$$x^1(z) := \sum_{m\in\mathbb{Z}} x_{m+\frac{1}{2}} z^{-m-1}, \quad x(z) := \sum_{m\in\mathbb{Z}} x_m z^{-m-1}.$$

The relations in Theorem 3 then can be rewritten succinctly as

$$[x(z), y(w)] = [xy](w)\delta(z/w) - (x,y)\omega_0\partial_w\delta(z/w), \tag{21}$$

$$[x^1(z), y^1(w)] = P(w)\left([x,y](w)\delta(z/w) - (x,y)\omega_0\partial_w\delta(z/w)\right) - \frac{1}{2}(x,y)(\partial P(w))\omega_0\delta(z/w), \tag{22}$$

$$\begin{aligned}[x(z), y^1(w)] &= [x,y]^1(w)\delta(z/w) - (x,y)(\partial_w\psi(c,w)\delta(z/w) - w\psi(c,w)\partial_w\delta(z/w)) \qquad (23)\\ &= [x^1(z), y(w)],\end{aligned}$$

where $x, y\in\{e, f, h\}$, $P(w) = w^4 - 2cw^2 + 1$ and $\psi(c,w) = \sum_{n\in\mathbb{Z}}\psi_n(c)w^n$ for $\psi_{i+j}(c) := \psi'_{ij}(c)$.

5 Oscillator Algebras

5.1 The $\beta - \gamma$ System

The $\beta - \gamma$ is the infinite-dimensional oscillator algebra $\hat{\mathfrak{a}}$ with generators $a_n, a_n^*, a_n^1, a_n^{1*}$, $n \in \mathbb{Z}$, together with $\mathbf{1}$ satisfying the relations

$$[a_n, a_m] = [a_m, a_n^1] = [a_m, a_n^{1*}] = [a_n^*, a_m^*] = [a_n^*, a_m^1] = [a_n^*, a_m^{1*}] = 0,$$

$$[a_n^1, a_m^1] = [a_n^{1*}, a_m^{1*}] = 0 = [\mathfrak{a}, \mathbf{1}],$$

$$[a_n, a_m^*] = \delta_{m+n,0}\mathbf{1} = [a_n^1, a_m^{1*}].$$

For $c = a, a^1$ and respectively $X = x, x^1$ with $r = 0$ or $r = 1$, sets $\mathbb{C}[\mathbf{x}] := \mathbb{C}[x_n, x_n^1 \,|\, n \in \mathbb{Z}$ and define $\rho : \hat{\mathfrak{a}} \to \mathfrak{gl}(\mathbb{C}[\mathbf{x}])$ by

$$\rho_r(c_m) := \begin{cases} \partial/\partial X_m & \text{if} \quad m \geq 0, \text{ and } r = 0 \\ X_m & \text{otherwise,} \end{cases} \tag{24}$$

$$\rho_r(c_m^*) := \begin{cases} X_{-m} & \text{if} \quad m \leq 0, \text{ and } r = 0 \\ -\partial/\partial X_{-m} & \text{otherwise,} \end{cases} \tag{25}$$

and $\rho_r(\mathbf{1}) = 1$. These two representations can be constructed using induction: For $r = 0$ the representation ρ_0 is the $\hat{\mathfrak{a}}$-module generated by $1 =: |0\rangle$, where

$$a_m|0\rangle = a_m^1|0\rangle = 0, \quad m \geq 0, \quad a_m^*|0\rangle = a_m^{1*}|0\rangle = 0, \quad m > 0.$$

For $r = 1$ the representation ρ_1 is the $\hat{\mathfrak{a}}$-module generated by $1 =: |0\rangle$, where

$$a_m^*|0\rangle = a_m^{1*}|0\rangle = 0, \quad m \in \mathbb{Z}.$$

If we write

$$\alpha(z) := \sum_{n\in\mathbb{Z}} a_n z^{-n-1}, \quad \alpha^*(z) := \sum_{n\in\mathbb{Z}} a_n^* z^{-n},$$

and

$$\alpha^1(z) := \sum_{n\in\mathbb{Z}} a_n^1 z^{-n-1}, \quad \alpha^{1*}(z) := \sum_{n\in\mathbb{Z}} a_n^{1*} z^{-n},$$

then

$$[\alpha(z), \alpha(w)] = [\alpha^*(z), \alpha^*(w)] = [\alpha^1(z), \alpha^1(w)] = [\alpha^{1*}(z), \alpha^{1*}(w)] = 0$$
$$[\alpha(z), \alpha^*(w)] = [\alpha^1(z), \alpha^{1*}(w)] = \mathbf{1}\delta(z/w).$$

Corresponding to these two representations there are two possible normal orderings: For $r = 0$ we use the usual normal ordering given by (18) and for $r = 1$ we define the *natural normal ordering* to be

$$\alpha(z)_+ = \alpha(z), \qquad \alpha(z)_- = 0$$
$$\alpha^1(z)_+ = \alpha^1(z), \qquad \alpha^1(z)_- = 0$$
$$\alpha^*(z)_+ = 0, \qquad \alpha^*(z)_- = \alpha^*(z),$$
$$\alpha^{1*}(z)_+ = 0, \qquad \alpha^{1*}(z)_- = \alpha^{1*}(z),$$

This means in particular that for $r = 0$ we get

$$\lfloor \alpha(z)\alpha^*(w) \rfloor = \textstyle\sum_{m\geq 0} \delta_{m+n,0} z^{-m-1} w^{-n} = \delta_-(z/w) = \iota_{z,w}\left(\frac{1}{z-w}\right) \quad (26)$$
$$\lfloor \alpha^*(z)\alpha(w) \rfloor = -\textstyle\sum_{m\geq 1} \delta_{m+n,0} z^{-m} w^{-n-1} = -\delta_+(w/z) = \iota_{z,w}\left(\frac{1}{w-z}\right) \quad (27)$$

(where $\iota_{z,w}$ Taylor series expansion in the " region" $|z| > |w|$), and for $r = 1$

$$\lfloor \alpha\alpha^* \rfloor = [\alpha(z)_-, \alpha^*(w)] = 0 \quad (28)$$
$$\lfloor \alpha^*\alpha \rfloor = [\alpha^*(z)_-, \alpha(w)] = -\sum_{\in\mathbb{Z}} \delta_{m+n,0} z^{-m} w^{-n-1} = -\delta(w/z), \quad (29)$$

where similar results hold for α^1. Notice that in both cases we have

$$[\alpha(z), \alpha^*(w)] = \lfloor \alpha(z)\alpha^*(w) \rfloor - \lfloor \alpha^*(w)\alpha(z) \rfloor = \delta(z/w).$$

The following two Theorems are needed for the proof of our main result:

Theorem 6 (Taylor's Theorem, [Kac98], 2.4.3). *Let $a(z)$ be a formal distribution. Then in the region $|z - w| < |w|$,*

$$a(z) = \sum_{j=0}^{\infty} \partial_w^{(j)} a(w)(z-w)^j. \quad (30)$$

Theorem 7 ([Kac98], Theorem 2.3.2). *Set $\mathbb{C}[\mathbf{x}] = \mathbb{C}[x_n, x_n^1 | n \in \mathbb{Z}]$ and $\mathbb{C}[\mathbf{y}] = C[y_m, y_m^1 | m \in \mathbb{N}^*]$. Let $a(z)$ and $b(z)$ be formal distributions with coefficients in the associative algebra $\mathrm{End}(\mathbb{C}[\mathbf{x}] \otimes \mathbb{C}[\mathbf{y}])$ where we are using the usual normal ordering. The following are equivalent:*

(i) There exists $c^j(w) \in \mathrm{End}(\mathbb{C}[\mathbf{x}] \otimes \mathbb{C}[\mathbf{y}])[[w, w^{-1}]]$ such that

$$[a(z), b(w)] = \sum_{j=0}^{N-1} \partial_w^{(j)} \delta(z-w) c^j(w).$$

(ii) $\lfloor ab \rfloor = \sum_{j=0}^{N-1} \iota_{z,w} \left(\frac{1}{(z-w)^{j+1}} \right) c^j(w).$

So the singular part of the *operator product expansion*

$$\lfloor ab \rfloor = \sum_{j=0}^{N-1} \iota_{z,w} \left(\frac{1}{(z-w)^{j+1}} \right) c^j(w)$$

completely determines the bracket of mutually local formal distributions $a(z)$ and $b(w)$. In the physics literature one writes

$$a(z)b(w) \sim \sum_{j=0}^{N-1} \frac{c^j(w)}{(z-w)^{j+1}}.$$

5.2 DJKM Heisenberg Algebra

Set

$$\psi'_{ij}(c) = \begin{cases} \mathbf{1}_{i+j-2} & \text{for} \quad i+j = 1, 0, -1, -2 \\ P_{-3,i+j-2}(c)(\mathbf{1}_{-3} + c\mathbf{1}_{-1}) & \text{for} \quad i+j = -1+2n \geq 3, \\ P_{-3,i+j-2}(c)(c\mathbf{1}_{-3} + \mathbf{1}_{-1}) & \text{for} \quad i+j = 1-2n \leq -3, \\ P_{-4,|i+j|-2}(c)\mathbf{1}_{-4} + P_{-2,|i+j|-2}(c)\mathbf{1}_{-2} & \text{for} \quad |i+j| = 2n \geq 2, \end{cases} \tag{31}$$

for $n \in \mathbb{Z}$ and $\psi'(c, w) = \sum_{n \in \mathbb{Z}} \psi'_n(c) w^n$ for $\psi'_{i+j}(c) := \psi'_{ij}(c)$.

The Cartan subalgebra $\mathfrak{h}$ tensored with $\mathcal{R}$ generates a subalgebra of $\hat{\mathfrak{g}}$ which is an extension of an oscillator algebra. This extension motivates the following definition: The Lie algebra with generators b_m, b_m^1, $m \in \mathbb{Z}$, $\mathbf{1}_i$, $i \in \{0, -1, -2, -3, -4\}$ and relations

$$[b_m, b_n] = (n-m)\,\delta_{m+n,0}\mathbf{1}_0 = -2m\,\delta_{m+n,0}\mathbf{1}_0 \tag{32}$$

$$[b^1_m, b^1_n] = 2\left((n+2)\delta_{m+n,-4} - 2c(n+1)\delta_{m+n,-2} + n\delta_{m+n,0}\right)\mathbf{1}_0 \tag{33}$$

$$[b^1_m, b_n] = 2n\psi'_{mn}(c) \tag{34}$$

$$[b_m, \mathbf{1}_i] = [b^1_m, \mathbf{1}_i] = 0, \tag{35}$$

give the appellation the *DJKM (affine) Heisenberg algebra* and denote it by $\hat{\mathfrak{b}}_3$.

If we introduce the formal distributions

$$\beta(z) := \sum_{n\in\mathbb{Z}} b_n z^{-n-1}, \quad \beta^1(z) := \sum_{n\in\mathbb{Z}} b^1_n z^{-n-1} = \sum_{n\in\mathbb{Z}} b_{n+\frac{1}{2}} z^{-n-1}. \tag{36}$$

(where $b_{n+\frac{1}{2}} := b^1_n$) then using calculations done earlier for DJKM Lie algebra we can see that the relations above can be rewritten in the form

$$[\beta(z), \beta(w)] = 2\mathbf{1}_0\partial_z\delta(z/w) = -2\partial_w\delta(z/w)\mathbf{1}_0$$

$$[\beta^1(z), \beta^1(w)] = -2\left(P(w)\partial_w\delta(z/w) + \frac{1}{2}\partial_w P(w)\delta(z/w)\right)\mathbf{1}_0$$

$$[\beta^1(z), \beta(w)] = 2\partial_w\psi'(c,w)\delta(z/w) - 2w\psi'(c,w)\partial_w\delta(z/w).$$

Set

$$\hat{\mathfrak{h}}_3^{\pm} := \sum_{n\gtrless 0}\left(\mathbb{C}b_n + \mathbb{C}b^1_n\right),$$

$$\hat{\mathfrak{h}}_3^0 := \mathbb{C}\mathbf{1}_0 \oplus \mathbb{C}\mathbf{1}_{-1} \oplus \mathbb{C}\mathbf{1}_{-2} \oplus \mathbb{C}\mathbf{1}_{-3} \oplus \mathbb{C}\mathbf{1}_{-4} \oplus \mathbb{C}b_0 \oplus \mathbb{C}b^1_0.$$

We introduce a Borel type subalgebra

$$\hat{\mathfrak{b}}_3 = \hat{\mathfrak{h}}_3^+ \oplus \hat{\mathfrak{h}}_3^0.$$

Due to the defining relations above one can see that $\hat{\mathfrak{b}}_3$ is a subalgebra.

Lemma 5.1 *Let $\mathcal{V} = \mathbb{C}\mathbf{v}_0 \oplus \mathbb{C}\mathbf{v}_1$ be a two-dimensional representation of $\hat{\mathfrak{h}}_3^+\mathbf{v}_i = 0$ for $i = 0, 1$. Suppose $\lambda, \mu, \nu, \varkappa, \chi_{-1}, \chi_{-2}, \chi_{-3}, \chi_{-4}, \kappa_0 \in \mathbb{C}$ are such that*

$$b_0\mathbf{v}_0 = \lambda\mathbf{v}_0, \qquad b_0\mathbf{v}_1 = \lambda\mathbf{v}_1$$

$$b^1_0\mathbf{v}_0 = \mu\mathbf{v}_0 + \nu\mathbf{v}_1, \qquad b^1_0\mathbf{v}_1 = \varkappa\mathbf{v}_0 + \mu\mathbf{v}_1$$

$$\mathbf{1}_j\mathbf{v}_i = \chi_j\mathbf{v}_i, \qquad \mathbf{1}_0\mathbf{v}_i = \kappa_0\mathbf{v}_i, \quad i = 0, 1, \quad j = -1, -2, -3, -4.$$

Then the above defines a representation of $\hat{\mathfrak{b}}_3$*. Not only that but also* $\chi_{-1} = \chi_{-2} = \chi_{-3} = \chi_{-4} = 0$ *and* $\psi'_{mn} = 0$*, for all* $m, n \in \mathbb{Z}$*.*

Proof. Since b_m acts by scalar multiplication for $m, n \geq 0$, the first defining relation (32) is satisfied for $m, n \geq 0$. The second relation (33) is also satisfied as the right-hand side is zero if $m \geq 0, n \geq 0$. If $n = 0$, then since b_0 acts by a scalar, the relation (34) leads to no condition on $\lambda, \mu, \nu, \varkappa, \chi_1, \kappa_0 \in \mathfrak{h}_3^0$. If $m \neq 0$ and $n \neq 0$, the third relation gives us

$$0 = b^1_m b_n \mathbf{v}_i - b_n b^1_m \mathbf{v}_i = [b^1_m, b_n]\mathbf{v}_i = 2n\psi'_{mn}\mathbf{v}_i = 0,$$

and then $\psi'_{mn} = 0$ for $n \neq 0$. Consequently $\chi_{-1} = \chi_{-2} = \chi_{-3} = \chi_{-4} = 0$ and $\psi'_{mn} = 0$ for all $m, n \in \mathbb{Z}$. □

Lemma 5.2. *The linear map* $\rho : \hat{\mathfrak{b}}_3 \to \mathrm{End}(\mathbb{C}[\mathbf{y}] \otimes \mathcal{V})$ *defined by*

$$\rho(b_n) = y_n \quad \text{for } n < 0 \tag{37}$$

$$\rho(b^1_n) = y^1_n + \delta_{n,-1}\partial_{y^1_{-3}}\chi_0 - \delta_{n,-3}\partial_{y^1_{-1}}\chi_0 \quad \text{for } n < 0 \tag{38}$$

$$\rho(b_n) = -2n\partial_{y_{-n}}\chi_0 \quad \text{for } n > 0 \tag{39}$$

$$\rho(b^1_n) = 2(n+2)\partial_{y^1_{-n-4}}\chi_0 - 4c(n+1)\partial_{y^1_{-n-2}}\chi_0 + 2n\partial_{y^1_{-n}}\chi_0 \quad \text{for } n > 0 \tag{40}$$

$$\rho(b_0) = \lambda \tag{41}$$

$$\rho(b^1_0) = 4\partial_{y^1_{-4}}\chi_0 - 2c\partial_{y^1_{-2}}\chi_0 + B^1_0. \tag{42}$$

is a representation of $\hat{\mathfrak{b}}_3$*.*

Proof. For $m, n > 0$, it is straight forward to see

$$[\rho(b_n), \rho(b_m)] = [\rho(b^1_n), \rho(b^1_m)] = 0,$$

and similarly for $m, n < 0$, $[\rho(b_n), \rho(b_m)] = 0$ and $[\rho(b^1_n), \rho(b^1_m)] = 0$ if $n \notin \{-1, -3\}$ and

$$\begin{aligned}
[\rho(b^1_{-1}), \rho(b^1_m)] &= [y^1_{-1} + \partial_{y^1_{-3}}\chi_0, y^1_m + \delta_{m,-1}\partial_{y^1_{-3}}\chi_0 - \delta_{m,-3}\partial_{y^1_{-1}}\chi_0] \\
&= -\delta_{m,-3}\chi_0[y^1_{-1}, \partial_{y^1_{-1}}]\chi_0 + \delta_{m,-3}[\partial_{y^1_{-3}}, y^1_{-3}]\chi_0 \\
&= -2\delta_{m,-3}\chi_0, \\
[\rho(b^1_{-3}), \rho(b^1_m)] &= [y^1_{-3} - \partial_{y^1_{-1}}\chi_0, y^1_m + \delta_{m,-1}\partial_{y^1_{-3}}\chi_0 - \delta_{m,-3}\partial_{y^1_{-1}}\chi_0] \\
&= \delta_{m,-1}\chi_0[y^1_{-3}, \partial_{y^1_{-3}}]\chi_0 - \delta_{m,-1}[\partial_{y^1_{-1}}, y^1_{-1}]\chi_0 \\
&= 2\delta_{m,-1}\chi_0,
\end{aligned}$$

$$[\rho(b_0^1), \rho(b_m^1)] = [4\partial_{y_{-4}^1}\chi_0 - 2c\partial_{y_{-2}^1}\chi_0, y_m^1 + \delta_{m,-1}\partial_{y_{-3}^1}\chi_0 - \delta_{m,-3}\partial_{y_{-1}^1}\chi_0]$$
$$= -4\delta_{m,-4}\chi_0 + 2c\delta_{m,-2}\chi_0.$$

For $m > 0$ and $n \leq 0$ we have

$$[\rho(b_m), \rho(b_n)] = [-2m\partial_{y_{-m}}\chi_0, y_n] = -2m\delta_{m,-n}\chi_0,$$
$$[\rho(b_m), \rho(b_n^1)] = [-2m\partial_{y_{-m}}\chi_0, y_n^1 + \delta_{n,-1}\partial_{y_{-3}^1}\chi_0 - \delta_{n,-3}\partial_{y_{-1}^1}\chi_0] = 0,$$
$$[\rho(b_m^1), \rho(b_n^1)] = [2(n+2)\partial_{y_{-n-4}^1}\chi_0 - 4c(n+1)\partial_{y_{-n-2}^1}\chi_0 + 2n\partial_{y_{-n}^1}\chi_0$$
$$, y_n^1 + \delta_{n,-1}\partial_{y_{-3}^1}\chi_0 - \delta_{n,-3}\partial_{y_{-1}^1}\chi_0]$$
$$= 2(n+2)\delta_{m+n,-4}\chi_0 - 4c(n+1)\delta_{m+n,-2}\chi_0 + 2n\delta_{m+n,0}\chi_0,$$
$$[\rho(b_m^1), \rho(b_n)] = [2(m+2)\partial_{y_{-m-4}^1}\chi_0 - 4c(m+1)\partial_{y_{-m-2}^1}\chi_0 + 2m\partial_{y_{-m}^1}\chi_0,$$
$$y_n + \delta_{n,-1}\partial_{y_{-3}^1}\chi_0 - \delta_{n,-3}\partial_{y_{-1}^1}\chi_0]$$
$$= 0.$$

□

6 Two Realizations of DJKM Algebra $\hat{\mathfrak{g}}$

Recall

$$P(w) = w^4 - 2cw^2 + 1. \tag{43}$$

Our main result is the following.

Theorem 8. *Fix $r \in \{0, 1\}$, which then fixes the corresponding normal ordering convention defined in the previous section. Set $\hat{\mathfrak{g}} = (\mathfrak{sl}(2,\mathbb{C}) \otimes \mathcal{R}) \oplus \mathbb{C}\omega_0 \oplus \mathbb{C}\omega_{-1} \oplus \mathbb{C}\omega_{-2} \oplus \mathbb{C}\omega_{-3} \oplus \mathbb{C}\omega_{-4}$ and assume that $\chi_0 \in \mathbb{C}$ and $\mathcal{V}$ as in Lemma 5.1. Then using (24), (25) and Lemma 5.2, the following defines a representation of DJKM algebra $\mathfrak{g}$ on $\mathbb{C}[\mathbf{x}] \otimes \mathbb{C}[\mathbf{y}] \otimes \mathcal{V}$:*

$$\tau(\omega_{-1}) = \tau(\omega_{-2}) = \tau(\omega_{-3}) = \tau(\omega_{-4}) = 0, \qquad \tau(\omega_0) = \chi_0 = \kappa_0 + 4\delta_{r,0},$$
$$\tau(f(z)) = -\alpha, \qquad \tau(f^1(z)) = -\alpha^1,$$
$$\tau(h(z)) = 2\left(: \alpha\alpha^* : + : \alpha^1\alpha^{1*} :\right) + \beta,$$
$$\tau(h^1(z)) = 2\left(: \alpha^1\alpha^* : + P(z) : \alpha\alpha^{1*} :\right) + \beta^1,$$
$$\tau(e(z)) =: \alpha(\alpha^*)^2 : + P(z) : \alpha(\alpha^{1*})^2 : + 2 : \alpha^1\alpha^*\alpha^{1*} : + \beta\alpha^* + \beta^1\alpha^{1*} + \chi_0\partial\alpha^*$$

Table 1 Defining relations

$[\cdot_\lambda \cdot]$	$f(w)$	$f^1(w)$	$h(w)$	$h^1(w)$	$e(w)$	$e^1(w)$
$f(z)$	0	0	$*$	$*$	$*$	$*$
$f^1(z)$		0	$*$	$*$	$*$	$*$
$h(z)$			$*$	$*$	$*$	$*$
$h^1(z)$				$*$	$*$	$*$
$e(z)$					0	0
$e^1(z)$						0

$$\tau(e^1(z)) = \alpha^1\alpha^*\alpha^* + P(z)\left(\alpha^1(\alpha^{1*})^2 + 2 : \alpha\alpha^*\alpha^{1*} :\right)$$
$$+ \beta^1\alpha^* + P(z)\beta\alpha^{1*} + \chi_0\left(P(z)\partial_z\alpha^{1*} + \frac{1}{2}\partial_z P(z)\alpha^{1*}\right).$$

Proof. The proof is very similar to the proof of Theorem 5.1 in [BCF09] and Theorem 5.1 in [CJ]. We need to check that the following table is preserved under τ.

Here $*$ indicates nonzero formal distributions that are obtained from the defining relations (21), (22), and (23). The proof is carried out using Wick's theorem and Taylor's theorem. We are going to make use of V. Kac's λ-notation (see [Kac98] section 2.2 for some of its properties) used in operator product expansions. If $a(z)$ and $b(w)$ are formal distributions, then

$$[a(z), b(w)] = \sum_{j=0}^{\infty} \frac{(a_{(j)}b)(w)}{(z-w)^{j+1}}$$

is transformed under the *formal Fourier transform*

$$F^{\lambda}_{z,w}a(z,w) = \mathrm{Res}_z e^{\lambda(z-w)}a(z,w),$$

into the sums

$$[a_\lambda b] = \sum_{j=0}^{\infty} \frac{\lambda^j}{j!} a_{(j)}b.$$

$$[\tau(f)_\lambda \tau(f)] = 0, \quad [\tau(f)_\lambda \tau(f^1)] = 0, \quad [\tau(f^1)_\lambda \tau(f^1)] = 0$$
$$[\tau(f)_\lambda \tau(h)] = -\left[\alpha_\lambda\left(2\left(\alpha\alpha^* + \alpha^1\alpha^{1*}\right) + \beta\right)\right] = -2\alpha = 2\tau(f),$$
$$[\tau(f)_\lambda \tau(h^1)] = -[\alpha_\lambda\left(2\left(\alpha^1\alpha^* + P\alpha\alpha^{1*}\right) + \beta^1\right)] = -2\alpha^1 = 2\tau(f^1),$$
$$[\tau(f)_\lambda \tau(e)] = -2\left(: \alpha\alpha^* : + : \alpha^1\alpha^{1*} :\right) - \beta - \chi_0\lambda = -\tau(h) - \chi_0\lambda$$

$$[\tau(f)_\lambda \tau(e^1)] = -2\left(:\alpha^1\alpha^*: + P:\alpha\alpha^{1*}:\right) - \beta^1 = -\tau(h^1).$$

$$[\tau(f^1)_\lambda \tau(h)] = -[\alpha^1_\lambda\left(2\left(:\alpha\alpha^*: + :\alpha^1\alpha^{1*}:\right) + \beta\right)] = -2\alpha^1 = 2\tau(f^1),$$

$$[\tau(f^1)_\lambda \tau(h^1)] = -[\alpha^1_\lambda\left(2\left(:\alpha^1\alpha^*: + P:\alpha\alpha^{1*}:\right) + \beta^1\right)] = -2P\alpha^1 = 2P\tau(f^1),$$

$$[\tau(f^1)_\lambda \tau(e)] = -\left(2P:\alpha\alpha^{1*}: + 2:\alpha^1\alpha^*: + \beta^1\right) = -\tau(h^1)$$

$$\begin{aligned}[\tau(f^1)_\lambda \tau(e^1)] &= -\left(P\left(2\left(:\alpha^1\alpha^{1*}: + :\alpha\alpha^*:\right) + \beta + \chi_0\lambda\right) + \frac{1}{2}\chi_0\partial P\right)\\ &= -\left(P\tau(h) + P\chi_0\lambda + \chi_0\frac{1}{2}\partial P\right).\end{aligned}$$

Note that $:a(z)b(z):$ and $:b(z)a(z):$ are usually not equal, but $:\alpha^1(w)\alpha^{*1}(w):=:\alpha^{1*}(w)\alpha^1(w):$ and $:\alpha(w)\alpha^*(w):=:\alpha^*(w)\alpha(w):$. Thus we calculate

$$\begin{aligned}[\tau(h)_\lambda \tau(h)] &= \left[\left(2\left(:\alpha\alpha^*: + :\alpha^1\alpha^{1*}:\right) + \beta\right)_\lambda \left(2\left(:\alpha\alpha^*: + :\alpha^1\alpha^{1*}:\right) + \beta\right)\right]\\ &= 4\left(-:\alpha\alpha^*: + :\alpha^*\alpha: - :\alpha^1\alpha^{1*}: + :\alpha^{1*}\alpha^1:\right) - 8\delta_{r,0}\lambda + [\beta_\lambda\beta]\\ &= -2(4\delta_{r,0} + \kappa_0)\lambda,\end{aligned}$$

which can be put into the form of (21):

$$\begin{aligned}[\tau(h(z)), \tau(h(w))] &= -2(4\delta_{r,0} + \kappa_0)\partial_w\delta(z/w) = -2\chi_0\partial_w\delta(z/w)\\ &= \tau\left(-2\omega_0\partial_w\delta(z/w)\right).\end{aligned}$$

Next we calculate

$$\begin{aligned}[\tau(h)_\lambda \tau(h^1)] &= \left[\left(2\left(:\alpha\alpha^*: + :\alpha^1\alpha^{1*}:\right) + \beta\right)_\lambda \left(2\left(:\alpha^1\alpha^*: + P:\alpha\alpha^{1*}:\right) + \beta^1\right)\right]\\ &= 4\left(\left(:\alpha^*\alpha^1: - :\alpha^1\alpha^*:\right) + P\left(-:\alpha\alpha^{1*}: + :\alpha^{1*}\alpha:\right)\right) + [\beta_\lambda\beta^1].\end{aligned}$$

Since $[a_n, a^{1*}_m] = [a^1_n, a^*_m] = 0$, we have

$$\left[\tau(h(z)), \tau(h^1(w))\right] = [\beta(z), \beta^1(w)] = 0.$$

As $\tau(\omega_1) = 0$, relation (23) is satisfied.

We continue with

$$[\tau(h^1)_\lambda \tau(h^1)] = 4P\left(-:\alpha\alpha^*:+:\alpha^{1*}\alpha^1:\right) + 4P\left(-:\alpha^1\alpha^{1*}:+:\alpha^*\alpha:\right)$$
$$- 8\delta_{r,0}P\lambda - 4\delta_{r,0}\partial P + [\beta^1_\lambda \beta^1]$$
$$= -8\delta_{r,0}P\lambda - 4\delta_{r,0}\partial P - 2\kappa_0(P\lambda + \frac{1}{2}\partial P),$$

yielding the relation

$$\left[\tau(h^1(z)), \tau(h^1(w))\right] = \tau(-(h,h)\omega_0 P\partial_w\delta(z/w) - \frac{1}{2}(h,h)\partial P\omega_0\delta(z/w))$$

Next we calculate the h's paired with the e's:

$$[\tau(h)_\lambda \tau(e)] = \Big[\left(2\left(:\alpha\alpha^*:+:\alpha^1\alpha^{1*}:\right)+\beta\right)_\lambda$$
$$:\alpha(\alpha^*)^2: + P:\alpha(\alpha^{1*})^2: + 2:\alpha^1\alpha^*\alpha^{1*}: + \beta\alpha^* + \beta^1\alpha^{1*} + \chi_0\partial\alpha^*\Big]$$
$$= 4:\alpha(\alpha^*)^2: - 2:\alpha(\alpha^*)^2: - 4\delta_{r,0}\alpha^*\lambda - 2P:\alpha(\alpha^{1*})^2: + 4:\alpha^*\alpha^1\alpha^{1*}:$$
$$+ 2\alpha^*\beta + 2\chi_0\alpha^*\lambda + 2\chi_0\partial\alpha^* + 4P:\alpha(\alpha^{1*})^2:$$
$$- 4\delta_{r,0}\alpha^*\lambda + 2\beta^1\alpha^{1*} - 2\lambda\alpha^*\kappa_0$$
$$= 2\tau(e)$$

and

$$[\tau(h^1)_\lambda \tau(e)] = 2:\alpha^1(\alpha^*)^2: + 2P:\alpha^1(\alpha^{1*})^2: + 4P:\alpha\alpha^*\alpha^{1*}: + 2\delta(z/w)\alpha^*\beta^1$$
$$+ 2P\beta\alpha^{1*} + 2P\chi_0\partial\alpha^{1*} + \partial P\alpha^{1*}\chi_0$$
$$= 2\tau(e^1)$$

Next we must calculate

$$[\tau(h)_\lambda \tau(e^1)] = 4:\alpha^1(\alpha^*)^2: - 4P\delta_{r,0}\alpha^{1*}\lambda + 2\alpha^*\beta^1 - 2\delta(z/w):\alpha^1\alpha^*\alpha^*:$$
$$- 4P\delta_{r,0}\alpha^{1*}\lambda + 2P:\alpha^{1*}\alpha^1\alpha^{1*}: + 4P:\alpha^{1*}\alpha\alpha^*: + 2P\beta\alpha^{1*}$$
$$+ 2\chi_0(P\alpha^{1*}\lambda + P\partial\alpha^{1*} + \frac{1}{2}\partial P\alpha^{1*}) - 2P\alpha^{1*}\kappa_0\lambda$$
$$= 2:\alpha^1(\alpha^*)^2: + 2P:\alpha^1(\alpha^{1*})^2: + 4P:\alpha\alpha^*\alpha^{1*}:$$
$$+ 2\beta^1\alpha^* + 2P\beta\alpha^{1*} + 2\chi_0\left(P\partial_w\alpha^{1*} + (1/2\partial P)\alpha^{1*}\right)$$
$$= 2\tau(e^1)$$

and the proof for $[\tau(h^1)_\lambda \tau(e^1)]$ is similar.

We prove the Serre relation for just one of the relations, $[\tau(e)_\lambda \tau(e^1)]$, and the proof of the others ($[\tau(e)_\lambda \tau(e)]$, $[\tau(e^1)_\lambda \tau(e^1)]$) are similar as the reader can verify.

$$\begin{aligned}
[\tau(e)_\lambda \tau(e^1)] = &\Big[:\alpha(\alpha^*)^2 :_\lambda \Big(:\alpha^1(\alpha^*)^2 : +2P :\alpha\alpha^*\alpha^{1*} : +\beta^1\alpha^*\Big)\Big] \\
&+ \Big[P:\alpha(\alpha^{1*})^2:_\lambda\Big(:\alpha^1(\alpha^*)^2 : +P\Big(:\alpha^1(\alpha^{1*})^2 : +2:\alpha\alpha^*\alpha^{1*}:\Big) + \beta^1\alpha^*\Big)\Big] \\
&+ \Big[2:\alpha^1\alpha^*\alpha^{1*}:_\lambda\Big(\alpha^1(\alpha^*)^2 + P\Big(:\alpha^1(\alpha^{1*})^2 : +2:\alpha\alpha^*\alpha^{1*}:\Big) + P\beta\alpha^{1*} \\
&\qquad + \chi_0\Big((w^4-2cw^2+1)\partial_w\alpha^{1*} + (2w^3-2cw)\alpha^{1*}\Big)\Big)\Big] \\
&+ \Big[\beta\alpha^*{}_\lambda\Big(2P:\alpha\alpha^*\alpha^{1*}: +\beta^1\alpha^* + P\beta\alpha^{1*}\Big)\Big] \\
&+ \Big[\beta^1\alpha^{1*}{}_\lambda\Big(:\alpha^1(\alpha^*)^2 : +P\alpha^1(\alpha^{1*})^2 + \beta^1\alpha^* + P\beta\alpha^{1*}\Big)\Big] \\
&+ \Big[\chi_0\partial\alpha^*{}_\lambda\Big(2P:\alpha\alpha^*\alpha^{1*}:\Big)\Big]
\end{aligned}$$

$$\begin{aligned}
= &\,2:\alpha^1\alpha^*(\alpha^*)^2: +2P:\alpha(\alpha^*)^2\alpha^{1*}: -4P:\alpha(\alpha^*)^2\alpha^{1*}: -4\delta_{r,0}P:\alpha^*\alpha^{1*}:\lambda \\
&\qquad - 4\delta_{r,0}P:\partial(\alpha^*)\alpha^{1*}: +\beta^1(\alpha^*)^2 \\
&- 2P:\alpha(\alpha^*)^2\alpha^{1*}: +2P\alpha^1\alpha^*(\alpha^{1*})^2 \\
&- 4\delta_{r,0}P:\alpha^{1*}\alpha^*:\lambda - 4\delta_{r,0}\partial P:\alpha^{1*}\alpha^*: -4\delta_{r,0}P:\partial\alpha^{1*}\alpha^*: \\
&- 2P^2:\alpha(\alpha^{1*})^3: +2P^2:\alpha(\alpha^{1*})^3: +P\beta^1(\alpha^{1*})^2 \\
&- 2:\alpha^1\alpha^*(\alpha^*)^2: +4P:\alpha^1\alpha^*(\alpha^{1*})^2: -2P:\alpha^1\alpha^*(\alpha^{1*})^2: -4\delta_{r,0}P:\alpha^*\alpha^{1*}:\lambda \\
&- 4\delta_{r,0}P:\partial(\alpha^*)\alpha^{1*}: +4P:\alpha(\alpha^*)^2\alpha^{1*}: -4P:\alpha^1\alpha^*(\alpha^{1*})^2: \\
&- 4\delta_{r,0}P:\alpha^*\alpha^{1*}:\lambda - 4\delta_{r,0}P:\alpha^*\partial\alpha^{1*}: +2P\beta:\alpha^*\alpha^{1*}: \\
&+ 2\chi_0\Big(P:\partial\alpha^*\alpha^{1*}: +P:\alpha^*\partial\alpha^{1*}: +P:\alpha^*\alpha^{1*}:\lambda + \frac{1}{2}(\partial P):\alpha^*\alpha^{1*}:\Big) \\
&- 2P\beta\alpha^*\alpha^{1*} - 2\kappa_0 P\alpha^*\alpha^{1*}\lambda - 2\kappa_0 P\partial\alpha^*\alpha^{1*} \\
&- \beta^1(\alpha^*)^2 - P\beta^1(\alpha^{1*})^2 - \kappa_0\Big(2P\alpha^*\alpha^{1*}\lambda + 2P\alpha^*\partial\alpha^{1*} + \partial P\alpha^*\alpha^{1*}\Big) \\
&+ 2\chi_0 P\alpha^*\alpha^{1*}\lambda
\end{aligned}$$

$$\begin{aligned}
= &-4\delta_{r,0}P:\alpha^*\alpha^{1*}:\lambda - 4\delta_{r,0}P:\partial(\alpha^*)\alpha^{1*}: \\
&\qquad + \chi_1\Big(2:\alpha^*\partial\alpha^*: + :(\alpha^*)^2:\lambda\Big) \\
&- 4\delta_{r,0}P:\alpha^{1*}\alpha^*:\lambda - 4\delta_{r,0}\partial P:\alpha^{1*}\alpha^*: -4\delta_{r,0}P:\partial\alpha^{1*}\alpha^*: \\
&- 4\delta_{r,0}P:\alpha^*\alpha^{1*}:\lambda - 4\delta_{r,0}P:\partial(\alpha^*)\alpha^{1*}:
\end{aligned}$$

$$
\begin{aligned}
&- 4\delta_{r,0}P : \alpha^*\alpha^{1*} : \lambda - 4\delta_{r,0}P : \alpha^*\partial\alpha^{1*} : \\
&+ 2\chi_0\Big(P : \partial\alpha^*\alpha^{1*} : + P : \alpha^*\partial\alpha^{1*} : + P : \alpha^*\alpha^{1*} : \lambda + \frac{1}{2}(\partial P) : \alpha^*\alpha^{1*} : \Big) \\
&- \kappa_0 P\alpha^*\alpha^{1*}\lambda - \kappa_0 P\partial\alpha^*\alpha^{1*} \\
&- \kappa_0\left(2P\alpha^*\alpha^{1*}\lambda + 2P\alpha^*\partial\alpha^{1*} + \partial P\alpha^*\alpha^{1*}\right) \\
&- 2\chi_0 P\alpha^*\alpha^{1*}\lambda \\
= {} & 0. \qquad \square
\end{aligned}
$$

Acknowledgements The first author would like to thank the other two authors and the University of São Paulo for hosting him while he visited Brazil in June of 2013 where part of this work was completed. The second author was partially supported by Fapesp (2010/50347-9) and CNPq (301743/2007-0). The third author was supported by FAPESP (2012/02459-8).

References

[BCF09] André Bueno, Ben Cox, and Vyacheslav Futorny, Free field realizations of the elliptic affine Lie algebra $\mathfrak{sl}(2, \mathbf{R}) \oplus (\Omega_R/d\mathrm{R})$. *J. Geom. Phys.*, 59(9):1258–1270, 2009.

[BI82] Joaquin Bustoz and Mourad E. H. Ismail, The associated ultraspherical polynomials and their q-analogues. *Canad. J. Math.*, 34(3):718–736, 1982.

[Bre94a] Murray Bremner, Generalized affine Kac-Moody Lie algebras over localizations of the polynomial ring in one variable. *Canad. Math. Bull.*, 37(1):21–28, 1994.

[Bre94b] Murray Bremner, Universal central extensions of elliptic affine Lie algebras. *J. Math. Phys.*, 35(12):6685–6692, 1994.

[Bre95] Murray Bremner, Four-point affine Lie algebras. *Proc. Amer. Math. Soc.*,123(7): 1981–1989, 1995.

[BS83] N. N. Bogoliubov and D. V. Shirkov, *Quantum Fields*. Benjamin/Cummings Publishing Co. Inc. Advanced Book Program, Reading, MA, 1983. Translated from the Russian by D. B. Pontecorvo.

[BT07] Georgia Benkart and Paul Terwilliger, The universal central extension of the three-point $\mathfrak{sl}_2$ loop algebra. *Proc. Amer. Math. Soc.*, 135(6):1659–1668, 2007.

[CF11] Ben Cox and Vyacheslav Futorny, DJKM algebras I: their universal central extension. *Proc. Amer. Math. Soc.*, 139(10):3451–3460, 2011.

[CFT13] Ben Cox, Vyacheslav Futorny, and Juan A. Tirao, DJKM algebras and non-classical orthogonal polynomials. *J. Differential Equations*, 255(9):2846–2870, 2013.

[CJ] Ben L. Cox and Elizabeth Jurisich, Realizations of the three point Lie algebra $\mathfrak{sl}(2, R) \oplus (\Omega_R/dR)$. arXiv:1303.6973.

[Cox08] Ben Cox, Realizations of the four point affine Lie algebra $\mathfrak{sl}(2, R)\oplus(\Omega_R/dR)$. *Pacific J. Math.*, 234(2):261–289, 2008.

[DJKM83] Etsurō Date, Michio Jimbo, Masaki Kashiwara, and Tetsuji Miwa, Landau-Lifshitz equation: solitons, quasiperiodic solutions and infinite-dimensional Lie algebras. *J. Phys. A*, 16(2):221–236, 1983.

[EFK98] Pavel I. Etingof, Igor B. Frenkel, and Alexander A. Kirillov, Jr., *Lectures on representation theory and Knizhnik-Zamolodchikov equations*, Volume 58 of Mathematical Surveys and Monographs. American Mathematical Society, Providence, RI, 1998.

[FBZ01] Edward Frenkel and David Ben-Zvi, *Vertex algebras and algebraic curves*, Volume 88, Mathematical Surveys and Monographs. American Mathematical Society, Providence, RI, 2001.

[FF90] Boris L. Feĭgin and Edward V. Frenkel, Affine Kac-Moody algebras and semi-infinite flag manifolds. *Comm. Math. Phys.*, 128(1):161–189, 1990.

[FF99] Boris Feigin and Edward Frenkel, Integrable hierarchies and Wakimoto modules. In *Differential topology, infinite-dimensional Lie algebras, and applications*, Volume 194, Amer. Math. Soc. Transl. Ser. 2, pages 27–60. Amer. Math. Soc., Providence, RI, 1999.

[Fre05] Edward Frenkel, Wakimoto modules, opers and the center at the critical level. *Adv. Math.*, 195(2):297–404, 2005.

[Fre07] Edward Frenkel, Langlands correspondence for loop groups, Volume 103, Cambridge Studies in Advanced Mathematics. Cambridge University Press, Cambridge, 2007.

[FS05] Alice Fialowski and Martin Schlichenmaier, Global geometric deformations of current algebras as Krichever-Novikov type algebras. *Comm. Math. Phys.*, 260(3):579–612, 2005.

[FS06] Alice Fialowski and Martin Schlichenmaier, Global geometric deformations of the virasoro algebra, current and affine algebras by Krichever-Novikov type algebra. *math.QA/0610851*, 2006.

[Hua98] Kerson Huang, *Quantum field theory*. John Wiley & Sons Inc., New York, 1998. From operators to path integrals.

[Ism05] Mourad E. H. Ismail, *Classical and quantum orthogonal polynomials in one variable*, Volume 98, Encyclopedia of Mathematics and its Applications. Cambridge University Press, Cambridge, 2005. With two chapters by Walter Van Assche, with a foreword by Richard A. Askey.

[JK85] H. P. Jakobsen and V. G. Kac, A new class of unitarizable highest weight representations of infinite-dimensional Lie algebras. In *Nonlinear equations in classical and quantum field theory* (Meudon/Paris, 1983/1984), pages 1–20. Springer, Berlin, 1985.

[Kac98] Victor Kac, *Vertex Algebras for Beginners*. American Mathematical Society, Providence, RI, second edition, 1998.

[Kas84] Christian Kassel, Kähler differentials and coverings of complex simple Lie algebras extended over a commutative algebra. In *Proceedings of the Luminy conference on algebraic K-theory* (Luminy, 1983), Volume 34, pp 265–275, 1984.

[KL82] C. Kassel and J.-L. Loday, Extensions centrales d'algèbres de Lie. *Ann. Inst. Fourier (Grenoble)*, 32(4):119–142 (1983), 1982.

[KL91] David Kazhdan and George Lusztig, Affine Lie algebras and quantum groups. *Internat. Math. Res. Notices*, (2):21–29, 1991.

[KL93] D. Kazhdan and G. Lusztig, Tensor structures arising from affine Lie algebras. I, II. *J. Amer. Math. Soc.*, 6(4):905–947, 949–1011, 1993.

[KN87a] Igor Moiseevich Krichever and S. P. Novikov, Algebras of Virasoro type, Riemann surfaces and strings in Minkowski space. *Funktsional. Anal. i Prilozhen.*, 21(4):47–61, 96, 1987.

[KN87b] Igor Moiseevich Krichever and S. P. Novikov, Algebras of Virasoro type, Riemann surfaces and the structures of soliton theory. *Funktsional. Anal. i Prilozhen.*, 21(2):46–63, 1987.

[KN89] Igor Moiseevich Krichever and S. P. Novikov, Algebras of Virasoro type, the energy-momentum tensor, and operator expansions on Riemann surfaces. *Funktsional. Anal. i Prilozhen.*, 23(1):24–40, 1989.

[MN99] Atsushi Matsuo and Kiyokazu Nagatomo, *Axioms for a vertex algebra and the locality of quantum fields*, Volume 4, *MSJ Memoirs*. Mathematical Society of Japan, Tokyo, 1999.

[Sch03a] Martin Schlichenmaier, Higher genus affine algebras of Krichever-Novikov type. *Mosc. Math. J.*, 3(4):1395–1427, 2003.

[Sch03b] Martin Schlichenmaier, Local cocycles and central extensions for multipoint algebras of Krichever-Novikov type. *J. Reine Angew. Math.*, 559:53–94, 2003.

[She03] O. K. Sheinman, Second-order Casimirs for the affine Krichever-Novikov algebras $\widehat{\mathfrak{gl}}_{g,2}$ and $\widehat{\mathfrak{sl}}_{g,2}$. In *Fundamental mathematics today (Russian)*, pp 372–404. Nezavis. Mosk. Univ., Moscow, 2003.

[She05] O. K. Sheinman, Highest-weight representations of Krichever-Novikov algebras and integrable systems. *Uspekhi Mat. Nauk*, 60(2(362)):177–178, 2005.

[SS98] M. Schlichenmaier and O. K. Scheinman, The Sugawara construction and Casimir operators for Krichever-Novikov algebras. *J. Math. Sci. (New York)*, 92(2):3807–3834, 1998. Complex analysis and representation theory, 1.

[SS99] M. Shlikhenmaier and O. K. Sheinman, The Wess-Zumino-Witten-Novikov theory, Knizhnik-Zamolodchikov equations, and Krichever-Novikov algebras. *Uspekhi Mat. Nauk*, 54(1(325)):213–250, 1999.

[SV90] V. V. Schechtman and A. N. Varchenko, Hypergeometric solutions of Knizhnik-Zamolodchikov equations. *Lett. Math. Phys.*, 20(4):279–283, 1990.

[Wak86] Minoru Wakimoto, Fock representations of the affine Lie algebra $A_1^{(1)}$. *Comm. Math. Phys.*, 104(4):605–609, 1986.

[Wim87] Jet Wimp, Explicit formulas for the associated Jacobi polynomials and some applications. *Canad. J. Math.*, 39(4):983–1000, 1987.

The Deformation Complex is a Homotopy Invariant of a Homotopy Algebra

Vasily Dolgushev and Thomas Willwacher

Well, mathematician X likes to write formulas...
Alexander Beilinson

Abstract To a homotopy algebra one may associate its deformation complex, which is naturally a differential graded Lie algebra. We show that ∞-quasi-isomorphic homotopy algebras have L_∞-quasiisomorphic deformation complexes by an explicit construction.

Key words Algebraic operads • Homotopy algebras

Mathematics Subject Classification (2010): 18D50.

1 Introduction

Given two homotopy algebras $\mathcal{A}$, $\mathcal{B}$ of a certain type (e.g., L_∞- or A_∞-algebras), we may define their deformation complexes Def($\mathcal{A}$) and Def($\mathcal{B}$) [11], which are differential graded Lie algebras. Suppose that $\mathcal{A}$ and $\mathcal{B}$ are quasiisomorphic. For example, there may be an L_∞- or A_∞-quasiisomorphism $\mathcal{A} \to \mathcal{B}$. It is natural to ask whether in this case the deformation complexes Def($\mathcal{A}$) and Def($\mathcal{B}$) are

V.D. acknowledges the NSF grant DMS-1161867 and the grant FASI RF 14.740.11.0347. T.W. thanks the Harvard Society of Fellows and the Swiss National Science Foundation (grant PDAMP2_137151) for their support.

V. Dolgushev (✉)
Department of Mathematics, Temple University, Wachman Hall Rm. 638, 1805 N. Broad St., Philadelphia, PA, 19122 USA
e-mail: vald@temple.edu

T. Willwacher
Department of Mathematics, ETH Zürich, Rämistrasse 101, 8092 Zürich, Switzerland
e-mail: thomas.willwacher@math.ethz.ch

G. Mason et al. (eds.), *Developments and Retrospectives in Lie Theory: Algebraic Methods*, Developments in Mathematics 38,
DOI 10.1007/978-3-319-09804-3_6

quasiisomorphic as L_∞-algebras, and whether a quasiisomorphism may be written down in a (sufficiently) functorial way. The answer to the above question is (not surprisingly) yes, as is probably known to the experts. However, the authors were not able to find a proof of this statement in the literature in the desired generality.

The modest purpose of this paper is fill in this gap by presenting the construction of an explicit sequence of quasiisomorphisms connecting $\mathrm{Def}(\mathcal{A})$ with $\mathrm{Def}(\mathcal{B})$.

This paper is organized as follows. After a brief description of our construction, we recall in Sect. 2, the necessary prerequisites about homotopy algebras. Section 3 is the core of this paper. In this section, we formulate the main statement (see Theorem 3.1), describe various auxiliary constructions, and finally prove Theorem 3.1 in Sect. 3.4. Section 4 is devoted to the notion of homotopy algebra and its deformation complex in the setting of dg sheaves on a topological space. In this section, we give a version of Theorem 3.1 (see Corollary 4.6) and describe its application.

1.1 The Construction in a Nutshell

For the reader who already knows some homotopy algebra, here is what we will do in this paper. First, the homotopy algebras of the type we consider are governed by some operad P. For example, for A_∞-algebras $\mathsf{P} = \mathsf{As}_\infty$ and for L_∞-algebras $\mathsf{P} = \mathsf{Lie}_\infty$. Providing P algebra structures on $\mathcal{A}$ and $\mathcal{B}$ is equivalent to providing operad maps $\mathsf{P} \to \mathsf{End}(\mathcal{A})$, $\mathsf{P} \to \mathsf{End}(\mathcal{B})$ into the endomorphism operads. The deformation complexes $\mathrm{Def}(\mathcal{A})$, $\mathrm{Def}(\mathcal{B})$ are by definition the deformation complexes of the operad maps $\mathrm{Def}(\mathcal{A}) = \mathrm{Def}(\mathsf{P} \to \mathsf{End}(\mathcal{A}))$, $\mathrm{Def}(\mathcal{B}) = \mathrm{Def}(\mathsf{P} \to \mathsf{End}(\mathcal{B}))$.

Similarly, one may define a two-colored operad HomP, whose algebras are triples $(\mathcal{A}, \mathcal{B}, F)$, where $\mathcal{A}$ and $\mathcal{B}$ are P algebras and F is a homotopy (∞-) morphism between them. Furthermore, given an ∞ quasiisomorphism $\mathcal{A} \rightsquigarrow \mathcal{B}$, we may build a colored operad map $\mathsf{HomP} \to \mathsf{End}(\mathcal{A}, \mathcal{B})$ into the colored endomorphism operad. One may build a deformation complex $\mathrm{Def}(\mathsf{HomP} \to \mathsf{End}(\mathcal{A}, \mathcal{B}))$, which is an L_∞-algebra. Furthermore, there are natural maps

$$\mathrm{Def}(\mathcal{A}) \leftarrow \mathrm{Def}(\mathsf{HomP} \to \mathsf{End}(\mathcal{A}, \mathcal{B})) \to \mathrm{Def}(\mathcal{B})$$

which one may check to be quasiisomorphisms. Hence this zigzag constitutes desired explicit and natural quasiisomorphisms of L_∞-algebras.

2 Preliminaries

The base field $\mathbb{K}$ has characteristic zero. The underlying symmetric monoidal category is the category of unbounded cochain complexes of $\mathbb{K}$-vector spaces. We will use the notation and conventions about labeled planar trees from [5]. In

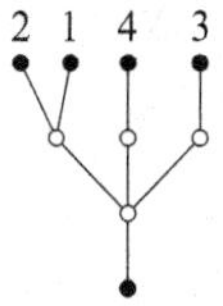

Fig. 1 An example of a pitchfork

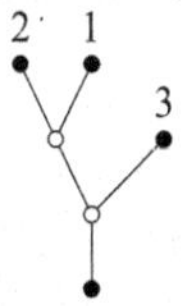

Fig. 2 This is not a pitchfork

particular, we denote by $\mathsf{Tree}(n)$ the groupoid of n-labeled planar trees. As in [5], we denote by $\mathsf{Tree}_2(n)$ the full subcategory of $\mathsf{Tree}(n)$ whose objects are n-labeled planar trees with exactly two nodal vertices. For a groupoid $\mathcal{G}$, the notation $\pi_0(\mathcal{G})$ is reserved for the set of isomorphism classes of objects in $\mathcal{G}$.

We say that an n-labeled planar tree $\mathbf{t}$ is a *pitchfork* if each leaf of $\mathbf{t}$ has height[1] 3. Figure 1 shows a pitchfork while Fig. 2 shows a tree that is not a pitchfork.

The notation $\mathsf{Tree}_{\pitchfork}(n)$ is reserved for the full sub-groupoid of $\mathsf{Tree}(n)$ whose objects are pitchforks.

Let C be a coaugmented dg cooperad satisfying the following technical condition:

Condition 2.1. *The cokernel $C_\circ$ of the coaugmentation carries an ascending exhaustive filtration*

$$\mathbf{0} = \mathcal{F}^0 C_\circ \subset \mathcal{F}^1 C_\circ \subset \mathcal{F}^2 C_\circ \subset \dots \tag{1}$$

which is compatible with the pseudo-cooperad structure on $C_\circ$.

For example, if the dg cooperad C has the properties

$$C(1) \cong \mathbb{K}, \qquad C(0) = \mathbf{0}, \tag{2}$$

then the filtration "by arity minus one" on $C_\circ$ satisfies the above technical condition.

For a cochain complex $\mathcal{V}$ we denote by

$$C(\mathcal{V}) := \bigoplus_{n \ge 1} \Big(C(n) \otimes \mathcal{V}^{\otimes n} \Big)_{S_n} \tag{3}$$

the "cofree" C-coalgebra cogenerated by $\mathcal{V}$.

[1] Recall that the height of a vertex v is the length of the (unique) path which connects v to the root vertex.

We denote by

$$\mathrm{coDer}\big(C(\mathcal{V})\big) \tag{4}$$

the cochain complex of coderivations of the cofree coalgebra $C(\mathcal{V})$ cogenerated by $\mathcal{V}$. In other words, $\mathrm{coDer}\big(C(\mathcal{V})\big)$ consists of $\mathbb{K}$-linear maps

$$\mathcal{D} : C(\mathcal{V}) \to C(\mathcal{V}) \tag{5}$$

which are compatible with the C-coalgebra structure on $C(\mathcal{V})$ in the following sense:

$$\Delta_n \circ \mathcal{D} = \sum_{i=1}^{n} \big(\mathsf{id}_C \otimes \mathsf{id}_{\mathcal{V}}^{\otimes(i-1)} \otimes \mathcal{D} \otimes \mathsf{id}_{\mathcal{V}}^{n-i}\big) \circ \Delta_n, \tag{6}$$

where Δ_n is the comultiplication map

$$\Delta_n : C(\mathcal{V}) \to \Big(C(n) \otimes \big(C(\mathcal{V})\big)^{\otimes n}\Big)_{S_n}.$$

The $\mathbb{Z}$-graded vector space (4) carries a natural differential ∂ induced by those on C and $\mathcal{V}$.

Since the commutator of two coderivations is again a coderivation, the cochain complex (4) is naturally a dg Lie algebra.

Recall that, since the C-coalgebra $C(\mathcal{V})$ is cofree, every coderivation $\mathcal{D}$: $C(\mathcal{V}) \to C(\mathcal{V})$ is uniquely determined by its composition $p_{\mathcal{V}} \circ \mathcal{D}$ with the canonical projection:

$$p_{\mathcal{V}} : C(\mathcal{V}) \to \mathcal{V}\,. \tag{7}$$

We denote by

$$\mathrm{coDer}'\big(C(\mathcal{V})\big) \tag{8}$$

the dg Lie subalgebra of coderivations $\mathcal{D} \in \mathrm{coDer}\big(C(\mathcal{V})\big)$ satisfying the additional technical condition

$$\mathcal{D}\Big|_{\mathcal{V}} = 0\,. \tag{9}$$

Due to [5, Proposition 4.2], the map

$$\mathcal{D} \mapsto p_{\mathcal{V}} \circ \mathcal{D}$$

induces an isomorphism of dg Lie algebras

$$\mathrm{coDer}'\big(C(\mathcal{V})\big) \cong \mathrm{Conv}(C_\circ, \mathsf{End}_{\mathcal{V}}), \tag{10}$$

where the differential ∂ on $\mathrm{Conv}(C_\circ, \mathsf{End}_{\mathcal{V}})$ comes solely from the differential on $C_\circ$ and $\mathcal{V}$. Here $\mathrm{Conv}(\cdots)$ denotes the convolution Lie algebra of ($\mathbb{S}$-module-) maps from a cooperad to an operad, cf. [10, Section 6.4.4].

Recall that [5, Proposition 5.2] $\mathrm{Cobar}(C)$-algebra structures on a cochain complex $\mathcal{V}$ are in bijection with Maurer–Cartan (MC) elements in $\mathrm{coDer}'\big(C(\mathcal{V})\big)$, i.e., with degree 1 coderivations

$$Q \in \mathrm{coDer}'\big(C(\mathcal{V})\big) \tag{11}$$

satisfying the Maurer–Cartan equation

$$\partial Q + \frac{1}{2}[Q, Q] = 0. \tag{12}$$

Hence, given a $\mathrm{Cobar}(C)$-algebra structure on $\mathcal{V}$, we may consider the dg Lie algebra (10) and the C-coalgebra $C(\mathcal{V})$ with the new differentials

$$\partial + [Q, \ ], \tag{13}$$

and

$$\partial + Q, \tag{14}$$

respectively.

In this text we use the following "pedestrian" definition of homotopy algebras:

Definition 2.2. Let C be a coaugmented dg cooperad satisfying Condition 2.1. *A homotopy algebra of type* C is a $\mathrm{Cobar}(C)$-algebra $\mathcal{V}$.

Using the above link between $\mathrm{Cobar}(C)$-algebra structures on $\mathcal{V}$ and Maurer–Cartan elements Q of $\mathrm{coDer}'\big(C(\mathcal{V})\big)$, we see that every homotopy algebra $\mathcal{V}$ of type C gives us a dg C-coalgebra $C(\mathcal{V})$ with the differential $\partial + Q$. This observation motivates our definition of an ∞-morphism between homotopy algebras:

Definition 2.3. Let $\mathcal{A}$, $\mathcal{B}$ be homotopy algebras of type C and let $Q_{\mathcal{A}}$ (resp. $Q_{\mathcal{B}}$) be the MC element of $\mathrm{coDer}'\big(C(\mathcal{A})\big)$ (resp. $\mathrm{coDer}'\big(C(\mathcal{B})\big)$) corresponding to the $\mathrm{Cobar}(C)$-algebra structure on $\mathcal{A}$ (resp. $\mathcal{B}$). Then an ∞*-morphism* from $\mathcal{A}$ to $\mathcal{B}$ is a homomorphism

$$F : C(\mathcal{A}) \to C(\mathcal{B})$$

of the dg C-coalgebras $C(\mathcal{A})$ and $C(\mathcal{B})$ with the differentials $\partial + Q_{\mathcal{A}}$ and $\partial + Q_{\mathcal{B}}$, respectively.

A homomorphism of dg C-coalgebras F is called an ∞ *quasiisomorphism* if the composition

$$\mathcal{A} \hookrightarrow C(\mathcal{A}) \xrightarrow{F} C(\mathcal{B}) \xrightarrow{p_{\mathcal{B}}} \mathcal{B}$$

is a quasiisomorphism of cochain complexes.

We say that two homotopy algebras $\mathcal{A}$ and $\mathcal{B}$ are *quasiisomorphic* if there exists a sequence of ∞ quasiisomorphisms connecting $\mathcal{A}$ with $\mathcal{B}$.

Definition 2.4. Let $\mathcal{A}$ be a homotopy algebra of type C and Q be the corresponding MC element of $\mathrm{coDer}'\big(C(\mathcal{A})\big)$. Then the cochain complex

$$\mathrm{Def}(\mathcal{A}) := \mathrm{coDer}'\big(C(\mathcal{A})\big) \tag{15}$$

with the differential $\partial + [Q, \ ]$ is called the *deformation complex* of the homotopy algebra $\mathcal{A}$.

3 The Main Statement

We observe that the deformation complex (15) of a homotopy algebra $\mathcal{A}$ is naturally a dg Lie algebra. We claim that

Theorem 3.1. *Let C be a coaugmented dg cooperad satisfying Condition 2.1. If $\mathcal{A}$ and $\mathcal{B}$ are quasiisomorphic homotopy algebras of type C, then the deformation complex* $\mathrm{Def}(\mathcal{A})$ *of $\mathcal{A}$ is L_∞-quasiisomorphic to the deformation complex* $\mathrm{Def}(\mathcal{B})$ *of $\mathcal{B}$.*

Remark 1. For A_∞-algebras this statement follows from the result [9] of B. Keller.

It is clearly sufficient to prove this theorem in the case when $\mathcal{A}$ and $\mathcal{B}$ are connected by a single ∞ quasiisomorphism $F : \mathcal{A} \rightsquigarrow \mathcal{B}$.

We will prove the theorem by constructing an L_∞-algebra $\mathrm{Def}(\mathcal{A} \overset{F}{\rightsquigarrow} \mathcal{B})$, together with quasiisomorphisms

$$\mathrm{Def}(\mathcal{A}) \leftarrow \mathrm{Def}(\mathcal{A} \overset{F}{\rightsquigarrow} \mathcal{B}) \rightarrow \mathrm{Def}(\mathcal{B}).$$

The next subsections are concerned with the definition of $\mathrm{Def}(\mathcal{A} \overset{F}{\rightsquigarrow} \mathcal{B})$. The proof of Theorem 3.1 is given in Sect. 3.4 below.

3.1 *The Auxiliary L_∞-Algebra* $\mathrm{Cyl}(C, \mathcal{A}, \mathcal{B})$

Let $\mathcal{A}$, $\mathcal{B}$ be cochain complexes. We consider the graded vector space

$$\mathrm{Cyl}(C, \mathcal{A}, \mathcal{B}) := \mathrm{Hom}(C_\circ(\mathcal{A}), \mathcal{A}) \oplus \mathbf{s}\mathrm{Hom}(C(\mathcal{A}), \mathcal{B}) \oplus \mathrm{Hom}(C_\circ(\mathcal{B}), \mathcal{B}) \quad (1)$$

with the differential coming from those on C, $\mathcal{A}$ and $\mathcal{B}$. Here we denote by $\mathbf{s}V$ the suspension of the graded vector space V. Concretely, if $v \in V$ has degree d, then the corresponding element $\mathbf{s}v \in \mathbf{s}V$ has degree $d+1$.

We equip the cochain complex $\mathrm{Cyl}(C, \mathcal{A}, \mathcal{B})$ with an L_∞-structure by declaring that

$$\{\mathbf{s}^{-1} P_1, \mathbf{s}^{-1} P_2, \dots, \mathbf{s}^{-1} P_n\}_n := \begin{cases} (-1)^{|P_1|+1}\mathbf{s}^{-1}[P_1, P_2] & \text{if } n = 2\,, \\ 0 & \text{otherwise}\,. \end{cases} \quad (2)$$

$$\{\mathbf{s}^{-1} R_1, \mathbf{s}^{-1} R_2, \dots, \mathbf{s}^{-1} R_n\}_n := \begin{cases} (-1)^{|R_1|+1}\mathbf{s}^{-1}[R_1, R_2] & \text{if } n = 2\,, \\ 0 & \text{otherwise}\,, \end{cases} \quad (3)$$

for $P_i \in \mathrm{Hom}(C_\circ(\mathcal{A}), \mathcal{A}) \cong \mathrm{Conv}(C_\circ, \mathsf{End}_{\mathcal{A}})$, and $R_i \in \mathrm{Hom}(C_\circ(\mathcal{B}), \mathcal{B}) \cong \mathrm{Conv}(C_\circ, \mathsf{End}_{\mathcal{B}})$, and $[\ ,\ ]$ is the Lie bracket on the convolution algebras $\mathrm{Conv}(C_\circ, \mathsf{End}_{\mathcal{A}})$ and $\mathrm{Conv}(C_\circ, \mathsf{End}_{\mathcal{B}})$, respectively.

Furthermore,

$$\{T, \mathbf{s}^{-1} P\}_2(X, a_1, \dots, a_n) = \sum_{\substack{0 \le p \le n \\ \sigma \in \mathrm{Sh}_{p,n-p}}} \sum_i (-1)^{|T|+|P|(|X'_{\sigma,i}|+1)} T\big(X'_{\sigma,i}, P(X''_{\sigma,i}; a_{\sigma(1)}, \dots, a_{\sigma(p)}), a_{\sigma(p+1)}, \dots, a_{\sigma(n)}\big)\,, \quad (4)$$

where $T \in \mathrm{Hom}(C(\mathcal{A}), \mathcal{B})$, $P \in \mathrm{Hom}(C_\circ(\mathcal{A}), \mathcal{A})$, $X \in C_\circ(n)$, $X'_{\sigma,i}$, $X''_{\sigma,i}$ are tensor factors in

$$\Delta_{\mathbf{t}_\sigma}(X) = \sum_i X'_{\sigma,i} \otimes X''_{\sigma,i}\,,$$

P is extended by zero to $\mathcal{A} \subset C(\mathcal{A})$, and $\mathbf{t}_\sigma$ is the n-labeled planar tree depicted in Fig. 1.

To define yet another collection of nonzero L_∞-brackets, denote by $\mathsf{Isom}_{\pitchfork}(m, r)$ the set of isomorphism classes of pitchforks $\mathbf{t} \in \mathsf{Tree}_{\pitchfork}(m)$ with

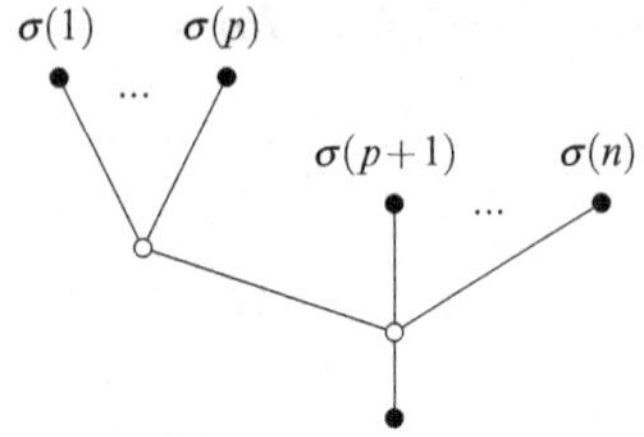

Fig. 1 Here σ is a $(p, n-p)$-shuffle

r nodal vertices of height 2. For every $z \in \mathsf{Isom}_{\pitchfork}(m, r)$ we choose a representative $\mathbf{t}_z$ and denote by $X^k_{z,i}$ the tensor factors in

$$\Delta_{\mathbf{t}_z}(X) = \sum_i X^0_{z,i} \otimes X^1_{z,i} \otimes \cdots \otimes X^r_{z,i}\,, \tag{5}$$

where $X \in C(m)$.

Finally, for vectors $T_j \in \mathrm{Hom}(C(\mathcal{A}), \mathcal{B})$ and $R \in \mathrm{Hom}(C_\circ(\mathcal{B}), \mathcal{B})$ we set

$$\{\mathbf{s}^{-1} R, T_1, \dots, T_r\}_{r+1}(X, a_1, \dots, a_m) =$$
$$\sum_{\sigma \in S_r} \sum_{z \in \mathsf{Isom}_{\pitchfork}(m,r)} \sum_i \pm (-1)^{|R|+1} R\big(X^0_{z,i}, T_{\sigma(1)}(X^1_{z,i}; a_{\lambda_z(1)}, \dots, a_{\lambda_z(n^z_1)}),$$
$$T_{\sigma(2)}(X^2_{z,i}; a_{\lambda_z(n^z_1+1)}, \dots, a_{\lambda_z(n^z_1+n^z_2)}), \dots, T_{\sigma(r)}(X^r_{z,i}; a_{\lambda_z(m-n^z_r+1)}, \dots, a_{\lambda_z(m)})\big), \tag{6}$$

where n^z_q is the number of leaves adjacent to the $(q+1)$-th nodal vertex of $\mathbf{t}_z$, $\lambda_z(l)$ is the label of the l-th leaf of $\mathbf{t}_z$, the map R is extended by zero to $\mathcal{B} \subset C(\mathcal{B})$ and the sign factor $\pm$ comes from the rearrangement of the homogeneous vectors

$$R, T_1, \dots, T_r, X^0_{z,i}, X^1_{z,i}, \dots, X^r_{z,i}, a_1, \dots, a_m \tag{7}$$

from their original positions in (7) to their positions in the right-hand side of (6).

We observe that, due to axioms of a cooperad, the right-hand side of (6) does not depend on the choice of representatives $\mathbf{t}_z \in \mathsf{Tree}_{\pitchfork}(m)$.

The remaining L_∞-brackets are either extended in the obvious way by symmetry or declared to be zero.

We claim that

Claim. The operations

$$\{\,,\,,\dots,\,\}_n : S^n(\mathbf{s}^{-1}\,\mathrm{Cyl}(C, \mathcal{A}, \mathcal{B})) \to \mathbf{s}^{-1}\,\mathrm{Cyl}(C, \mathcal{A}, \mathcal{B}), \qquad n \geq 2 \tag{8}$$

defined above, have degree 1 and satisfy the desired L_∞-identities:

$$\partial\{f_1, f_2, \dots, f_n\}_n + \sum_{i=1}^{n}(-1)^{|f_1|+\dots+|f_{i-1}|}\{f_1, \dots, f_{i-1}, \partial(f_i), f_{i+1}, \dots, f_n\}_n$$

$$\sum_{p=2}^{n-1} \sum_{\sigma\in \mathrm{Sh}_{p,n-p}} \pm\{\{f_{\sigma(1)}, f_{\sigma(2)}, \dots, f_{\sigma(p)}\}_p, f_{\sigma(p+1)}, \dots, f_{\sigma(n)}\}_{n-p+1}\,, \qquad (9)$$

where $f_j \in \mathbf{s}^{-1}\,\mathrm{Cyl}(C, \mathcal{A}, \mathcal{B})$ and the usual Koszul rule of signs is applied.

Before proving Claim 3.1, we would like to show that

Claim. The MC equation for the L_∞-algebra $\mathrm{Cyl}(C, \mathcal{A}, \mathcal{B})$ is well defined. Moreover, MC elements of the L_∞-algebra $\mathrm{Cyl}(C, \mathcal{A}, \mathcal{B})$ are triples:

- A $\mathrm{Cobar}(C)$-algebra structure on $\mathcal{A}$,
- A $\mathrm{Cobar}(C)$-algebra structure on $\mathcal{B}$, and
- An ∞-morphism from $\mathcal{A}$ to $\mathcal{B}$.

Proof. Let U be a degree 1 element in $\mathrm{Cyl}(C, \mathcal{A}, \mathcal{B})$.

We observe that the components of

$$\{\mathbf{s}^{-1}\,U, \mathbf{s}^{-1}\,U, \dots, \mathbf{s}^{-1}\,U\}_n$$

in $\mathrm{Hom}(C_\circ(\mathcal{A}), \mathcal{A})$ and $\mathrm{Hom}(C_\circ(\mathcal{B}), \mathcal{B})$ are zero for all $n \ge 3$. Furthermore, for every $(X, a_1, \dots, a_k) \in C(\mathcal{A})$

$$\{\mathbf{s}^{-1}\,U, \mathbf{s}^{-1}\,U, \dots, \mathbf{s}^{-1}\,U\}_n(X, a_1, \dots, a_k) = 0 \qquad \forall\ n \ge k+1\,.$$

Therefore the infinite sum

$$[\partial, U] + \sum_{n=2}^{\infty} \frac{1}{n!}\{\mathbf{s}^{-1}\,U, \mathbf{s}^{-1}\,U, \dots, \mathbf{s}^{-1}\,U\}_n \qquad (10)$$

makes sense for every degree 1 element U in $\mathrm{Cyl}(C, \mathcal{A}, \mathcal{B})$ and we can talk about MC elements of $\mathrm{Cyl}(C, \mathcal{A}, \mathcal{B})$.

To prove the second statement, we split the degree 1 element $U \in \mathrm{Cyl}(C, \mathcal{A}, \mathcal{B})$ into a sum

$$U = Q_{\mathcal{A}} + \mathbf{s}\,U_F + Q_{\mathcal{B}}\,,$$

where $Q_{\mathcal{A}} \in \mathrm{Conv}(C_\circ, \mathsf{End}_{\mathcal{A}})$, $Q_{\mathcal{B}} \in \mathrm{Conv}(C_\circ, \mathsf{End}_{\mathcal{B}})$, and $U_F \in \mathrm{Hom}(C(\mathcal{A}), \mathcal{B})$.

Then the MC equation for U is equivalent to the following three equations:

$$\partial Q_{\mathcal{A}} + \frac{1}{2}[Q_{\mathcal{A}}, Q_{\mathcal{A}}] = 0 \quad \text{in} \quad \mathrm{Conv}(C_{\circ}, \mathsf{End}_{\mathcal{A}}), \tag{11}$$

$$\partial Q_{\mathcal{B}} + \frac{1}{2}[Q_{\mathcal{B}}, Q_{\mathcal{B}}] = 0 \quad \text{in} \quad \mathrm{Conv}(C_{\circ}, \mathsf{End}_{\mathcal{B}}), \tag{12}$$

and

$$[\partial, U_F] + \{U_F, \mathbf{s}^{-1}\, Q_{\mathcal{A}}\}_2 + \sum_{r=1}^{\infty} \frac{1}{r!} \{\mathbf{s}^{-1}\, Q_{\mathcal{B}}, U_F, U_F, \dots, U_F\}_{r+1} = 0\,. \tag{13}$$

Equations (11) and (12) imply that $Q_{\mathcal{A}}$ (resp. $Q_{\mathcal{B}}$) gives us a Cobar(C)-algebra structure on $\mathcal{A}$ (resp. $\mathcal{B}$). Furthermore, Eq. (13) means that U_F is an ∞-morphism from $\mathcal{A}$ to $\mathcal{B}$. □

3.1.1 Proof of Claim 3.1

The most involved identity on L_∞-brackets defined above is

$$\begin{aligned}
&\{\{\mathbf{s}^{-1}\, R_1, \mathbf{s}^{-1}\, R_2\}_2, T_1, \dots, T_n\}_{n+1} + \\
&\sum_{\substack{1 \le p \le n-1 \\ \sigma \in \mathrm{Sh}_{p,n-p}}} \pm \{\mathbf{s}^{-1}\, R_1, \{\mathbf{s}^{-1}\, R_2, T_{\sigma(1)}, \dots, T_{\sigma(p)}\}_{p+1}, T_{\sigma(p+1)}, \dots, T_{\sigma(n)}\}_{n-p+2} + \\
&\sum_{\substack{1 \le p \le n-1 \\ \sigma \in \mathrm{Sh}_{p,n-p}}} \pm \{\mathbf{s}^{-1}\, R_2, \{\mathbf{s}^{-1}\, R_1, T_{\sigma(1)}, \dots, T_{\sigma(p)}\}_{p+1}, T_{\sigma(p+1)}, \dots, T_{\sigma(n)}\}_{n-p+2} = 0\,.
\end{aligned} \tag{14}$$

This identity is a consequence of a combinatorial fact about certain isomorphism classes in the groupoid $\mathsf{Tree}(n)$. To formulate this fact, we recall that the set of isomorphism classes of r-labeled planar trees with two nodal vertices are in bijection with the set of shuffles

$$\bigsqcup_{p=0}^{r} \mathrm{Sh}_{p,r-p}\,. \tag{15}$$

This bijection assigns to a shuffle $\sigma \in \mathrm{Sh}_{p,r-p}$ the r-labeled planar tree $\mathbf{t}_\sigma$ shown in Fig. 2.

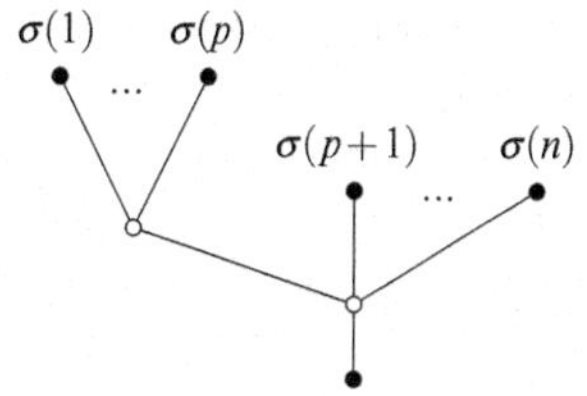

Fig. 2 Here σ is a $(p, r-p)$-shuffle

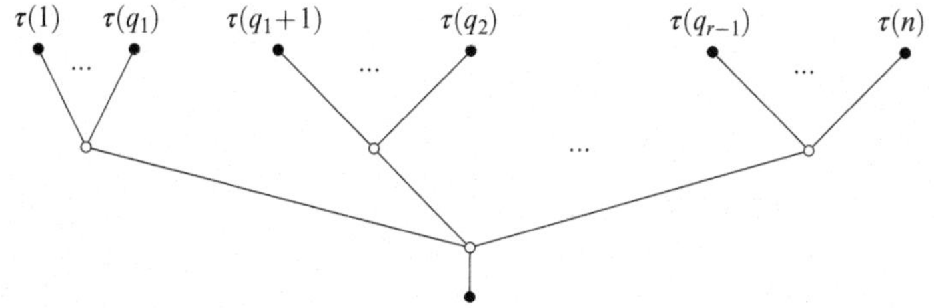

Fig. 3 The pitchfork $\mathbf{t}_\tau^{\pitchfork}$

Next, we observe that $\pi_0(\mathsf{Tree}_{\pitchfork}(n))$ is in bijection with the set

$$\bigsqcup_{r\ge 1} \mathfrak{S}\mathfrak{H}_{n,r}, \tag{16}$$

where[2]

$$\mathfrak{S}\mathfrak{H}_{n,r} = \bigsqcup_{1\le q_1<q_2<\cdots<q_{r-1}<q_r=n} \{\tau \in \mathrm{Sh}_{q_1,q_2-q_1,\dots,n-q_{r-1}} \,|\, \tau(1) < \tau(q_1+1) < \tau(q_2+1) < \cdots < \tau(q_{r-1}+1)\}. \tag{17}$$

This bijection assigns to a shuffle τ in the set (16) the isomorphism class of the pitchfork $\mathbf{t}_\tau^{\pitchfork}$ depicted in Fig. 3.

Note that, in the degenerate cases $r = 1$ and $r = n$, $\mathfrak{S}\mathfrak{H}_{n,r}$ is the one-element set consisting of the identity permutation $\mathsf{id} \in S_n$. The corresponding pitchforks are shown in Figs. 4 and 5, respectively.

For every permutation $\tau \in \mathfrak{S}\mathfrak{H}_{n,r}$ and a shuffle $\sigma \in \mathrm{Sh}_{p,r-p}$ we can form the following n-labeled planar tree:

$$\mathbf{t}_\tau^{\pitchfork} \bullet_1 \mathbf{t}_\sigma\,, \tag{18}$$

where $\mathbf{t} \bullet_j \mathbf{t}'$ denotes the insertion of the tree $\mathbf{t}'$ into the j-th nodal vertex of the tree $\mathbf{t}$ (see Section 2.2 in [5]).

[2]It is obvious that for every $\tau \in \mathfrak{S}\mathfrak{H}_{n,r}$, $\tau(1) = 1$.

Fig. 4 The pitchfork for $r = 1$

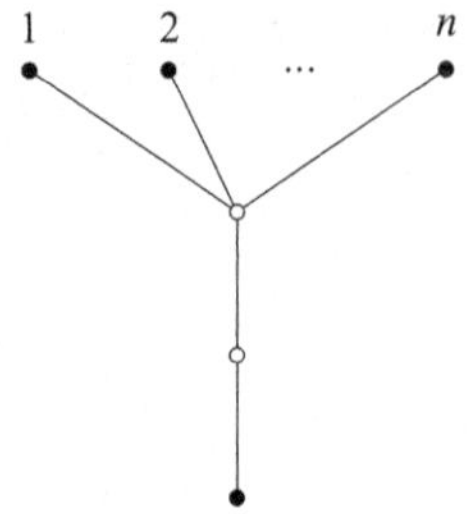

Fig. 5 The pitchfork for $r = n$

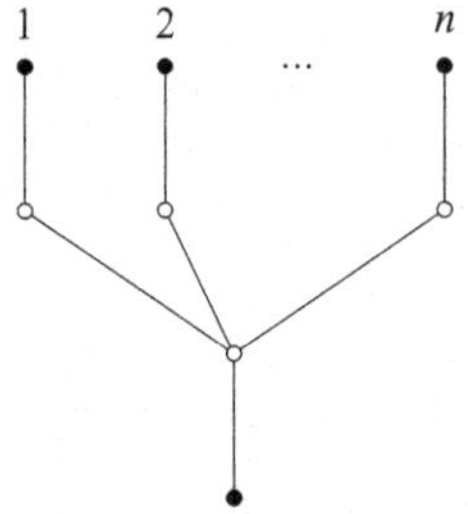

It is clear that for distinct pairs $(\tau, \sigma) \in \mathfrak{S}\mathfrak{H}_{n,r} \times \mathrm{Sh}_{p,r-p}$, we get mutually non-isomorphic labeled planar trees.

Let $\tau' \in \mathfrak{S}\mathfrak{H}_{n,r'}$ and m_i be the number of edges which terminate at the $(i+1)$-th nodal vertex of $\mathbf{t}^{\pitchfork}_{\tau'}$. For every $\tau'' \in \mathfrak{S}\mathfrak{H}_{m_i,r''}$, we may form the n-labeled planar tree

$$\mathbf{t}^{\pitchfork}_{\tau'} \bullet_{i+1} \mathbf{t}^{\pitchfork}_{\tau''} . \tag{19}$$

It is clear that for distinct triples $(\tau', i, \tau'') \in \mathfrak{S}\mathfrak{H}_{n,r'} \times \{1, 2, \ldots, r'\} \times \mathfrak{S}\mathfrak{H}_{m_i,r''}$, the corresponding labeled planar trees (19) are mutually nonisomorphic. Furthermore, every tree of the form (19) is isomorphic to exactly one tree of the form (18) and vice versa. This is precisely the combinatorial fact that is need to prove that identity (14) holds.

Indeed, the terms in the expression

$$\{\{\mathbf{s}^{-1} R_1, \mathbf{s}^{-1} R_2\}_2, T_1, \ldots, T_n\}_{n+1}$$

involve trees of the form (18) and the terms in the expressions

$$\sum_{\substack{1 \le p \le n-1 \\ \sigma \in \mathrm{Sh}_{p,n-p}}} \pm\{\mathbf{s}^{-1} R_1, \{\mathbf{s}^{-1} R_2, T_{\sigma(1)}, \ldots, T_{\sigma(p)}\}_{p+1}, T_{\sigma(p+1)}, \ldots, T_{\sigma(n)}\}_{n-p+2}$$

and

$$\sum_{\substack{1 \le p \le n-1 \\ \sigma \in \mathrm{Sh}_{p,n-p}}} \pm\{\mathbf{s}^{-1} R_2, \{\mathbf{s}^{-1} R_1, T_{\sigma(1)}, \ldots, T_{\sigma(p)}\}_{p+1}, T_{\sigma(p+1)}, \ldots, T_{\sigma(n)}\}_{n-p+2}$$

involve trees of the form (19).

Thus it only remains to check that the sign factors match.

The remaining identities on L_∞-brackets are simpler and we leave their verification to the reader.

Claim 3.1 is proved. □

3.2 The L_∞-Algebra $\mathrm{Cyl}(C, \mathcal{A}, \mathcal{B})^{\mathbf{s}F_1}$ and Its MC Elements

Let

$$F_1 : \mathcal{A} \to \mathcal{B} \tag{20}$$

be a map of cochain complexes.

We may view $\mathbf{s}F_1$ as a degree 1 element in $\mathrm{Cyl}(C, \mathcal{A}, \mathcal{B})$:

$$\mathbf{s}F_1 \in \mathbf{s}\mathrm{Hom}(\mathcal{A}, \mathcal{B}) \subset \mathbf{s}\mathrm{Hom}(C(\mathcal{A}), \mathcal{B}) \subset \mathrm{Cyl}(C, \mathcal{A}, \mathcal{B})\,.$$

Since F_1 is compatible with the differentials on $\mathcal{A}$ and $\mathcal{B}$, $\mathbf{s}F_1$ is obviously a MC element of $\mathrm{Cyl}(C, \mathcal{A}, \mathcal{B})$ and, in view of Claim 3.1, $\mathbf{s}F_1$ corresponds to the triple:

- The trivial $\mathrm{Cobar}(C)$-algebra structure on $\mathcal{A}$,
- The trivial $\mathrm{Cobar}(C)$-algebra structure on $\mathcal{B}$, and
- A strict[3] ∞-morphism F_1 from $\mathcal{A}$ to $\mathcal{B}$.

Let $Q_1, Q_2, \dots, Q_m$ be vectors in $\mathrm{Cyl}(C, \mathcal{A}, \mathcal{B})$. We recall that the components of

$$\{\underbrace{F_1, F_1, \dots, F_1}_{n \text{ times}}, \mathbf{s}^{-1}Q_1, \mathbf{s}^{-1}_{}Q_2, \dots, \mathbf{s}^{-1}Q_m\}_{n+m}$$

in $\mathrm{Hom}(C_\circ(\mathcal{A}), \mathcal{A})$ and $\mathrm{Hom}(C_\circ(\mathcal{B}), \mathcal{B})$ are zero if $n + m > 2$. Furthermore, for every $(X, a_1, \dots, a_k) \in C(\mathcal{A})$

$$\{\underbrace{F_1, F_1, \dots, F_1}_{n \text{ times}}, \mathbf{s}^{-1}Q_1, \mathbf{s}^{-1}Q_2, \dots, \mathbf{s}^{-1}Q_m\}_{n+m}(X, a_1, \dots, a_k) = 0 \; (\in \mathcal{B})$$

provided $n + m \geq k + 2$.

Therefore we may twist (see [7, Remark 3.11.]) the L_∞-algebra on $\mathrm{Cyl}(C, \mathcal{A}, \mathcal{B})$ by the MC element $\mathbf{s}F_1$. We denote by

$$\mathrm{Cyl}(C, \mathcal{A}, \mathcal{B})^{\mathbf{s}F_1} \tag{21}$$

the L_∞-algebra obtained in this way.

---

[3] i.e., an ∞-morphism $F : \mathcal{A} \rightsquigarrow \mathcal{B}$ whose all higher structure maps are zero.

It is not hard to see that[4]

$$\mathrm{Cyl}_{\circ}(C,\mathcal{A},\mathcal{B})^{\mathbf{s}F_1} := \mathrm{Hom}(C_\circ(\mathcal{A}),\mathcal{A}) \oplus \mathbf{s}\mathrm{Hom}(C_\circ(\mathcal{A}),\mathcal{B}) \oplus \mathrm{Hom}(C_\circ(\mathcal{B}),\mathcal{B}) \tag{22}$$

is an L_∞-subalgebra of $\mathrm{Cyl}(C,\mathcal{A},\mathcal{B})^{\mathbf{s}F_1}$. Furthermore, Claim 3.1 implies that

Claim. MC elements of the L_∞-algebra (22) are triples:

- A $\mathrm{Cobar}(C)$-algebra structure on $\mathcal{A}$,
- A $\mathrm{Cobar}(C)$-algebra structure on $\mathcal{B}$,
- An ∞-morphism $F:\mathcal{A}\rightsquigarrow\mathcal{B}$ for which the composition

$$\mathcal{A}\hookrightarrow C(\mathcal{A})\stackrel{F}{\longrightarrow}C(\mathcal{B})\stackrel{p_{\mathcal{B}}}{\longrightarrow}\mathcal{B}$$

coincides with F_1 .

□

Remark 2. Using the ascending filtration (1) on the pseudo-operad $C_\circ$, we equip the L_∞-algebra $\mathrm{Cyl}(C,\mathcal{A},\mathcal{B})$ with the complete descending filtrations:

$$\mathrm{Cyl}(C,\mathcal{A},\mathcal{B}) = \mathcal{F}_0\mathrm{Cyl}(C,\mathcal{A},\mathcal{B}) \supset \mathcal{F}_1\mathrm{Cyl}(C,\mathcal{A},\mathcal{B}) \supset \mathcal{F}_2\mathrm{Cyl}(C,\mathcal{A},\mathcal{B}) \supset \dots, \tag{23}$$

where (for $m\ge 1$)

$$\begin{gathered}\mathcal{F}_m\mathrm{Cyl}(C,\mathcal{A},\mathcal{B}) := \\ \big\{Q'\oplus F\oplus Q'' \in \mathrm{Hom}(C_\circ(\mathcal{A}),\mathcal{A})\oplus \mathbf{s}\mathrm{Hom}(C(\mathcal{A}),\mathcal{B})\oplus\mathrm{Hom}(C_\circ(\mathcal{B}),\mathcal{B}) \\ Q'(X,a_1,\dots,a_k)=0\,,\quad F(X,a_1,\dots,a_k)=0\,,\quad Q''(X,b_1,\dots,b_k)=0 \quad \forall\, X\in\mathcal{F}^{m-1}C(k)\big\}\,.\end{gathered} \tag{24}$$

The same formulas define a complete descending filtration on the L_∞-algebras $\mathrm{Cyl}(C,\mathcal{A},\mathcal{B})^{\mathbf{s}F_1}$ and $\mathrm{Cyl}_{\circ}(C,\mathcal{A},\mathcal{B})^{\mathbf{s}F_1}$.

We observe that

$$\mathrm{Cyl}_{\circ}(C,\mathcal{A},\mathcal{B})^{\mathbf{s}F_1} = \mathcal{F}_1\mathrm{Cyl}_{\circ}(C,\mathcal{A},\mathcal{B})^{\mathbf{s}F_1} \tag{25}$$

and hence $\mathrm{Cyl}_{\circ}(C,\mathcal{A},\mathcal{B})^{\mathbf{s}F_1}$ is pro-nilpotent. Later, we will use this advantage of $\mathrm{Cyl}_{\circ}(C,\mathcal{A},\mathcal{B})^{\mathbf{s}F_1}$ over $\mathrm{Cyl}(C,\mathcal{A},\mathcal{B})^{\mathbf{s}F_1}$.

[4]In $\mathrm{Cyl}_{\circ}(C,\mathcal{A},\mathcal{B})^{\mathbf{s}F_1}$, we have $\mathbf{s}\mathrm{Hom}(C_\circ(\mathcal{A}),\mathcal{B})$ instead of $\mathbf{s}\mathrm{Hom}(C(\mathcal{A}),\mathcal{B})$.

3.3 What If F_1 Is a Quasiisomorphism?

Starting with a chain map (20) we define two maps of cochain complexes:

$$P \mapsto f(P) = F_1 \circ P \; : \; \mathrm{Hom}(C_\circ(\mathcal{A}), \mathcal{A}) \to \mathrm{Hom}(C_\circ(\mathcal{A}), \mathcal{B}) \tag{26}$$

$$R \mapsto \tilde{f}(R) = R \circ C_\circ(F_1) \; : \; \mathrm{Hom}(C_\circ(\mathcal{B}), \mathcal{B}) \to \mathrm{Hom}(C_\circ(\mathcal{A}), \mathcal{B}) \tag{27}$$

and observe that the cochain complex $\mathrm{Cyl}_\circ(C, \mathcal{A}, \mathcal{B})^{\mathbf{s}F_1}$ is precisely the cochain complex $\mathrm{Cyl}(f, \tilde{f})$ defined in (9), (10) in the Appendix.

Hence, using Lemma 1, we deduce the following statement:

Proposition 3.2. *If the chain map $F_1 : \mathcal{A} \to \mathcal{B}$ induces an isomorphism on the level of cohomology, then so do the following canonical projections:*

$$\pi_{\mathcal{A}} : \mathrm{Cyl}_\circ(C, \mathcal{A}, \mathcal{B})^{\mathbf{s}F_1} \to \mathrm{Hom}(C_\circ(\mathcal{A}), \mathcal{A})\,, \tag{28}$$

$$\pi_{\mathcal{B}} : \mathrm{Cyl}_\circ(C, \mathcal{A}, \mathcal{B})^{\mathbf{s}F_1} \to \mathrm{Hom}(C_\circ(\mathcal{B}), \mathcal{B})\,. \tag{29}$$

The maps $\pi_{\mathcal{A}}$ and $\pi_{\mathcal{B}}$ are strict homomorphisms of L_∞-algebras.

Proof. Since we work over a field of characteristic zero, the functors Hom, $\otimes$, as well as the functors of taking (co)invariants with respect to actions of symmetric groups preserve quasiisomorphisms. Therefore the maps (26) and (27) are quasiisomorphisms of cochain complexes.

Thus the first statement follows directly from Lemma 1.

The second statement is an obvious consequence of the definition of L_∞-brackets on $\mathrm{Cyl}_\circ(C, \mathcal{A}, \mathcal{B})^{\mathbf{s}F_1}$. □

3.4 Proof of Theorem 3.1

We will now give a proof of Theorem 3.1

Let $\mathcal{A}$ and $\mathcal{B}$ be homotopy algebras of type C. As said above, we may assume, without loss of generality, that $\mathcal{A}$ and $\mathcal{B}$ are connected by a single ∞ quasiisomorphism:

$$F : \mathcal{A} \rightsquigarrow \mathcal{B}\,. \tag{30}$$

We denote by α^{Cyl} the MC element of $\mathrm{Cyl}_\circ(C, \mathcal{A}, \mathcal{B})^{\mathbf{s}F_1}$ which corresponds to the triple

- The homotopy algebra structure on $\mathcal{A}$,
- The homotopy algebra structure on $\mathcal{B}$, and
- The ∞-morphism F.

Due to (25), we may twist (see [7, Remark 3.11.]) the L_∞-algebra $\mathrm{Cyl}_\circ(C, \mathcal{A}, \mathcal{B})^{\mathbf{s}F_1}$ by the MC element α^{Cyl} . We denote by

$$\mathrm{Def}(\mathcal{A} \overset{F}{\rightsquigarrow} \mathcal{B}) \tag{31}$$

the L_∞-algebra which is obtained from $\mathrm{Cyl}_\circ(C, \mathcal{A}, \mathcal{B})^{\mathbf{s}F_1}$ via twisting by the MC element α^{Cyl}.

We also denote by $Q_{\mathcal{A}}$ (resp. $Q_{\mathcal{B}}$) the MC element of $\mathrm{Conv}(C_\circ, \mathsf{End}_{\mathcal{A}})$ (resp. $\mathrm{Conv}(C_\circ, \mathsf{End}_{\mathcal{B}})$) corresponding to the homotopy algebra structure on $\mathcal{A}$ (resp. $\mathcal{B}$) and recall that $\mathrm{Def}(\mathcal{A})$ (resp. $\mathrm{Def}(\mathcal{B})$) is obtained from $\mathrm{Conv}(C_\circ, \mathsf{End}_{\mathcal{A}})$ (resp. $\mathrm{Conv}(C_\circ, \mathsf{End}_{\mathcal{B}})$) via twisting by the MC element $Q_{\mathcal{A}}$ (resp. $Q_{\mathcal{B}}$).

It is easy to see that

$$\pi_{\mathcal{A}}(\alpha^{\mathrm{Cyl}}) = Q_{\mathcal{A}}, \qquad \pi_{\mathcal{B}}(\alpha^{\mathrm{Cyl}}) = Q_{\mathcal{B}}. \tag{32}$$

Since $\pi_{\mathcal{A}}$ (28) and $\pi_{\mathcal{B}}$ (29) are strict L_∞-morphisms, they do not change under twisting by MC elements. Thus, we conclude that the same maps $\pi_{\mathcal{A}}$ and $\pi_{\mathcal{B}}$ give us (strict) L_∞-morphisms

$$\pi_{\mathcal{A}} : \mathrm{Def}(\mathcal{A} \overset{F}{\rightsquigarrow} \mathcal{B}) \to \mathrm{Def}(\mathcal{A}),$$
$$\pi_{\mathcal{B}} : \mathrm{Def}(\mathcal{A} \overset{F}{\rightsquigarrow} \mathcal{B}) \to \mathrm{Def}(\mathcal{B}). \tag{33}$$

According to [7, Proposition 6.2], twisting preserves quasiisomorphisms. Thus, due to Proposition 3.2, the two arrows in (33) are (strict) L_∞-quasiisomorphisms, as desired.

Theorem 3.1 is proven. □

4 Sheaves of Homotopy Algebras

For a topological space X we consider the category dgSh_X of dg sheaves (i.e., sheaves of unbounded cochain complexes of $\mathbb{K}$-vector spaces). We recall that dgSh_X is a symmetric monoidal category for which the monoidal product is the tensor product followed by sheafification.

Given coaugmented dg cooperad C (satisfying condition (2.1)) one may give the following naive definition of a homotopy algebra of type C in the category dgSh_X:

Definition 4.1 (Naive!). We say that a dg sheaf $\mathcal{A}$ on X carries a structure of *a homotopy algebra of type* C if $\mathcal{A}$ is an algebra over the dg operad $\mathrm{Cobar}(C)$.

One can equivalently define a homotopy algebra of type C by considering coderivations of the cofree C-coalgebra (in the category dgSh_X)

$$C(\mathcal{A}) := \bigoplus_n \left(C(n) \otimes \mathcal{A}^{\otimes n}\right)_{S_n}. \tag{1}$$

In other words, a homotopy algebra of type C on $\mathcal{A}$ is a degree 1 coderivation $\mathcal{Q}$ of $C(\mathcal{A})$ satisfying the MC equation and the additional condition

$$\mathcal{Q}\Big|_{\mathcal{A}} = 0 .$$

Given such a coderivation $\mathcal{Q}$, it is natural to consider the C-coalgebra (1) with the new differential

$$\partial + \mathcal{Q} \tag{2}$$

where ∂ comes from the differentials on C and $\mathcal{A}$.

This observation motivates the following naive definition of ∞-morphism of homotopy algebra in dgSh_X:

Definition 4.2 (Naive!). Let $\mathcal{A}$ and $\mathcal{B}$ be homotopy algebras of type C in dgSh_X and let $\mathcal{Q}_{\mathcal{A}}$ and $\mathcal{Q}_{\mathcal{B}}$ be the corresponding coderivations of $C(\mathcal{A})$ and $C(\mathcal{B})$ respectively. An ∞-*morphism* $F : \mathcal{A} \rightsquigarrow \mathcal{B}$ is a map of sheaves

$$F : C(\mathcal{A}) \to C(\mathcal{B})$$

which is compatible with the C-coalgebra structure and the differentials $\partial + \mathcal{Q}_{\mathcal{A}}$, $\partial + \mathcal{Q}_{\mathcal{B}}$.

An important disadvantage of the above naive definitions is that they do not admit an analogue of the homotopy transfer theorem [10, Theorem 10.3.2]. For this reason we propose "more mature" definitions based on the use of the Thom–Sullivan normalization [1], [12, Appendix A].

Let $\mathfrak{U}$ be a covering of X and $\mathcal{A}$ be a dg sheaf on X. The associated cosimplicial set $\mathfrak{U}(\mathcal{A})$ is naturally a cosimplicial cochain complex. So, applying the Thom–Sullivan functor N^{TS} to $\mathfrak{U}(\mathcal{A})$, we get a cochain complex

$$N^{\mathrm{TS}}\mathfrak{U}(\mathcal{A}) \tag{3}$$

which computes the Cech hypercohomology of $\mathcal{A}$ with respect to the cover $\mathfrak{U}$.

Let us assume, for simplicity, that there exists an acyclic covering $\mathfrak{U}$ for $\mathcal{A}$. In particular, $\check{H}_{\mathfrak{U}}(\mathcal{A}) \cong H(\mathcal{A})$ agrees with the sheaf cohomology of $\mathcal{A}$.

Then we have the following definition:

Definition 4.3. *A homotopy algebra structure of type* C on a dg sheaf $\mathcal{A}$ is Cobar(C)-algebra structure on the cochain complex (3).

Remark 3. Since the Thom–Sullivan normalization N^{TS} is a symmetric monoidal functor from cosimplicial cochain complexes into cochain complexes, a homotopy algebra structure on $\mathcal{A}$ in the sense of naive Definition 4.1 is a homotopy algebra structure on $\mathcal{A}$ in the sense of Definition 4.3.

Remark 4. Let $\mathfrak{U}'$ be another acyclic covering of X and $\mathfrak{V}$ be a common acyclic refinement of $\mathfrak{U}$ and $\mathfrak{U}'$. Since the functor N^{TS} preserves quasiisomorphisms, the cochain complexes $N^{\mathrm{TS}}\mathfrak{U}(\mathcal{A})$ and $N^{\mathrm{TS}}\mathfrak{U}'(\mathcal{A})$ are connected by the following pair of quasiisomorphisms:

$$N^{\mathrm{TS}}\mathfrak{U}(\mathcal{A}) \stackrel{\sim}{\longrightarrow} N^{\mathrm{TS}}\mathfrak{V}(\mathcal{A}) \stackrel{\sim}{\longleftarrow} N^{\mathrm{TS}}\mathfrak{U}'(\mathcal{A})\,. \tag{4}$$

Hence, using the usual homotopy transfer theorem [10, Theorem 10.3.2], we conclude that the notion of homotopy algebra structure on a dg sheaf $\mathcal{A}$ is, in some sense, independent of the choice of acyclic covering.

Proceeding further in this fashion, we give the definition of an ∞-morphism (and ∞ quasiisomorphism) in the setting of sheaves:

Definition 4.4. Let $\mathcal{A}$ and $\mathcal{B}$ be dg sheaves on X equipped with structures of homotopy algebras of type C. *An ∞-morphism F* from $\mathcal{A}$ to $\mathcal{B}$ is an ∞-morphism

$$F : N^{\mathrm{TS}}\mathfrak{U}(\mathcal{A}) \rightsquigarrow N^{\mathrm{TS}}\mathfrak{U}(\mathcal{B}) \tag{5}$$

of the corresponding homotopy algebras (in the category of cochain complexes) for some acyclic cover $\mathfrak{U}$. If (5) is an ∞ quasiisomorphism then, we say that, F is an *∞ quasiisomorphism* from $\mathcal{A}$ to $\mathcal{B}$.

Remark 5. Again, since the Thom–Sullivan normalization N^{TS} is a symmetric monoidal functor from cosimplicial cochain complexes into cochain complexes, an ∞-morphism in the sense of naive Definition 4.2 gives us an ∞-morphism in the of Definition 4.4.

4.1 The Deformation Complex in the Setting of Sheaves

Let X be a topological space and let $\mathcal{A}$ be a dg sheaf on X. Let us assume that $\mathfrak{U}$ is an acyclic (for $\mathcal{A}$) cover of X and $\mathcal{A}$ carries a homotopy algebra of type C defined in terms of this cover $\mathfrak{U}$.

Definition 4.5. *The deformation complex* of the sheaf of homotopy algebras $\mathcal{A}$ is

$$\mathrm{Def}(\mathcal{A}) := \mathrm{Def}(N^{\mathrm{TS}}\mathfrak{U}(\mathcal{A})).$$

Remark 6. The above definition of the deformation complex is independent on the choice of the acyclic cover in the following sense: Let $\mathfrak{U}'$ be another acyclic cover of X. Since the cochain complexes $N^{\mathrm{TS}}\mathfrak{U}(\mathcal{A})$ and $N^{\mathrm{TS}}\mathfrak{U}'(\mathcal{A})$ are connected by the pair of quasiisomorphisms (4), Theorem 3.1 and the homotopy transfer theorem imply that the deformation complexes corresponding to different acyclic coverings are connected by a sequence of quasiisomorphisms of dg Lie algebras.

Theorem 3.1 has the following obvious implication.

Corollary 4.6. *Let $\mathcal{A}$ and $\mathcal{B}$ be dg sheaves on X equipped with structures of homotopy algebras of type C. If $\mathcal{A}$ and $\mathcal{B}$ are connected by a sequence of ∞ quasiisomorphisms, then* $\mathrm{Def}(\mathcal{A})$ *and* $\mathrm{Def}(\mathcal{B})$ *are quasiisomorphic dg Lie algebras.*

□

4.2 An Application of Corollary 4.6

In applications we often deal with honest (versus ∞) algebraic structures on sheaves and maps of sheaves which are compatible with these algebraic structures on the nose (not up to homotopy). Here we describe a setting of this kind in which Corollary 4.6 can be applied.

Let O be a dg operad and $\mathrm{Cobar}(C)$ be a resolution of O for which the cooperad C satisfies condition (2.1).

Every dg sheaf of O-algebras $\mathcal{A}$ is naturally a sheaf of $\mathrm{Cobar}(C)$-algebras. Hence, $\mathcal{A}$ carries a structure of homotopy algebra of type C and we define the deformation complex of $\mathcal{A}$ as

$$\mathrm{Def}(\mathcal{A}) := \mathrm{Def}(N^{\mathrm{TS}}\mathfrak{U}(\mathcal{A}))\,.$$

Theorem 4.7. *Let $\mathcal{A}$ and $\mathcal{B}$ be dg sheaves of O-algebras on a topological space X. If there exists a sequence of quasiisomorphisms of dg sheaves of O-algebras*

$$\mathcal{A} \stackrel{\sim}{\leftarrow} \mathcal{A}_1 \stackrel{\sim}{\rightarrow} \mathcal{A}_2 \stackrel{\sim}{\leftarrow} \cdots \stackrel{\sim}{\rightarrow} \mathcal{A}_n \stackrel{\sim}{\rightarrow} \mathcal{B}$$

then the dg Lie algebras $\mathrm{Def}(\mathcal{A})$ *and* $\mathrm{Def}(\mathcal{B})$ *are quasiisomorphic.*

Proof. It is suffices to prove this theorem in the case when $\mathcal{A}$ and $\mathcal{B}$ are connected by a single quasiisomorphism

$$f : \mathcal{A} \stackrel{\sim}{\rightarrow} \mathcal{B} \tag{6}$$

of dg sheaves of O-algebras.

Since the functor N^{TS} preserves quasiisomorphisms, f induces a quasiisomorphism

$$f_* : N^{\mathrm{TS}}\mathfrak{U}(\mathcal{A}) \stackrel{\sim}{\rightarrow} N^{\mathrm{TS}}\mathfrak{U}(\mathcal{B}) \tag{7}$$

for any acyclic cover $\mathfrak{U}$.

Furthermore, since N^{TS} is compatible with the symmetric monoidal structure, the map f_* is compatible with the O-algebra structures on $N^{\mathrm{TS}}\mathfrak{U}(\mathcal{A})$ and $N^{\mathrm{TS}}\mathfrak{U}(\mathcal{B})$.

Therefore, f_* may be viewed as an ∞ quasiisomorphism from $\mathcal{A}$ to $\mathcal{B}$. Thus Corollary 4.6 implies the desired statement. □

4.3 A Concluding Remark About Definitions 4.3–4.5

For certain applications, Definitions 4.3–4.5 may still be naive. One may ask about the possibility to extend the notion of homotopy algebras to the setting of twisted complexes [2–4, 8]. For some application one may need a universal way of keeping track of "dependencies on covers" by using the notion of hypercover. For other applications one may need a notion of deformation complex which would also govern deformations of $\mathcal{A}$ as a sheaf or possibly as a (higher) stack.

However, for applications considered in [6], the framework of Definitions 4.3–4.5 is sufficient.

Appendix: Cylinder Type Construction

Given a pair $(f, \tilde{f})$ of maps of cochain complexes

$$V \stackrel{f}{\longrightarrow} W \stackrel{\tilde{f}}{\longleftarrow} \tilde{V}\,, \tag{8}$$

we form another cochain complex $\mathrm{Cyl}(f, \tilde{f})$. As a graded vector space

$$\mathrm{Cyl}(f, \tilde{f}) := V \oplus \mathbf{s}W \oplus \tilde{V} \tag{9}$$

and the differential ∂^{Cyl} is defined by the formula

$$\partial^{\mathrm{Cyl}}(v + \mathbf{s}w + \tilde{v}) := \partial v + \mathbf{s}(f(v) - \partial w + \tilde{f}(\tilde{v})) + \partial \tilde{v}\,. \tag{10}$$

The equation

$$\partial^{\mathrm{Cyl}} \circ \partial^{\mathrm{Cyl}} = 0$$

is a consequence of $\partial^2 = 0$ and the compatibility of f (resp. $\tilde{f}$) with the differentials[5] on V, W, and $\tilde{V}$.

We have the obvious pair of maps of cochain complexes:

$$V \stackrel{\pi_V}{\longleftarrow} \mathrm{Cyl}(f, \tilde{f}) \stackrel{\pi_{\tilde{V}}}{\longrightarrow} \tilde{V} \tag{11}$$

$$\pi_V(v + \mathbf{s}w + \tilde{v}) = v, \tag{12}$$

$$\pi_{\tilde{V}}(v + \mathbf{s}w + \tilde{v}) = \tilde{v}. \tag{13}$$

We claim that

[5]By abuse of notation, we denote by the same letter ∂ the differential on V, W, and $\tilde{V}$.

Lemma 1. *If f and $\tilde{f}$ are quasiisomorphisms of cochain complexes, then so are π_V and $\pi_{\tilde{V}}$.*

Proof. Let us prove that π_V is surjective on the level of cohomology.

For this purpose, we observe that for every cocycle $v \in V$ its image $f(v)$ in W is cohomologous to some cocycle of the form $\tilde{f}(v')$, where v' is a cocycle in $\tilde{V}$. The latter follows easily from the fact that f and $\tilde{f}$ are quasiisomorphisms.

In other words, for every degree n cocycle $v \in V$ there exists a degree n cocycle $v' \in V$ and a degree $(n-1)$ vector $w \in W$ such that

$$f(v) - \tilde{f}(v') - \partial w = 0 . \tag{14}$$

Hence, $v + \mathbf{s}w - v'$ is a cocycle in $\mathrm{Cyl}(f, \tilde{f})$ such that $\pi(v + \mathbf{s}w - v') = v$.

Let us now prove that π_V is injective on the level of cohomology.

For this purpose, we observe that the cocycle condition for $v + \mathbf{s}w + v' \in \mathrm{Cyl}(f, \tilde{f})$ is equivalent to the three equations:

$$\partial v = 0 , \tag{15}$$

$$\partial v' = 0 , \tag{16}$$

and

$$f(v) + \tilde{f}(v') - \partial w = 0 . \tag{17}$$

Therefore, for every cocycle $v + \mathbf{s}w + v' \in \mathrm{Cyl}(f, \tilde{f})$, the vectors v and v' are cocycles in V and $\tilde{V}$, respectively, and the cocycles $f(v)$ and $-\tilde{f}(v')$ in W are cohomologous.

Hence, $v + \mathbf{s}w + v' \in \mathrm{Cyl}(f, \tilde{f})$ is a cocycle and v is exact, then so is v', i.e., there exist vectors $v_1 \in V$ and $v'_1 \in \tilde{V}$ such that

$$v = \partial v_1 , \qquad v' = \partial v'_1 .$$

Subtracting the coboundary of $v_1 \oplus \mathbf{s}0 \oplus v'_1$ from $v \oplus \mathbf{s}w \oplus v'$ we get a cocycle in $\mathrm{Cyl}(f, \tilde{f})$ of the form

$$0 \oplus \mathbf{s}(w - f(v_1) - \tilde{f}(v'_1)) \oplus 0. \tag{18}$$

Since $w - f(v_1) - \tilde{f}(v'_1)$ is a cocycle on W and $\tilde{f}$ is a quasiisomorphism, there exists a cocycle $\tilde{v} \in \tilde{V}$ and a vector $w_1 \in W$ such that

$$w - f(v_1) - \tilde{f}(v'_1) - \tilde{f}(\tilde{v}) - \partial(w_1) = 0 . \tag{19}$$

Hence the cocycle (18) is the coboundary of

$$0 \oplus (-\mathbf{s}w_1) \oplus \tilde{v} \in \mathrm{Cyl}(f, \tilde{f}).$$

Thus π_V is indeed injective on the level of cohomology.

Switching the roles $V \leftrightarrow \tilde{V}$, $f \leftrightarrow \tilde{f}$, and $\pi_V \leftrightarrow \pi_{\tilde{V}}$ we also prove the desired statement about $\pi_{\tilde{V}}$. □

Acknowledgement We would like to thank Bruno Vallette for useful discussions.

References

1. A. K. Bousfield and V. K. A. M. Gugenheim, *On PL de Rham theory and rational homotopy type*, Mem. Amer. Math. Soc. **8**, 179 (1976) ix+94 pp.
2. N. O'Brian, *Geometry of twisting cochains*, Compositio Math. **63**, 1 (1987) 41–62.
3. N. O'Brian, D. Toledo, and Y. L. Tong, *Hierzebruch-Riemann-Roch for coherent sheaves*, Amer. J. Math. **103**, 2 (1981) 253–271.
4. P. Bressler, A. Gorokhovsky, R. Nest and B. Tsygan, *Chern character for twisted complexes*, Geometry and dynamics of groups and spaces, Progr. Math., 265, Birkhäuser, Basel, 2008, pp. 309–324.
5. V.A. Dolgushev and C.L. Rogers, Notes on algebraic operads, graph complexes, and Willwacher's construction, *Mathematical aspects of quantization*, 25–145, Contemp. Math., **583**, AMS., Providence, RI, 2012; arXiv:1202.2937.
6. V.A. Dolgushev, C. L. Rogers, and T.H. Willwacher, *Kontsevich's graph complex, GRT, and the deformation complex of the sheaf of polyvector fields*, arXiv:1211.4230.
7. V.A. Dolgushev and T. H. Willwacher, *Operadic Twisting – with an application to Deligne's conjecture*, arXiv:1207.2180.
8. H. Gillet, *K-theory of twisted complexes*, Contemporary Mathematics, **55**, 1 (1986) 159–191.
9. B. Keller, *Hochschild cohomology and derived Picard groups*, Journal of Pure and Applied Algebra, **190** (2004) 177–196.
10. J.-L. Loday and B. Vallette, *Algebraic operads*, Grundlehren der Mathematischen Wissenschaften **346**, Springer, Heidelberg, 2012. xxiv+634 pp.
11. S. Merkulov and B. Vallette, *Deformation theory of representations of prop(erad)s. I and II*, J. Reine Angew. Math. **634**, **636** (2009) 51–106, 123–174.
12. M. Van den Bergh, *On global deformation quantization in the algebraic case*, J. Algebra **315**, 1 (2007), 326–395.

Invariants of Artinian Gorenstein Algebras and Isolated Hypersurface Singularities

Michael Eastwood and Alexander Isaev

Abstract We survey our recently proposed method for constructing biholomorphic invariants of quasihomogeneous isolated hypersurface singularities and, more generally, invariants of graded Artinian Gorenstein algebras. The method utilizes certain polynomials associated to such algebras, called nil-polynomials, and we compare them with two other classes of polynomials that have also been used to produce invariants.

Key words Artinian Gorenstein algebras • Isolated hypersurface singularities

Mathematics Subject Classification (2010): Primary 13H10, Secondary 13E10, 32S25, 13A50.

1 Introduction

On 27th October 2001, one of us (MGE) gave a talk entitled 'Invariants of isolated hypersurface singularities' at the West Coast Lie Theory Workshop held at the University of California, Berkeley. The purpose of this talk was to propose, at that time only by means of examples, a method for extracting invariants of isolated hypersurface singularities. Since classical invariant theory was the key ingredient in the method and since classical invariants may be derived from representation theory of the general linear group, this seemed an appropriate topic for the West Coast Lie Theory Series. The examples from this talk were presented in [5]. Since that time, the method from [5] has been considerably improved (by authors other than MGE)

Work supported by the Australian Research Council.

M. Eastwood (✉) • A. Isaev
Mathematical Sciences Institute, The Australian National University, Canberra, ACT 0200, Australia
e-mail: michael.eastwood@anu.edu.au; alexander.isaev@anu.edu.au

G. Mason et al. (eds.), *Developments and Retrospectives in Lie Theory: Algebraic Methods*, Developments in Mathematics 38,
DOI 10.1007/978-3-319-09804-3_7

and related to other methods. Also, it has been realized that the proper framework in which to formulate the results is within the theory of Gorenstein algebras. Although this now seems quite far from Lie theory, we believe it is the natural development and we thank the editors of this volume for the opportunity to present this survey, which might otherwise appear out of place.

Let V be a complex hypersurface germ with an isolated singularity at $0 \in \mathbb{C}^n$, $n \geq 2$. It follows that the complex structure of V is reduced, i.e., defined by the ideal $I(V)$ in the algebra $\mathcal{O}_n$ of germs of holomorphic functions at the origin that consists of all elements of $\mathcal{O}_n$ vanishing on V. The singularity of V is called *quasihomogeneous* if some (hence every) generator of $I(V)$ in some coordinates $z_1, \ldots, z_n$ near the origin is the germ of a quasihomogeneous polynomial $Q(z_1, \ldots, z_n)$, i.e., a polynomial satisfying $Q(t^{p_1}z_1, \ldots, t^{p_n}z_n) \equiv t^q Q(z_1, \ldots, z_n)$ for fixed positive integers $p_1, \ldots, p_n, q$ and all $t \in \mathbb{C}$, where p_j is called the weight of z_j and q the weight of Q. The singularity is called *homogeneous* if one can choose Q to be a form (homogeneous polynomial). This paper concerns the biholomorphic invariants of quasihomogeneous singularities introduced in our recent article [6]. Various invariants of hypersurface singularities have been extensively studied by many authors (see Chapter 1 in [13] for an account of some of the results). The objective of [6] was to construct, in the quasihomogeneous case, numerical invariants that: (i) are easy to compute, and (ii) form a complete system with respect to the biholomorphic equivalence problem for hypersurface germs. Although we succeed only in a very limited setting (concerning binary quintics and sextics, whose classical invariants are well understood), to the best of our knowledge, no such system of invariants had been previously known.

Our approach utilizes the *Milnor algebra* of V, which is the complex local commutative associative algebra $M(V) := \mathcal{O}_n/J(f)$, where f is any generator of $I(V)$ and $J(f)$ is the ideal in $\mathcal{O}_n$ generated by all first-order partial derivatives of f calculated with respect to some coordinate system near the origin. It is easy to observe that the above definition is independent of the choice of f as well as the coordinate system, and that the Milnor algebras of biholomorphically equivalent singularities are isomorphic. Furthermore, the dimension n and the isomorphism class of $M(V)$ determine V up to biholomorphic equivalence (see [23] and a more general result in [18]). Thus, any quantity that depends only on the isomorphism class of $M(V)$ is a biholomorphic invariant of V, and any collection of quantities of this kind uniquely characterizing the isomorphism class of every Milnor algebra is a complete system of biholomorphic invariants for hypersurface germs of fixed dimension.

In order to produce invariants of Milnor algebras of quasihomogeneous singularities, we focus on three important properties of $M(V)$. First, since the singularity of V is isolated, one has $\dim_{\mathbb{C}} M(V) < \infty$ (see, e.g., Chapter 1 in [13]), that is, the algebra $M(V)$ is *Artinian*. It then follows that the first-order partial derivatives of f form a regular sequence in $\mathcal{O}_n$ (see Theorem 2.1.2 in [3]), hence, by [2], the algebra $M(V)$ is *Gorenstein*. (Recall that a local commutative associative algebra A of finite vector space dimension greater than 1 is Gorenstein if the annihilator $\mathrm{Ann}(\mathfrak{m}) := \{x \in \mathfrak{m} : x\mathfrak{m} = 0\}$ of the maximal ideal $\mathfrak{m} \subset A$ is

1-dimensional – see, e.g., [16].) Finally, $M(V)$ is *(nonnegatively) graded*, i.e., it can be represented as a direct sum $M(V) = \bigoplus_{i\geq 0} L_i$, where L_i are subspaces such that $L_0 \simeq \mathbb{C}$ and $L_i L_j \subset L_{i+j}$ for all i, j. Indeed, choosing coordinates near the origin in which f is the germ of a quasihomogeneous polynomial, we set L_i to be the subspace of elements of $M(V)$ represented by germs of quasihomogeneous polynomials of weight i.

Rather than focussing on invariants of Milnor algebras of quasihomogeneous singularities, one can take a broader viewpoint and introduce certain invariants of general complex graded Artinian Gorenstein algebras. Our method for constructing invariants is based on associating to every algebra A a form P_A of degree d_A on a complex vector space W_A of dimension N_A, such that for any pair of isomorphic algebras A, $\tilde{A}$, there exists a linear isomorphism $\varphi : W_A \to W_{\tilde{A}}$ with $P_A = P_{\tilde{A}} \circ \varphi$. Then, upon identification of W_A with $\mathbb{C}^{N_A}$, for any absolute classical invariant $\mathbf{I}$ of forms of degree d_A on $\mathbb{C}^{N_A}$, the quantity $\mathbf{I}(P_A)$ is invariantly defined. Observe that for a given choice of P_A all invariants of this kind are easy to calculate using computer algebra.

The idea of building invariants by the above method goes back at least as far as article [9] (see also [7] and references therein for more detail), where it was briefly noted for the case of *standard graded* Artinian Gorenstein algebras, in which case P_A is a *Macaulay inverse system* for A. For instance, if the singularity of V is homogeneous, the algebra $M(V)$ is standard graded. For general quasihomogeneous singularities the idea was explored in [5], with P_A being a certain form a on $\mathfrak{m}/\mathfrak{m}^2$. Finally, in [6] we utilized homogeneous components of *nil-polynomials* introduced in [10] to construct a large number of invariants of arbitrary graded Artinian Gorenstein algebras. Relationships among the above three choices of P_A are discussed in Sect. 2, where we will see, in particular, that nil-polynomials can be regarded as certain extensions of both inverse systems and the form a. Hence, the invariants produced from nil-polynomials incorporate those arising from the other two possibilities for P_A. The construction of this most general system of invariants and results concerning its completeness are surveyed in Sect. 3.

2 Polynomials Associated to Artinian Gorenstein Algebras

In this section we establish relationships among three kinds of polynomials arising from Artinian Gorenstein algebras. As mentioned in the introduction, we will consider inverse systems, the form a introduced in [5], and nil-polynomials introduced in [10]. For expository purposes, it is convenient for us to start with nil-polynomials.

Let A be an Artinian Gorenstein algebra over a field $\mathbb{F}$ of characteristic zero, with $\dim_{\mathbb{F}} A > 2$ and maximal ideal $\mathfrak{m}$. Define a map $\exp : \mathfrak{m} \to \mathbf{1} + \mathfrak{m}$ by the formula

$$\exp(x) := \sum_{s=0}^{\infty} \frac{1}{s!} x^s,$$

where $x^0 := \mathbf{1}$, with $\mathbf{1}$ being the identity element of A. By Nakayama's lemma, $\mathfrak{m}$ is a nilpotent algebra, and therefore the above sum is in fact finite, with the highest-order term corresponding to $s = \nu$, where $\nu \geq 2$ is the nil-index of $\mathfrak{m}$ (i.e., the largest of all integers μ for which $\mathfrak{m}^\mu \neq 0$). Fix a hyperplane Π in $\mathfrak{m}$ complementary to $\mathrm{Ann}(\mathfrak{m}) = \mathfrak{m}^\nu$. An $\mathbb{F}$-valued polynomial P on Π is called a nil-polynomial if there exists a linear form $\omega : A \to \mathbb{F}$ such that $\ker \omega = \langle \Pi, \mathbf{1} \rangle$ and

$$P = \omega \circ \exp|_\Pi, \ \text{i.e.,}\ P(x) = \omega\left(\sum_{s=2}^{\nu} \frac{1}{s!} x^s\right), \quad x \in \Pi,$$

where $\langle \cdot \rangle$ denotes linear span.

As shown in [10, 11, 17], nil-polynomials solve the isomorphism problem for Artinian Gorenstein algebras as follows: A and $\tilde{A}$ are isomorphic if and only if the graphs $\Gamma \subset \Pi \times \mathbb{F}$, $\tilde{\Gamma} \subset \tilde{\Pi} \times \mathbb{F}$ of any nil-polynomials P, $\tilde{P}$ arising from A, $\tilde{A}$, respectively, are affinely equivalent, that is, there exists a bijective affine map $\psi : \Pi \times \mathbb{F} \to \tilde{\Pi} \times \mathbb{F}$ such that $\psi(\Gamma) = \tilde{\Gamma}$. Furthermore, if both A and $\tilde{A}$ are graded, then Γ and $\tilde{\Gamma}$ are affinely equivalent if and only if P and $\tilde{P}$ are linearly equivalent up to scale, i.e., there exist $c \in \mathbb{F}^*$ and a linear isomorphism $\varphi : \Pi \to \tilde{\Pi}$ with $cP = \tilde{P} \circ \varphi$. As will be seen in Sect. 3, this is exactly the property that allows one to use nil-polynomials for producing invariants of graded Artinian Gorenstein algebras.

Further, any nil-polynomial $P = \omega \circ \exp|_\Pi$ arising from a Gorenstein algebra A extends to the polynomial $\hat{P} := \omega \circ \exp$ on all of $\mathfrak{m}$. Let

$$\hat{P}^{[s]}(x) := \frac{1}{s!}\omega(x^s), \quad x \in \mathfrak{m},$$

be the homogeneous component of $\hat{P}$ of degree s, with $s = 2, \dots, \nu$. One has $\hat{P}^{[s]}(y) = 0$, $\hat{P}^{[s]}(x + y) = \hat{P}^{[s]}(x)$ for all $x \in \mathfrak{m}$, $y \in \mathfrak{m}^{\nu+2-s}$. Thus, $\hat{P}^{[s]}$ gives rise to a form $\mathbf{P}^{[s]}$ on the quotient $\mathfrak{m}/\mathfrak{m}^{\nu+2-s}$. The forms $\mathbf{P}^{[s]}$ will be used in Sect. 3 for constructing the invariants mentioned above. Here we only observe that the highest-degree form $\mathbf{P}^{[\nu]}$ defined on $\mathfrak{m}/\mathfrak{m}^2$ is special. Indeed, for any other nil-polynomial P' arising from A, the corresponding form $\mathbf{P}'^{[\nu]}$ coincides with $\mathbf{P}^{[\nu]}$ up to scale. Moreover, it is clear from the definition of the form a given on p. 305 in [5] that it is equal, up to scale, to $\mathbf{P}^{[\nu]}$. Thus, loosely speaking, a can be regarded, up to proportionality, as the highest-degree homogeneous component of any nil-polynomial.

Next, let $k := \mathrm{emb}\dim A := \dim_{\mathbb{F}} \mathfrak{m}/\mathfrak{m}^2 \geq 1$ be the embedding dimension of A. Since $\dim_{\mathbb{F}} A > 2$, the hyperplane Π contains a k-dimensional subspace that forms a complement to $\mathfrak{m}^2$ in $\mathfrak{m}$. Fix any such subspace L, choose a basis $e_1, \dots, e_k$

in it, and let $y_1, \dots, y_k$ be the coordinates with respect to this basis. Denote by $R \in \mathbb{F}[y_1, \dots, y_k]$ the restriction of the nil-polynomial P to L expressed in these coordinates. Clearly, one has

$$R(y_1, \dots, y_k) = \sum_{j=0}^{\nu} \frac{1}{j!} \omega\Big((y_1 e_1 + \cdots + y_k e_k)^j\Big),$$

and the homogeneous component $R^{[\nu]}$ of degree ν of R is given by

$$R^{[\nu]}(y_1, \dots, y_k) = \frac{1}{\nu!} \omega\Big((y_1 e_1 + \cdots + y_k e_k)^\nu\Big).$$

Thus, identifying L with $\mathfrak{m}/\mathfrak{m}^2$, we see that $R^{[\nu]}$ is a coordinate representation of the form $\mathbf{P}^{[\nu]}$ and therefore that of the form a up to a scaling factor.

Further, the elements $e_1, \dots, e_k$ generate A as an algebra, hence A is isomorphic to $\mathbb{F}[x_1, \dots, x_k]/I$, where I is the ideal of all relations among $e_1, \dots, e_k$, i.e., polynomials $f \in \mathbb{F}[x_1, \dots, x_k]$ with $f(e_1, \dots, e_k) = 0$. Observe that I contains the monomials $x_1^{\nu+1}, \dots, x_k^{\nu+1}$, and therefore A is also isomorphic to $\mathbb{F}[[x_1, \dots, x_k]]/\mathbb{F}[[x_1, \dots, x_k]]I$. It is well known that since the quotient $\mathbb{F}[x_1, \dots, x_k]/I$ is Gorenstein, there is a polynomial $g \in \mathbb{F}[y_1, \dots, y_k]$ of degree ν satisfying $\mathrm{Ann}(g) = I$, where

$$\mathrm{Ann}(g) := \left\{ f \in \mathbb{F}[x_1, \dots, x_k] : f\left(\frac{\partial}{\partial y_1}, \dots, \frac{\partial}{\partial y_k}\right)(g) = 0 \right\}$$

is the annihilator of g (see, e.g., [7] and references therein). The freedom in choosing g with $\mathrm{Ann}(g) = I$ is fully understood, and any such polynomial is called a Macaulay inverse system for the Gorenstein quotient $\mathbb{F}[x_1, \dots, x_k]/I$. The classical correspondence $I \leftrightarrow g$ can be also derived from the *Matlis duality* (see Section 5.4 in [22]).

Inverse systems can be used for solving the isomorphism problem for quotients of this kind. Namely, two Gorenstein quotients are isomorphic if and only if their inverse systems are equivalent in a certain sense (see Proposition 16 in [9] and a more explicit formulation in Proposition 2.2 in [7]). Observe that in general the equivalence relation for inverse systems is harder to analyze than the affine equivalence of graphs of nil-polynomials mentioned above, and therefore the criterion for isomorphism of Artinian Gorenstein algebras in terms of inverse systems seems to be less convenient in applications than that in terms of nil-polynomials. There is one case, however, when the criterion in terms of inverse systems is rather useful. It is discussed in Remark 2 at the end of this section.

The following theorem provides a connection between nil-polynomials and inverse systems.

Theorem 1. *The polynomial R is an inverse system for the quotient $\mathbb{F}[x_1, \dots, x_k]/I$.*

Proof. Fix any polynomial $f \in \mathbb{F}[x_1, \ldots, x_k]$

$$f = \sum_{0 \le i_1, \ldots, i_k \le N} a_{i_1, \ldots, i_k} x_1^{i_1} \ldots x_k^{i_k}$$

and calculate

$$\begin{aligned}
&f\left(\frac{\partial}{\partial y_1}, \ldots, \frac{\partial}{\partial y_k}\right)(R) \\
&= \sum_{0 \le i_1, \ldots, i_k \le N} a_{i_1, \ldots, i_k} \sum_{j = i_1 + \cdots + i_k}^{\nu} \frac{1}{(j - (i_1 + \cdots + i_k))!} \\
&\qquad \times \omega\Big((y_1 e_1 + \cdots + y_k e_k)^{j - (i_1 + \cdots + i_k)} e_1^{i_1} \ldots e_k^{i_k}\Big) \\
&= \sum_{m=0}^{\nu} \frac{1}{m!} \omega\Big((y_1 e_1 + \cdots + y_k e_k)^m \times \sum_{\substack{0 \le i_1, \ldots, i_k \le N, \\ i_1 + \cdots + i_k \le \nu - m}} a_{i_1, \ldots, i_k} e_1^{i_1} \ldots e_k^{i_k}\Big) \\
&= \sum_{m=0}^{\nu} \frac{1}{m!} \omega\Big((y_1 e_1 + \cdots + y_k e_k)^m f(e_1, \ldots, e_k)\Big).
\end{aligned} \tag{1}$$

Formula (1) immediately implies $I \subset \mathrm{Ann}(R)$.

Conversely, let $f \in \mathbb{F}[x_1, \ldots, x_k]$ be an element of $\mathrm{Ann}(R)$. Then (1) yields

$$\sum_{m=0}^{\nu} \frac{1}{m!} \omega\Big((y_1 e_1 + \cdots + y_k e_k)^m f(e_1, \ldots, e_k)\Big) = 0. \tag{2}$$

Collecting the terms containing $y_1^{i_1} \ldots y_k^{i_k}$ in (2) we obtain

$$\omega\Big(e_1^{i_1} \ldots e_k^{i_k} f(e_1, \ldots, e_k)\Big) = 0 \tag{3}$$

for all indices $i_1, \ldots, i_k$. Since $e_1, \ldots, e_k$ generate A, identities (3) yield

$$\omega\Big(A\, f(e_1, \ldots, e_k)\Big) = 0. \tag{4}$$

Further, since the bilinear form $(a, b) \mapsto \omega(ab)$ is nondegenerate on A (see, e.g., p. 11 in [14]), identity (4) implies $f(e_1, \ldots, e_k) = 0$. Therefore $f \in I$, which shows that $I = \mathrm{Ann}(R)$ as required. □

Remark 1. Theorem 1 easily generalizes to the case of Artinian Gorenstein quotients $\mathbb{F}[x_1, \ldots, x_m]/I$, where I lies in the ideal generated by $x_1, \ldots, x_m$ and m is not necessarily the embedding dimension of the quotient. Indeed, let $e_1, \ldots, e_m$

be the elements of $\mathbb{F}[x_1, \dots, x_m]/I$ represented by $x_1, \dots, x_m$, respectively, and consider the polynomial in $\mathbb{F}[y_1, \dots, y_m]$ defined as follows:

$$S(y_1, \dots, y_m) := \sum_{j=0}^{\nu} \frac{1}{j!} \omega\Big((y_1 e_1 + \cdots + y_m e_m)^j\Big),$$

where ω is a linear form on $\mathbb{F}[x_1, \dots, x_m]/I$ with kernel complementary to $\mathrm{Ann}(\mathfrak{m})$ and ν is the nil-index of the maximal ideal of $\mathbb{F}[x_1, \dots, x_m]/I$. Then arguing as in the proof of Theorem 1, we see that S is an inverse system for $\mathbb{F}[x_1, \dots, x_m]/I$. However, if $m > \mathrm{emb\,dim}\mathbb{F}\,[x_1, \dots, x_m]/I$, this inverse system does not come from restricting a nil-polynomial to a subspace of $\mathfrak{m}$ complementary to $\mathfrak{m}^2$.

We will now give an example illustrating the relationships among nil-polynomials, inverse systems and the form a established above.

Example 1. Consider the following one-parameter family of algebras:

$$A_t := \mathbb{F}[x_1, x_2]/(2x_1^3 + t x_1 x_2^3, t x_1^2 x_2^2 + 2x_2^5), \quad t \in \mathbb{F},\ t \neq \pm 2.$$

It is straightforward to verify that every A_t is a Gorenstein algebra of dimension 15 with $\nu = 7$ for $t \neq 0$ and $\nu = 6$ for $t = 0$.

Consider the following monomials in $\mathbb{F}[x_1, x_2]$:

$$x_1,\ x_2,\ x_1^2,\ x_1x_2,\ x_2^2,\ x_1^2x_2,\ x_1x_2^2,\ x_2^3,\ x_1x_2^3,\ x_1^2x_2^2,\ x_2^4,\ x_1^2x_2^3,\ x_1x_2^4,\ x_1^2x_2^4.$$

Let $e_1, \dots, e_{14}$, respectively, be the elements of A_t represented by these monomials. They form a basis of $\mathfrak{m}$. The hyperplane $\Pi := \langle e_1, \dots, e_{13} \rangle$ in $\mathfrak{m}$ is complementary to $\mathrm{Ann}(\mathfrak{m}) = \langle e_{14} \rangle$, and we denote by $y_1, \dots, y_{13}$ the coordinates in Π with respect to $e_1, \dots, e_{13}$. Further, define $\omega : A_t \to \mathbb{F}$ to be the linear form such that $\ker \omega = \langle \Pi, \mathbf{1} \rangle$ and $\omega(e_{14}) = 1$. Then the nil-polynomial $P := \omega \circ \exp|_\Pi$ is expressed in the coordinates $y_1, \dots, y_{13}$ as follows:

$$\begin{aligned} P(y_1, \dots, y_{13}) = {} & \frac{t}{10080} y_2^7 - \frac{1}{48} y_2^4 \left(y_1^2 - \frac{t}{5} y_2 y_5 \right) + \frac{t}{48} y_1^4 y_2 - \frac{1}{4} y_1^2 y_2^2 y_5 \\ & - \frac{1}{6} y_1 y_2^3 y_4 + \frac{t}{24} y_2^3 y_5^2 + \frac{t}{48} y_2^4 y_8 - \frac{1}{24} y_2^4 y_3 + \text{terms of deg} \leq 4. \end{aligned}$$

Further, setting $L := \langle e_1, e_2 \rangle$ and restricting P to L, we arrive at an inverse system of A_t:

$$R(y_1, y_2) = \frac{t}{10080} y_2^7 - \frac{1}{48} y_1^2 y_2^4 + \frac{t}{48} y_1^4 y_2$$

(note that all terms of deg ≤ 4 in P vanish for $y_3 = \cdots = y_{13} = 0$). Then, choosing the highest-degree terms in R and identifying L with $\mathfrak{m}/\mathfrak{m}^2$, we obtain the following coordinate representation of the form a:

$$a(y_1, y_2) = \begin{cases} \dfrac{t}{10080} y_2^7 \text{ if } t \neq 0, \\ \\ -\dfrac{1}{48} y_1^2 y_2^4 \text{ if } t = 0 \end{cases} \qquad \text{up to scale.}$$

Remark 2. Let now A be a standard graded algebra, i.e., an algebra that can be represented as a direct sum $A = \bigoplus_{i \geq 0} L_i$, with $L_0 \simeq \mathbb{F}$, $L_i L_j \subset L_{i+j}$ for all i, j, and $L_l = (L_1)^l$ for all $l \geq 1$. Choose

$$\Pi := \bigoplus_{i=1}^{\nu-1} L_i, \quad L := L_1.$$

In this case for any choice of the basis $e_1, \ldots, e_k$ in L the ideal I is homogeneous, i.e., generated by homogeneous relations. For an arbitrary nil-polynomial P on Π its restriction to L expressed in the corresponding coordinates is

$$R(y_1, \ldots, y_k) = \frac{1}{\nu!} \omega\Big((y_1 e_1 + \cdots + y_k e_k)^\nu \Big).$$

Identifying L with $\mathfrak{m}/\mathfrak{m}^2$, we observe that the homogeneous inverse system R is a coordinate representation of the form $\mathbf{P}^{[\nu]}$ and therefore that of the form a up to scale.

Theorem 1 yields the well-known fact (see, e.g., Proposition 7 in [9]) that a standard graded Artinian Gorenstein algebra, when written as a quotient by a homogeneous ideal, admits a homogeneous inverse system (note that any two such systems are proportional). In this situation the criterion for isomorphism of algebras in terms of inverse systems stated in Proposition 16 in [9] becomes rather simple: two quotients are isomorphic if and only if their homogeneous inverse systems are linearly equivalent up to scale (see Proposition 17 in [9] and Proposition 2.2 in [7]). We note that this classical criterion can be easily derived from Theorem 1 as well.

3 The System of Invariants

In this section we survey the construction and properties of the system of invariants introduced in [6]. Here we assume that $\mathbb{F} = \mathbb{C}$, although much of what follows works for any algebraically closed field of characteristic zero.

Let A and $\tilde{A}$ be graded Artinian Gorenstein algebras of vector space dimension greater than 2, and $P : \Pi \to \mathbb{C}$, $\tilde{P} : \tilde{\Pi} \to \mathbb{C}$ some nil-polynomials arising from A, $\tilde{A}$, respectively. Assume that A and $\tilde{A}$ are isomorphic. As stated in Sect. 2, in this case P and $\tilde{P}$ are linearly equivalent up to scale, i.e., there exist $c \in \mathbb{C}^*$ and a linear isomorphism $\varphi : \Pi \to \tilde{\Pi}$ with $cP = \tilde{P} \circ \varphi$. Moreover, as shown in [10, 11, 17], the map

$$\hat{\varphi} : \mathfrak{m} \to \tilde{\mathfrak{m}}, \quad x + y \mapsto \varphi(x) + c\,\tilde{\omega}_0^{-1}(\omega(y)), \quad x \in \Pi,\ y \in \mathrm{Ann}(\mathfrak{m}),$$

is an algebra isomorphism, where $\tilde{\omega}_0 := \tilde{\omega}|_{\mathrm{Ann}(\tilde{\mathfrak{m}})}$.

Further, for $s = 2, \ldots, \nu$ consider the forms $\mathbf{P}^{[s]}$ and $\tilde{\mathbf{P}}^{[s]}$ on $\mathfrak{m}/\mathfrak{m}^{\nu+2-s}$ and $\tilde{\mathfrak{m}}/\tilde{\mathfrak{m}}^{\nu+2-s}$ arising from P and $\tilde{P}$, respectively, as explained in Sect. 2. Since the map $\hat{\varphi}$ is an algebra isomorphism, there exist algebra isomorphisms $\varphi^{[s]} : \mathfrak{m}/\mathfrak{m}^{\nu+2-s} \to \tilde{\mathfrak{m}}/\tilde{\mathfrak{m}}^{\nu+2-s}$ such that $c\mathbf{P}^{[s]} = \mathbf{P}^{[s]} \circ \varphi^{[s]}$. This fact allows one to utilize classical invariant theory for constructing numerical invariants of graded Artinian Gorenstein algebras. We will now recall the definitions of relative and absolute classical invariants (see, e.g., [20] for details). These definitions can be given in a coordinate-free setting.

Let W be a finite-dimensional complex vector space and $\mathcal{Q}_W^m$ the linear space of holomorphic forms of degree m on W, with $m \geq 2$. Define an action of $\mathrm{GL}(W)$ on $\mathcal{Q}_W^m$ by the formula

$$(C, Q) \mapsto Q_C, \; Q_C(w) := Q(C^{-1}w), \; \text{where } C \in \mathrm{GL}(W),\, Q \in \mathcal{Q}_W^m,\, w \in W.$$

If two forms lie in the same $\mathrm{GL}(W)$-orbit, they are called linearly equivalent. A relative invariant (or relative classical invariant) of forms of degree m on W is a polynomial $\mathcal{I} : \mathcal{Q}_W^m \to \mathbb{C}$, such that for any $Q \in \mathcal{Q}_W^m$ and any $C \in \mathrm{GL}(W)$, one has $\mathcal{I}(Q) = (\det C)^{\ell}\mathcal{I}(Q_C)$, where ℓ is a nonnegative integer called the weight of $\mathcal{I}$. It follows that $\mathcal{I}$ is in fact homogeneous of degree $\ell \dim_{\mathbb{C}} W/m$. Finite sums of relative invariants comprise the algebra of polynomial $\mathrm{SL}(W)$-invariants of $\mathcal{Q}_W^m$, called the algebra of invariants (or algebra of classical invariants) of forms of degree m on W. As shown by Hilbert in [15], this algebra is finitely generated. For any two invariants $\mathcal{I}$ and $\mathcal{J}$, with $\mathcal{J} \not\equiv 0$, the ratio $\mathcal{I}/\mathcal{J}$ yields a rational function on $\mathcal{Q}_W^m$ that is defined, in particular, at the points where $\mathcal{J}$ does not vanish. If $\mathcal{I}$ and $\mathcal{J}$ have equal weights, this function does not change under the action of $\mathrm{GL}(W)$, and we say that $\mathcal{I}/\mathcal{J}$ is an absolute invariant (or absolute classical invariant) of forms of degree m on W.

If one fixes coordinates $z_1, \ldots, z_n$ in W, then W is identified with $\mathbb{C}^n$, $\mathrm{GL}(W)$ with $\mathrm{GL}(n, \mathbb{C})$, and any element $Q \in \mathcal{Q}_W^m$ is written as a homogeneous polynomial of degree m in $z_1, \ldots, z_n$. Invariants are usually defined in terms of the coefficients of the polynomial in $z_1, \ldots, z_n$ representing Q. Observe, however, that the value of any absolute invariant at Q is independent of the choice of coordinates in W.

The above discussion yields the following result.

Theorem 2 ([6]). *Let A be a graded Artinian Gorenstein algebra with* $\dim_{\mathbb{C}} A > 2$, *and P a nil-polynomial arising from A. Further, for a fixed $s \in \{2, \dots, \nu\}$, let W be a complex vector space isomorphic to $\mathfrak{m}/\mathfrak{m}^{\nu+2-s}$ by means of a linear map $\psi : W \to \mathfrak{m}/\mathfrak{m}^{\nu+2-s}$. Fix an absolute invariant $\mathbf{I}$ of forms of degree s on W. Then the value $\mathbf{I}(\psi^* \mathbf{P}^{[s]})$ depends only on the isomorphism class of A.*

For each s, let $\psi_s : \mathbb{C}^{N_s} \to \mathfrak{m}/\mathfrak{m}^{\nu+2-s}$ be some linear isomorphism, where $N_s := \dim_{\mathbb{C}} \mathfrak{m}/\mathfrak{m}^{\nu+2-s}$. We have thus constructed the following system of invariants:

$$\mathfrak{I} := \bigsqcup_{s=2}^{\nu} \mathfrak{I}_s,$$

where

$$\mathfrak{I}_s := \left\{ \mathbf{I}(\psi_s^* \mathbf{P}^{[s]}),\ \mathbf{I} \text{ is an absolute invariant of forms of degree } s \text{ on } \mathbb{C}^{N_s} \right\}.$$

The highest stratum $\mathfrak{I}_\nu$ of $\mathfrak{I}$ was introduced in [5] by means of the form a, which, as we noted in Sect. 2, coincides with any $\mathbf{P}^{[\nu]}$ up to scale. For standard graded Artinian Gorenstein algebras the idea of building invariants by the above method was briefly indicated in [9] in relation to homogeneous inverse systems. Any homogeneous inverse system in $\mathbb{C}[z_1, \dots, z_k]$ arising from a given algebra of embedding dimension k is calculated as explained in Remark 2 and therefore is proportional to a coordinate representation of any $\mathbf{P}^{[\nu]}$. Hence, the invariants resulting from the idea expressed in [9] also comprise the stratum $\mathfrak{I}_\nu$. This stratum will play an important role below.

We now return to quasihomogeneous singularities, which are the main motivation of this study, and denote by $\mathfrak{I}^M$ the restriction of the system $\mathfrak{I}$ to the Milnor algebras of such singularities. In the remainder of the paper we will discuss the completeness property of $\mathfrak{I}^M$. At this stage, all our completeness results only concern homogeneous singularities, and this is the case that we will consider from now on. We thus further restrict $\mathfrak{I}^M$ to the class of Milnor algebras of homogeneous singularities and denote the restriction by $\mathfrak{I}^{MH}$.

Let Q be a nonzero element of $\mathcal{Q}^m_{\mathbb{C}^n}$ with $n \geq 2$, $m \geq 3$, and V_Q the germ at the origin of the hypersurface $\{Q = 0\}$. Then the singularity of V_Q is isolated if and only if $\Delta(Q) \neq 0$, where Δ is the discriminant (see Chapter 13 in [12]). Define

$$X_n^m := \{Q \in \mathcal{Q}^m_{\mathbb{C}^n} : \Delta(Q) \neq 0\}.$$

Any hypersurface germ V at the origin in $\mathbb{C}^n$ with homogeneous singularity and $\dim_{\mathbb{C}} M(V) > 1$ is biholomorphic to some V_Q with $Q \in X_n^m$, $m \geq 3$.

Next, for $Q, \tilde{Q} \in X_n^m$ the germs V_Q, $V_{\tilde{Q}}$ are biholomorphically equivalent if and only if Q and $\tilde{Q}$ are linearly equivalent. On the other hand, we have the following fact.

Proposition 1 ([6][1]). *The orbits of the* $\mathrm{GL}(n, \mathbb{C})$*-action on* X_n^m *are separated by invariant regular functions on the affine algebraic variety* X_n^m*, i.e., by absolute invariants of the form* I/Δ^p*, where* p *is a nonnegative integer and* I *a relative invariant.*

Let $\mathcal{I}_n^m$ be the algebra of invariant regular functions on X_n^m. This algebra is finitely generated. The above discussion yields that the completeness of the system $\mathfrak{I}^{MH}$ will follow if one shows that the algebra $\mathcal{I}_n^m$ can be somehow "extracted" from $\mathfrak{I}^{MH}$.

Observe that emb dim $M(V_Q) = n$ and $M(V_Q)$ is canonically isomorphic to the quotient $\mathbb{M}(V_Q) := \mathbb{C}[z_1, \ldots, z_n]/\mathbb{J}(Q)$, where $\mathbb{J}(Q)$ is the ideal in $\mathbb{C}[z_1, \ldots, z_n]$ generated by all first-order partial derivatives of Q. From now on, we will only consider algebras of this form. Notice that the ideal $\mathbb{J}(Q)$ is homogeneous, and therefore, as stated in Remark 2, the isomorphism class of $\mathbb{M}(V_Q)$ is determined by the linear equivalence class of any of its homogeneous inverse systems. Hence, it is a reasonable idea to explore whether $\mathcal{I}_n^m$ can be derived from the highest stratum $\mathfrak{I}_\nu^{MH}$ of $\mathfrak{I}^{MH}$.

By Lemma 3.3 in [21], the annihilator $\mathrm{Ann}(\mathfrak{m})$ of the maximal ideal $\mathfrak{m}$ of $\mathbb{M}(V_Q)$ is generated by the element represented by the Hessian of Q, which implies that the nil-index of $\mathfrak{m}$ is found from the formula $\nu = n(m-2)$. Therefore, every nil-polynomial P arising from $\mathbb{M}(V_Q)$ has degree $n(m-2)$, and, since emb dim $\mathbb{M}(V_Q) = n$, the corresponding highest-degree form $\mathbf{P}^{[n(m-2)]}$ is a form on an n-dimensional vector space. We say that any such form $\mathbf{P}^{[n(m-2)]}$ is *associated* to Q (recall that all these forms are proportional to each other). Thus, upon identification of $\mathfrak{m}/\mathfrak{m}^2$ with $\mathbb{C}^n$, every element of $\mathfrak{I}_{n(m-2)}^{MH}$ calculated for the algebra $\mathbb{M}(V_Q)$ is given as $\mathbf{I}(\mathbf{Q})$, where $\mathbf{I}$ is an absolute invariant of forms of degree $n(m-2)$ on $\mathbb{C}^n$ and $\mathbf{Q}$ is any form associated to Q.

Computing associated forms is quite easy for any realization of the algebra. Choose a basis $e_1, \ldots, e_n$ in a complement to $\mathfrak{m}^2$ in $\mathfrak{m}$, and let f_j be the element of $\mathfrak{m}/\mathfrak{m}^2$ represented by e_j for $j = 1, \ldots, n$. Denote by $w_1, \ldots, w_n$ the coordinates in $\mathfrak{m}/\mathfrak{m}^2$ with respect to the basis $f_1, \ldots, f_n$. Further, choose a vector v spanning $\mathrm{Ann}(\mathfrak{m})$. If $k_1, \ldots, k_n$ are nonnegative integers such that $k_1 + \cdots + k_n = n(m-2)$, the product $e_1^{k_1} \ldots e_n^{k_n}$ is an element of $\mathrm{Ann}(\mathfrak{m})$, and thus we have $e_1^{k_1} \ldots e_n^{k_n} = \mu_{k_1,\ldots,k_n} v$ for some $\mu_{k_1,\ldots,k_n} \in \mathbb{C}$. Then the form

$$\sum_{k_1+\cdots+k_n=n(m-2)} \mu_{k_1,\ldots,k_n} \binom{n(m-2)}{k_1, \ldots, k_n} w_1^{k_1} \ldots w_n^{k_n} \tag{5}$$

[1]The proof of this proposition given in [6] was suggested to us by A. Gorinov.

is a coordinate representation of a form associated to Q, where

$$\binom{n(m-2)}{k_1, \ldots, k_n} := \frac{(n(m-2))!}{k_1! \ldots k_n!}$$

is a multinomial coefficient.

We now propose a conjecture.

Conjecture 1. For any $\mathrm{I} \in \mathcal{I}_n^m$ there exists an absolute invariant $\mathbf{I}$ of forms of degree $n(m-2)$ on $\mathbb{C}^n$, such that for all $Q \in X_n^m$, the invariant $\mathbf{I}$ is defined at some (hence at every) form $\mathbf{Q}$ associated to Q and $\mathbf{I}(\mathbf{Q}) = \mathrm{I}(Q)$.

For binary quartics ($n = 2, m = 4$) and ternary cubics ($n = 3, m = 3$) the conjecture was essentially verified in [5] (see also [6] and Example 2 below). Furthermore, in [6] we showed that the conjecture holds for binary quintics ($n = 2$, $m = 5$) and binary sextics ($n = 2$, $m = 6$) as well. If the conjecture were confirmed in full generality, it would imply that the stratum $\mathfrak{I}_{n(m-2)}^{MH}$ is a complete system of invariants for homogeneous hypersurface singularities defined by forms in X_n^m.

We will now mention another interesting consequence of Conjecture 1. First, we observe that $\mathbf{I}(\mathbf{Q})$ is rational when regarded as a function of Q, provided $\mathbf{I}$ is defined for at least one associated form. In order to see this, we choose a particular (canonical) identification of $\mathfrak{m}/\mathfrak{m}^2$ with $\mathbb{C}^n$. Namely, for $j = 1, \ldots, n$, let e_j be the element of $\mathfrak{m}$ represented by the coordinate function z_j. Also, we let v be the element represented by the Hessian of Q. Denote by $\mathbf{Q}^c$ the corresponding associated form found from formula (5). As follows from Remark 2, the form $\mathbf{Q}^c$ is an inverse system for $\mathbb{M}(V_Q)$. It is clear that $\mu_{k_1,\ldots,k_n}$ that occur in (5) are rational functions of the coefficients of Q, and therefore for an absolute invariant $\mathbf{I}$ of forms of degree $n(m-2)$ on $\mathbb{C}^n$, the expression $\mathbf{I}(\mathbf{Q}^c)$ is an invariant rational function of Q if it is defined at least at one point of X_n^m.

Let $\mathcal{R}_n^m$ denote the collection of all invariant rational functions on X_n^m obtained in this way. Further, let $\hat{\mathcal{I}}_n^m$ be the algebra of restrictions to X_n^m of all absolute invariants of forms of degree m on $\mathbb{C}^n$. Note that $\mathcal{R}_n^m$ lies in $\hat{\mathcal{I}}_n^m$ (see Proposition 1 in [4]). We claim that Conjecture 1 implies

$$\mathcal{R}_n^m = \hat{\mathcal{I}}_n^m. \tag{6}$$

Indeed, since every element of $\hat{\mathcal{I}}_n^m$ can be represented as a ratio of two elements of $\mathcal{I}_n^m$ (see Proposition 6.2 in [19]), identity (6) is equivalent to the inclusion $\mathcal{I}_n^m \subset \mathcal{R}_n^m$, which clearly follows from Conjecture 1 (cf. Conjecture 3.2 in [6]).

We remark that identity (6) is interesting from the invariant-theoretic point of view, since it means that the invariant theory of forms of degree m can be completely recovered from that of forms of degree $n(m-2)$. Observe that (6) is *a priori* weaker than Conjecture 1. Indeed, it may potentially happen that for some $\mathrm{I} \in \mathcal{I}_n^m$

there exists an absolute invariant **I** of forms of degree $n(m-2)$ on $\mathbb{C}^n$, such that $\mathbf{I}(\mathbf{Q}) \equiv \mathtt{I}(Q)$, where $\mathbf{I}(\mathbf{Q})$ is regarded as a function of Q, but for some $Q_0 \in X_n^m$ the invariant **I** is not defined at the forms associated to Q_0. In the recent paper [1], identity (6) was shown to always hold.

We will now illustrate Conjecture 1 with the example of simple elliptic singularities of type $\tilde{E}_6$.

Example 2. Simple elliptic $\tilde{E}_6$-singularities form a family V_t parametrized by $t \in \mathbb{C}$ satisfying $t^3 + 27 \neq 0$. Namely, for every such t let $V_t := V_{Q_t}$, where Q_t is the following cubic on $\mathbb{C}^3$:

$$Q_t(z_1, z_2, z_3) := z_1^3 + z_2^3 + z_3^3 + t z_1 z_2 z_3.$$

Since $n = m = 3$, we have $\nu = n(m-2) = 3$; thus any form associated to Q_t is again a ternary cubic. To compute such a form using formula (5), set e_j to be the element of $\mathfrak{m}$ represented by z_j for $j = 1, 2, 3$ and v the element represented by $z_1 z_2 z_3$. Then for the coefficients in formula (5) we have

$$\mu_{3,0,0} = \mu_{0,3,0} = \mu_{0,0,3} = -\frac{t}{3}, \quad \mu_{1,1,1} = 1,$$

with all the remaining μ_{k_1,k_2,k_3} being zero. These coefficients yield the following associated form:

$$\mathbf{Q}_t := -\frac{t}{3}(w_1^3 + w_2^3 + w_3^3) + 6 w_1 w_2 w_3.$$

The form $\mathbf{Q}_t$ is an inverse system for $\mathbb{M}(V_t)$ and has been known for a long time (see [5, 9]). For $t \neq 0$, $t^3 - 216 \neq 0$ one has $\Delta(\mathbf{Q}_t) \neq 0$, in which case the original cubic Q_t is associated to $\mathbf{Q}_t$ and thus there is a natural duality between Q_t and $\mathbf{Q}_t$.

Further, the algebra of classical invariants of ternary cubics is generated by certain invariants $\mathcal{I}_4$ and $\mathcal{I}_6$, where the subscripts indicate the degrees (see pp. 381–389 in [8]). For a ternary cubic of the form

$$Q(z) = a z_1^3 + b z_2^3 + c z_3^3 + 6 d z_1 z_2 z_3$$

the values of $\mathcal{I}_4$ and $\mathcal{I}_6$ are computed as follows:

$$\mathcal{I}_4(Q) = abcd - d^4, \quad \mathcal{I}_6(Q) = a^2 b^2 c^2 - 20abcd^3 - 8d^6,$$

and $\Delta(Q) = \mathcal{I}_6^2 + 64\mathcal{I}_4^3$.[2]

[2]This formula for the discriminant of a ternary cubic differs from the general formula given in [12] by a scalar factor.

Consider the j-invariant of ternary cubics, which is the absolute invariant defined as follows:

$$j := \frac{64\,\mathcal{I}_4^3}{\Delta}.$$

It is easy to see that the restriction $j|_{X_3^3}$ generates the algebra $\mathcal{I}_3^3$. In particular, any two nonequivalent ternary cubics with nonvanishing discriminant are distinguished by j (see Proposition 1). Further, we have

$$j(Q_t) = -\frac{t^3(t^3-216)^3}{1728(t^3+27)^3}.$$

Details on computing $j(Q)$ for any ternary cubic Q with $\Delta(Q) \neq 0$ can be found, for example, in [5].

Next, consider the following absolute invariant of ternary cubics:

$$\mathbf{j} := \frac{1}{j}.$$

A straightforward calculation shows that for any $Q \in X_3^3$ the absolute invariant $\mathbf{j}$ is defined at $\mathbf{Q}_t$ and $\mathbf{j}(\mathbf{Q}_t) = j(Q_t)$, which demonstrates that Conjecture 1 is indeed valid for $n = m = 3$.

References

1. Alper, J., Isaev, A., *Associated forms in classical invariant theory and their applications to hypersurface singularities*, Math. Ann., published online, DOI 10.1007/s00208-014-1054-2.
2. Bass, H., *On the ubiquity of Gorenstein rings.* Math. Z. **82** (1963), 8–28.
3. Bruns, W., Herzog, J., *Cohen-Macaulay Rings.* Cambridge Studies in Advanced Mathematics 39. Cambridge University Press, Cambridge (1993).
4. Dieudonné, J. A., Carrell, J. B., *Invariant theory, old and new.* Adv. Math. **4** (1970), 1–80.
5. Eastwood, M. G., *Moduli of isolated hypersurface singularities.* Asian J. Math. **8** (2004), 305–313.
6. Eastwood, M. G., Isaev, A. V., *Extracting invariants of isolated hypersurface singularities from their moduli algebras.* Math. Ann. **356** (2013), 73–98.
7. Elias, J., Rossi, M. E., *Isomorphism classes of short Gorenstein local rings via Macaulay's inverse system.* Trans. Amer. Math. Soc. **364** (2012), 4589–4604.
8. Elliott, E. B., *An Introduction to the Algebra of Quantics.* Oxford University Press (1895).
9. Emsalem, J.: *Géométrie des points épais.* Bull. Soc. Math. France **106** (1978), 399–416.
10. Fels, G., Isaev, A., Kaup, W., Kruzhilin, N., *Isolated hypersurface singularities and special polynomial realizations of affine quadrics.* J. Geom. Analysis **21** (2011), 767–782.
11. Fels, G., Kaup, W., *Nilpotent algebras and affinely homogeneous surfaces.* Math. Ann. **353** (2012), 1315–1350

12. Gelfand, I. M., Kapranov, M. M., Zelevinsky, A. V., *Discriminants, Resultants and Multidimensional Determinants*. Modern Birkhäuser Classics. Birkhäuser Boston, Inc., Boston, MA (2008).
13. Greuel, G.-M., Lossen, C., Shustin, E., *Introduction to Singularities and Deformations*. Springer Monographs in Mathematics. Springer, Berlin (2007).
14. Hertling, C., *Frobenius Manifolds and Moduli Spaces for Singularities*. Cambridge Tracts in Mathematics 151. Cambridge University Press, Cambridge (2002).
15. Hilbert, D., *Ueber die Theorie der algebraischen Formen*. Math. Ann. **36** (1890), 473–534.
16. Huneke, C., Hyman Bass and ubiquity: Gorenstein rings. In: *Algebra, K-theory, Groups, and Education* (New York, 1997). Contemp. Math. 243, pp. 55–78. Amer. Math. Soc., Providence, RI (1999).
17. Isaev, A. V., *On the affine homogeneity of algebraic hypersurfaces arising from Gorenstein algebras*. Asian J. Math. **15** (2011), 631–640.
18. Mather, J., Yau, S. S.-T., *Classification of isolated hypersurface singularities by their moduli algebras*. Invent. Math. **69** (1982), 243–251.
19. Mukai, S., *An Introduction to Invariants and Moduli*. Cambridge Studies in Advanced Mathematics 81. Cambridge University Press, Cambridge (2003).
20. Olver, P., *Classical Invariant Theory*. London Mathematical Society Student Texts 44. Cambridge University Press, Cambridge (1999).
21. Saito, K., *Einfach-elliptische Singularitäten*. Invent. Math. **23** (1974), 289–325.
22. Sharpe, D. W., Vámos, P., *Injective Modules*. Cambridge Tracts in Mathematics and Mathematical Physics 62. Cambridge University Press, London–New York (1972).
23. Shoshitaishvili, A. N., *Functions with isomorphic Jacobian ideals*. Funct. Anal. Appl. **10** (1976), 128–133.

Generalized Loop Modules for Affine Kac–Moody Algebras

Vyacheslav Futorny and Iryna Kashuba

Abstract We construct new families of irreducible modules for any affine Kac–Moody algebra by considering the parabolic induction from irreducible modules over the Heisenberg subalgebra with a nonzero central charge.

Key words Kac–Moody algebra • Loop module • Parabolic induction • Heisenberg subalgebra

Mathematics Subject Classification (2010): 17B67.

1 Introduction

In the recent paper [BBFK] parabolic induction was used to construct irreducible modules for affine Lie algebras from certain modules over the Heisenberg subalgebra. In particular, when the central charge is nonzero the induced module is always irreducible. Previously this result was also known for highest weight modules (with respect to nonstandard Borel subalgebras) with nonzero central charge [C, FS, F1, F4]. We extend this result to any irreducible $\mathbb{Z}$-graded module over the Heisenberg subalgebra.

Let $\mathfrak{G}$ be an affine Lie algebra with a 1-dimensional center $Z = \mathbb{C}c$, that is, a Kac–Moody algebra corresponding to an affine generalized Cartan matrix. Let $\mathfrak{H}$ be a Cartan subalgebra of $\mathfrak{G}$, $\mathfrak{H} = \mathfrak{h} \oplus \mathbb{C}c \oplus \mathbb{C}d$, where d is the degree derivation and

V. Futorny (✉) • I. Kashuba
Institute of Mathematics, University of São Paulo, Caixa Postal 66281, CEP 05314-970, São Paulo, Brazil
e-mail: futorny@ime.usp.br; kashuba@ime.usp.br

G. Mason et al. (eds.), *Developments and Retrospectives in Lie Theory: Algebraic Methods*, Developments in Mathematics 38,
DOI 10.1007/978-3-319-09804-3_8

$\mathfrak{h}$ is a Cartan subalgebra of underlined simple finite-dimensional Lie subalgebra $\mathfrak{g}$ of $\mathfrak{G}$. Then $\mathfrak{G}$ has the following root decomposition:

$$\mathfrak{G} = \mathfrak{H} \oplus (\oplus_{\alpha \in \Delta} \mathfrak{G}_\alpha),$$

where $\mathfrak{G}_\alpha := \{x \in \mathfrak{G} \mid [h, x] = \alpha(h)x \text{ for every } h \in \mathfrak{H}\}$ and Δ is the root system of $\mathfrak{G}$.

Denote by $U(\mathfrak{G})$ the universal enveloping algebra of $\mathfrak{G}$.

A $\mathfrak{G}$-module V is a *weight* module if $V = \oplus_{\mu \in \mathfrak{H}^*} V_\mu$, where

$$V_\mu = \{v \in V | hv = \mu(h)v, \forall h \in \mathfrak{H}\}.$$

We will denote by $\mathscr{W}$ the category of all weight $\mathfrak{G}$-modules.

Classification of irreducible weight modules is only known for modules with finite-dimensional weight spaces [FT] (nonzero central charge), [DG] (zero central charge) and in certain subcategories of induced modules with infinite-dimensional weight spaces [BBFK, F2, FK].

Let G be the Heisenberg subalgebra of $\mathfrak{G}$ generated by the root spaces $\mathfrak{G}_{k\delta}$, $k \in \mathbb{Z}\setminus\{0\}$, i.e., $G = \sum_{k \in \mathbb{Z}\setminus\{0\}} \mathfrak{G}_{k\delta} \oplus \mathbb{C}c$. Hence G has a natural $\mathbb{Z}$-grading. We will denote by $\mathscr{K}$ the category of all $\mathbb{Z}$-graded G-modules with the grading compatible with the above grading of G. If V is a G-module (respectively $\mathfrak{G}$-module) with a scalar action of c, then this scalar is called the *l*evel of V.

Let $\mathscr{P} = \mathscr{P}_0 \oplus \mathfrak{N}$ be a parabolic subalgebra of $\mathfrak{G}$ with the Levi subalgebra $\mathscr{P}_0 = G + \mathfrak{H}$. If N is a module from $\mathscr{K}$, then one defines on it a structure of a $\mathscr{P}_0$-module by choosing any $\lambda \in \mathfrak{H}^*$ and setting $hv = \lambda(h)v$ for any $h \in \mathfrak{H}$ and any $v \in N$. Moreover, N can be viewed as a $\mathscr{P}$-module with a trivial action of $\mathfrak{N}$.

Consider the induced $\mathfrak{G}$-module $\text{ind}_\lambda(\mathscr{P}, \mathfrak{G}; N) = U(\mathfrak{G}) \otimes_{U(\mathscr{P})} N$. Hence

$$\text{ind}_\lambda(\mathscr{P}, \mathfrak{G}) : N \longmapsto \text{ind}_\lambda(\mathscr{P}, \mathfrak{G}; N)$$

defines a functor from the category $\mathscr{K}$ of G-modules to the category $\mathscr{W}$ of weight $\mathfrak{G}$-modules.

Denote by $\tilde{\mathscr{K}}$ the full subcategory of $\mathbb{Z}$-graded G-modules on which the central element c acts injectively. The main result of [BBFK] shows that this functor preserves irreducibility when applied to a particular class of irreducible G-modules. Our main result is the following theorem which shows that the same result holds for any irreducible G-module in $\tilde{\mathscr{K}}$. The proof of the result in [BBFK] depends on the property of a $\mathbb{Z}$-graded G-module to have a special countable basis while our proof is basis-free.

Theorem 1. *$\mathscr{P} = \mathscr{P}_0 \oplus \mathfrak{N}$ a parabolic subalgebra of $\mathfrak{G}$, where $\mathscr{P}_0 = G + \mathfrak{H}$. Let $\lambda \in \mathfrak{H}^*$ such that $\lambda(c) \neq 0$. Then the functor* $\text{ind}_\lambda(\mathscr{P}, \mathfrak{G})$ *preserves the irreducibles.*

Thus any irreducible $\mathbb{Z}$-graded module V over the Heisenberg subalgebra and any $\lambda \in \mathfrak{H}^*$ with $\lambda(c) \neq 0$ induce the irreducible $\mathfrak{G}$-module $\mathrm{ind}_\lambda(\mathscr{P}, \mathfrak{G})(V)$. Moreover,

$$\mathrm{ind}_\lambda(\mathscr{P}, \mathfrak{G})(V) \simeq \mathrm{ind}_{\lambda'}(\mathscr{P}, \mathfrak{G})(V')$$

if and only if $V \simeq V'$ and $\lambda = \lambda'$.

In [FK] a similar reduction theorem was shown for pseudo-parabolic subalgebras. We explain the difference between our result and the main result in [FK].

Let $\mathscr{P} = \mathscr{P}_0 \oplus \mathfrak{N}$ be a nonsolvable parabolic subalgebra of $\mathfrak{G}$ with infinite-dimensional Levi factor $\mathscr{P}_0$. Then $\mathscr{P}_0 = [\mathscr{P}_0, \mathscr{P}_0] \oplus G(\mathscr{P}) + \mathfrak{H}$, where $G(\mathscr{P}) \subset G$ is the orthogonal completion (with respect to the Killing form) of the Heisenberg subalgebra of $[\mathscr{P}_0, \mathscr{P}_0]$. We have $G(\mathscr{P}) + ([\mathscr{P}_0, \mathscr{P}_0] \cap G) = G$ and $G(\mathscr{P}) \cap [\mathscr{P}_0, \mathscr{P}_0] = \mathbb{C}c$. Here $[\mathscr{P}_0, \mathscr{P}_0]$ is a sum of affine subalgebras of $\mathfrak{G}$ whose intersection equals $\mathbb{C}c$. Consider a triangular decomposition $G(\mathscr{P}) = G(\mathscr{P})_- \oplus \mathbb{C}c \oplus G(\mathscr{P})_+$ of $G(\mathscr{P})$. Then a pseudo-parabolic subalgebra is defined as $\mathscr{P}^{ps} = \mathscr{P}_0^{ps} \oplus \mathfrak{N}^{ps}$, where $\mathscr{P}_0^{ps} = [\mathscr{P}_0, \mathscr{P}_0] + \mathfrak{H}$, $\mathfrak{N}^{ps} = \mathfrak{N} \oplus G(\mathscr{P})_+$ and $\mathscr{P} = \mathscr{P}^{ps} \oplus G(\mathscr{P})_-$. One can start with an irreducible weight module over $[\mathscr{P}_0, \mathscr{P}_0]$, extend it naturally to a module over $\mathscr{P}_0^{ps}$, and then induce up to a $\mathfrak{G}$-module letting $\mathfrak{N}^{ps}$ to act trivially. The main result of [FK] establishes the irreducibility of such induced module when the central element acts nonzero. It is essential for the proof that $\mathscr{P}$ is not solvable Lie algebra, that is, $\dim[\mathscr{P}_0, \mathscr{P}_0] > 1$. On the other, the parabolic subalgebra $\mathscr{P} = G + \mathfrak{H}$ considered in this paper is solvable and the proof in this case requires a different argument.

All results in the paper hold for both *untwisted* and *twisted* affine Lie algebras of rank greater than 1.

2 Parabolic Induction

Let π (respectively $\dot{\pi}$) be a basis of the root system Δ (respectively $\dot{\Delta}$) of $\mathfrak{G}$ (respectively $\mathfrak{g}$), $\Delta_+(\pi)$ (respectively $\dot{\Delta}_+(\dot{\pi})$) the set of positive roots with respect to π (respectively $\dot{\pi}$). Denote by $\delta \in \Delta_+(\pi)$ the indivisible positive imaginary root. Then $\Delta^{\mathrm{im}} = \{k\delta | k \in \mathbb{Z} \setminus \{0\}\}$ is the set of imaginary roots.

A closed subset $P \subset \Delta$ is called a *partition* if $P \cap (-P) = \emptyset$ and $P \cup (-P) = \Delta$. In the case of finite-dimensional simple Lie algebras every partition corresponds to a choice of positive roots and all partitions are conjugate by the Weyl group. For affine root systems the partitions were classified in [JK, F3] and in this case there exist a finite family of nonconjugate partitions. A *parabolic* subset $P \in \Delta$ is a closed subset in Δ such that $P \cup (-P) = \Delta$. Parabolic subsets were classified in [F3]. Given a parabolic subset P, one defines a parabolic subalgebra $\mathfrak{G}_P$ of $\mathfrak{G}$ generated by $\mathfrak{H}$ and all the root spaces $\mathfrak{G}_\alpha$, $\alpha \in P$.

Set $P_0 = P \cap -P$. Then $\mathfrak{G}$ has the following triangular decomposition

$$\mathfrak{G} = \mathfrak{G}_P^- \oplus \mathfrak{G}_P^0 \oplus \mathfrak{G}_P^+,$$

where $\mathfrak{G}_P^{\pm} = \sum_{\alpha\in P\setminus(-P)} \mathfrak{G}_{\pm\alpha}$ and the Levi factor $\mathfrak{G}^0_{\mathscr{P}}$ is generated by $\mathfrak{H}$ and the subspaces $\mathfrak{G}_\alpha$, $\alpha \in P_0$. We have $\mathfrak{G}_P = \mathfrak{G}^0_P \oplus \mathfrak{G}^+_P$.

Let $\mathscr{P} = \mathscr{P}_0 \oplus \mathfrak{N}$ be a parabolic subalgebra of $\mathfrak{G}$ with Levi factor $\mathscr{P}_0$. Then it has *type* I if $\mathscr{P}_0$ is a finite-dimensional reductive Lie algebra and it has *type II* if $\mathscr{P}_0$ contains the Heisenberg subalgebra G. In the latter case if $\mathfrak{G}_{\alpha+k\delta} \subset \mathfrak{N}$ for some real root α and some $k \in \mathbb{Z}$, then the same holds for $\mathfrak{G}_{\alpha+s\delta}$ for all $s \in \mathbb{Z}$.

The smallest Levi factor of a parabolic subalgebra of type II is $\mathscr{P}_0 = G + \mathfrak{H}$. Then $\mathfrak{N}$ is generated by $\mathfrak{G}_{\alpha+n\delta}$, $\alpha \in \dot{\Delta}_+(\dot{\pi})$, $n \in \mathbb{Z}$ for some choice of $\dot{\pi}$. Such parabolic subalgebras will be considered in this paper.

3 $\mathbb{Z}$-Graded Modules over a Heisenberg Subalgebra

Here we construct a series of simple modules for the infinite Heisenberg Lie algebras which are $\mathbb{Z}$-graded with infinite-dimensional homogeneous components and have a nonzero central charge.

Let $H = \mathbb{C}c \oplus \bigoplus_{i\in\mathbb{Z}\setminus\{0\}} \mathbb{C}e_i$ be an infinite-dimensional Lie algebra of Heisenberg type, where $[e_i, e_j] = \delta_{i,-j}c$, $[e_j, c] = 0$ for all $i \geq 1$ and all j. Note that $H \simeq G$ if $\mathfrak{g} = sl(2)$. If $\mathfrak{g} \neq sl(2)$, then $U(G)$ is isomorphic to a tensor product of finitely many copies of $U(H)$. Hence irreducible G-modules can be constructed by taking a tensor product of irreducible H-modules.

Denote by $\mathscr{K}_H$ the category of all $\mathbb{Z}$-graded H-modules V such that $V = \bigoplus_{i\in\mathbb{Z}} V_i$ and $e_i V_j \subset V_{i+j}$.

If $V \in \mathscr{K}_H$ is irreducible, then the action of c is scalar, the *central charge* of V. All irreducible modules with a zero central charge were classified by Chari [Ch]. Irreducible weight module with a nonzero finite-dimensional weight space and with a nonzero central charge were described in [F2]. A class of admissible diagonal modules in $\mathscr{K}_H$ was constructed in [BBFK] using the classification of simple weight admissible modules over the infinite rank Weyl algebra A_∞ [BBF] (roughly admissibility means the existence of a maximal set for any increasing chain of annihilators in A_∞ of elements of the module, see [BBFK] for a precise definition). After rescaling the generators of H, we may assume that an irreducible module $V \in \mathscr{K}_H$ has central charge 1 and hence it becomes a module for $A = A_\infty$, since $U(H)$ modulo the ideal generated by $c-1$ is isomorphic to A_∞ by identifying $\partial_i = e_i$ and $x_i = e_{-i}$ for all $i > 0$. These H-modules have a nonzero central charge and infinite-dimensional weight subspaces (with respect to the $\mathbb{Z}$-grading). All simple weight modules over the infinite rank Weyl algebra A_∞ were classified in [FGM] providing a classification of all irreducible diagonal H-modules. Taking any such module and applying a parabolic induction we obtain new families of irreducible modules (after taking a quotient by a maximal submodule if necessary) for the affine $sl(2)$ algebra.

In order to generalize this to any affine Kac–Moody algebra we need to extend the above mentioned result in [FGM] for the Heisenberg algebra G. For each $k \in \mathbb{Z}$, $k \neq 0$, let $\mathsf{d}_k = \dim\mathfrak{G}_{k\delta}$. Set $[\mathsf{d}_k] = \{1, \ldots, \mathsf{d}_k\}$. Choose a basis $\{x_{k,i} \mid i \in [\mathsf{d}_k]\}$

for $\mathfrak{G}_{k\delta}$ such that $[x_{k,i}, x_{-k,j}] = \delta_{i,j}kc, i, j$. Then G can be written as direct sum of finitely many Lie subalgebras isomorphic to H (possibly with a shift of gradation). Hence, taking a tensor product of irreducible modules from [FGM] and applying a parabolic induction, we obtain new families of irreducible modules for arbitrary affine Lie algebra.

4 Generalized Loop Modules

Let $\mathscr{P} = \mathscr{P}_0 \oplus \mathfrak{N}$ be a parabolic subalgebra of $\mathfrak{G}$ such that $\mathscr{P}_0 = G + \mathfrak{H}$. Let N be an irreducible G-module from $\mathscr{K}$ and $\lambda \in \mathfrak{H}^*$. Define $\mathscr{P}_0$-module structure on N by $hv = \lambda(h)v$ for all $v \in N$. Next define on N a weight (with respect to $\mathfrak{H}$) module over the parabolic subalgebra $\mathscr{P}$ with a trivial action of $\mathfrak{N}$ and let

$$M_{\mathscr{P}}(N, \lambda) = \mathrm{ind}(\mathscr{P}, \mathfrak{G}; N)$$

be the induced weight $\mathfrak{G}$-module. If N is irreducible, then $M_{\mathscr{P}}(N, \lambda)$ has a unique irreducible quotient $L_{\mathscr{P}}(N, \lambda)$. Extending [BBFK] we call $M_{\mathscr{P}}(N, \lambda)$ the *generalized loop module* associated with N and λ.

A nonzero element v of a $\mathfrak{G}$-module V is called *$\mathscr{P}$-primitive* if $\mathfrak{N}v = 0$.

Proposition 1. *Let V be an irreducible weight $\mathfrak{G}$-module with a $\mathscr{P} = \mathscr{P}_0 \oplus \mathfrak{N}$-primitive element of weight λ, $\mathscr{P}_0 = G + \mathfrak{H}$, $N = \sum_{k\in\mathbb{Z}} V_{\lambda+k\delta}$. Then N is an irreducible $\mathscr{P}$-module and V is isomorphic to $L_{\mathscr{P}}(N, \lambda)$.*

Proof. The proof is standard. □

The following theorem implies the main result.

Theorem 2. *Let $\mathscr{P} = G+\mathfrak{H}$ be a parabolic subalgebra of $\mathfrak{G}$, $V \in \mathscr{K}$ a G-module, $\lambda \in \mathfrak{H}^*$ with $\lambda(c) \neq 0$. Then for any submodule U of $M_{\mathscr{P}}(V, \lambda)$ we have $U \cap (1 \otimes V) \neq 0$. In particular, $M_{\mathscr{P}}(V, \lambda)$ is irreducible if and only if V is irreducible.*

Recall that the algebra H is $\mathbb{Z}$-graded. Below we will use a change of grading on H by interchanging of some positive and negative components of the original grading. Similar change of grading can be done for the algebra G. Any such change of grading corresponds to an automorphism of H (respectively G).

Lemma 1. *Let V be an arbitrary nonzero level $\mathbb{Z}$-graded module over G. Fix $k_1, \ldots, k_s \in \mathbb{Z}$, $k_i \neq k_j$ if $i \neq j$, and $f_i \in G_{k_i\delta}$, $i = 1, \ldots, s$ and set $f_0 = 1$. Also fix nonzero elements $x_k \in G_{k\delta}$, $k \in \mathbb{Z} \setminus \{0\}$ such that $[x_k, x_{-k}] \neq 0$ for any k. Then for any nonzero homogeneous element $v \in V$ there exist a change of $\mathbb{Z}$-grading on G and $n \in \mathbb{Z}$, $n >> 0$, such that for any $N \geq n$,*

$$\sum_{i=0}^{s} x_{N-i} f_i v \neq 0.$$

Proof. Let $f_i v = 0$ for all $i = 1, \dots, s$. Then for any $N \in \mathbb{Z}$, $N \neq 0$ and for any nonzero $x \in G_{N\delta}$, either $xv \neq 0$ or there exists $x' \in G_{-N\delta}$ such that $x'v \neq 0$. Then the statement follows after a suitable change of the $\mathbb{Z}$-grading on G. Consider now the general case. Assume $f_i v \neq 0$ for some i. Without loss of generality, we may assume that $f_i v \neq 0$ for all $i = 1, \dots, s$. Suppose that for some $N \in \mathbb{Z}$ and for all nonzero elements $x_i \in G_{(N-k_i)\delta}$ and $x_i' \in G_{(-N+k_i)\delta}$, $i = 0, \dots, s$, such that $[x_i, x_i'] \neq 0$ for all i,

$$\sum_{i=0}^{s} x_i f_i v = \sum_{i=0}^{s} x_i' f_i v = 0.$$

Then $x_0 v = -\sum_{i=1}^{s} x_i f_i v$ and $x_0' v = -\sum_{i=1}^{s} x_i' f_i v$. If N is sufficiently large, then we will have $[x_i, f_j] = [x_i', f_j] = 0$ for all i, j. Then

$$[\sum_{i=1}^{s} x_i f_i, \sum_{i=1}^{s} x_i' f_i]$$

$$= \sum_{i,j=1}^{s} (x_i x_j' f_i f_j - x_j' x_i f_j f_i)$$

$$= \sum_{i=1}^{s} [x_i, x_i'] f_i^2.$$

Hence

$$[x_0, x_0']v = \sum_{i=1}^{s} [x_i, x_i'] f_i^2 v,$$

and thus v can be written as an integral linear combination of $f_1^2 v, \dots, f_s^2 v$. Due to the $\mathbb{Z}$-grading on V, this can only happen when $f_1^2 v = \dots = f_s^2 v = v = 0$ which is a contradiction. Therefore the statement is shown for all sufficiently large N after a suitable change of the $\mathbb{Z}$-grading on G. □

The universal enveloping algebra $U(G)$ inherits the $\mathbb{Z}$-grading from G, $U(G) = \oplus_{i \in \mathbb{Z}} U(G)_i$. One can easily generalize Lemma 1 for the case when f_i's are arbitrary elements of $U(G)$. Namely, we have

Corollary 1. *Let V be an arbitrary nonzero level $\mathbb{Z}$-graded module over G, $u_i \in U(G)_i$, $i \in I \subset \mathbb{Z}$. Fix nonzero elements $x_k \in G_{k\delta}$, $k \in \mathbb{Z} \backslash \{0\}$ such that $[x_k, x_{-k}] \neq 0$ for any k. Then for any nonzero homogeneous element $v \in V$ such that $u_0 v + v \neq 0$*

(if $0 \in I$) there exist a change of $\mathbb{Z}$-grading on G and $n \in \mathbb{Z}$, $n >> 0$, such that for any $N \geq n$,

$$\sum_{i \in I} x_{N-i} u_i v + x_N v \neq 0.$$

Proof. The proof is similar to the proof of Lemma 1. It is sufficient to compare the top graded components of u_i's and use the fact that $[x_i, u_j] = [x_i', u_j] = 0$ for all i, j when N is sufficiently large. Here x_i's and x_i''s are as in Lemma 1. □

Proof of Theorem 2. The proof is analogous to the proof of Lemma 5.3 in [BBFK]. Let $\mathscr{P} = \mathscr{P}_0 \oplus \mathfrak{N}$, $\mathscr{P}_0 = G + \mathfrak{H}$. Consider a Lie subalgebra $\mathfrak{N}^-$ of $\mathfrak{G}$ obtained from $\mathfrak{N}$ applying the Chevalley involution. Then $\mathscr{P} \oplus \mathfrak{N}^- = \mathfrak{G}$. Let $M_{\mathscr{P}}(\lambda, V) = \mathrm{ind}_\lambda(\mathscr{P}, \mathfrak{G}; V)$ where V is a G-module, $\lambda \in \mathfrak{H}^*$ with $\lambda(c) \neq 0$. Denote $\hat{M}_{\mathscr{P}}(\lambda, V) = 1 \otimes V$. Then $\hat{M}_{\mathscr{P}}(\lambda, V)$ is a $\mathscr{P}$-submodule of $M_{\mathscr{P}}(\lambda, V)$ isomorphic to V, which consists of $\mathscr{P}$-primitive elements. Let U be a nonzero submodule of $M_{\mathscr{P}}(\lambda, V)$. We will show that $\hat{U} = U \cap \hat{M}_{\mathscr{P}}(\lambda, V) \neq 0$ and that U is generated by $\hat{U}$ which implies the statement. Let $v \in U$ be a nonzero element. One can assume that u is a weight vector. Then

$$v = \sum_{i \in I} u_i v_i,$$

where $u_i \in U(\mathfrak{N}^-)$ are linearly independent homogeneous elements, $i \in I$, $v_i \in \hat{M}_{\mathscr{P}}(\lambda, V)$. First we assume that $\mathfrak{G}$ is not of type $\mathsf{A}_{2\ell}^{(2)}$.

Let $\dot{\pi} = \{\alpha_1, \ldots, \alpha_n\}$ be a suitable base of simple roots for $\dot{\Delta}$ such that $\mathfrak{N} = \oplus_{\alpha \in \dot{\Delta}_+, k \in \mathbb{Z}} \mathfrak{G}_{\alpha + k\delta}$, where $\dot{\Delta}_+$ is the set of positive roots with respect to $\dot{\pi}$ and $\alpha + k\delta \in \Delta$. Then

$$\mathfrak{N}^- = \oplus_{\alpha \in \dot{\Delta}_+, k \in \mathbb{Z}} \mathfrak{G}_{-\alpha + k\delta},$$

as long as $\alpha + k\delta \in \Delta$. If $\alpha = \sum_{j=1}^n k_j \alpha_j$, where each k_j is in $\mathbb{Z}_{\geq 0}$, then we set $\mathrm{ht}(\alpha) = \sum_{j=1}^n k_j$, the *height* of α.

Since v is a weight vector then each u_i is a homogeneous element of $U(\mathfrak{G})$ of degree

$$-\varphi_i = -\sum_{j=1}^n k_{ij} \alpha_j + m_i \delta,$$

where $k_{ij} \in \mathbb{Z}_{\geq 0}$, $m_i \in \mathbb{Z}$. We have that all φ_i's have the same height. Assume first that $\mathrm{ht}(\varphi_i) = 1$ for all i. Using the fact that all u_i's are linearly independent, all v_i's are $\mathscr{P}$-primitive elements and Corollary 1, one can show that there exists a nonzero $x \in \mathfrak{N}$ such that $[x, u_i] \in U(G)$ for all i and $0 \neq xv \in \hat{M}_{\mathscr{P}}(\lambda, V)$. Thus, we obtain a nonzero element which belongs to $U \cap \hat{M}_{\mathscr{P}}(\lambda, V)$. This completes the proof in the case $\mathrm{ht}(\varphi) = 1$. The case $\mathrm{ht}(\varphi) > 1$ is considered similarly. Using the fact that u_i's are linearly independent it is easy to find an element y of $\mathfrak{N}$ of height 1 which produces a nonzero element yv and reduces the height by 1. Then induction implies the statement of the theorem. In particular, if V is irreducible, then $M_{\mathscr{P}}(\lambda, V)$ is irreducible. If $\mathfrak{G}$ is of type $\mathsf{A}_{2\ell}^{(2)}$, then for any $\alpha \in (\dot{\Delta}_l)_+$, $\frac{1}{2}(\alpha + (2n-1)\delta) \in \Delta$. Let α_0 be a root of Δ such that $\{\alpha_0, \ldots, \alpha_n\}$ is a base of simple roots for Δ. Then $\alpha_0 = \frac{1}{2}(-\alpha_1 - 2\alpha_2 - \ldots - 2\alpha_n + \delta)$. Using this, one can write any root as $\alpha + k\delta$ for some integer k where $\alpha = k_0\alpha_0 + \sum_{j=2}^{n} k_j\alpha_j$, where each k_j is in $\mathbb{Z}_{\geq 0}$. Setting $\mathsf{ht}(\alpha) = k_0 + \sum_{j=2}^{n} k_j$ and using induction on $\mathsf{ht}(\alpha)$ as above, we complete the proof in the case of $\mathsf{A}_{2\ell}^{(2)}$. □

We also conjecture that the structure of $M_{\mathscr{P}}(\lambda, V)$ is completely determined by $\hat{M}_{\mathscr{P}}(\lambda, V)$ when $\lambda(c) \neq 0$. Namely, that any submodule U of $M_{\mathscr{P}}(V, \lambda)$ is generated by $V_U = U \cap \hat{M}_{\mathscr{P}}(\lambda, V)$ as a $\mathfrak{G}$-module. Therefore we have

$$U \simeq \mathrm{ind}_{\lambda}(\mathscr{P}, \mathfrak{G}; V_U).$$

This is known to be true in different particular cases, see [FK].

Theorem 2 immediately implies our main result about the structure of generalized loop modules. As a consequence of this result, any irreducible module V from the category $\mathscr{K}$ with a nonzero central charge $a \neq 0$ and any $\lambda \in \mathfrak{H}^*$, such that $\lambda(c) = a$ will determine an irreducible module $M_{\mathscr{P}}(\lambda, V)$ for the affine Lie algebra $\mathfrak{G}$. If V and V' are irreducible modules in $\mathscr{K}$ with central charge $a \neq 0$, and $\lambda, \mu \in \mathfrak{H}^*$ are such that $\lambda(c) = \mu(c) = a$, then $M_{\mathscr{P}}(\lambda, V)$ and $M_{\mathscr{P}}(\mu, V')$ are isomorphic if and only if V and V' are isomorphic as G-modules (up to a shift of gradation) and $\lambda = \mu$.

Suppose H_1 and H_2 are Lie algebras isomorphic to H. Consider H_i-module V^i, $i = 1, 2$, in $\mathscr{K}$ and the $H_1 \otimes H_2$-module $W = V^1 \otimes V^2$. If $V^1 = \oplus_{i\in\mathbb{Z}} V_i^1$ and $V^2 = \oplus_{i\in\mathbb{Z}} V_i^2$, then $W \in \mathscr{K}$ where $V_i^1 \otimes V_j^2 \subset W_{i+j}$. If both V^i, $i = 1, 2$ are also modules in $\mathscr{W}(\mathscr{A})$, that is they are weight modules over the infinite rank Weyl algebra, and all weight spaces are 1-dimensional, then W is a weight module over $\mathscr{A} \otimes \mathscr{A}$ with 1-dimensional weight spaces. Moreover, if both V^1 and V^2 are irreducible, then W is irreducible $H_1 \otimes H_2$-module.

Corollary 2. *Let $\mathfrak{G}$ be an affine Lie algebra with the Heisenberg subalgebra G. Write G as a direct sum of Lie subalgebras $G = \oplus_{i=1}^{m} G_i$ where $G_i \simeq H$ for all i. For each i, let V_i be an irreducible G_i-module with a nonzero central charge $a \in \mathbb{C}$. Choose $\lambda \in \mathfrak{H}^*$ such that $\lambda(c) = a$ and define a $G + \mathfrak{H}$-module structure on $V = \otimes_{i=1}^{m} V_i$ by $hv = \lambda(h)v$ for all $h \in \mathfrak{H}$, $v \in V$. Consider any parabolic subalgebra $\mathscr{P}$ of $\mathfrak{G}$ such that $\mathscr{P}_0 = G + \mathfrak{H}$. Then $M_{\mathscr{P}}(\lambda, V)$ is irreducible.*

Proof. Since $V = \otimes_i V_i$ is irreducible G-module, then it is irreducible $\mathscr{P}$-module and the statement follows immediately from Theorem 2. □

As a particular case we obtain the following statement.

Corollary 3. *Let V be a highest weight G-module with central charge $a \in \mathbb{C}$. Choose $\lambda \in \mathfrak{H}^*$ such that $\lambda(c) = a$ and define a $G + \mathfrak{H}$-module structure on $V = \otimes_{i=1}^{m} V_i$ by $hv = \lambda(h)v$ for all $h \in \mathfrak{H}$, $v \in V$. Consider any parabolic subalgebra $\mathscr{P}$ of $\mathfrak{G}$ such that $\mathscr{P}_0 = G + \mathfrak{H}$. Then $M_{\mathscr{P}}(\lambda, V)$ is irreducible.*

Acknowledgements The first author was supported in part by the CNPq grant (301320/2013-6) and by the Fapesp grant (2014/09310-5). The second author was supported by the CNPq grant (309742/2013-7).

References

[BBF] V. Bekkert, G. Benkart, V. Futorny, *Weyl algebra modules, Kac-Moody Lie algebras and related topics*, 17–42, Contemp. Math., 343, Amer. Math. Soc., Providence, RI, 2004.

[BBFK] G. Benkart, V. Bekkert, V. Futorny, I. Kashuba, *New irreducible modules for Heisenberg and Affine Lie algebras*, Journal of Algebra. **373** (2013), 284–298.

[Ch] V. Chari, *Integrable representations of affine Lie algebras*, Invent. Math. **85** (1986), no.2, 317–335.

[C] B. Cox, *Verma modules induced from nonstandard Borel subalgebras*, Pacific J. Math. **165** (1994), 269–294.

[DG] I. Dimitrov, D. Grantcharov, *Classification of simple weight modules over affine Lie algebras*, arXiv:0910.0688v1.

[F1] V. Futorny, Representations of affine Lie algebras, *Queen's Papers in Pure and Applied Math.*, v. 106 (1997), Kingston, Ont., Canada.

[F2] V. Futorny, *Irreducible non-dense $A_1^{(1)}$-modules*, Pacific J. of Math. **172** (1996), 83–99.

[F3] V. Futorny, *The parabolic subsets of root systems and corresponding representations of affine Lie algebras*, Contemporary Math., 131 (1992), part 2, 45–52.

[F4] V. Futorny, *Imaginary Verma modules for affine Lie algebras*, Canad. Math. Bull., v. **37**(2), 1994, 213–218.

[FGM] V. Futorny, D. Grantcharov, V. Mazorchuk, *Representations of Weyl algebras*, Proceedings of AMS, to appear.

[FK] V. Futorny and I. Kashuba, *Induced modules for Kac-Moody Lie algebras*, SIGMA - Symmetry, Integrability and Geometry: Methods and Applications **5** (2009), Paper 026.

[FS] V. Futorny and H. Saifi, *Modules of Verma type and new irreducible representations for Affine Lie Algebras*, CMS Conference Proceedings, v.14 (1993), 185–191.

[FT] V. Futorny, A. Tsylke, *Classification of irreducible nonzero level modules with finite-dimensional weight spaces for affine Lie algebras*, J. Algebra **238** (2001), 426–441.

[JK] H. Jakobsen and V. Kac, *A new class of unitarizable highest weight representations of infinite dimensional Lie algebras*, Lecture Notes in Physics **226** (1985), Springer-Verlag, 1–20.

Twisted Localization of Weight Modules

Dimitar Grantcharov

Abstract We discuss how the twisted localization functor leads to a classification of the simple objects and a description of the injectives in various categories of weight modules. The article is a survey on existing results for finite-dimensional simple Lie algebras and superalgebras, affine Lie algebras, and algebras of differential operators.

Key words Lie algebra • Lie superalgebra • Weyl algebra • Weight module • Localization

Mathematics Subject Classification (2010): 17B10.

1 Introduction

The study of weight modules of Lie algebras and superalgebras have attracted considerable attenion in the last 30 years. Examples of weight modules include parabolically induced modules (in particular, modules in the category $\mathscr{O}$), $\mathscr{D}$-modules, generalized Harish-Chandra modules, among others. The first major steps in the systematic study of weight modules were made in the 1980s and 1990s by G. Benkart, D. Britten, S. Fernando, V. Futorny, A. Joseph, F. Lemire, and others, [3, 6, 19, 20, 30]. Two remarkable results for simple finite-dimensional Lie algebras include the Fernando–Futorny parabolic induction theorem, [19, 20], reducing the classification of all simple weight modules with finite weight multiplicities to the classification of all simple cuspidal modules (defined in Sect. 3.5); and the result of Fernando [19] that cuspidal modules exist only for Lie algebras of type A

Partially supported by NSA grant H98230-13-1-0245.

D. Grantcharov (✉)
Department of Mathematics, University of Texas at Arlington, Arlington, TX 76019, USA
e-mail: grandim@uta.edu

G. Mason et al. (eds.), *Developments and Retrospectives in Lie Theory: Algebraic Methods*, Developments in Mathematics 38,
DOI 10.1007/978-3-319-09804-3_9

and C. In 2000 O. Mathieu [32], classified the simple cuspidal modules, and in this way completed the classification of all simple weight modules with finite weight multiplicities of all finite-dimensional reductive Lie algebras.

An essential part of Mathieu's classification result is the application of the twisted localization functor: the result states that every cuspidal module is a twisted localization of a highest weight module. The definition of a twisted localization functor for Lie algebras in [32] generalizes other fundamental constructions: Deodhar's localization functors, and in particular, Enright's completions, see [11] and [18]. The twisted localization functor is especially convenient on categories of bounded weight modules, i.e., those weight modules whose sets of weight multiplicities are uniformly bounded. The functor admits several important properties: it preserves central characters; it commutes with the parabolic induction and the translation functors; it generally maps injectives and projectives to injectives and projectives, respectively. After its inception in [32], it was obvious that the application of the functor is not limited only to finite-dimensional Lie algebras. For the last 10 years the twisting localization functor was used in numerous other important classification results and in particular for:

- Description of injective modules in categories of bounded and cuspidal weight modules [25–27];
- Classification of simple weight modules with finite weight multiplicities of affine Lie algebras and Schrödinger algebras [13, 17];
- Classification of simple and injective weight modules of algebras of twisted differential operators on the projective space [27].

A detailed account on the twisted localization functor can be found in Chapter 3 of [33].

The main goal of this paper is to present a survey on the applications of the twisted localization functor in the classification of the simple and injective objects of various categories of weight modules. Due to the technical nature of the classification results, the content of the paper is limited to the presentation of simple and indecomposable injective objects in terms of twisted localization, but it does not address the uniqueness of this presentation. One should note that such uniqueness is present for all results collected in the paper and the reader is referred to the corresponding references for details.

The content of the paper is as follows. Section 3 is devoted to the background material on weight modules of associative algebras. The notations and definitions of all categories of weight modules is introduced in this section. The cases of simple finite-dimensional Lie algebras and superalgebras, Weyl algebras of differential operators, and affine Lie algebras are considered as separate subsections. In Sect. 4 we first introduce the twisted localization functor in its most general setting and then following the subsection order of Sect. 3, we present the classification results on the simple and injective weight modules on a case-by-case basis.

2 Index of Notation

Below we list some notation that are frequently used in the paper under the section number they are first introduced.

3.1 supp, M^λ, $M^{(\lambda)}$, $(\mathscr{U}, \mathscr{H})$-mod, ${}^{\mathrm{f}}(\mathscr{U}, \mathscr{H})$-mod, ${}^{\mathrm{w}}(\mathscr{U}, \mathscr{H})$-mod, ${}^{\mathrm{wf}}(\mathscr{U}, \mathscr{H})$-mod.

3.2 $\mathscr{D}(n)$, $\mathscr{D}$, $(\mathscr{D}, \mathscr{H})$-mod, ${}^{\mathrm{b}}(\mathscr{D}, \mathscr{H})$-mod, $(\mathscr{D}, \mathscr{H})_\nu$-mod, ${}^{\mathrm{b}}(\mathscr{D}, \mathscr{H})_\nu$-mod.

3.3 $|\nu|$, E, $(\mathscr{D}^E, \mathscr{H})^a$-mod, $(\mathscr{D}^E, \mathscr{H})^a_\nu$-mod, ${}_{\mathrm{s}}(\mathscr{D}, \mathscr{H})$-mod, ${}_{\mathrm{s}}(\mathscr{D}^E, \mathscr{H})^a$-mod, ${}^{\mathrm{b}}_{\mathrm{s}}(\mathscr{D}^E, \mathscr{H})^a$-mod, ${}_{\mathrm{s}}(\mathscr{D}^E, \mathscr{H})^a_\nu$-mod, ${}^{\mathrm{b}}_{\mathrm{s}}(\mathscr{D}^E, \mathscr{H})^a_\nu$-mod, $S_{\mathscr{H}'}$.

3.4 $\mathscr{D}(\infty)$

3.5 $\mathscr{B}$, $\mathscr{G}\mathscr{B}$, $\overline{\mathscr{B}}$, $\overline{\mathscr{G}\mathscr{B}}$, $\mathscr{B}_\nu$, $\mathscr{G}\mathscr{B}_\nu$, $\mathscr{B}^\lambda$, $\mathscr{G}\mathscr{B}^\lambda$, $\mathscr{B}^\lambda_\nu$, $\mathscr{G}\mathscr{B}^\lambda_\nu$, $\overline{\mathscr{B}}^\lambda_\nu$, $\overline{\mathscr{G}\mathscr{B}}^\lambda_\nu$.

3.6 $\mathfrak{h}^{ss}_{\bar{0}}$.

3.7 $\mathfrak{G}$, $\mathfrak{H}$, $\mathscr{L}(\mathfrak{g})$, $\mathscr{L}(\mathfrak{g}, \sigma)$, $\mathscr{A}(\mathfrak{g})$, $\mathscr{A}(\mathfrak{g}, \sigma)$.

4.1 D_F, $\Theta^{\mathbf{x}}_F$, $\Theta^{\mathbf{x}}_F$, $D^{\mathbf{x}}_F$.

4.2 D_i^+, D_i^-, $D_i^{x,+}$, $D_i^{x,-}$, $D_{i,j}$, $D^x_{i,j}$.

4.4 D_Γ, $D^{\mathbf{x}}_\Gamma$, γ, $\mathscr{F}_\nu$, $\mathscr{F}_\nu^{\log}$, σ_J, $\mathscr{F}_\nu^{\log}(J)$, $\mathrm{Int}(\nu)$, Γ_a, Φ, ψ_A, Ψ_A, ψ_C, Ψ_C, $S_{\mathfrak{h}}$.

4.6 $\mathscr{L}_{a_1,\dots,a_k}(Y_1 \otimes \ldots \otimes Y_k)$, $\mathscr{L}^\sigma_{a_1,\dots,a_k}(Y_1 \otimes \ldots \otimes Y_k)$, $V_{a_1,\dots,a_k}(Y_1 \otimes \ldots \otimes Y_k)$.

3 Background

In this paper the ground field is $\mathbb{C}$ and $\mathbb{N}$ stands for the set of positive integers. All vector spaces, algebras, and tensor products are assumed to be over $\mathbb{C}$ unless otherwise stated.

3.1 Categories of Weight Modules of Associative Algebras

Let $\mathscr{U}$ be an associative unital algebra and let $\mathscr{H} \subset \mathscr{U}$ be a commutative subalgebra. We assume in addition that $\mathscr{H}$ is a polynomial algebra identified with the symmetric algebra of a vector space $\mathfrak{h}$, and that we have a decomposition

$$\mathscr{U} = \bigoplus_{\mu \in \mathfrak{h}^*} \mathscr{U}^\mu,$$

where

$$\mathscr{U}^\mu = \{x \in \mathscr{U} \,|\, [h, x] = \mu(h)x, \forall h \in \mathfrak{h}\}.$$

Let $Q_{\mathscr{U}} = \mathbb{Z}\Delta_{\mathscr{U}}$ be the $\mathbb{Z}$-lattice in $\mathfrak{h}^*$ generated by $\Delta_{\mathscr{U}} = \{\mu \in \mathfrak{h}^* \mid \mathscr{U}^\mu \neq 0\}$. We obviously have $\mathscr{U}^\mu \mathscr{U}^\nu \subset \mathscr{U}^{\mu+\nu}$.

We call a $\mathscr{U}$-module M *a generalized weight* $(\mathscr{U}, \mathscr{H})$*-module* if $M = \bigoplus_{\lambda\in\mathfrak{h}^*} M^{(\lambda)}$, where

$$M^{(\lambda)} = \{m \in M | (h - \lambda(h)\mathrm{Id})^N m = 0 \text{ for some } N > 0 \text{ and all } h \in \mathfrak{h}\}.$$

Equivalently, generalized weight modules are those on which $\mathfrak{h}$ acts locally finitely. We call $M^{(\lambda)}$ the generalized weight space of M and $\dim M^{(\lambda)}$ the weight multiplicity of the weight λ. Note that

$$\mathscr{U}^{\mu} M^{(\lambda)} \subset M^{(\mu+\lambda)}. \tag{1}$$

A generalized weight module M is called a *weight module* if $M^{(\lambda)} = M^{\lambda}$, where

$$M^{\lambda} = \{m \in M | (h - \lambda(h)\mathrm{Id})m = 0 \text{ for all } h \in \mathfrak{h}\}.$$

Equivalently, weight modules are those on which $\mathfrak{h}$ acts semisimply. Weight and generalized weight modules in similar situations and general setting are also studied in [15].

By $(\mathscr{U}, \mathscr{H})$-mod and ${}^{\mathrm{w}}(\mathscr{U}, \mathscr{H})$-mod we denote the category of generalized weight modules and weight modules, respectively. Furthermore, by ${}^{\mathrm{f}}(\mathscr{U}, \mathscr{H})$-mod and ${}^{\mathrm{b}}(\mathscr{U}, \mathscr{H})$-mod we denote the subcategories of $(\mathscr{U}, \mathscr{H})$-mod consisting of modules with finite weight multiplicities and bounded set of weight multiplicities, respectively. By ${}^{\mathrm{wf}}(\mathscr{U}, \mathscr{H})$-mod and ${}^{\mathrm{wb}}(\mathscr{U}, \mathscr{H})$-mod we denote the subcategories of ${}^{\mathrm{w}}(\mathscr{U}, \mathscr{H})$-mod that are in ${}^{\mathrm{f}}(\mathscr{U}, \mathscr{H})$-mod and ${}^{\mathrm{b}}(\mathscr{U}, \mathscr{H})$-mod, respectively.

For any module M in $(\mathscr{U}, \mathscr{H})$-mod we set

$$\operatorname{supp} M := \{\lambda \in \mathfrak{h}^* \mid M^{(\lambda)} \neq 0\}$$

to be the *support* of M. It is clear from (1) that $\mathrm{Ext}^1_{\mathscr{A}}(M, N) = 0$ if $(\operatorname{supp} M + Q_U) \cap \operatorname{supp} N = \emptyset$, where $\mathscr{A}$ is any of the categories of generalized weight modules or weight modules defined above. Then we have

$$(\mathscr{U}, \mathscr{H})\text{-mod} = \bigoplus_{\overline{\mu}\in\mathfrak{h}^*/Q} (\mathscr{U}, \mathscr{H})_{\overline{\mu}}\text{-mod}, \tag{2}$$

where $(\mathscr{U}, \mathscr{H})_{\overline{\mu}}$-mod denotes the subcategory of $(\mathscr{D}, \mathscr{H})$-mod consisting of modules M with $\operatorname{supp} M \subset \overline{\mu} = \mu + Q$. We similarly define ${}^{\mathrm{w}}(\mathscr{U}, \mathscr{H})_{\overline{\mu}}$-mod, ${}^{\mathrm{f}}(\mathscr{U}, \mathscr{H})_{\overline{\mu}}$-mod, ${}^{\mathrm{b}}(\mathscr{U}, \mathscr{H})_{\overline{\mu}}$-mod, ${}^{\mathrm{wf}}(\mathscr{U}, \mathscr{H})_{\overline{\mu}}$-mod, and ${}^{\mathrm{wb}}(\mathscr{U}, \mathscr{H})_{\overline{\mu}}$-mod, and obtain the corresponding support composition where the direct summands are parametrized by elements of $\mathfrak{h}^*/Q$. With a slight abuse of notation, for $\mu \in \mathfrak{h}^*$ we set $(\mathscr{U}, \mathscr{H})_{\mu}$-mod $= (\mathscr{U}, \mathscr{H})_{\overline{\mu}}$-mod, etc.

3.2 Weight $\mathscr{D}(n)$-Modules

Let $\mathscr{D}(n)$ be the Weyl algebra, i.e., the algebra of polynomial differential operators of the ring $\mathbb{C}[t_1, \ldots, t_n]$ and consider $\mathscr{U} = \mathscr{D}(n)$. When $n \geq 1$ is fixed, we use the notation $\mathscr{D}$ for $\mathscr{D}(n)$. Let $\mathfrak{h} = \mathrm{Span}\{t_1\partial_1, \ldots, t_n\partial_n\}$ and hence $\mathscr{H} = \mathbb{C}[t_1\partial_1, \ldots, t_n\partial_n]$ is a maximal commutative subalgebra of $\mathscr{D}$. Note that the adjoint action of the abelian Lie subalgebra $\mathfrak{h}$ on $\mathscr{D}$ is semisimple. We identify $\mathbb{C}^n$ with the dual space of $\mathfrak{h} = \mathrm{Span}\{t_1\partial_1, \ldots, t_n\partial_n\}$, and fix $\{\varepsilon_1, \ldots, \varepsilon_n\}$ to be the standard basis of this space, i.e., $\varepsilon_i(t_j\partial_j) = \delta_{ij}$. Then $Q = \bigoplus_{i=1}^n \mathbb{Z}\varepsilon_i$ is identified with $\mathbb{Z}^n$, and

$$\mathscr{D} = \bigoplus_{\mu\in\mathbb{Z}^n} \mathscr{D}^\mu.$$

Here $\mathscr{D}^0 = \mathscr{H}$ and each $\mathscr{D}^\mu$ is a free left $\mathscr{H}$-module of rank 1 with generator $\prod_{\mu_i\geq 1} t_i^{\mu_i} \prod_{\mu_j<0} \partial_j^{-\mu_j}$.

From this we see that the simple objects of ${}^{\mathrm{f}}(\mathscr{D}, \mathscr{H})$-mod (equivalently, of ${}^{\mathrm{wf}}(\mathscr{D}, \mathscr{H})$-mod) are in ${}^{\mathrm{b}}(\mathscr{D}, \mathscr{H})$-mod, i.e., have bounded sets of weight multiplicities. Using the description of these simple objects (see for example Theorem 4.5), we obtain that ${}^{\mathrm{b}}(\mathscr{D}, \mathscr{H})$-mod $=$ ${}^{\mathrm{f}}(\mathscr{D}, \mathscr{H})$-mod and ${}^{\mathrm{wb}}(\mathscr{D}, \mathscr{H})$-mod $=$ ${}^{\mathrm{wf}}(\mathscr{D}, \mathscr{H})$-mod. The latter category was studied in [2] and [25] and the former in [1].

The support of every $(\mathscr{D}, \mathscr{H})$-module will be considered as a subset of $\mathbb{C}^n$ and we have a natural decomposition

$$(\mathscr{D}, \mathscr{H})\text{-mod} = \bigoplus_{\overline{\nu}\in\mathbb{C}^n/\mathbb{Z}^n} (\mathscr{D}, \mathscr{H})_{\overline{\nu}}\text{-mod},$$

As before, for $\nu \in \mathbb{C}^n$ we write $(\mathscr{D}, \mathscr{H})_\nu$-mod $= (\mathscr{D}, \mathscr{H})_{\overline{\nu}}$-mod. The same applies for the subcategories ${}^{\mathrm{w}}(\mathscr{D}, \mathscr{H})_{\overline{\nu}}$-mod, ${}^{\mathrm{b}}(\mathscr{D}, \mathscr{H})_{\overline{\nu}}$-mod $=$ ${}^{\mathrm{f}}(\mathscr{D}, \mathscr{H})_{\overline{\nu}}$-mod, and ${}^{\mathrm{wb}}(\mathscr{D}, \mathscr{H})_{\overline{\nu}}$-mod $=$ ${}^{\mathrm{wf}}(\mathscr{D}, \mathscr{H})_{\overline{\nu}}$-mod.

3.3 Weight $\mathscr{D}^E$-Modules

In this subsection $\mathscr{D} = \mathscr{D}(n+1)$ and we assume $n \geq 1$. Let $E = \sum_{i=1}^{n+1} t_i\partial_i$ be the Euler vector field. Denote by $\mathscr{D}^E$ the centralizer of E in $\mathscr{D}$. Note that $\mathscr{D}$ has a $\mathbb{Z}$-grading $\mathscr{D} = \bigoplus_{m\in\mathbb{Z}} \mathscr{D}^m$, where $\mathscr{D}^m = \{d \in \mathscr{D} | [E, d] = md\}$. It is not hard to see that the center of $\mathscr{D}^E$ is generated by E. The quotient algebra $\mathscr{D}^E/(E-a)$ is the algebra of global sections of twisted differential operators on $\mathbb{P}^n$.

Let $a \in \mathbb{C}$, let $(\mathscr{D}^E, \mathscr{H})^a$-mod be the category of generalized weight $\mathscr{D}^E$-modules with locally nilpotent action of $E - a$ and ${}^{\mathrm{b}}(\mathscr{D}^E, \mathscr{H})^a$-mod be the subcategory of ${}^{\mathrm{b}}(\mathcal{D}^E, \mathcal{H})^a$-mod consisting of modules with finite weight multiplicities. We have again a decomposition

$$(\mathcal{D}^E, \mathcal{H})^a\text{-mod} = \bigoplus_{|\nu|=a} (\mathcal{D}^E, \mathcal{H})^a_\nu\text{-mod},$$

where $(\mathcal{D}^E, \mathcal{H})^a_\nu$-mod is the subcategory of modules with support in $\nu + \sum_{i=1}^n \mathbb{Z}(\varepsilon_i - \varepsilon_{i+1})$ and $|\nu| := \sum_{i=1}^{n+1} \nu_i$.

Let $\mathcal{H}'$ be the subalgebra of $\mathcal{D}$ generated by $t_i\partial_i - t_j\partial_j$. We denote by ${}_{\mathrm{s}}(\mathcal{D}, \mathcal{H})$-mod (respectively, ${}_{\mathrm{s}}(\mathcal{D}^E, \mathcal{H})$-mod , ${}^{\mathrm{b}}_{\mathrm{s}}(\mathcal{D}^E, \mathcal{H})$-mod) the subcategory of $(\mathcal{D}, \mathcal{H})$-mod (resp., the subcategory of $(\mathcal{D}^E, \mathcal{H})$-mod , ${}^{\mathrm{b}}(\mathcal{D}^E, \mathcal{H})$-mod) consisting of all modules semisimple over $\mathcal{H}'$. Similarly we define the categories ${}_{\mathrm{s}}(\mathcal{D}^E, \mathcal{H})^a$-mod, ${}^{\mathrm{b}}_{\mathrm{s}}(\mathcal{D}^E, \mathcal{H})^a$-mod, ${}_{\mathrm{s}}(\mathcal{D}^E, \mathcal{H})^a_\nu$-mod, and ${}^{\mathrm{b}}_{\mathrm{s}}(\mathcal{D}^E, \mathcal{H})^a_\nu$-mod. These categories are studied in [27] in detail. In this paper we will limit our attention to the role that the simples and injectives of these categories play in the classification of the simples and injectives in categories of bounded and generalized bounded $\mathfrak{sl}(n+1)$-modules (see Proposition 4.10).

Define the left exact functor $S_{\mathcal{H}'} : (\mathcal{D}, \mathcal{H})\text{-mod} \to {}_{\mathrm{s}}(\mathcal{D}, \mathcal{H})$-mod to be the one that maps M to its submodule consisting of all $\mathcal{H}'$-eigenvectors. One can easily verify that $S_{\mathcal{H}'}$ maps injectives to injectives and blocks to blocks. For the injectives, note that if $\alpha : X \to I$ is a homomorphism of $\mathcal{D}$-modules, X is in ${}_{\mathrm{s}}(\mathcal{D}, \mathcal{H})$-mod, and I is in $(\mathcal{D}, \mathcal{H})$-mod, then $\alpha(X)$ is in ${}_{\mathrm{s}}(\mathcal{D}, \mathcal{H})$-mod, hence a submodule of $S_{\mathcal{H}'}(I)$.

Finally, we define two functors between the categories of weight $\mathcal{D}$-modules and weight $\mathcal{D}^E$-modules. For any $M \in (\mathcal{D}, \mathcal{H})$-mod, let

$$\Gamma_a(M) = \bigcup_{l>0} \mathrm{Ker}(E-a)^l.$$

Then Γ_a is an exact functor from the category $(\mathcal{D}, \mathcal{H})$-mod to the category $(\mathcal{D}^E, \mathcal{H})^a$-mod. The induction functor

$$\Phi(X) = \mathcal{D} \otimes_{\mathcal{D}^E} X$$

is its left adjoint.

3.4 Weight $\mathcal{D}(\infty)$-Modules

Denote by $\mathbb{C}[t_i]_{i\in\mathbb{N}}$ the polynomial algebra in t_i, $i \in \mathbb{Z}$. In this section $\mathcal{U} = \mathcal{D}(\infty)$ is the infinite Weyl algebra, where $\mathcal{D}(\infty)$ is defined as the subalgebra of $\mathrm{End}(\mathbb{C}[t_i]_{i\in\mathbb{N}})$ generated by the operators t_i (multiplication by t_i) and ∂_i (derivative with respect to t_i). We also let $\mathfrak{h}$ to be the space spanned by $t_i\partial_i$, $i \in \mathbb{N}$, and hence $\mathcal{H} = S(\mathfrak{h}) = \mathbb{C}[t_i\partial_i]_{i\in\mathbb{N}}$. All other definitions from Sect. 3.2 transfer trivially to the infinite case.

3.5 Weight Modules of Lie Algebras

Let $\mathfrak{g}$ be a simple finite-dimensional Lie algebra and let $U = U(\mathfrak{g})$ be its universal enveloping algebra. We fix a Cartan subalgebra $\mathfrak{h}$ of $\mathfrak{g}$ and denote by $(\ ,\ )$ the Killing form on $\mathfrak{g}$. We apply the setting of Sect. 3.1 with $\mathscr{U} = U$ and $\mathscr{H} = S(\mathfrak{h})$. We will use the following notation: $\mathscr{G}\mathscr{B} = {}^{\mathrm{b}}(\mathscr{U}, \mathscr{H})$-mod, $\mathscr{B} = {}^{\mathrm{wb}}(\mathscr{U}, \mathscr{H})$-mod, $\mathscr{G}\mathscr{B}_\mu = \mathscr{G}\mathscr{B}_{\overline{\mu}} = {}^{\mathrm{b}}(\mathscr{U}, \mathscr{H})_{\overline{\mu}}$-mod, and $\mathscr{B}_\mu = \mathscr{B}_{\overline{\mu}} = {}^{\mathrm{wb}}(\mathscr{U}, \mathscr{H})_{\overline{\mu}}$-mod. By a result of Fernando [19] and Benkart, Britten and Lemire [3], infinite-dimensional simple objects of $\mathscr{B}$ and $\mathscr{G}\mathscr{B}$ exist only for Lie algebras of type A and C. The modules in $\mathscr{B}$ and $\mathscr{G}\mathscr{B}$ are called bounded and generalized bounded modules, respectively.

A generalized weight module M with finite weight multiplicities will be called a *generalized cuspidal module* if the nonzero elements of the root space $\mathfrak{g}^\alpha$ act injectively (and hence bijectively) on M for all roots α of $\mathfrak{g}$. If M is a weight cuspidal module we will call it simply *cuspidal module*. By $\mathscr{G}\mathscr{C}$ and $\mathscr{C}$ we will denote the categories of generalized cuspidal and cuspidal modules, respectively, and the corresponding subcategories defined by the supports will be denoted by $\mathscr{G}\mathscr{C}_{\overline{\mu}}$ and $\mathscr{C}_{\overline{\mu}}$. One should note that the simple objects of $\mathscr{B}$ and $\mathscr{G}\mathscr{B}$ (as well as those of $\mathscr{C}$ and $\mathscr{G}\mathscr{C}$) coincide. The category $\mathscr{C}$ was described in [26] and [34] for $\mathfrak{g} = \mathfrak{sl}(n+1)$, and in [7] for $\mathfrak{g} = \mathfrak{sp}(2n)$. The category $\mathscr{G}\mathscr{C}$ was described in [34] for $\mathfrak{g} = \mathfrak{sl}(n+1)$, and in [35] for $\mathfrak{g} = \mathfrak{sp}(2n)$.

The induced form on $\mathfrak{h}^*$ will be denoted by $(\ ,\ )$ as well. In this case $Q \subset \mathfrak{h}^*$ is the root lattice. By W we denote the Weyl group of $\mathfrak{g}$. Denote by $Z := Z(U)$ the center of U and let $Z' := \mathrm{Hom}(Z, \mathbb{C})$ be the set of all central characters (here Hom stands for homomorphisms of unital $\mathbb{C}$-algebras). By $\chi_\lambda \in Z'$ we denote the central character of the irreducible highest weight module with highest weight λ. Recall that $\chi_\lambda = \chi_\mu$ iff $\lambda + \rho = w(\mu + \rho)$ for some element w of the Weyl group W, where, as usual, ρ denotes the half-sum of positive roots. Finally, recall that λ is *dominant integral* if $(\lambda, \alpha) \in \mathbb{Z}_{\geq 0}$ for all positive roots α.

One should note that every generalized bounded module has finite Jordan–Hölder series (see Lemma 3.3 in [32]). Since the center Z of U preserves weight spaces, it acts locally finitely on the generalized bounded modules. For every central character $\chi \in Z'$ let $\mathscr{G}\mathscr{B}^\chi$ (respectively, $\mathscr{B}^\chi, \mathscr{G}\mathscr{C}^\chi, \mathscr{C}^\chi$) denote the category of all generalized bounded modules (respectively, bounded, generalized cuspidal, cuspidal) modules M with generalized central character χ, i.e., such that for some $n(M)$, $(z - \chi(z))^{n(M)} = 0$ on M for all $z \in Z$. It is clear that every generalized bounded module M is a direct sum of finitely many $M_i \in \mathscr{G}\mathscr{B}^{\chi_i}$. Thus, one can write

$$\mathscr{G}\mathscr{B} = \bigoplus_{\substack{\chi \in Z' \\ \bar{\mu} \in \mathfrak{h}^*/Q}} \mathscr{G}\mathscr{B}^\chi_{\bar{\mu}},\ \mathscr{G}\mathscr{C} = \bigoplus_{\substack{\chi \in Z' \\ \bar{\mu} \in \mathfrak{h}^*/Q}} \mathscr{G}\mathscr{C}^\chi_{\bar{\mu}},\ \mathscr{B} = \bigoplus_{\substack{\chi \in Z' \\ \bar{\mu} \in \mathfrak{h}^*/Q}} \mathscr{B}^\chi_{\bar{\mu}},\ \mathscr{C} = \bigoplus_{\substack{\chi \in Z' \\ \bar{\mu} \in \mathfrak{h}^*/Q}} \mathscr{C}^\chi_{\bar{\mu}},$$

where $\mathscr{G}\mathscr{B}^\chi_{\bar{\mu}} = \mathscr{G}\mathscr{B}^\chi \cap \mathscr{G}\mathscr{B}_{\bar{\mu}}$, etc. Note that many of the direct summands above are trivial.

By χ_λ we denote the central character of the simple highest weight $\mathfrak{g}$-module with highest weight λ. For simplicity we put $\mathscr{GB}^\lambda := \mathscr{GB}^{\chi_\lambda}$, $\mathscr{GB}^\lambda_{\bar{\mu}} := \mathscr{GB}^{\chi_\lambda}_{\bar{\mu}}$, etc.

Let $\overline{\mathscr{B}}$ (respectively, $\overline{\mathscr{GB}}$) be the full subcategory of all weight modules (respectively, generalized weight modules) consisting of $\mathfrak{g}$-modules M with countable dimensional weight spaces whose finitely generated submodules belong to $\mathscr{B}$ (respectively, $\mathscr{G}B$). With the aid of (2), it is not hard to see that every such M is a direct limit $\varinjlim M_i$ for some directed system $\{M_i \mid i \in I\}$ such that each $M_i \in \mathscr{GB}$ (respectively, $M_i \in \mathscr{B}$). It implies that the action of the center Z of the universal enveloping algebra U on M is locally finite and we have decompositions

$$\overline{\mathscr{B}} = \bigoplus_{\substack{\chi \in Z', \\ \bar{\mu} \in \mathfrak{h}^*/Q}} \overline{\mathscr{B}}^\chi_{\bar{\mu}}, \quad \overline{\mathscr{GB}} = \bigoplus_{\substack{\chi \in Z' \\ \bar{\mu} \in \mathfrak{h}^*/Q}} \overline{\mathscr{GB}}^\chi_{\bar{\mu}}.$$

In a similar way we define $\overline{\mathscr{C}}$ and $\overline{\mathscr{GC}}$ and obtain their block decompositions. Finally, we set $\overline{\mathscr{GB}}^\lambda_{\bar{\mu}} := \overline{\mathscr{GB}}^{\chi_\lambda}_{\bar{\mu}}$, $\overline{\mathscr{B}}^\lambda_{\bar{\mu}} := \overline{\mathscr{B}}^{\chi_\lambda}_{\bar{\mu}}$, and so on.

3.6 Weight Modules of Lie Superalgebras

In this case $\mathfrak{g}$ is a simple finite-dimensional Lie superalgebra, $\mathscr{U} = U(\mathfrak{g})$. Let $\mathfrak{h} = \mathfrak{h}_{\bar{0}} \oplus \mathfrak{h}_{\bar{1}}$ be a Cartan subalgebra of $\mathfrak{g}$, i.e., a self-normalizing nilpotent Lie subsuperalgebra of $\mathfrak{g}$. In particular (see [36, 37]), $\mathfrak{h}_{\bar{0}}$ is a Cartan subalgebra of $\mathfrak{g}_{\bar{0}}$ and $\mathfrak{h}_{\bar{1}}$ is the generalized weight space of weight 0 of the $\mathfrak{h}_{\bar{0}}$-module $\mathfrak{g}_{\bar{1}}$. We fix a Levi subalgebra $\mathfrak{h}^{ss}_{\bar{0}}$ of $\mathfrak{h}_{\bar{0}}$ and set $\mathscr{H} = S(\mathfrak{h}^{ss}_{\bar{0}})$. Note that in this case the vector space $\mathfrak{h}$ from Sect. 3.1 is $\mathfrak{h}^{ss}_{\bar{0}}$. If $\mathfrak{g}$ is a classical Lie superalgebra, then $\mathfrak{h}^{ss}_{\bar{0}} = \mathfrak{h}_{\bar{0}}$. For the four Cartan type series $\mathfrak{g}$, a list of the corresponding $\mathfrak{h}^{ss}_{\bar{0}}$ can be found for example in the appendix of [24]. Note that while the definitions of bounded and generalized bounded modules can be easily transferred from the Lie algebra case to the Lie superalgebra case (by replacing $\mathfrak{h}$ with $\mathfrak{h}^{ss}_{\bar{0}}$), the definitions of cuspidal and generalized cuspidal modules for Lie superalgbras require some additional conditions (see §1.5 in [14]).

3.7 Weight Modules of Affine Lie Algebras

In this section $\mathscr{U} = U(\mathfrak{G})$ and $\mathscr{H} = S(\mathfrak{H})$ where $\mathfrak{G}$ is an affine Lie algebra and $\mathfrak{H}$ is a Cartan subalgebra of $\mathfrak{G}$. For the reader's convenience, we recall the construction of affine Lie algebras and fix notation. For more detail, see [31].

Let $\mathfrak{g}$ be a simple finite-dimensional Lie algebra with a nondegenerate invariant symmetric bilinear form $(\ ,\ )$. Denote by $\mathscr{L}(\mathfrak{g})$ the loop algebra $\mathfrak{g} \otimes \mathbb{C}[t, t^{-1}]$. The affine Lie algebra $\mathscr{A}(\mathfrak{g}) = \mathscr{L}(\mathfrak{g}) \oplus \mathbb{C}D \oplus \mathbb{C}K$ has commutation relations

$$[x \otimes t^m, y \otimes t^n] = [x, y] \otimes t^{m+n} + \delta_{m,-n} m\,(x, y)K, \quad [D, x \otimes t^m] = mx \otimes t^m, \quad [K, \mathscr{A}(\mathfrak{g})] = 0,$$

where $x, y \in \mathfrak{g}$, $m, n \in \mathbb{Z}$, and $\delta_{i,j}$ is Kronecker's delta. The form $(\ ,\ )$ extends to a nondegenerate invariant symmetric bilinear form on $\mathscr{A}(\mathfrak{g})$, still denoted by $(\ ,\ ) : \mathscr{A}(\mathfrak{g}) \times \mathscr{A}(\mathfrak{g}) \to \mathbb{C}$, via

$$(x \otimes t^m, y \otimes t^n) = \delta_{m,-n}(x, y), \quad (D, K) = 1,$$
$$(x \otimes t^m, D) = (x \otimes t^m, K) = (K, K) = (D, D) = 0.$$

If σ is a diagram automorphism of $\mathfrak{g}$ of order s, then σ extends to an automorphism of $\mathscr{A}(\mathfrak{g})$, still denoted by σ, via

$$\sigma(x \otimes t^m) = \zeta^m \sigma(x) t^m, \quad \sigma(D) = D, \quad \sigma(K) = K,$$

where ζ is a fixed primitive sth root of unity. The twisted affine Lie algebra $\mathscr{A}(\mathfrak{g}, \sigma)$ is the Lie algebra $\mathscr{A}(\mathfrak{g})^\sigma$ of σ-fixed points of $\mathscr{A}(\mathfrak{g})$. Note that

$$\mathscr{A}(\mathfrak{g}, \sigma) = \mathscr{L}(\mathfrak{g}, \sigma) \oplus \mathbb{C}D \oplus \mathbb{C}K,$$

where $\mathscr{L}(\mathfrak{g}, \sigma) = \mathscr{L}(\mathfrak{g})^\sigma$ is the subalgebra of σ-fixed points of $\mathscr{L}(\mathfrak{g})$. One has that

$$\mathscr{L}(\mathfrak{g}, \sigma) = \bigoplus_{j \in \mathbb{Z}} \mathfrak{g}_{\bar{j}} \otimes t^j,$$

where

$$\mathfrak{g} = \bigoplus_{\bar{j} \in \mathbb{Z}/s\mathbb{Z}} \mathfrak{g}_{\bar{j}}$$

is the decomposition of $\mathfrak{g}$ into σ-eigenspaces. The restriction of $(\ ,\ )$ to $\mathscr{A}(\mathfrak{g}, \sigma)$ is a nondegenerate invariant symmetric bilinear form for which we will use the same notation.

If $\mathfrak{h}$ is a Cartan subalgebra of $\mathfrak{g}$, then $\mathfrak{h} \oplus \mathbb{C}D \oplus \mathbb{C}K$ is a Cartan subalgebra of $\mathscr{A}(\mathfrak{g})$. Furthermore, σ preserves $\mathfrak{h}$ and $\mathfrak{h}^\sigma \oplus \mathbb{C}D \oplus \mathbb{C}K$ is a Cartan subalgebra of $\mathscr{A}(\mathfrak{g}, \sigma)$. For the rest of the paper $\mathfrak{h}$ denotes a fixed Cartan subalgebra of $\mathfrak{g}$, $\mathfrak{G}$ denotes $\mathscr{A}(\mathfrak{g})$ or $\mathscr{A}(\mathfrak{g}, \sigma)$, and $\mathfrak{H}$ denotes the corresponding Cartan subalgebra of $\mathfrak{G}$.

The Lie algebra $\mathfrak{G}$ admits a root decomposition

$$\mathfrak{G} = \mathfrak{H} \oplus \left(\bigoplus_{\alpha \in \Delta} \mathfrak{G}^\alpha \right).$$

To describe the root system Δ of $\mathfrak{G}$, let $\delta \in \mathfrak{H}^*$ denote the element with

$$\delta(D) = 1, \quad \delta(h) = \delta(K) = 0 \quad \text{for every} \quad h \in \mathfrak{h}.$$

If $\mathfrak{G} = \mathscr{A}(\mathfrak{g})$, denote the root system of $\mathfrak{g}$ by $\mathring{\Delta}$. If $\mathfrak{G} = \mathscr{A}(\mathfrak{g}, \sigma)$, denote the nonzero weights of the $\mathfrak{g}_{\bar{0}}$-module $\mathfrak{g}_{\bar{j}}$ by $\mathring{\Delta}_{\bar{j}}$ and set $\mathring{\Delta} = \cup_{\bar{j} \in \mathbb{Z}/s\mathbb{Z}} \mathring{\Delta}_{\bar{j}}$. Note that for $\mathfrak{G} = \mathscr{A}(\mathfrak{g}, \sigma) \not\cong A_{2l}^{(2)}$, $\mathring{\Delta} = \mathring{\Delta}_{\bar{0}}$ is the root system of $\mathfrak{g}_{\bar{0}}$ and for $\mathfrak{G} \cong A_{2l}^{(2)}$, $\mathring{\Delta}$ is the non–reduced root system BC_l.

The decomposition $\mathfrak{H} = \mathfrak{h} \oplus \mathbb{C}D \oplus \mathbb{C}K$ (respectively, $\mathfrak{H} = \mathfrak{h}^{\sigma} \oplus \mathbb{C}D \oplus \mathbb{C}K$) allows us to consider $\mathring{\Delta}$ as a subset of $\mathfrak{H}^*$. The root system Δ decomposes as

$$\Delta = \Delta^{\text{re}} \sqcup \Delta^{\text{im}},$$

where

$$\Delta^{\text{im}} = \{n\delta \mid n \in \mathbb{Z} \backslash \{0\}\}$$

are the *imaginary roots* of $\mathfrak{G}$ and the real roots Δ^{re} are given as follows.

(i) If $\mathfrak{G} = \mathscr{A}(\mathfrak{g})$, then

$$\Delta^{\text{re}} = \{\alpha + n\delta \mid n \in \mathbb{Z},\ \alpha \in \mathring{\Delta}\}.$$

(ii) If $\mathfrak{G} = \mathscr{A}(\mathfrak{g}, \sigma)$, then

$$\Delta^{\text{re}} = \{\alpha + n\delta \mid n \in \mathbb{Z},\ \alpha \in \mathring{\Delta}_{\bar{n}}\}.$$

3.8 Parabolically Induced Module

In this subsection we have that $\mathscr{U} = U(\mathfrak{g})$ or $\mathscr{U} = U(\mathfrak{G})$, where $\mathfrak{g}$ is a simple finite dimensional Lie algebra or superalgebras, and $\mathfrak{G}$ is an affine Lie algebra, i.e., we have the setting of Sects. 3.5, 3.6, or 3.7. In this subsection, for simplicity, Δ will denote the root system of $\mathfrak{g}$ or $\mathfrak{G}$.

If Δ is symmetric (i.e., $\Delta = -\Delta$), then we call a proper subset P of Δ a *parabolic set in* Δ if

$$\Delta = P \cup (-P) \quad \text{and} \quad \alpha, \beta \in P \quad \text{with} \quad \alpha + \beta \in \Delta \quad \text{implies} \quad \alpha + \beta \in P.$$

A detailed treatment of parabolic subsets of symmetric root systems can be found in [12]. If $\Delta \neq -\Delta$, then $P \subsetneq \Delta$ will be called *parabolic* if $P = \tilde{P} \cap \Delta$ for some parabolic subset $\tilde{P}$ of $\Delta \cup (-\Delta)$.

For a symmetric root systems Δ and a parabolic subset of roots P of Δ, we call $L := P \cap (-P)$ the *Levi component* of P, $N^+ := P \backslash (-P)$ the *nilradical* of P, and

$P = L \sqcup N^+$ the Levi decomposition of P. If $\Delta \neq -\Delta$, then we choose a parabolic subset $\tilde{P}$ of $\Delta \cap (-\Delta)$ such that $P = \tilde{P} \cap \Delta$, and define $\tilde{L} = \tilde{P} \cap (-\tilde{P})$ and $\tilde{N}^+ = \tilde{P} \backslash (-\tilde{P})$. We call $L := \tilde{L} \cap P$ a Levi component of P, $N^+ = \tilde{N}^+ \cap P$ a nilradical of P, and $P = L \sqcup N^+$ a Levi decomposition of P. We note that in the nonsymmetric case the definition of a Levi component and nilradical of P essentially depends on the choice of a parabolic subset $\tilde{P}$ of $\Delta \cap (-\Delta)$. We refer the reader to Remarks 1.7 and 3.3 in [28] for examples.

We will call a subalgebra $\mathfrak{p}$ of $\mathfrak{g}$ (respectively, $\mathfrak{P}$ of $\mathfrak{G}$) *parabolic* if it is of the form $\mathfrak{p}_P = \mathfrak{h} \oplus \left(\bigoplus_{\mu \in P} \mathfrak{g}^\mu\right)$ (respectively, $\mathfrak{P}_P = \mathfrak{H} \oplus \left(\bigoplus_{\mu \in P} \mathfrak{G}^\mu\right)$) for some parabolic subset P of Δ. If L and N^+ are Levi component and a nilradical of a parabolic set P, respectively, then $\mathfrak{l} = \mathfrak{h} \oplus \left(\bigoplus_{\mu \in L} \mathfrak{g}^\mu\right)$ and $\mathfrak{n} = \mathfrak{h} \oplus \left(\bigoplus_{\mu \in N^+} \mathfrak{g}^\mu\right)$ are called a Levi subalgebra, and a nilradical of $\mathfrak{p}_P$, respectively. Similarly, we define a Levi subalgebra, and a nilradical of a parabolic subalgebra $\mathfrak{P}_P$ of $\mathfrak{G}$.

Let $\mathfrak{l}$ and $\mathfrak{n}$ be a Levi subalgebra and a nilradical of a parabolic subalgebra $\mathfrak{p}$. Every $\mathfrak{l}$-module S can be considered as a $\mathfrak{p}$-module with the trivial action of $\mathfrak{n}$. We define $M_{\mathfrak{p}}(S) = U(\mathfrak{g}) \otimes_{U(\mathfrak{p})} S$. Among the submodules of $M_{\mathfrak{p}}(S)$ which intersect trivially with $1 \otimes S$ there is a unique maximal one $Z_{\mathfrak{p}}(S)$. Set $V_{\mathfrak{p}}(S) = M_{\mathfrak{p}}(S)/Z_{\mathfrak{p}}(S)$. In this paper we will call a $\mathfrak{g}$-module *parabolically induced* if it isomorphic to $V_{\mathfrak{p}}(S)$ for some parabolic subalgebra $\mathfrak{p}$ of $\mathfrak{g}$ and some module S over a Levi component $\mathfrak{l}$ of $\mathfrak{p}$. We similarly define parabolically induced modules of $\mathfrak{G}$. For properties of the version of the parabolic induction functor used in the paper the reader is referred to [14].

4 Twisted Localization

4.1 Twisted Localization in General Setting

Retain the notation of Sect. 3.1. Namely, $\mathscr{U}$ is an associative unital algebra, and $\mathscr{H} = S(\mathfrak{h})$ is a commutative subalgebra of $\mathscr{U} = \bigoplus_{\mu \in Q_{\mathscr{U}}} U^\mu$, in particular, $\mathrm{ad}(h)$ is semisimple on $\mathscr{U}$ for every $h \in \mathfrak{h}$. Now let $F = \{f_j \mid j \in J\}$ be a subset of commuting elements of $\mathscr{U}$ such that $\mathrm{ad}\,(f)$, $f \in F$, are locally nilpotent endomorphisms of $\mathscr{U}$. In addition, we assume that for every $u \in \mathscr{U}$, $uf = fu$, for all but finitely many $f \in F$. Let $\langle F \rangle$ be the multiplicative subset of $\mathscr{U}$ generated by F, i.e., the $\langle F \rangle$ consists of the elements $f_1^{n_1} \dots f_k^{n_k}$ for $f_i \in F$, $n_i \in \mathbb{N}$. By $D_F \mathscr{U}$ we denote the localization of $\mathscr{U}$ relative to $\langle F \rangle$. Note that $\langle F \rangle$ satisfies Ore's localizability condition due to the fact that f_i are locally ad-nilpotent. The proof of that and more details on the $\langle F \rangle$-localization can be found in §4 of [32].

For a $\mathscr{U}$-module M, by $D_F M = D_F \mathscr{U} \otimes_{\mathscr{U}} M$ we denote the localization of M relative to $\langle F \rangle$. We will consider $D_F M$ both as a $\mathscr{U}$-module and as a $D_F \mathscr{U}$-module. By $\theta_F : M \to D_F M$ we denote the localization map defined by $\theta_F(m) = 1 \otimes m$. Then

$$\mathrm{ann}_M F := \{m \in M \mid sm = 0 \text{ for some } s \in F\}$$

is a submodule of M (often called, the torsion submodule with respect to F). Note that if $\mathrm{ann}_M F = 0$, then θ_F is an injection. In the latter case, we will say that M is F-torsion free, and M will be considered naturally as a submodule of $D_F M$. Note also that if $F = F_1 \cup F_2$, then $D_{F_1} D_{F_2} M \simeq D_{F_2} D_{F_1} M \simeq D_F M$.

It is well known that D_F is a functor from the category of $\mathscr{U}$-modules to the category of $D_F\mathscr{U}$-modules. For any category $\mathscr{A}$ of $\mathscr{U}$-modules, by $\mathscr{A}_F$ we denote the category of $D_F\mathscr{U}$-modules considered as $\mathscr{U}$-modules are in $\mathscr{A}$. Some useful properties of the localization functor D_F are listed in the following lemma.

Lemma 4.1. (i) *If $\varphi : M \to N$ is a homomorphism of $\mathscr{U}$-modules, then $D_F(\varphi)\theta_F = \theta_F \varphi$.*
(ii) *D_F is an exact functor.*
(iii) *If N is a $D_F\mathscr{U}$-module and $\varphi : M \to N$ is a homomorphism of $\mathscr{U}$-modules, then there exists a unique homomorphism of $D_F\mathscr{U}$-modules $\overline{\varphi} : D_F M \to N$ such that $\overline{\varphi}\theta_F = \varphi$. If we identify N with $D_F N$, then $\overline{\varphi} = D_F(\varphi)$.*
(iv) *Let $\mathscr{A}$ be any category of U-modules. If I is an injective module in $\mathscr{A}_F$, then I (considered as an $\mathscr{U}$-module) is injective in $\mathscr{A}$ as well.*

We now introduce the "generalized conjugation" in $D_F\mathscr{U}$ following §4 of [32]. For $\mathbf{x} \in \mathbb{C}^J$ define the automorphism $\Theta_F^{\mathbf{x}}$ of $D_F\mathscr{U}$ in the following way. For $u \in D_F U$, $f \in F$, and $x \in \mathbb{C}$ set

$$\Theta_{\{f\}}^{x}(u) := \sum_{i \geq 0} \binom{x}{i} \mathrm{ad}(f)^i(u)\, f^{-i},$$

where $\binom{x}{i} := x(x-1)\ldots(x-i+1)/i!$ for $x \in \mathbb{C}$ and $i \in \mathbb{Z}_{\geq 0}$. Note that the sum on the right-hand side is well defined since f is ad-nilpotent on $\mathscr{U}$. Now, for $J \subset I$ and $\mathbf{x} = (x_j)_{j \in J} \in \mathbb{C}^J$ define

$$\Theta_F^{\mathbf{x}}(u) := \prod_{j \in J} \Theta_{\{f_j\}}^{x_j}(u).$$

The product above is in fact finite since $\mathrm{ad}(f_j)(u) = 0$ for all but finitely many f_j. Note that if $J = \{1, 2, \ldots, k\}$ and $\mathbf{x} \in \mathbb{Z}^k$, we have $\Theta_F^{\mathbf{x}}(u) = \mathbf{f}^{\mathbf{x}} u \mathbf{f}^{-\mathbf{x}}$, where $\mathbf{f}^{\mathbf{x}} := f_1^{x_1} \ldots f_k^{x_k}$.

For a $D_F\mathscr{U}$-module N by $\Phi_F^{\mathbf{x}} N$ we denote the $D_F\mathscr{U}$-module N twisted by $\Theta_F^{\mathbf{x}}$. The action on $\Phi_F^{\mathbf{x}} N$ is given by

$$u \cdot v^{\mathbf{x}} := (\Theta_F^{\mathbf{x}}(u) \cdot v)^{\mathbf{x}},$$

where $u \in D_F\mathscr{U}$, $v \in N$, and $w^{\mathbf{x}}$ stands for the element w considered as an element of $\Phi_F^{\mathbf{x}} N$. In the case $J = \{1, 2, \ldots, k\}$ and $\mathbf{x} \in \mathbb{Z}^k$, there is a natural isomorphism of $D_F\mathscr{U}$-modules $M \to \Phi_F^{\mathbf{x}} M$ given by $m \mapsto (\mathbf{f}^{\mathbf{x}} \cdot m)^{\mathbf{x}}$ with inverse map defined by $n^{\mathbf{x}} \mapsto \mathbf{f}^{-\mathbf{x}} \cdot n$. In view of this isomorphism, for $\mathbf{x} \in \mathbb{Z}^k$, we will identify M with $\Phi_F^{\mathbf{x}} M$, and for any $\mathbf{x} \in \mathbb{C}^k$ we will write $\mathbf{f}^{\mathbf{x}} \cdot m$ (or simply $\mathbf{f}^{\mathbf{x}} m$) for $m^{-\mathbf{x}}$ whenever

$m \in M$. The action of $\Phi_F^{\mathbf{x}}$ on a homomorphism $\alpha : M \to N$ of $D_F\mathscr{U}$-modules is defined by $\Phi_F^{\mathbf{x}}(\alpha)(\mathbf{f}^{\mathbf{x}} \cdot m) = \mathbf{f}^{\mathbf{x}} \cdot (\alpha(m))$.

The basic properties of the twisting functor $\Phi_F^{\mathbf{x}}$ on $D_F\mathscr{U}$-mod are summarized in the following lemma. The proofs are straightforward.

Lemma 4.2. *Let $F = \{f_1, \ldots, f_k\}$ be a set of locally* ad*-nilpotent commuting elements of $\mathscr{U}$, M and N be $D_F\mathscr{U}$-modules, $m \in M$, $u \in \mathscr{U}$, and $\mathbf{x}, \mathbf{y} \in \mathbb{C}^k$.*

(i) $\Theta_F^{\mathbf{x}} \circ \Theta_F^{\mathbf{y}} = \Theta_F^{\mathbf{x}+\mathbf{y}}$, *in particular* $\mathbf{f}^{\mathbf{x}} \cdot (\mathbf{f}^{\mathbf{y}} \cdot m) = \mathbf{f}^{\mathbf{x}+\mathbf{y}} \cdot m$;
(ii) $\Phi_F^{\mathbf{x}}\Phi_F^{\mathbf{y}} = \Phi_F^{\mathbf{x}+\mathbf{y}}$, *in particular,* $\Phi_F^{\mathbf{x}}\Phi_F^{-\mathbf{x}} = \mathrm{Id}$ *on the category of $D_F\mathscr{U}$-modules;*
(iii) $\mathbf{f}^{\mathbf{x}} \cdot (u \cdot (\mathbf{f}^{-\mathbf{x}} \cdot m)) = \Theta_F^{\mathbf{x}}(u) \cdot m$;
(iv) $\Phi_F^{\mathbf{x}}$ *is an exact functor;*
(v) *M is simple (respectively, injective) if and only if $\Phi_F^{\mathbf{x}}M$ is simple (respectively, injective);*
(vi) $\mathrm{Hom}_{\mathscr{U}}(M, N) = \mathrm{Hom}_{\mathscr{U}}(\Phi_F^{\mathbf{x}}M, \Phi_F^{\mathbf{x}}N)$.

For any $\mathscr{U}$-module M, and $\mathbf{x} \in \mathbb{C}^J$ we define the *twisted localization $D_F^{\mathbf{x}}M$ of M relative to F and* $\mathbf{x}$ by $D_F^{\mathbf{x}}M := \Phi_F^{\mathbf{x}}D_FM$. The twisted localization is a exact functor from $\mathscr{U}$-mod to $D_F\mathscr{U}$-mod.

Some properties of the functor $D_F^{\mathbf{x}}$ on the category of generalized weight $(\mathscr{U}, \mathscr{H})$-modules are described in the following lemma.

Lemma 4.3. *Assume that $f_i \in \mathscr{U}^{\mathbf{a_i}}$ for $\mathbf{a_i} \in Q_{\mathscr{U}}$.*

(i) *If M is a generalized weight $(\mathscr{U}, \mathscr{H})$-module, then D_FM is a generalized weight $(D_F\mathscr{U}, \mathscr{H})$-module.*
(ii) *If N is a generalized weight $(D_F\mathscr{U}, \mathscr{H})$-module, then $\mathbf{f}^{\mathbf{x}}m \in N^{(\lambda+\mathbf{xa})}$ whenever $m \in N^{(\lambda)}$, where $\mathbf{xa} = x_1\mathbf{a_1} + \ldots + x_k\mathbf{a_k}$. In particular, $\Phi_F^{\mathbf{x}}N$ is a generalized weight $(D_F\mathscr{U}, \mathscr{H})$-module.*

In what follows we will treat each of the special cases of $\mathscr{U}$ and $\mathscr{H}$ considered in the previous section separately. We will show in particular that in all cases every simple object of ${}^{\mathrm{f}}(\mathscr{U}, \mathscr{H})$-mod (equivalently, in ${}^{\mathrm{wf}}(\mathscr{U}, \mathscr{H})$-mod) is either parabolically induced or it is isomorphic to a twisted localization of a well understood module (for example, highest weight module, loop module, etc.).

4.2 Twisted Localization for $\mathscr{D}(n)$

Consider now $\mathscr{U} = \mathscr{D}(n)$ and $\mathscr{H} = \mathbb{C}[t_1\partial_1, \ldots, t_n\partial_n]$. Obvious choices for F are $\{t_i\}$, $\{\partial_i\}$, $\{t_i\partial j\}$, $i \neq j$. We set for convenience $D_i^+ = D_{\{\partial_i\}}$, $D_i^- = D_{\{t_i\}}$, $D_i^{x,-} := D_{\{t_i\}}^x$, $D_i^{x,+} = D_{\{\partial_i\}}^x$, and $D_{i,j}^x = D_{\{t_i\partial_j\}}^x$ for $x \in \mathbb{C}$ and $i \neq j$.

Let us first focus on the case $n = 1$, i.e., $\mathscr{D} = \mathscr{D}(1)$. We fix $t_1 = t$ and $\partial_1 = \partial$. For $\nu \in \mathbb{C}$ we set

$$\mathscr{F}_\nu = t^\nu\mathbb{C}[t, t^{-1}]$$

and consider $\mathcal{F}_\nu$ as a $\mathcal{D}$-module with the natural action of $\mathcal{D}$. It is easy to check that $\mathcal{F}_\nu \in {}^{\mathrm{b}}(\mathcal{D}, \mathcal{H})_\mu$−mod and $\mathcal{F}_\nu$ is simple iff $\nu \notin \mathbb{Z}$. By definition, $\mathcal{F}_\nu \simeq \mathcal{F}_\mu$ if and only if $\mu - \nu \in \mathbb{Z}$. So, if $\nu \in \mathbb{Z}$ we may assume $\nu = 0$. One easily checks that $\mathcal{F}_0$ has length 2 and one has the following nonsplit exact sequence

$$0 \to \mathcal{F}_0^+ \to \mathcal{F}_0 \to \mathcal{F}_0^- \to 0,$$

where $\mathcal{F}_0^+ = \mathbb{C}[t]$ and $\mathcal{F}_0^-$ is a simple quotient. Moreover, if σ denotes the automorphism of $\mathcal{D}$ defined by $\sigma(t) = \partial, \sigma(\partial) = -t$, then $\mathcal{F}_0^- \simeq (\mathcal{F}_0^+)^\sigma$. As follows for instance from [25], any simple object in $(\mathcal{D}, \mathcal{H})$-mod is isomorphic to $\mathcal{F}_\nu$ for some non-integer ν, $\mathcal{F}_0^-$ or $\mathcal{F}_0^+$. A classification of all simple $\mathcal{D}(1)$-modules (not necessarily generalized weight ones) can be found in [5]. Also, an alternative approach that leads to a classification of the simple weight modules of a more general class of Weyl type algebras can be found in [16].

Now let $\mathcal{D} = \mathcal{D}(n)$. To describe the simple and injective modules in the category $(\mathcal{D}, \mathcal{H})$-mod we use the fact that $\mathcal{D}(n)$ is the tensor product of n copies of $\mathcal{D}(1)$. With this in mind, for $\nu \in \mathbb{C}^n$, we set $\mathcal{F}_\nu = \mathcal{F}_{\nu_1} \otimes \ldots \otimes \mathcal{F}_{\nu_n}$, $\mathcal{F}^+ = \mathcal{F}_0^+ \otimes \ldots \otimes \mathcal{F}_0^+$, $\mathcal{F}_\nu^{\log} = \mathcal{F}_{\nu_1}^{\log} \otimes \ldots \otimes \mathcal{F}_{\nu_n}^{\log}$, and $\mathcal{F}_0^+ = \mathbb{C}[t_1, \ldots, t_n]$. If μ has the property that $\mu_i \geq 0$ whenever $\mu_i \in \mathbb{Z}$, by $\mathcal{F}_\mu^+$ we denote the submodule of $\mathcal{F}_\mu$ generated by t^μ. One can check that $\mathcal{F}_\mu^+$ is the unique simple submodule of $\mathcal{F}_\mu$. For a nonempty subset J of $\{1, 2, \ldots., n\}$ let σ_J be the automorphism of $\mathcal{D}(n)$ defined as σ on the j-th $\mathcal{D}(1)$-component of $\mathcal{D}(n)$ for $j \in J$, and as identity for all other $\mathcal{D}(1)$-components. We set $\mathcal{F}_\nu^{\log}(J) = (\mathcal{F}_\nu^{\log})^{\sigma_J}$ and $\mathcal{F}_\nu(J) = (\mathcal{F}_\nu)^{\sigma_J}$.

The application of the twisted localization functors $D_i^{x,\pm}$ on the modules $\mathcal{F}_\nu^{\log}(J)$ is described in the following lemma. Recall that $\varepsilon_i \in \mathbb{C}^n$ are defined by $(\varepsilon_i)_j = \delta_{ij}$.

Lemma 4.4. *The following isomorphisms hold.*

$$D_i^{x,+}(\mathcal{F}_\nu^{\log}(J)) \simeq \mathcal{F}_{\nu - x\varepsilon_i}^{\log}(J \cup i), \; D_i^{x,-}(\mathcal{F}_\nu^{\log}(J)) \simeq \mathcal{F}_{\nu + x\varepsilon_i}^{\log}(J \setminus i).$$

One can show that for $\nu = (\nu_1, \ldots, \nu_n)$ every simple module in $(\mathcal{D}, \mathcal{H})_\nu$-mod is the tensor product $S_1 \otimes \cdots \otimes S_n$ where each S_i is a simple module in $(\mathcal{D}(1), \mathcal{H}(1))_{\nu_i}$-mod (see for example [25]).

Let $\mathrm{Int}(\nu)$ be the set of all i such that $\nu_i \in \mathbb{Z}$ and $\mathcal{P}(\nu)$ be the power set of $\mathrm{Int}(\nu)$. For every $J \in \mathcal{P}(\nu)$ set

$$\mathcal{S}_\nu(J) := S_1 \otimes \cdots \otimes S_n,$$

where $S_i = \mathcal{F}_{\nu_i}$ if $i \notin \mathrm{Int}(\nu)$, $S_i = \mathcal{F}_0^+$ if $i \in \mathrm{Int}(\nu) \setminus J$ and $S_i = \mathcal{F}_0^-$ if $i \in J$. As discussed above, $\mathcal{S}_\nu(J)$ are exactly the simple objects of $(\mathcal{D}, \mathcal{H})$-mod. On the other hand, as follows from Proposition 5.2 in [27], $\mathcal{F}_\nu^{\log}(J)$ is injective in $(\mathcal{D}, \mathcal{H})$-mod. Combining these results with the lemma above, one has the following theorem.

Theorem 4.5 *Every simple object M in $(\mathscr{D}, \mathscr{H})$-mod (or, equivalently, in ${}^{\mathrm{w}}(\mathscr{D}, \mathscr{H})$-mod, in ${}^{\mathrm{f}}(\mathscr{D}, \mathscr{H})$-mod, and in ${}^{\mathrm{wf}}(\mathscr{D}, \mathscr{H})$-mod) is isomorphic to some $\mathscr{S}_\nu(J)$ and the injective envelope of $\mathscr{S}_\nu(J)$ in $(\mathscr{D}, \mathscr{H})$-mod is isomorphic to $\mathscr{F}_\nu^{\log}(J)$. In particular, every such simple module M is isomorphic to the σ_J-twist of a twisted localized module $D_F^{\mathbf{x}}\mathscr{F}_0^+$, and the injective envelope of M in $(\mathscr{D}, \mathscr{H})$-mod is isomorphic to the σ_J-twist of $D_F^{\mathbf{x}}\mathscr{F}_0^{\log}$.*

4.3 Twisted Localization for $\mathscr{D}(\infty)$

Recall that $\mathbb{N}$ stands for the set of positive integers. The $\mathscr{D}(\infty)$-analogs of the $\mathscr{D}(n)$-modules $\mathscr{F}_\nu$ can be defined with the aid of twisted localization. Indeed, for $\nu \in \mathbb{C}^{\mathbb{N}}$ and $T = \{t_i \mid i \in \mathbb{N}\}$, set $\mathscr{F}_\nu = D_T^\nu \mathscr{F}_0^+$, where, as before, $\mathscr{F}_0^+ = \mathbb{C}[t_i]_{i\in\mathbb{N}}$. In particular, one can show that $\mathscr{F}_0 \simeq \mathbb{C}[t_i^{\pm 1}]_{i\in\mathbb{N}}$. One way to think of $\mathscr{F}_\nu$ is as the space with basis $\mathbf{t}^\nu \mathbf{t}^{\mathbf{z}}$, where $\mathbf{z} = (z_1, z_2, \dots)$ runs through all sequences of integers, such that all but finitely many z_i are zero, and $\mathbf{t}^{\mathbf{z}} = \prod_{i>0} t_i^{z_i}$. As in the case of $\mathscr{D}(n)$, for any $J \subset \mathbb{N}$, we may define an automorphism σ_J of $\mathscr{D}(\infty)$. The following theorem is proved in [21].

Theorem 4.6 (i) *For every $\nu \in \mathbb{C}^{\mathbb{N}}$, the module $\mathscr{F}_\nu$ has a unique simple submodule $\mathscr{F}_\nu^+$. If ν is such that $\nu_i \geq 0$ whenever $\nu_i \in \mathbb{Z}$, then the module $\mathscr{F}_\nu^+$ is the submodule of $\mathscr{F}_\nu$ generated by $\mathbf{t}^\nu$.*

(ii) *Every simple module in ${}^{\mathrm{w}}(\mathscr{D}(\infty), \mathscr{H})$-mod is isomorphic to the σ_J-twist $(\mathscr{F}_\nu^+)^{\sigma_J}$ of some $\mathscr{F}_\nu^+$. Here J is a subset of* Int(ν) *and ν can be chosen so that $\nu_i \geq 0$ whenever $\nu_i \in \mathbb{Z}$.*

(iii) The injective envelope of $(\mathscr{F}_\nu^+)^{\sigma_J}$ in ${}^{\mathrm{w}}(\mathscr{D}(\infty), \mathscr{H})$-mod is $(\mathscr{F}_\nu)^{\sigma_J}$.

Remark 4.7 The above theorem can be extended to the category $(\mathscr{D}(\infty), \mathscr{H})$-mod. Namely, every simple object in $(\mathscr{D}(\infty), \mathscr{H})$-mod is also of the form $(\mathscr{F}_\nu^+)^{\sigma_J}$ and the injective envelope of $(\mathscr{F}_\nu^+)^{\sigma_J}$ in $(\mathscr{D}(\infty), \mathscr{H})$-mod is isomorphic to $(\mathscr{F}_\nu^{\log})^{\sigma_J}$. Here $\mathscr{F}_\nu^{\log} = D_T^\nu \mathscr{F}_0^{\log}$, and $\mathscr{F}_0^{\log} = \mathbb{C}[t_i^{\pm 1}, \log t_i]_{i\in\mathbb{N}}$. To prove that $(\mathscr{F}_\nu^{\log})^{\sigma_J}$ is injective in $(\mathscr{D}(\infty), \mathscr{H})$-mod, it is sufficient to prove it for $J = \emptyset$, and for the latter we follow the steps of the proof of the same statement in $(\mathscr{D}(n), \mathscr{H})$-mod (see Proposition 5.2 in [27]).

4.4 Twisted Localization for Finite-Dimensional Lie Algebras

In this case we consider a simple finite-dimensional Lie algebra $\mathfrak{g}$ with a fixed Cartan subalgebra $\mathfrak{h}$, $\mathscr{U} = U(\mathfrak{g})$ and $\mathscr{H} = S(\mathfrak{h})$. The multiplicative sets will be always of the form $F = \langle e_\alpha \mid \alpha \in \Gamma \rangle$, where Γ is a set of k commuting roots and e_α is in the α-root space of $\mathfrak{sl}(n+1)$. For $\mathbf{x} \in \mathbb{C}^k$, we write D_Γ and $D_\Gamma^{\mathbf{x}}$ for D_F and $D_F^{\mathbf{x}}$, respectively. The following theorem is proved in [32].

Proposition 4.8 *Every simple module in* ${}^{\mathrm{wf}}(\mathscr{U}, \mathscr{H})$ *(equivalently, in* ${}^{\mathrm{f}}(\mathscr{U}, \mathscr{H})$*) is either parabolically induced, or it is cuspidal. Every cuspidal module is isomorphic to* $D_{\Gamma}^{\mathbf{x}} L$ *for some simple highest weight bounded module* L*, a set* Γ *of* n *commuting roots, and* $\mathbf{x} \in \mathbb{C}^n$*.*

For the classification of the simple highest weight bounded modules we refer the reader to Sects. 8 and 9 in [32]. Below we describe the injectives in the categories of bounded and generalized bounded modules using equivalence of categories in each of the two cases $\mathfrak{g} = \mathfrak{sl}(n+1)$ and $\mathfrak{g} = \mathfrak{sp}(2n)$.

4.4.1 The Case $\mathfrak{g} = \mathfrak{sl}(n+1)$

In this case $\mathfrak{h}^*$ is identified with the subspace of $\mathbb{C}^{n+1}$ spanned by the simple roots $\varepsilon_1 - \varepsilon_2, \ldots, \varepsilon_n - \varepsilon_{n+1}$. By γ we denote the projection $\mathbb{C}^{n+1} \to \mathfrak{h}^*$ with one-dimensional kernel $\mathbb{C}(\varepsilon_1 + \cdots + \varepsilon_{n+1})$.

Consider the homomorphism $\psi_A : U(\mathfrak{sl}(n+1)) \to \mathscr{D}(n+1)$ defined by $\psi(E_{ij}) = t_i \partial_j$, $i \neq j$, where E_{ij} is the elementary (ij)th matrix in $\mathfrak{sl}(n+1)$. The image of ψ_A is contained in $\mathscr{D}^E$. Using lift by ψ_A any $\mathscr{D}^E$-module becomes $\mathfrak{sl}(n+1)$-module. Since $\psi_A(U(\mathfrak{h})) \subset \mathscr{H}$, one has a functor $\Psi_A : (\mathscr{D}^E, \mathscr{H})\text{-mod} \to \overline{\mathscr{G}\mathscr{B}}$. One can check that Ψ_A is exact and that we have the following commutative diagram.

$$\begin{array}{ccccc}
(\mathscr{D},\mathscr{H})\text{-mod} & \underset{\Phi}{\overset{\Gamma_a}{\rightleftarrows}} & (\mathscr{D}^E,\mathscr{H})^a\text{-mod} & \xrightarrow{\Psi_A} & \overline{\mathscr{G}\mathscr{B}}^{\gamma(a\varepsilon_1)} \\
\downarrow S_{\mathscr{H}'} & & \downarrow S_{\mathscr{H}'} & & \downarrow S_{\mathfrak{h}} \\
{}_{\mathrm{s}}(\mathscr{D},\mathscr{H})\text{-mod} & \underset{\Phi}{\overset{\Gamma_a}{\rightleftarrows}} & {}_{\mathrm{s}}(\mathscr{D}^E,\mathscr{H})^a\text{-mod} & \xrightarrow{\Psi_A} & \overline{\mathscr{B}}^{\gamma(a\varepsilon_1)}
\end{array}$$

where $S_{\mathfrak{h}}$ stands for the functor $\mathscr{G}\mathscr{B}^{\gamma(a\varepsilon_1)} \to \mathscr{B}^{\gamma(a\varepsilon_1)}$ mapping a module to its submodule consisting of all $\mathfrak{h}$-eigenvectors.

Using translation functors (in terms of [4]), one can show that every block $\mathscr{G}\mathscr{B}^{\mu}$ and $\mathscr{B}^{\mu}$, of $\mathscr{G}\mathscr{B}$ and $\mathscr{B}$, respectively, is equivalent to $\mathscr{G}\mathscr{B}^{\gamma(a\varepsilon_1)}$ and $\mathscr{B}^{\gamma(a\varepsilon_1)}$, respectively, for some $a \in \mathbb{C}$. The case $a \notin \mathbb{Z}$ corresponds to a nonintegral central character, $a = -1, \ldots, -n$ to singular central character, and all remaining a to a regular integral central character.

The following theorem is proved in [27].

Theorem 4.9 *Assume that* $a \notin \mathbb{Z}$ *or* $a = -1, \ldots, -n$*. Then* Ψ_A *provides an equivalence between* ${}^{\mathrm{b}}(\mathscr{D}^E, \mathscr{H})^a$-mod *and* $\mathscr{G}\mathscr{B}^{\gamma(a\varepsilon_1)}$ *and between* $(\mathscr{D}^E, \mathscr{H})^a$-mod *and* $\overline{\mathscr{G}\mathscr{B}}^{\gamma(a\varepsilon_1)}$*.*

Moreover, Ψ_A *provides an equivalence between* ${}^{\mathrm{b}}_{\mathrm{s}}(\mathscr{D}^E, \mathscr{H})^a$-mod *and* $\mathscr{B}^{\gamma(a\varepsilon_1)}$*, as well as, between* ${}_{\mathrm{s}}(\mathscr{D}^E, \mathscr{H})^a$-mod *and* $\overline{\mathscr{B}}^{\gamma(a\varepsilon_1)}$*.*

An important part of the proof of the above proposition is the description of the injectives and the application of the twisted localization functor. More precisely, we have the following.

Proposition 4.10 *Assume that $a \notin \mathbb{Z}$ or $a = -1, \ldots, -n$, and that $\nu \in \mathbb{C}^{n+1}$. Every indecomposable injective in $\overline{\mathcal{GB}}_{\gamma(\nu)}^{\gamma(a\varepsilon_1)}$ (respectively, in $\overline{\mathcal{B}}_{\gamma(\nu)}^{\gamma(a\varepsilon_1)}$) is isomorphic to $\Psi_A(\Gamma_a(\mathcal{F}_\nu^{\log}(J)))$ (respectively, to $\Psi_A(\Gamma_a(S_{\mathcal{H}'}(\mathcal{F}_\nu^{\log}(J)))))$ for some subset J of $\{1, 2, \ldots, n+1\}$. In particular, if $\nu = 0$ and $J = \emptyset$, the injective envelope of $\Psi_A(\Gamma_a(\mathcal{F}_0^+))$ in $\overline{\mathcal{GB}}^{\gamma(a\varepsilon_1)}$ is $\Psi_A(\Gamma_a(\mathcal{F}_0^{\log}))$, while the injective envelope of $\Psi_A(\Gamma_a(\mathcal{F}_0^+))$ in $\overline{\mathcal{B}}^{\gamma(a\varepsilon_1)}$ is $\Psi_A(\Gamma_a(S_{\mathcal{H}'}(\mathcal{F}_0^{\log})))$.*

It is not hard to see that

$$S_{\mathcal{H}'}(\mathcal{F}_\nu) = \mathcal{F}_\nu, \; S_{\mathcal{H}'}(\mathcal{F}_\nu^{\log}) = \mathcal{F}_\nu \otimes \mathbb{C}[u],$$

where $u = \log(t_1 t_2 \cdots t_{n+1})$.

One can easily show that the twisted localization functor commutes with the functors Γ_a, Φ, Ψ_A, more precisely, that the following diagram is commutative.

$$\begin{array}{ccccc}
(\mathcal{D}, \mathcal{H})_\nu\text{-mod} & \underset{\Phi}{\overset{\Gamma_a}{\rightleftarrows}} & (\mathcal{D}^E, \mathcal{H})_\nu^a\text{-mod} & \xrightarrow{\Psi_A} & \overline{\mathcal{GB}}_{\gamma(\nu)}^{\gamma(a\varepsilon_1)} \\
\downarrow D_{i,j}^x & & \downarrow D_{i,j}^x & & \downarrow D_{\varepsilon_i-\varepsilon_j}^x \\
(\mathcal{D}, \mathcal{H})_\mu\text{-mod} & \underset{\Phi}{\overset{\Gamma_a}{\rightleftarrows}} & (\mathcal{D}^E, \mathcal{H})_\mu^a\text{-mod} & \xrightarrow{\Psi_A} & \overline{\mathcal{GB}}_{\gamma(\mu)}^{\gamma(a\varepsilon_1)}
\end{array}$$

where $\mu = \nu + x(\varepsilon_i - \varepsilon_j)$. With this and Lemma 4.2. (v) in mind, one can describe the injectives $\overline{\mathcal{B}}$ and $\overline{\mathcal{GB}}$ (for singular and nonintegral central characters) as twisted localization of the injectives in $\overline{\mathcal{C}}$ and $\overline{\mathcal{GC}}$. Note that $\mathcal{B}$ and $\mathcal{GB}$ do not have injectives for $\mathfrak{g} = \mathfrak{sl}(n+1)$, and this is the main reason one needs to introduce the categories $\overline{\mathcal{B}}$ and $\overline{\mathcal{GB}}$.

4.4.2 The Case $\mathfrak{g} = \mathfrak{sp}(2n)$

In contrast with $\mathfrak{g} = \mathfrak{sl}(n+1)$, for $\mathfrak{g} = \mathfrak{sp}(2n)$, the category $\mathcal{B}$ does have enough injectives. In fact one has (as proved in [25]) the following.

Theorem 4.11 *The injective envelope of a simple object L in $\mathcal{B}$ is isomorphic to $D_F^{\mathbf{x}} L$ for some set of n commuting long roots F and $\mathbf{x} \in \mathbb{C}^n$.*

To describe the injectives in $\overline{\mathcal{GB}}$ (note that $\mathcal{GB}$ does not have injectives) one needs to use an equivalence Ψ_C of categories analogous to the functor Ψ_A described above. To define Ψ_C, we use the presentation of every element $X \in \mathfrak{g}$ as a block matrix of the form

$$\begin{bmatrix} A & B \\ C & -A^t \end{bmatrix}$$

where A is an arbitrary $n \times n$-matrix, and B and C are symmetric $n \times n$-matrices. Then the maps

$$B \mapsto \sum_{i \leq j} b_{ij} t_i t_j, \; C \mapsto \sum_{i \leq j} c_{ij} \partial_i \partial_j$$

extend to a homomorphism of Lie algebras

$$\mathfrak{g} \to \mathscr{D}(n)$$

which induces a homomorphism of associative algebras

$$\omega \colon U(\mathfrak{g}) \to \mathscr{D}(n).$$

The image of ω coincides with the $\mathscr{D}(n)^{\text{ev}} := \bigoplus_{|\nu| \in 2\mathbb{Z}} \mathscr{D}(n)^{\nu}$ (recall that $|\nu| = \nu_1 + \ldots + \nu_n$). For convenience in this subsubsection we set $\mathscr{D} = \mathscr{D}(n)$ and $\mathscr{D}^{\text{ev}} = \mathscr{D}(n)^{\text{ev}}$. With the aid of the homomorphism ω we obtain equivalence of categories ${}^{\text{wb}}(\mathscr{D}^{\text{ev}}, \mathscr{H})$-mod $\to$ $\mathscr{B}^{\chi}$ and $(\mathscr{D}^{\text{ev}}, \mathscr{H})$-mod $\to$ $\overline{\mathscr{G}B}^{\chi}$, where χ stands for any central character of a bounded $\mathfrak{g}$-module (translation functors provided equivalence between all blocks $\mathscr{B}^{\chi}$). By definition, the functor Ψ_C is the one providing the second equivalence, i.e., $(\mathscr{D}^{\text{ev}}, \mathscr{H})$-mod $\to \overline{\mathscr{G}B}^{\chi}$. The first equivalence is established in §5 of [25]. For the second equivalence one needs to prove the $\mathfrak{sp}(2n)$-analogs of Lemmas 8.8 and 8.9 in [27]. More precisely, one first proves that Ψ_C is an equivalence on the cuspidal blocks by using the description of the cuspidal blocks in the category $\overline{\mathscr{G}\mathscr{B}}$ provided by [35]. Then applying twisted localization functors to the cuspidal injectives we obtain all injectives in $\overline{\mathscr{G}\mathscr{B}}$. To describe the injectives in $(\mathscr{D}^{\text{ev}}, \mathscr{H})$-mod we introduce the modules $\mathscr{F}_{\nu}^{\text{ev}}$, $\mathscr{F}_{\nu}^{\text{ev},+}$, $\mathscr{F}_{\nu}^{\text{ev,log}}$, and $\mathscr{F}_{\nu}^{\text{ev,log}}(J)$ which are the even degree components of $\mathscr{F}_{\nu}$, $\mathscr{F}_{\nu}^{+}$, $\mathscr{F}_{\nu}^{\text{log}}$, and $\mathscr{F}_{\nu}^{\text{log}}(J)$, respectively. Let χ^+ be the central character of the module $\Psi_C(\mathscr{F}_0^{\text{ev},+})$.

Theorem 4.12 *Every indecomposable injective object in* $\overline{\mathscr{G}B}^{\chi^+}$ *is isomorphic to* $\Psi_C(\mathscr{F}_{\nu}^{\text{ev,log}}(J))$ *for some* ν *and a subset* J *of* $\{1, 2, \ldots, n\}$. *In particular, the injective envelope of* $\Psi_C(\mathscr{F}_0^{\text{ev},+})$ *in* $\overline{\mathscr{G}B}^{\chi^+}$ *is* $\Psi_C(\mathscr{F}_0^{\text{ev,log}})$.

4.5 *Twisted Localization for Finite-Dimensional Lie Superalgebras*

In this case we consider a simple finite-dimensional Lie superalgebra $\mathfrak{g}$, $\mathscr{U} = U(\mathfrak{g})$ and $\mathscr{H} = S(\mathfrak{h}_{\bar{0}}^{ss})$. We consider multiplicative subsets consisting of root elements e_α in $\mathfrak{g}^\alpha$ for even roots α.

The following theorem is proved in [23].

Proposition 4.13 *Let $\mathfrak{g}$ be a classical Lie superalgebra. Every simple module in ${}^{\mathrm{f}}(\mathscr{U}, \mathscr{H})$ (equivalently, in ${}^{\mathrm{wf}}(\mathscr{U}, \mathscr{H})$) is isomorphic to $\mathrm{D}_\Gamma^{\mathbf{x}} L$ for some simple highest weight module L, a set Γ of n even commuting roots, and $\mathbf{x} \in \mathbb{C}^n$.*

The parabolic induction theorem for any $\mathfrak{g}$ (including the Cartan type series) is proved in [14], where a classification of all cuspidal modules of $\mathfrak{g}$ is provided for all $\mathfrak{g}$ except for $\mathfrak{g} = \mathfrak{osp}(m|2n)$, $m = 1, 3, 4, 5, 6$, $\mathfrak{g} = \mathfrak{psq}(n)$, $\mathfrak{g} = D(\alpha)$, and the Cartan type series. The simple cuspidal modules of $\mathfrak{psq}(n)$ are classified in [22], while those of $\mathfrak{g} = D(\alpha)$, are classified in [29]. The classification of the simple cuspidal modules for $\mathfrak{g} = \mathfrak{osp}(m|2n)$, $m = 1, 3, 4, 5, 6$, and for the Cartan type series remains an open question.

4.6 *Twisted Localization for Affine Lie Algebras*

In this case, we consider commuting subsets of real roots $\Gamma \subset \Delta^{\mathrm{re}}$ and the multiplicative subsets of $U(\mathfrak{G})$ are $F = \{e_\alpha \mid \alpha \in \Gamma\}$, where $e_\alpha \in \mathfrak{g}^\alpha$. As in the Lie algebra case we will write D_Γ for D_F. An important part of the classification of the simple objects of ${}^{\mathrm{wf}}(\mathscr{U}, \mathscr{H})$ achieved in [13] is the constructions of loop modules and their relations with twisted localization.

We recall the definition and some properties of loop modules. For more detail, see [8–10]. Let $\mathfrak{G} = \mathscr{A}(\mathfrak{g})$, let $Y_1, \ldots, Y_k$ be weight $\mathfrak{g}$-modules, and let $a_1, \ldots, a_k$ be nonzero scalars. Following [9], we define the loop module $\mathscr{L}_{a_1,\ldots,a_k}(Y_1 \otimes \ldots \otimes Y_k)$ in the following way: the underlining vector space of $\mathscr{L}_{a_1,\ldots,a_k}(Y_1 \otimes \ldots \otimes Y_k)$ is

$$(Y_1 \otimes \ldots \otimes Y_k) \otimes \mathbb{C}[t, t^{-1}],$$

$X \otimes t^n \in \mathscr{A}(\mathfrak{g})$ acts as

$$(X \otimes t^n) \cdot ((v_1 \otimes \ldots \otimes v_k) \otimes t^s) = \sum_{i=1}^{k} a_i^n (v_1 \otimes \ldots \otimes X \cdot v_i \otimes \ldots \otimes v_k) \otimes t^{n+s},$$

D acts as $t\frac{d}{dt}$, and K acts trivially. If the scalars $a_1, \ldots, a_k$ are distinct, then $\mathscr{L}_{a_1,\ldots,a_k}(Y_1 \otimes \ldots \otimes Y_k)$ is completely reducible with finitely many simple components. Furthermore, the simple components of $\mathscr{L}_{a_1,\ldots,a_k}(Y_1 \otimes \ldots \otimes Y_k)$ are isomorphic simple $\mathscr{L}(\mathfrak{g})$-modules. Denote by $V_{a_1,\dots,a_k}(Y_1 \otimes \dots \otimes Y_k)$ the simple $\mathscr{L}(\mathfrak{g})$-module which is a component of $\mathscr{L}_{a_1,\dots,a_k}(Y_1 \otimes \dots \otimes Y_k)$. Considered as $\mathscr{A}(\mathfrak{g})$-modules, the constituents of $\mathscr{L}_{a_1,\dots,a_k}(Y_1 \otimes \dots \otimes Y_k)$ differ only by a shift of the action of D. By a slight abuse of notation we denote by $V_{a_1,\dots,a_k}(Y_1 \otimes \dots \otimes Y_k)$ any shift of a simple $\mathscr{A}(\mathfrak{g})$–component of $\mathscr{L}_{a_1,\dots,a_k}(Y_1 \otimes \dots \otimes Y_k)$. A relation between loop modules and twisted localization is provided in the following result in [13].

Proposition 4.14 *Let $\mathfrak{G} = \mathscr{L}(\mathfrak{g})$ be an untwisted affine Lie algebra, let $\alpha \in \mathring{\Delta}$, and let $\mathscr{L}_{a_0,\dots,a_k}(S \otimes F_1 \otimes \dots \otimes F_k)$ be the loop $\mathfrak{G}$-module for which $a_i \in \mathbb{C}$, F_i are simple finite-dimensional $\mathfrak{g}$-modules, and S is a simple e_α-torsion free $\mathfrak{g}$-module. Then $\mathscr{L}_{a_0,\dots,a_k}(S \otimes F_1 \otimes \dots \otimes F_k)$ is $e_{\alpha+r\delta}$-torsion free and*

$$D_{\alpha+r\delta}\mathscr{L}_{a_0,\dots,a_k}(S \otimes F_1 \otimes \dots \otimes F_k) \simeq \mathscr{L}_{a_0,\dots,a_k}(D_\alpha S \otimes F_1 \otimes \dots \otimes F_k)$$

for every integer r.

If $\mathfrak{G} = \mathscr{A}(\mathfrak{g}, \sigma)$, then $\mathscr{L}_{a_1,\dots,a_k}(Y_1 \otimes \dots \otimes Y_k)$ admits an endomorphism compatible with σ if and only if the modules $Y_1, \dots, Y_k$ come in r-tuples of isomorphic modules and for each r-tuple the corresponding scalars $a_1, \dots, a_r$ are multiples (with the same scalar) of the rth roots of unity. We denote the corresponding endomorphism of $\mathscr{L}_{a_1,\dots,a_k}(Y_1 \otimes \dots \otimes Y_k)$ by σ again and let $\mathscr{L}^\sigma_{a_1,\dots,a_k}(Y_1 \otimes \dots \otimes Y_k)$ denote the fixed points of σ. Similarly, we can define $V^\sigma_{a_1,\dots,a_k}(Y_1 \otimes \dots \otimes Y_k)$.

The following is the main theorem in [13].

Theorem 4.15 *Every simple module M of $\mathfrak{G}$ in ${}^{\mathrm{f}}(\mathscr{U}, \mathscr{H})$ (equivalently, in ${}^{\mathrm{wf}}(\mathscr{U}, \mathscr{H})$) either is parabolically induced or is isomorphic (up to a shift) to*

$$V_{a_0,a_1,\dots,a_k}(N \otimes F_1 \otimes \dots \otimes F_k),$$

where $a_0, \dots, a_k$ are distinct nonzero scalars, N is a cuspidal $\mathfrak{g}$-module, and $F_1, \dots, F_k$ are finite-dimensional $\mathfrak{g}$-modules. In the latter case $\mathfrak{G} \cong A_l^{(1)}$ or $\mathfrak{G} \cong C_l^{(1)}$. In particular, M is isomorphic to $D_F^{\mathbf{x}} V_{a_0,a_1,\dots,a_k}(L \otimes F_1 \otimes \dots \otimes F_k)$, for some bounded highest weight $\mathfrak{g}$-module L, commuting set of ℓ roots F, and $x \in \mathbb{C}^\ell$.

The study of the category of bounded modules of $\mathfrak{G}$, and in particular, describing the injectives in that category, is a largely open question.

Acknowledgements I would like to thank the referee for the helpful suggestions. Also, special thanks are due to Geoff Mason, Ivan Penkov, and Joe Wolf for organizing the workshop series "Lie Groups, Lie Algebras and their Representations" where I had the opportunity to report some of the results in this paper.

References

1. V. Bavula and V. Bekkert, Indecomposable representations of generalized Weyl algebras, *Comm. Algebra* **28** (2000), 5067–5100.
2. V. Bekkert, G. Benkart, V. Futorny, Weight modules for Weyl algebras, in "*Kac-Moody Lie algebras and related topics*", Contemp. Math. **343** (2004) 17–42.
3. G. Benkart, D. Britten, F. Lemire, Modules with bounded weight multiplicities for simple Lie algebras, *Math. Z.* **225** (1997), 333–353.
4. J. Bernstein, S. Gelfand, Tensor products of finite and infinite-dimensional representations of semisimple Lie algebras, *Compositio Math.* **41** (1980), 245–285.
5. R. Block, The irreducible representations of the Lie algebra $sl(2)$ and of the Weyl algebra, *Adv. in Math.* **39** (1981), 69–110.
6. D. Britten, F. Lemire, A classification of simple Lie modules having a 1-dimensional weight space. *Trans. Amer. Math. Soc.* **299** (1987), 683–697.
7. D. Britten, O. Khomenko, F. Lemire, V. Mazorchuk, Complete reducibility of torsion free C_n-modules of finite degree, *J. Algebra* **276** (2004), 129–142.
8. V. Chari, Integrable representations of affine Lie-algebras, *Invent. Math.* **85** (1986), 317–335.
9. V. Chari and A. Pressley, New unitary representations of loop groups, *Math. Ann.* **275** (1986), 87–104.
10. V. Chari and A. Pressley, Integrable representations of twisted affine Lie algebras, *J. Algebra* **113** (1988), 438–464.
11. V. Deodhar, On a construction of representations and a problem of Enright, *Invent. Math.* **57** (1980), 101–118.
12. I. Dimitrov, V. Futorny, D. Grantcharov, Parabolic sets of roots, *Contemp. Math.* **499** (2009), 61–74.
13. I. Dimitrov, D. Grantcharov, *Simple weight modules of affine Lie algebras*, arXiv:0910.0688.
14. I. Dimitrov, O. Mathieu, I. Penkov, On the structure of weight modules, *Trans. Amer. Math. Soc.* **352** (2000), 2857–2869.
15. Yu. Drozd, V. Futorny, and S. Ovsienko, Harish-Chandra subalgebras and Gelfand-Zetlin modules, Finite-dimensional algebras and related topics, NATO Adv. Sci. Inst. Ser. C Math. Phys. Sci., Kluwer Acad. Publ., Dordrecht. **424** (1994) 79–93.
16. Yu. Drozd, B. Guzner, S. Ovsienko, Weight modules over generalized Weyl algebras, *J. Algebra* **184** (1996), 491–504.
17. B. Dubsky, *Classification of simple weight modules with finite-dimensional weight spaces over the Schrödinger algebra*, arXiv:1309.1346.
18. T. J. Enright, On the fundamental series of a real semisimple Lie algebra: their irreducibility, resolutions and multiplicity formulae, *Ann. of Math.* **2** (1979), 1–82.
19. S. Fernando, Lie algebra modules with finite-dimensional weight spaces I, *Trans. Amer. Math. Soc.* **322** (1990), 757–781.
20. V. Futorny, The weight representations of semisimple finite-dimensional Lie algebras, Ph. D. Thesis, Kiev University, 1987.
21. V. Futorny, D. Grantcharov, V. Mazorchuk, Weight modules over infinite dimensional Weyl algebras, to appear in *Proc. Amer. Math. Soc.* **142** (2014), 3049–3057.
22. M. Gorelik, D. Grantcharov, Bounded highest weight modules of $\mathfrak{q}(n)$, to appear in *Int. Math. Res. Not.* (2013), doi: 10.1093/imrn/rnt147.
23. D. Grantcharov, Explicit realizations of simple weight modules of classical Lie superalgebras, *Cont. Math.* **499** (2009), 141–148.
24. D. Grantcharov, A. Pianzola, Automorphisms of toroidal Lie superalgebras, *J. Algebra* **319** (2008), 4230–4248.
25. D. Grantcharov, V. Serganova, Category of $\mathfrak{sp}(2n)$-modules with bounded weight multiplicities, *Mosc. Math. J.* **6** (2006), 119–134.
26. D. Grantcharov, V. Serganova, Cuspidal representations of $\mathfrak{sl}(n+1)$, *Adv. Math.* **224** (2010), 1517–1547.

27. D. Grantcharov, V. Serganova, *On weight modules of algebras of twisted differential operators on the projective space*, arXiv:1306.3673.
28. D. Grantcharov, M. Yakimov, Cominiscule parabolic subalgebras of simple finite dimensional Lie superalgebras, *J. Pure Appl. Algebra* **217** (2013), 1844–1863.
29. C. Hoyt, *Weight modules of* $D(2, 1, a)$, arXiv:1307.8006.
30. A. Joseph, The primitive spectrum of an enveloping algebra, *Astérisque* **173–174** (1989), 13–53.
31. V. Kac, Infinite Dimensional Lie Algebras. Third edition. Cambridge University Press, Cambridge, 1990. xxii+400 pp.
32. O. Mathieu, Classification of irreducible weight modules, *Ann. Inst. Fourier* **50** (2000), 537–592.
33. V. Mazorchuk, Lectures on $sl_2(\mathbb{C})$-modules, Imperial College Press, London, 2010.
34. V. Mazorchuk, C. Stroppel, Cuspidal $\mathfrak{sl}_n$-modules and deformations of certain Brauer tree algebras, *Adv. Math.* **228** (2011), 1008–1042.
35. V. Mazorchuk, C. Stroppel, Blocks of the category of cuspidal $\mathfrak{sp}(2n)$-modules, *Pacif. J. Math.* **251** (2011), 183–196
36. I. Penkov, V. Serganova, Generic irreducible representations of finite-dimensional Lie superalgebras, *Internat. J. Math.* **5** (1994), 389–419.
37. M. Scheunert, Invariant supersymmetric multilinear forms and the Casimir elements of P-type Lie superalgebras, *J. Math. Phys.* **28** (1987), 1180–1191.

Dirac Cohomology and Generalization of Classical Branching Rules

Jing-Song Huang

Abstract We generalize certain classical results on branching rules such as the Littlewood restriction formulae. Our formulae are expressed in terms of a linear integral combination of the Littlewood–Richardson coefficients and in terms of Dirac cohomology.

Key words Branching rule • Dirac cohomology • Littlewood–Richardson coefficient • Harish-Chandra module • Holomorphic representation

Mathematics Subject Classification (2010): 22E46, 22E 47

1 Introduction

Let G and H be complex algebraic groups with an embedding $H \hookrightarrow G$. Let V and W be completely reducible representations of G and H respectively. We let

$$[V, W] = \dim \mathrm{Hom}_H(W, V),$$

where V is regarded as a representation of H by restriction. A description of this number $[V, W]$ as the multiplicity of W in the restriction of V to H is referred to as a branching rule.

The Littlewood–Richardson rule which describes the decomposition of tensor product of two irreducible representations of the general linear group is exactly the branching rule with $H = GL_n(\mathbb{C})$ diagonally embedded into

The research in this paper is supported by grants from Research Grant Council of HKSAR and the National Science Foundation of China.

J.-S. Huang (✉)
Department of Mathematics, Hong Kong University of Science and Technology, Clear Water Bay, Kowloon, Hong Kong SAR, China
e-mail: mahuang@ust.hk

G. Mason et al. (eds.), *Developments and Retrospectives in Lie Theory: Algebraic Methods*, Developments in Mathematics 38,
DOI 10.1007/978-3-319-09804-3_10

$G = GL_n(\mathbb{C}) \times GL_n(\mathbb{C})$. The multiplicity of an irreducible representation ν in the tensor product of two other irreducible representations λ and μ is denoted $c^{\nu}_{\lambda,\mu}$ [Kn]. The numbers $c^{\nu}_{\lambda,\mu}$ are called Littlewood–Richardson coefficients. The Littlewood–Richardson rule plays an important role in representation theory and it has been a topic of intense study in the theory of combinatorics with connection to many other branches of mathematics. For a new proof the Littlewood–Richardson rule and the reference to some of related work, we refer to the recent article by Howe and Lee [HL].

Let G be a classical reductive algebraic group over $\mathbb{C}$: $G = GL_n(\mathbb{C}) = GL_n$, the general linear group, or $G = O_n(\mathbb{C}) = O_n$, the orthogonal group or $G = SO_n(\mathbb{C}) = SO_n$, the connected component of O_n; or $G = Sp_{2n}(\mathbb{C}) = Sp_{2n}$ the symplectic group. There are a total of four kinds of simple classical symmetric pairs. We list them here:

(1) Diagonal: $GL_n \subset GL_n \times GL_n$, $O_n \subset O_n \times O_n$, $Sp_n \subset Sp_n \times Sp_n$;
(2) Direct sum: $GL_n \times GL_m \subset GL_{n+m}$, $O_n \times O_m \subset O_{n+m}$, $Sp_n \times Sp_m \subset Sp_{n+m}$;
(3) Polarization: $GL_n \subset SO_{2n}$ and $GL_n \subset Sp_{2n}$;
(4) Bilinear form: $O_n \subset GL_n$ and $Sp_{2n} \subset GL_{2n}$.

In 1940s, D. E. Littlewood [L1, L2] gave a formula for the decomposition of some representations of GL_n restricted to O_n, and GL_{2n} restricted to Sp_{2n}. Those branching rules are in terms of Littlewood–Richardson coefficients and for the classical symmetric pairs associated to the bilinear form

$$O_n \subset GL_n \text{ and } Sp_{2n} \subset GL_{2n}.$$

They are well known as the Littlewood Restriction Formulae. Inspired by the work of Enright and Willenbring [EW], we obtained a generalization of the Littlewood Restriction Formulae which are expressed in terms of the Littlewood–Richardson coefficients and Dirac cohomology [HPZ].

The Dirac cohomology is a new tool in representation theory. It appears to be a basic invariant of irreducible unitary representations and more general admissible representations. The main result of [HP1], conjectured by Vogan [V], says that the standard parameters of the infinitesimal character of a Harish-Chandra module X and the infinitesimal character of its Dirac cohomology $H_D(X)$ are conjugated. The Dirac cohomology $H_D(X)$ has been completely determined for finite-dimensional $\mathfrak{g}$-modules and unitary irreducible $A_{\mathfrak{q}}(\lambda)$-modules [HKP], which includes all discrete series and tempered representations. Another important family of irreducible unitary representations is the unitary highest weight modules. The results of [HPR] ensure that their Dirac cohomology is equal to the $\mathfrak{p}^+$ cohomology (up to a 1-dimensional character), which had been determined in [E]. Still, it remains to be an interesting open problem to find all irreducible unitary representations with nonzero Dirac cohomology and to determine their Dirac cohomology.

Kostant extended Vogan's conjecture to the setting of the cubic Dirac operator and proved a nonvanishing result on Dirac cohomology for highest weight modules in the most general setting [K2]. He also determined the Dirac cohomology of finite-dimensional modules in the equal rank case. The Dirac cohomology for all irreducible highest weight modules was determined in [HX] in terms of coefficients of Kazhdan–Lusztig polynomials.

The purpose of this paper is to describe the branching rules for the symmetric pairs of classical algebraic groups associated to the polarization

$$GL_n \subset SO_{2n} \text{ and } GL_n \subset Sp_{2n}.$$

Our branching rules cover not only finite-dimensional representations, but also infinite-dimensional lowest (or highest) weight modules. We also recall the generalized Littlewood Restriction Formulae that were obtained in [HPZ]. These formulae describe the branching rules for finite-dimensional representations of GL_n restricted to O_n and GL_{2n} to Sp_{2n}. We note that the formulae obtained in [HPZ] are inspired by the work of [EW] in terms of nilpotent Lie algebra cohomology.

We also note that the stable branching rules for all classical symmetric pairs were studied by Howe, Tan and Willenbring [HTW]. Their formulae are also in terms of the Littlewood–Richardson coefficients, but are involved with combining products of Littlewood–Richardson coefficients. In all of our branching formula there is only a linear integral combination of Littlewood–Richardson coefficients.

The paper is organized as follows. In Sect. 2 we define the holomorphic representations for simple Lie groups of Hermitian symmetric type and recall the definition of the related category $\mathcal{O}^{\mathfrak{q}}$. In Sect. 3 we recall the definition and basic properties of the Dirac cohomology. In Sect. 4 we prove a K-character formula for the holomorphic representations. In Sect. 5 we prove the branching rules for $GL_n \hookrightarrow SO_{2n}$ and $GL_n \hookrightarrow Sp_{2n}$. In Sect. 6 we prove the branching rules for $O_n \hookrightarrow GL_n$ and $Sp_{2n} \hookrightarrow GL_{2n}$.

2 Hermitian Symmetric Pairs and Holomorphic Representations

We write $U(n)$, $SO(n)$ and $Sp(2n)$ for the compact real forms of GL_n, SO_n and Sp_{2n} respectively. The two compact symmetric spaces $SO(2n)/U(n)$ and $Sp(2n)/U(n)$ associated to polarization are Hermitian symmetric. Since some of our branching rules are naturally extended to the infinite-dimensional representations, we will include the consideration of the Hermitian symmetric spaces $SO^*(2n)/U(n)$ and $Sp(2n,\mathbb{R})/U(n)$ of noncompact type. The branching laws can be regarded the description of the multiplicities of K-types of $(\mathfrak{g}, K)$-modules for the noncompact simple Lie groups of Hermitian type.

Therefore in this section we assume that G is a simple noncompact Lie group with a maximal compact subgroup such that the pair (G, K) is Hermitian symmetric. This assumption is equivalent to the condition that K has a one-dimensional center. We denote by $\mathfrak{g}_0$, $\mathfrak{k}_0$ the Lie algebras of G and K, and by $\mathfrak{g}$, $\mathfrak{k}$ their complexifications. The $\mathfrak{k}$-module $\mathfrak{g}$ decomposes as $\mathfrak{g} = \mathfrak{k} \oplus \mathfrak{p} = \mathfrak{k} \oplus \mathfrak{p}^+ \oplus \mathfrak{p}^-$.

An irreducible unitary representation π on a Hilbert space is said to be *holomorphic* if it has nonzero vectors that are annihilated by $\mathfrak{p}^-$. We will work with the space X_π of K-finite vectors of the unitary representation (π, H). The space X_π is called the Harish-Chandra module of the representation π. All of our considerations can be extended to the more general admissible $(\mathfrak{g}, K)$-modules. An irreducible $(\mathfrak{g}, K)$-module X is holomorphic if it has nonzero vectors that are annihilated by $\mathfrak{p}^-$. Equivalently, an irreducible holomorphic $(\mathfrak{g}, K)$-module is defined to be an irreducible $(\mathfrak{g}, K)$-module that is also a lowest weight module.

Fix a Cartan subalgebra $\mathfrak{h}$ of $\mathfrak{k}$. Then $\mathfrak{h}$ is also a Cartan subalgebra of $\mathfrak{g}$. The roots $\Delta = \Delta(\mathfrak{g}, \mathfrak{h})$ decompose as $\Delta = \Delta_c \cup \Delta_n$, where $\Delta_c = \Delta(\mathfrak{k}, \mathfrak{h})$ (the compact roots), and Δ_n consists of the $\mathfrak{h}$-weights of $\mathfrak{p}$ (the noncompact roots). We fix a positive root system Δ_c^+ for $\mathfrak{k}$. Let Δ_n^+ be the set of roots corresponding to $\mathfrak{p}^+$. Then $\Delta^+ = \Delta_c^+ \cup \Delta_n^+$ is a positive root system for $\mathfrak{g}$. Let ρ (resp. ρ_c, ρ_n) equal one half the sum of the roots in Δ^+ (resp. Δ_c^+, Δ_n^+).

Let F^μ denote the irreducible finite-dimensional representation of $\mathfrak{k}$ with highest weight μ. One can consider F^μ as a module for the Lie algebra $\mathfrak{k} \oplus \mathfrak{p}^-$, by letting $\mathfrak{p}^-$ act by zero. The module

$$N(\mu) = U(\mathfrak{g}) \otimes_{U(\mathfrak{k} \oplus \mathfrak{p}^-)} F^\mu$$

is called a (lowest weight) generalized Verma module. This module has a unique irreducible quotient denoted by $L(\mu)$. Conversely, any irreducible $(\mathfrak{g}, \mathfrak{k})$-module which is also a lowest weight $\mathfrak{g}$-module must be $L(\mu)$ for some μ. Note that the lowest weight of $N(\mu)$ and $L(\mu)$ is the $\mathfrak{k}$-lowest weight of F^μ, which we denote by μ^-. Thus the infinitesimal character of $N(\mu)$ and $L(\mu)$ is $\mu^- - \rho$, up to conjugacy by the Weyl group $W = W_{\mathfrak{g}}$ of $\mathfrak{g}$. The problem of classifying unitary highest (or lowest) weight modules was solved by Enright, Howe and Wallach [EHW].

Note that $\mathfrak{q} = \mathfrak{k} + \mathfrak{p}^+$ is a parabolic subalgebra of $\mathfrak{g}$. Let τ be the Chevalley automorphism of $\mathfrak{g}$. This is the automorphism of $\mathfrak{g}$ induced by multiplying -1 on the Cartan subalgebra. It takes positive roots to their negatives and the corresponding positive root vectors to negative root vectors. Under the action of τ, the holomorphic or the lowest weight $(\mathfrak{g}, K)$-modules become antiholomorphic or highest weight $(\mathfrak{g}, K)$-modules. This family of $\mathfrak{g}$-modules fall into the category $\mathcal{O}^{\mathfrak{q}}$ associated with a parabolic subalgebra $\mathfrak{q} = \mathfrak{k} + \mathfrak{p}^+$.

In the remaining part of this section we recall the definition and some of the basic properties of the category $\mathcal{O}^{\mathfrak{q}}$ associated with an arbitrary parabolic subalgebra $\mathfrak{q}$ of $\mathfrak{g}$ [BGG]. As a matter of fact, we only need the relevant concepts and results in the next section for the special case when the parabolic subalgebra $\mathfrak{q}$ is equal to $\mathfrak{k} + \mathfrak{p}^+$. But the general case is as easy as (or as hard as) the special case.

Let $\mathfrak{g}$ be a complex semisimple Lie algebra. Let $\mathfrak{h}$ be a Cartan subalgebra of $\mathfrak{g}$. Denote by $\Phi \subseteq \mathfrak{h}^*$ the root system of $(\mathfrak{g}, \mathfrak{h})$. For $\alpha \in \Phi$, let $\mathfrak{g}_\alpha$ be the root subspace of $\mathfrak{g}$ corresponding to α. We fix a choice of the set of positive roots Φ^+ and let Δ be the corresponding subset of simple roots in Φ^+. Note that each subset $I \subset \Delta$ generates a root system $\Phi_I \subset \Phi$, with positive roots $\Phi_I^+ = \Phi_I \cap \Phi^+$.

The parabolic subalgebras of $\mathfrak{g}$ up to conjugation are in one-to-one correspondence with the subsets in Δ. We let

$$\mathfrak{l}_I = \mathfrak{h} \oplus \sum_{\alpha \in \Phi_I} \mathfrak{g}_\alpha$$

be the Levi subalgebra and let

$$\mathfrak{u}_I = \sum_{\alpha \in \Phi^+ \backslash \Phi_I^+} \mathfrak{g}_\alpha, \; \bar{\mathfrak{u}}_I = \sum_{\alpha \in \Phi^+ \backslash \Phi_I^+} \mathfrak{g}_{-\alpha}$$

be the nilpotent radical and its dual space with respect to the Killing form B. Then $\mathfrak{q}_I = \mathfrak{l}_I \oplus \mathfrak{u}_I$ is the standard parabolic subalgebra associated with I. We set

$$\rho = \rho(\mathfrak{g}) = \frac{1}{2} \sum_{\alpha \in \Phi^+} \alpha, \; \rho(\mathfrak{l}_I) = \frac{1}{2} \sum_{\alpha \in \Phi_I^+} \alpha, \text{ and } \rho(\mathfrak{u}_I) = \frac{1}{2} \sum_{\alpha \in \Phi^+ \backslash \Phi_I^+} \alpha.$$

Then we have $\rho(\bar{\mathfrak{u}}_I) = -\rho(\mathfrak{u}_I)$. We note that once I is fixed there is little use for other subsets of Δ. We will omit the subscript if a subalgebra is clearly associated with I.

Definition 2.1. The category $\mathcal{O}^{\mathfrak{q}}$ is defined to be the full subcategory of $U(\mathfrak{g})$-modules M that satisfy the following conditions:

(i) M is a finitely generated $U(\mathfrak{g})$-module;
(ii) M is a direct sum of finite-dimensional simple $U(\mathfrak{l})$-modules;
(iii) M is locally finite as a $U(\mathfrak{q})$-module.

We adopt notations in [Hum]. Let Λ_I^+ be the set of Φ_I^+-dominant integral weights in $\mathfrak{h}^*$, namely,

$$\Lambda_I^+ := \{\lambda \in \mathfrak{h}^* \mid \langle \lambda, \alpha^\vee \rangle \in \mathbb{Z}^{\geq 0} \text{ for all } \alpha \in \Phi_I^+\}.$$

Here $\langle , \rangle$ is the bilinear form on $\mathfrak{h}^*$ (induced from the Killing form B) and $\alpha^\vee = 2\alpha/\langle \alpha, \alpha \rangle$.

Let $F(\lambda)$ be the finite-dimensional simple $\mathfrak{l}$-module with highest weight λ. Then $\lambda \in \Lambda_I^+$. We consider $F(\lambda)$ as a $\mathfrak{q}$-module by letting $\mathfrak{u}$ act trivially on it. Then the *parabolic Verma module* with highest weight λ is the induced module

$$M_I(\lambda) = U(\mathfrak{g}) \otimes_{U(\mathfrak{q})} F(\lambda).$$

The module $M_I(\lambda)$ is a quotient of the ordinary Verma module $M(\lambda)$. Using Theorem 1.2 in [Hum], we can write unambiguously $L(\lambda)$ for the unique simple quotient of $M_I(\lambda)$ and $M(\lambda)$. Furthermore, since every nonzero module in $\mathcal{O}^{\mathfrak{q}}$ has at least one nonzero vector of maximal weight, Proposition 9.3 in [Hum] implies that every simple module in $\mathcal{O}^{\mathfrak{q}}$ is isomorphic to $L(\lambda)$ for some $\lambda \in \Lambda_I^+$ and is therefore determined uniquely up to isomorphism by its highest weight.

Let $Z(\mathfrak{g})$ be the center of $U(\mathfrak{g})$. Recall that $M_I(\lambda)$ and all its subquotients including $L(\lambda)$ have the same infinitesimal character χ_λ. Here χ_λ is an algebra homomorphism $Z(\mathfrak{g}) \to \mathbb{C}$ such that $z \cdot v = \chi_\lambda(z)v$ for all $z \in Z(\mathfrak{g})$ and all $v \in M(\lambda)$.

It follows from Corollary 1.2 in [Hum] that every nonzero module $M \in \mathcal{O}^{\mathfrak{q}}$ has a finite filtration with nonzero quotients each of which is a highest weight module in $\mathcal{O}^{\mathfrak{q}}$. Then the action of $Z(\mathfrak{g})$ on M is finite. We set

$$M^\chi := \{v \in M \mid (z - \chi(z))^n v = 0 \text{ for some } n > 0 \text{ depending on } z\}.$$

Then $z - \chi(z)$ acts locally nilpotently on M^χ for all $z \in Z(\mathfrak{g})$ and M^χ is a $U(\mathfrak{g})$-submodule of M. Let $\mathcal{O}^{\mathfrak{q}}_\chi$ denote the full subcategory of $\mathcal{O}^{\mathfrak{q}}$ whose objects are the modules M for which $M = M^\chi$. By the above discussion we have the following direct sum decomposition:

$$\mathcal{O}^{\mathfrak{q}} = \bigoplus_{\chi} \mathcal{O}^{\mathfrak{q}}_\chi,$$

where χ is of the form $\chi = \chi_\lambda$ for some $\lambda \in \mathfrak{h}^*$.

Let W be the Weyl group associated to the root system Φ. We define the dot action of W on $\mathfrak{h}^*$ by $w \cdot \lambda = w(\lambda + \rho) - \rho$ for $\lambda \in \mathfrak{h}^*$. Then $\chi_\lambda = \chi_\mu$ if and only if $\lambda \in W \cdot \mu$ by the Harish-Chandra isomorphism $Z(\mathfrak{g}) \to S(\mathfrak{h})^W$. An element $\lambda \in \mathfrak{h}^*$ is called *regular* if the isotropy group of λ in W is trivial. In other words, λ is regular if $\langle \lambda + \rho, \alpha^\vee \rangle \neq 0$ for all $\alpha \in \Phi$. A nonregular element in $\mathfrak{h}^*$ will be called *singular*.

Denote by Γ the set of all $\mathbb{Z}^{\geq 0}$-linear combinations of simple roots in Δ. Let $\mathcal{X}$ be the additive group of functions $f : \mathfrak{h}^* \to \mathbb{Z}$ whose support lies in a finite union of sets of the form $\lambda - \Gamma$ for $\lambda \in \mathfrak{h}^*$. Define the convolution product on $\mathcal{X}$ by

$$(f * g)(\lambda) := \sum_{\mu+\nu=\lambda} f(\mu)g(\nu).$$

We regard $e(\lambda)$ as a function in $\mathcal{X}$ which takes value 1 at λ and value 0 at $\mu \neq \lambda$. Then $e(\lambda) * e(\mu) = e(\lambda + \mu)$. It is clear that $\mathcal{X}$ is a commutative ring under convolution, with $e(0)$ as its multiplicative identity. Let

$$M_\lambda := \{v \in M \mid h \cdot v = \lambda(h)v \text{ for all } h \in \mathfrak{h}\}.$$

We say that a weight module (semisimple $\mathfrak{h}$-module) M *has a character* if

$$\operatorname{ch} M := \sum_{\lambda\in\mathfrak{h}^*} \dim M_\lambda \, e(\lambda) \tag{2.2}$$

is contained in $\mathcal{X}$. In this case $\operatorname{ch} M$ is called the *formal character* of M. Notice that all the modules in $\mathcal{O}^{\mathfrak{q}}$ have characters, as do all the finite-dimensional semisimple $\mathfrak{h}$-modules. In particular, if M has a character and $\dim L < \infty$, then $M \otimes L$ has a character

$$\operatorname{ch}(M \otimes L) = \operatorname{ch} M * \operatorname{ch} L.$$

In addition, for semisimple $\mathfrak{h}$-modules which have characters, their direct sums, submodules and quotients also have characters.

3 Dirac Cohomology of Holomorphic $(\mathfrak{g}, K)$-Modules

In this section we first recall the definition of Kostant's cubic Dirac and the basic properties of Dirac cohomology. Then we focus on the case of the Hermitian symmetric pairs and the holomorphic representations.

Let $\mathfrak{r} \subset \mathfrak{g}$ be any reductive Lie subalgebra of the semisimple Lie algebra $\mathfrak{g}$. Let B denote the Killing form of $\mathfrak{g}$. We assume that the restriction $B|_{\mathfrak{r}}$ of B to $\mathfrak{r}$ is nondegenerate. Let $\mathfrak{g} = \mathfrak{r} \oplus \mathfrak{s}$ be the orthogonal decomposition with respect to B. Then the restriction $B|_{\mathfrak{s}}$ is also nondegenerate. Denote by $C(\mathfrak{s})$ the Clifford algebra of $\mathfrak{s}$ with

$$uu' + u'u = -2B(u, u')$$

for all $u, u' \in \mathfrak{s}$. The above choice of sign is the same as in [HP1], but it is different from [K1] or [HPR]. We note that the two conventions have no essential difference since the two bilinear forms are equivalent over $\mathbb{C}$. Now fix an orthonormal basis $Z_1, \ldots, Z_m$ of $\mathfrak{s}$. Kostant [K1] defines the cubic Dirac operator D by

$$D = \sum_{i=1}^{m} Z_i \otimes Z_i + 1 \otimes v \in U(\mathfrak{g}) \otimes C(\mathfrak{s}).$$

Here $v \in C(\mathfrak{s})$ is the image of the fundamental 3-form $w \in \bigwedge^3(\mathfrak{s}^*)$,

$$w(X, Y, Z) = \frac{1}{2} B(X, [Y, Z]),$$

under the Chevalley map $\bigwedge(\mathfrak{s}^*) \to C(\mathfrak{s})$ and the identification of $\mathfrak{s}^*$ with $\mathfrak{s}$ by the Killing form B. Explicitly,

$$v = \frac{1}{2} \sum_{1 \le i < j < k \le m} B([Z_i, Z_j], Z_k) Z_i Z_j Z_k.$$

The cubic Dirac operator has a good square similar to the case of symmetric pairs [P]. To explain this, we start with a Lie algebra map

$$\alpha : \mathfrak{r} \to C(\mathfrak{s})$$

which is defined by the adjoint map ad : $\mathfrak{r} \to \mathfrak{so}(\mathfrak{s})$ composed with the embedding of $\mathfrak{so}(\mathfrak{s})$ into $C(\mathfrak{s})$ using the identification $\mathfrak{so}(\mathfrak{s}) \simeq \bigwedge^2 \mathfrak{s}$. The explicit formula for α is

$$\alpha(X) = \frac{1}{2} \sum_{i<j} B(X, [Z_i, Z_j]) Z_i Z_j, \qquad X \in \mathfrak{r}. \tag{3.1}$$

Using α we can embed the Lie algebra $\mathfrak{r}$ diagonally into $U(\mathfrak{g}) \otimes C(\mathfrak{s})$ by

$$X \to X_\Delta = X \otimes 1 + 1 \otimes \alpha(X).$$

This embedding extends to $U(\mathfrak{r})$. We denote the image of $\mathfrak{r}$ by $\mathfrak{r}_\Delta$, and then the image of $U(\mathfrak{r})$ is the enveloping algebra $U(\mathfrak{r}_\Delta)$ of $\mathfrak{r}_\Delta$. Let $\Omega_\mathfrak{g}$ (resp. $\Omega_\mathfrak{r}$) be the Casimir elements for $\mathfrak{g}$ (resp. $\mathfrak{r}$). The image of $\Omega_\mathfrak{r}$ under Δ is denoted by $\Omega_{\mathfrak{r}_\Delta}$.

Let $\mathfrak{h}_\mathfrak{r}$ be a Cartan subalgebra of $\mathfrak{r}$ which is contained in $\mathfrak{h}$. It follows from Kostant [K1, Theorem 2.16] that

$$D^2 - \Omega_\mathfrak{g} \otimes 1 + \Omega_{\mathfrak{r}_\Delta} = (\|\rho\|^2 - \|\rho_\mathfrak{r}\|^2) 1 \otimes 1, \tag{3.2}$$

where $\rho_\mathfrak{r}$ denote the half sum of positive roots for $(\mathfrak{r}, \mathfrak{h}_\mathfrak{r})$.

Definition 3.3. Let S be a spin module of $C(\mathfrak{s})$. Consider the action of D on $V \otimes S$

$$D : V \otimes S \to V \otimes S \tag{3.4}$$

with $\mathfrak{g}$ acting on V and $C(\mathfrak{s})$ on S. The *Dirac cohomology* of V is defined to be the $\mathfrak{r}$-module

$$H_D(V) := \operatorname{Ker} D / \operatorname{Ker} D \cap \operatorname{Im} D.$$

We denote by $W_\mathfrak{r}$ the Weyl group associated to the root system $\Phi(\mathfrak{r}, \mathfrak{h}_\mathfrak{r})$. The following theorem due to Kostant is an extension of Vogan's conjecture on the symmetric pair case which is proved in [HP1]. (See [K2] Theorems 4.1 and 4.2 or [HP2] Theorem 4.1.4).

Theorem 3.5. *There is an algebra homomorphism* $\zeta : Z(\mathfrak{g}) \to Z(\mathfrak{r}) \cong Z(\mathfrak{r}_\Delta)$ *such that for any* $z \in Z(\mathfrak{g})$ *one has*

$$z \otimes 1 - \zeta(z) = Da + aD \text{ for some } a \in U(\mathfrak{g}) \otimes C(\mathfrak{s}).$$

Moreover, ζ *is determined by the following commutative diagram:*

$$\begin{array}{ccc} Z(\mathfrak{g}) & \xrightarrow{\zeta} & Z(\mathfrak{r}) \\ \eta \downarrow & & \eta_{\mathfrak{r}} \downarrow \\ S(\mathfrak{h})^W & \xrightarrow{\text{Res}} & S(\mathfrak{h}_{\mathfrak{r}})^{W_{\mathfrak{r}}}. \end{array}$$

Here the vertical maps η *and* $\eta_{\mathfrak{r}}$ *are Harish-Chandra isomorphisms.*

We will focus on the case when $\mathfrak{r} = \mathfrak{l}$ and $\mathfrak{s} = \mathfrak{u} + \bar{\mathfrak{u}}$. Applying the above theorem we have the following theorem on generalized infinitesimal characters, which is a generalization of Vogan's conjecture on infinitesimal characters [HP1].

Theorem 3.6. ([DH], **Theorem 4.3**) *Let* V *be a* $\mathfrak{g}$*-module with a generalized* $Z(\mathfrak{g})$ *infinitesimal character* χ_μ*. Suppose that an* $\mathfrak{l}$*-module* N *with a generalized* $Z(\mathfrak{l})$ *infinitesimal character* $\chi_\lambda^{\mathfrak{l}}$ *is contained in the Dirac cohomology* $H_D(V)$*. Then* $\lambda + \rho_{\mathfrak{l}} = w(\mu + \rho)$ *for some* $w \in W$.

Remark 3.7. We note that the spin action $\alpha(\mathfrak{l})$ on S makes it a finite-dimensional $\mathfrak{l}$-module. If $V \in \mathcal{O}^{\mathfrak{q}}$, then $V \otimes S$ is a direct sum of finite-dimensional simple $\mathfrak{l}$-modules. Hence any submodule, quotient or subquotient of $V \otimes S$ is also a direct sum of finite-dimensional simple $\mathfrak{l}$-modules.

As a consequence we have the following proposition (See also [DH] Theorem 4.3).

Proposition 3.8. *Suppose that* V *is in* $\mathcal{O}^{\mathfrak{q}}_{\chi_\mu}$*. Then the Dirac cohomology* $H_D(V)$ *is a completely reducible finite-dimensional* $\mathfrak{l}$*-module. Moreover, if the finite-dimensional* $\mathfrak{l}$*-module* $F(\lambda)$ *is contained in* $H_D(V)$*, then* $\lambda + \rho_{\mathfrak{l}} = w(\mu + \rho)$ *for some* $w \in W$.

When G is simple and Hermitian symmetric with the maximal compact subgroup K, the $\mathfrak{k}$-module $\mathfrak{g}$ decomposes as $\mathfrak{g} = \mathfrak{k} \oplus \mathfrak{p} = \mathfrak{k} \oplus \mathfrak{p}^+ \oplus \mathfrak{p}^-$. In this case the fundamental 3-form $\omega = 0$ and therefore the corresponding cubic term vanishes in Kostant's Dirac operator. We can choose the basis b_i of $\mathfrak{p}$ in the following special way. Let $\Delta_n^+ = \{\beta_1, \ldots, \beta_m\}$. For each β_i we choose a root vector $e_i \in \mathfrak{p}^+$. Let $f_i \in \mathfrak{p}^-$ be the root vector for the root $-\beta_i$ such that $B(e_i, f_i) = 1$. Then for the basis b_i of $\mathfrak{p}$ we choose $e_1, \ldots, e_m; f_1, \ldots, f_m$. The dual basis is then $f_1, \ldots, f_m; e_1, \ldots, e_m$. Thus the Dirac operator is

$$D = \sum_{i=1}^{m} e_i \otimes f_i + f_i \otimes e_i.$$

We also note that in the Hermitian case $\mathfrak{p}$ is even-dimensional, there is a unique irreducible $C(\mathfrak{p})$-module, the spin module S, which we choose to construct as $S = \bigwedge \mathfrak{p}^+$. It is also a module for the double $\tilde{K}$ of K. Since $\mathfrak{p}^+ \cong (\mathfrak{p}^-)^*$, we have

$$X \otimes S \cong X \otimes \bigwedge \mathfrak{p}^+ \cong \mathrm{Hom}(\bigwedge \mathfrak{p}^-, X) \tag{3.9}$$

as vector spaces. Note that the underlying vector space $\bigwedge \mathfrak{p}^+$ of the spin module S carries the adjoint action of $\mathfrak{k}$, but the relevant $\mathfrak{k}$-action on S is the spin action defined using the map (3.1). The spin action is equal to the adjoint action shifted by the character $-\rho_n$ of $\mathfrak{k}$ (see [K1, Proposition 3.6]). So as a $\mathfrak{k}$-module, $X \otimes S$ differs from $X \otimes \bigwedge \mathfrak{p}^+$ and $\mathrm{Hom}(\bigwedge \mathfrak{p}^-, X)$ by a twist of the 1-dimensional $\mathfrak{k}$-module $\mathbb{C}_{-\rho_n}$.

Let $C = \sum_{i=1}^m f_i \otimes e_i$ and $C^- = \sum_{i=1}^m e_i \otimes f_i$; so $D = C + C^-$. Then, under the identifications (3.9), C acts on $X \otimes S$ as the $\mathfrak{p}^-$-cohomology differential, while C^- acts by 2 times the $\mathfrak{p}^+$-homology differential (see [HP2, Proposition 9.1.6] or [HPR]). Furthermore, C and C^- are adjoints of each other with respect to the Hermitian inner product on $X \otimes S$ mentioned above (see [HP2, Lemma 9.3.1] or [HPR]). It was proved that Dirac cohomology is isomorphic to nilpotent Lie algebra cohomology up to a one-dimensional character by using a version of Hodge decomposition.

Theorem 3.10. [[HPR], Theorem 7.11] *Let X be a unitary $(\mathfrak{g}, K)$-module. Then*

$$H_D(X) \cong H^*(\mathfrak{p}^-, X) \otimes \mathbb{C}_{-\rho_n} \cong H_*(\mathfrak{p}^+, X) \otimes \mathbb{C}_{-\rho_n}$$

as $\mathfrak{k}$-modules.

Note that we may use $\bigwedge \mathfrak{p}^-$ instead of $\bigwedge \mathfrak{p}^+$ to construct the spin module S. Then we have

$$H_D(X) \cong H^*(\mathfrak{p}^+, X) \otimes \mathbb{C}_{\rho_n} \cong H_*(\mathfrak{p}^-, X) \otimes \mathbb{C}_{\rho_n}. \tag{3.11}$$

Namely, the Dirac operator is independent of the choice of positive roots. Thus, we also have

$$H^*(\mathfrak{p}^+, X) \otimes \mathbb{C}_{\rho_n} \cong H^*(\mathfrak{p}^-, X) \otimes \mathbb{C}_{-\rho_n}; \; H_*(\mathfrak{p}^+, X) \otimes \mathbb{C}_{-\rho_n} \cong H_*(\mathfrak{p}^-, X) \otimes \mathbb{C}_{\rho_n}.$$

It also follows that we know the Dirac cohomology of all irreducible unitary highest weight modules explicitly from Enright's calculation of $\mathfrak{p}^+$-cohomology [E].

Now we discuss the action of an automorphism on Dirac cohomology. Let τ be an automorphism of G preserving K. Then $\tau\big|_K$ is an automorphism of K. Also, τ induces automorphisms of $\mathfrak{g}_0$ and $\mathfrak{g}$, denoted again by τ, and τ preserves the Cartan decomposition $\mathfrak{g} = \mathfrak{k} \oplus \mathfrak{p}$. Finally, $\tau\big|_{\mathfrak{p}}$ extends to an automorphism of the Clifford algebra $C(\mathfrak{p})$, denoted again by τ.

For any $(\mathfrak{g}, K)$-module (π, X), if we set $X^\tau = X$, then $(\pi \circ \tau, X^\tau)$ is also $(\mathfrak{g}, K)$-module. Similarly, for any K-module (φ, V), if we set $V^\tau = V$, then $(\varphi \circ \tau, V^\tau)$ is also a K-module. The same is true if we replace K by $\tilde{K}$. The following property of

Dirac cohomology was proved for any unitary $(\mathfrak{g}, K)$-module in [HPZ] (Prop. 5.1 of [HPZ]). The proof is extended easily to any $(\mathfrak{g}, K)$-modules.

Proposition 3.12. *Let (π, X) be a $(\mathfrak{g}, K)$-module. Then*

$$H_D(X^\tau) \cong (H_D(X))^\tau.$$

Proof. Denote by $c : C(\mathfrak{p}) \to \operatorname{End} S$ the map given by the action of $C(\mathfrak{p})$ on the spin module S. Setting $S^\tau = S$ we see that $c \circ \tau : C(\mathfrak{p}) \to \operatorname{End} S^\tau$ makes S^τ into a $C(\mathfrak{p})$-module. This module has to be isomorphic to S, since it is of the same dimension as S, and any $C(\mathfrak{p})$-module is isomorphic to a multiple of S.

Now to define $H_D(X^\tau)$, we can use $(c \circ \tau, S^\tau)$ instead of (c, S). Then the action of the Dirac operator on $X^\tau \otimes S^\tau$ is given by

$$D^\tau = \sum_i \pi(\tau(Z_i)) \otimes c(\tau(Z_i)).$$

Here $\{Z_i\}$ is an orthonormal basis of $\mathfrak{p}$. Since $\tau(Z_i)$ is another orthonormal basis of $\mathfrak{p}$, and since D does not depend on the choice of the orthonormal basis Z_i, we see that

$$D^\tau = \sum_i \pi(Z_i) \otimes c(Z_i) = D.$$

This implies that $H_D(X^\tau) = H_D(X) = \operatorname{Ker} D / \operatorname{Ker} D \cap \operatorname{Im} D$ as vector spaces. It remains to see that the $\mathfrak{k}$-action on $H_D(X^\tau)$ is the same as the τ-twist of the $\mathfrak{k}$-action on $H_D(X^\tau)$. These actions are induced by the diagonal $\mathfrak{k}$-actions on $X^\tau \otimes S^\tau$, respectively $X \otimes S$. So it is enough to show that

$$(\pi \circ \tau \otimes c \circ \tau) \circ (\mathrm{id} \otimes 1 + 1 \otimes \alpha) = (\pi \otimes c) \circ (\mathrm{id} \otimes 1 + 1 \otimes \alpha) \circ \tau,$$

where $\alpha : \mathfrak{k} \to U(\mathfrak{g}) \otimes C(\mathfrak{p})$ is the map defined in (3.1). This is obvious from the formula (3.1). □

The above proposition enables us to obtain the Dirac cohomology of lowest weight modules from that of highest weight modules. Let X be an irreducible holomorphic $(\mathfrak{g}, K)$-module. If τ is the Chevalley automorphism, then $V = X^\tau$ is in $\mathcal{O}^\mathfrak{q}$ as we saw in Sect. 2. Then the Casimir element $\Omega_\mathfrak{g}$ acts semisimply on V. We have shown that $H_D(V)$ is isomorphic to the nilpotent Lie algebra cohomology up to a character in [HX]. We recall here the main steps of the proof of this isomorphism ([HX] Theorem 5.12). The nilpotent Lie algebra homology is $\mathbb{Z}_2$-graded as follows:

$$H_+(\mathfrak{u}, V) = \bigoplus_{i=0} H_{2i}(\mathfrak{u}, V) \text{ and } H_-(\mathfrak{u}, V) = \bigoplus_{i=0} H_{2i+1}(\mathfrak{u}, V).$$

Then there are injective $\mathfrak{l}$-module homomorphisms ([HX] Proposition 4.8):

$$H_D^{\pm}(V) \to H_{\pm}(\mathfrak{u}, V) \otimes \mathbb{C}_{\rho(\bar{\mathfrak{u}})}.$$

Note that we also have ([HX] Proposition 5.2),

$$\operatorname{ch} H_D^{+}(V) - \operatorname{ch} H_D^{-}(V) = (\operatorname{ch} H_{+}(\mathfrak{u}, V) - \operatorname{ch} H_{-}(\mathfrak{u}, V)) * \operatorname{ch} \mathbb{C}_{\rho(\bar{\mathfrak{u}})}.$$

The key lemma in [HX] (Lemma 5.11) states that $H_{+}(\mathfrak{u}, V)$ and $H_{-}(\mathfrak{u}, V)$ have no common $\mathfrak{l}$-submodules, namely,

$$\operatorname{Hom}_{\mathfrak{l}}(H_{+}(\mathfrak{u}, V), H_{-}(\mathfrak{u}, V)) = 0.$$

It follows that $H_D^{\pm}(V) \to H_{\pm}(\mathfrak{u}, V) \otimes \mathbb{C}_{\rho(\bar{\mathfrak{u}})}$ are isomorphisms.

For a Harish-Chandra module V, we have similar injective homomorphisms of $H_D^{\pm}(V)$ into $H^{\pm}(\mathfrak{p}^{+}, V) \otimes \mathbb{C}_{-\rho(\mathfrak{p}^{+})}$. We conjecture that these injective homomorphisms are actually isomorphisms for any simple Harish-Chandra module V.

It is shown in [HX] that determining $H_D(L(\lambda))$ is equivalent to determining $\operatorname{ch} L(\lambda)$ in term of $\operatorname{ch} M_I(\mu)$, which is solved by the Kazhdan–Lusztig algorithm. Namely, if

$$\operatorname{ch} L(\lambda) = \sum (-1)^{\epsilon(\lambda,\mu)} m(\lambda, \mu) \operatorname{ch} M_I(\mu),$$

then we have

$$H_D(L(\lambda)) = \bigoplus m(\lambda, \mu) F(\mu) \otimes \mathbb{C}_{\rho(\mathfrak{u})}.$$

By using the known results on Kazhdan–Lusztig polynomials we can calculate explicitly the Dirac cohomology of all irreducible highest weight modules (and lowest weight modules).

In case $L(\lambda)$ is a finite-dimensional representation V_λ with highest weight $\lambda \in \mathfrak{h}^*$, Kostant [K2] calculated the Dirac cohomology of V_λ with respect to any equal rank quadratic subalgebra $\mathfrak{r}$ of $\mathfrak{g}$. Suppose that $\mathfrak{h} \subset \mathfrak{r} \subset \mathfrak{g}$ is the Cartan subalgebra for both $\mathfrak{r}$ and $\mathfrak{g}$. Define $W(\mathfrak{g}, \mathfrak{h})^1$ to be the subset of the Weyl group $W(\mathfrak{g}, \mathfrak{h})$ by

$$W(\mathfrak{g}, \mathfrak{h})^1 = \{w \in W(\mathfrak{g}, \mathfrak{h}) \,|w(\rho) \text{ is } \Delta^{+}(\mathfrak{r}, \mathfrak{h}) - \text{dominant}\}.$$

This is the same as the subset of element $w \in W(\mathfrak{g}, \mathfrak{h})$ that maps the positive Weyl $\mathfrak{g}$-chamber into the positive $\mathfrak{r}$-chamber. There is a bijection $W(\mathfrak{r}, \mathfrak{h}) \times W(\mathfrak{g}, \mathfrak{h})^1 \to W(\mathfrak{g}, \mathfrak{h})$ given by $(w, \tau) \mapsto w\tau$. Kostant proved [K2] that

$$H_D(V_\lambda) = \bigoplus_{w \in W(\mathfrak{g}, \mathfrak{h})^1} E_{w(\lambda+\rho)-\rho(\mathfrak{r})}.$$

Kostant's theorem implies

Proposition 3.13. *Let G be a semisimple Lie group of Hermitian symmetric type. Let V_λ be a finite-dimensional representation with highest weight λ of G. Then*

$$H_D(V_\lambda) = \bigoplus_{w \in W(\mathfrak{g},\mathfrak{h})^1} E_{w(\lambda+\rho)-\rho_c}.$$

We note that the above formula also follows from the calculation of Dirac cohomology for the more general case of all finite-dimensional Harish-Chandra modules in [HKP].

4 K-Characters of Holomorphic $(\mathfrak{g}, K)$-Modules

In this section we prove a K-character formula for holomorphic or irreducible lowest weight $(\mathfrak{g}, K)$-modules. This is a straightforward extension of the K-character formula for unitary lowest weight modules in [HPZ].

Recall that if V is an admissible $(\mathfrak{g}, K)$-module with K-type decomposition $V = \bigoplus_\lambda m_\lambda F^\lambda$, then the K-character of V is the formal series

$$\operatorname{ch} V = \sum_\lambda m_\lambda \operatorname{ch} F^\lambda,$$

where $\operatorname{ch} F^\lambda$ is the character of the irreducible K-module F^λ. Moreover, this definition makes sense also for virtual $(\mathfrak{g}, K)$-modules V; in that case, the integers m_λ can be negative. In the following we will often deal with representations of the spin double cover $\tilde{K}$ of K, and not K, but we will still denote the corresponding character by ch.

We keep the notation of the previous section. Since $\mathfrak{p}$ is even-dimensional, the spin module S decomposes as $S^+ \oplus S^-$, with the $\mathfrak{k}$-submodules $S^\pm$ being the even respectively odd part of $S \cong \bigwedge \mathfrak{p}^+$. For any irreducible $(\mathfrak{g}, K)$-module X, we consider the $\tilde{K}$-equivariant operators

$$D^\pm : X \otimes S^\pm \to X \otimes S^\mp$$

defined by the restrictions of D.

Recall that ([HX] Proposition 5.2) for any X in $\mathcal{O}^\mathfrak{q}$ such that given the Casimir element $\Omega_\mathfrak{g}$ acting semisimply, one has

$$\operatorname{ch} X * (\operatorname{ch} S^+ - \operatorname{ch} S^-) = \operatorname{ch} H_D^+ - \operatorname{ch} H_D^-. \tag{4.1}$$

By the same argument in the proof of Proposition 5.2 of [HX] we can prove the same formula for K-characters of admissible $(\mathfrak{g}, K)$-modules such that $\Omega_\mathfrak{g}$ acts semisimply. Consider the exact sequence

$$0 \to \operatorname{Ker} D^+ \to V \otimes S^+ \to V \otimes S^- \to \operatorname{Coker} D^+ \to 0,$$

where $\operatorname{Coker} D^+ = V \otimes S^- / \operatorname{Im} D^+$. It follows that

$$\operatorname{ch}(V \otimes S^+) - \operatorname{ch}(V \otimes S^-) = \operatorname{ch} \operatorname{Ker} D^+ - \operatorname{ch} \operatorname{Coker} D^+.$$

Since D is an isomorphism on $\operatorname{Im} D^2$, the calculation of $\operatorname{ch} \operatorname{Ker} D^+ - \operatorname{ch} \operatorname{Coker} D^+$ can be reduced to the subspace $U = \operatorname{Ker} D^2$. We write D' for the restriction of D to U. Then $(D')^2 = 0$ and $\operatorname{Im} D'^- \cong U / \operatorname{Ker} D'^-$. It follows that

$$\begin{aligned} \operatorname{ch} \operatorname{Ker} D^+ - \operatorname{ch} \operatorname{Coker} D^+ &= \operatorname{ch} \operatorname{Ker} D'^+ - \operatorname{ch} \operatorname{Coker} D'^+ \\ &= \operatorname{ch} \operatorname{Ker} D'^+ / \operatorname{Im} D'^- - \operatorname{ch} \operatorname{Ker} D'^- / \operatorname{Im} D'^+ = \operatorname{ch} H_D^+(V) - \operatorname{ch} H_D^-(V). \end{aligned}$$

To obtain K-character formulae for generalized Verma modules we need the following lemma. Let $S(\mathfrak{p}^+)$ denote the symmetric algebra of $\mathfrak{p}^+$.

Lemma 4.2. *Let $\mathbb{C}_{-\rho_n}$ be the 1-dimensional $\mathfrak{k}$-module with weight $-\rho_n$. Then*

$$ch\, S(\mathfrak{p}^+)(ch\, S^+ - ch\, S^-) = ch\, \mathbb{C}_{-\rho_n}.$$

Proof. Note that $S(\mathfrak{p}^+) \otimes S \cong S(\mathfrak{p}^+) \otimes \bigwedge \mathfrak{p}^+$ is the Koszul complex of $\mathfrak{p}^+$. By the Euler–Poincaré principle, $\operatorname{ch} S(\mathfrak{p}^+)(\operatorname{ch} S^+ - \operatorname{ch} S^-)$ is the Euler characteristic of this complex, i.e., the alternating sum of its cohomology modules. It is well known (see for example [HP2, Proposition 3.3.5]) that the cohomology of the Koszul complex is spanned by the vector $1 \otimes 1$. Since $\mathbb{C} \cdot 1 \otimes 1 = \mathbb{C}_{-\rho_n}$ as a $\mathfrak{k}$-module, the lemma follows. □

Since $N(\mu) = S(\mathfrak{p}^+) \otimes F^\mu$ as a K-module, Lemma 4.2 immediately implies

$$\operatorname{ch} N(\mu)(\operatorname{ch} S^+ - \operatorname{ch} S^-) = \operatorname{ch} F^\mu \operatorname{ch} \mathbb{C}_{-\rho_n}. \tag{4.3}$$

This leads to a Blattner-type formula for $\operatorname{ch} N(\mu)$; we will give more explicit expression for some special but very interesting cases at the end of this section. Before that, we want to use (4.3) to express $\operatorname{ch} L(\mu)$ in terms of its Dirac cohomology (or rather, Dirac index). For this we need the following lemma.

Lemma 4.4. **([HPZ] Lemma 3.4)** *Let V be a virtual $(\mathfrak{g}, K)$-module, with $ch\, V = \sum_{i=1}^{\infty} n_i\, ch\, F^{\mu_i}$ for some $n_i \in \mathbb{Z}$ and some distinct μ_i. Assume that the numbers (μ_i, ρ_n), $i \geq 1$, are bounded from below. Then the identity $ch\, V(ch\, S^+ - ch\, S^-) = 0$ implies that $V = 0$.*

Proof. We include here the proof given in [HPZ]. The arguments were adopted from [HS] Sect. 4. We enclose the proof here for the convenience of readers. We assume that our positive root system Δ^+ is chosen so that the simple roots are compact roots $\alpha_1, \dots, \alpha_{r-1}$ and a noncompact root α_r. Now any positive noncompact root β is of the form $\sum m_i \alpha_i$ with all $m_i \geq 0$ and $m_r > 0$. It is clear that $(\alpha_i, \rho_n) = 0$

for $i \le r-1$. Moreover, $(\alpha_r, \rho_n) = (\alpha_r, \rho - \rho_{\mathfrak{k}}) > 0$, since α_r has negative inner product with compact simple roots and hence also with $\rho_{\mathfrak{k}}$. It follows that $(\beta, \rho_n) > 0$ for any positive noncompact root β.

If $V \neq 0$, we can assume that $n_1 \neq 0$ and that

$$(\mu_1, \rho_n) \le (\mu_i, \rho_n) \tag{4.5}$$

for all i. Since $F^{\mu_1} \otimes S$ contains $F^{\mu_1} \otimes 1 \cong F^{\mu_1 - \rho_n}$ with multiplicity 1, ch $F^{\mu_1-\rho_n}$ appears in ch F^{μ_1}(ch S^+ − ch S^-) with coefficient 1.

In order to cancel this contribution to ch V(ch S^+ − ch S^-), ch $F^{\mu_1-\rho_n}$ must appear in some ch F^{μ_i}(ch S^+ − ch S^-), and hence also in $F^{\mu_i} \otimes S$, for some $i > 1$. It is well known that all weights of S are of the form $-\rho_n + \sum_j \beta_j$ with β_j distinct noncompact positive roots (the sum can be empty). So we must have $\mu_i - \rho_n + \sum_j \beta_j = \mu_1 - \rho_n$, i.e., $\mu_1 = \mu_i + \sum_j \beta_j$. Since $\mu_i \neq \mu_1$, the sum is nonempty. Note that $(\beta_j, \rho_n) > 0$ for each j. It follows that $(\mu_1, \rho_n) > (\mu_i, \rho_n)$, which contradicts (4.5). □

Proposition 4.6. *Let $L(\mu)$ be an irreducible lowest weight* $(\mathfrak{g}, K)$*-module. Assume that $H_D^+(L(\mu)) = \sum_\xi F^\xi$ and $H_D^-(L(\mu)) = \sum_\eta F^\eta$. Then*

$$ch\, L(\mu) = \sum_\xi ch\, N(\xi + \rho_n) - \sum_\eta ch\, N(\eta + \rho_n). \tag{4.7}$$

Proof. Using (4.1) for $X = L(\mu)$, (4.3), and the obvious fact ch $N(\nu)$ ch $\mathbb{C}_{\rho_n}$ = ch $N(\nu + \rho_n)$, we see that

$$\begin{aligned} \operatorname{ch} L(\mu)(\operatorname{ch} S^+ - \operatorname{ch} S^-) &= \operatorname{ch} H_D^+(L(\mu)) - \operatorname{ch} H_D^-(L(\mu)) \\ &= \sum_\xi \operatorname{ch} F^\xi - \sum_\eta \operatorname{ch} F^\eta \\ &= \Big(\sum_\xi \operatorname{ch} N(\xi + \rho_n) - \sum_\eta \operatorname{ch} N(\eta + \rho_n)\Big)(\operatorname{ch} S^+ - \operatorname{ch} S^-). \end{aligned}$$

Thus we have

$$\Big(\operatorname{ch} L(\mu) - \sum_\xi \operatorname{ch} N(\xi + \rho_n) + \sum_\eta \operatorname{ch} N(\eta + \rho_n)\Big)(\operatorname{ch} S^+ - \operatorname{ch} S^-) = 0.$$

The assertion now follows from Lemma 4.4 applied to $V = L(\mu) - \sum_\xi N(\xi + \rho_n) + \sum_\eta N(\eta + \rho_n)$. Namely, if F^ν is a K-type appearing in either $N(\gamma)$ or $L(\gamma)$ for some γ, then $\nu = \gamma + \sum_j \beta_j$ for some positive noncompact roots β_j. In the proof of Lemma 4.4 we saw that $(\beta_j, \rho_n) > 0$, so it follows that $(\nu, \rho_n) \ge (\gamma, \rho_n)$ (equality is attained when the sum $\sum_j \beta_j$ is empty). It follows that the K-types appearing in V are bounded from below and so Lemma 4.4 applies. □

5 Branching Rules for $GL_n \hookrightarrow Sp_{2n}$ and $GL_n \hookrightarrow SO_{2n}$

The branching formulae of finite-dimensional representations for $GL_n \hookrightarrow Sp_{2n}$ and $GL_n \hookrightarrow SO_{2n}$ can be regarded as K-type multiplicities for noncompact Lie groups of $Sp(2n,\mathbb{R})$ and $SO^*(2n)$. In this section we show how to obtain the K-multiplicity formulae from the K-character formulae that are proved in the previous section. The formulae for K-multiplicity naturally extend to infinite-dimensional holomorphic $(\mathfrak{g}, K)$-modules or the lowest weight modules.

We start with the symplectic case, for which the Lie algebra for the noncompact Lie group $Sp(2n,\mathbb{R})$ is $\mathfrak{g}_0 = \mathfrak{sp}(2n,\mathbb{R})$. In this case, both $\mathfrak{g} = \mathfrak{sp}(2n,\mathbb{C})$ and $\mathfrak{k} = \mathfrak{gl}(n,\mathbb{C})$ have rank n. So if $\mathfrak{t}$ is a common Cartan subalgebra of $\mathfrak{g}$ and $\mathfrak{k}$, both $\mathfrak{t}$ and its dual can be identified with $\mathbb{C}^n$. It is standard to choose the positive compact roots to be $e_i - e_j$ for $1 \leq i < j \leq n$, and the positive noncompact roots to be $e_i + e_j$ for $i < j$ and $2e_i$, $i = 1,\ldots,n$. The simple roots corresponding to our choice of positive roots are $e_i - e_{i+1}$, $i = 1,\ldots,n-1$, and $2e_n$. If we as usual denote by ρ and ρ_c the half sums of the positive roots for $\mathfrak{g}$ respectively for $\mathfrak{k}$, and by $\rho_n = \rho - \rho_c$ the half sum of the noncompact positive roots, then we see that

$$\rho = (n,\ldots,1); \qquad \rho_c = (\frac{n-1}{2},\ldots,-\frac{n-1}{2}); \qquad \rho_n = (\frac{n+1}{2},\ldots,\frac{n+1}{2}).$$

The Weyl group W_K consists of permutations of the variables, while W_G also contains arbitrary sign changes of the variables. The fundamental chamber for $\mathfrak{g}$ is given by the inequalities $x_1 \geq x_2 \geq \cdots \geq x_n \geq 0$, while the fundamental chamber for $\mathfrak{k}$ is given by $x_1 \geq x_2 \geq \cdots \geq x_n$. (These are the closed fundamental chambers; the open ones are given by strict inequalities.)

We now consider the orthogonal case, for which the Lie algebra of $SO^*(2n,\mathbb{R})$ is $\mathfrak{g}_0 = \mathfrak{so}^*(2n,\mathbb{R})$. The Lie algebras $\mathfrak{g} = \mathfrak{so}(2n,\mathbb{C})$ and $\mathfrak{k} = \mathfrak{gl}(n,\mathbb{C})$ both have rank n, and we choose a common Cartan subalgebra $\mathfrak{t}$ in both of them. Both $\mathfrak{t}$ and $\mathfrak{t}^*$ are identified with $\mathbb{C}^n$. We choose the positive compact roots to be $e_i - e_j$ for $1 \leq i < j \leq n$, and the noncompact positive roots to be $e_i + e_j$ for $1 \leq i < j \leq n$. The simple roots corresponding to our choice of positive roots are $e_i - e_{i+1}$, $i = 1,\ldots,n-1$, and $e_{n-1} + e_n$. In this case,

$$\rho = (n-1,\ldots,0); \quad \rho_c = (\frac{n-1}{2},\ldots,-\frac{n-1}{2}); \quad \rho_n = (\frac{n-1}{2},\ldots,\frac{n-1}{2}).$$

(The entries of ρ_c and ρ decrease by one, while the entries of ρ_n are constant.)

The Weyl group W_K consists of permutations of the variables, while W_G also contains arbitrary sign changes of even number of the variables. The fundamental chamber for $\mathfrak{g}$ is given by the inequalities $x_1 \geq x_2 \geq \cdots \geq x_{n-1} \geq |x_n|$, while the fundamental chamber for $\mathfrak{k}$ is given by $x_1 \geq x_2 \geq \cdots \geq x_n$. (These are the closed fundamental chambers; the open ones are given by strict inequalities.)

We now assume $\mathfrak{g}_0$ is either $\mathfrak{sp}(2k,\mathbb{R})$ or $\mathfrak{so}^*(2k)$. Then $\mathfrak{k}_0 = \mathfrak{u}(k)$. Let $\tau \in \mathrm{Aut}(\mathfrak{g}_0)$ be defined by $\tau(X) = -X^T$, where T denotes the matrix transpose. It is easy to see that τ fixes $\mathfrak{u}(k)$ and that the extension of τ to $\mathfrak{g}$ interchanges $\mathfrak{p}^+$ and $\mathfrak{p}^-$. The following lemma is straightforward.

Lemma 5.1. *Let τ be as above. For $\sigma = (\sigma_1, \dots, \sigma_k)$, let $\sigma' = (-\sigma_k, \dots, -\sigma_1)$.*

(i) *For any irreducible representation (φ, F^σ) of $\mathfrak{u}(k)$, $\varphi \circ \tau$ is the dual representation $F^{\sigma'}$ of F^σ.*

(ii) *Let $(\pi, N(\sigma))$ be a lowest weight generalized Verma module for $(\mathfrak{g}, K)$. Then $(\pi \circ \tau, N(\sigma)^\tau)$ is the highest weight generalized Verma module $\overline{N}(\sigma') = U(\mathfrak{g}) \otimes_{U(\mathfrak{k}\oplus\mathfrak{p}^+)} F^{\sigma'}$.*

Now we turn to the parametrization of the irreducible finite-dimensional representations for the classical groups. A partition σ is a finite sequence of weakly decreasing positive integers, $\sigma_1 \geq \sigma_2 \geq \dots \geq \sigma_k$. For such a partition σ, we denote by $l(\sigma)(= k)$, the length of σ and $|\sigma|(= \sum_i \sigma_i)$, the size of σ. Given a partition σ, we denote the conjugate partition to σ by σ'. That is, σ' is the partition obtained by flipping the Young diagram of σ over the main diagonal. Equivalently, $(\sigma')_i = |\{j : \sigma_j \geq i\}|$. Note that $|\sigma| = |\sigma'|$ and $l(\sigma) = (\sigma')_1$.

For each partition σ with at most k parts, let F^σ denote the irreducible (finite-dimensional) representation of GL_k with highest weight $\sigma_1\epsilon_1 + \sigma_2\epsilon_2 + \cdots + \sigma_k\epsilon_k$. Here we are using the standard coordinates for the dual Cartan algebra of GL_k; ϵ_i denotes the i-th projection of $\mathfrak{h} \cong \mathbb{C}^k$. Similarly, for k even, let V^σ be the irreducible Sp_k representation indexed by σ, and for σ with $(\sigma')_1 + (\sigma')_2 \leq k$ let E^σ be the irreducible O_k representation indexed by σ (see [GW], page 420).

We remark that not all representations of GL_k correspond to nonnegative integer partitions as above; those that do are called polynomial representations, since their matrix coefficients turn out to be polynomial functions. However, we do not lose generality by studying only polynomial representations, because any representation can be twisted to a polynomial one by a sufficiently large one-dimensional character.

Given nonnegative integer partitions σ, μ and ν, each with at most k parts, the classical Littlewood–Richardson coefficient $c^\sigma_{\mu\nu}$ is defined by

$$c^\sigma_{\mu\nu} = \dim \mathrm{Hom}_{GL_k}(F^\sigma, F^\mu \otimes F^\nu).$$

Let P denote the set of partitions. We define

$$P_R = \{\sigma \in P : \sigma_i \in 2\mathbb{N} \text{ for all } i\},$$
$$P_C = \{\sigma \in P : (\sigma')_i \in 2\mathbb{N} \text{ for all } i\}. \tag{5.2}$$

The set P_R (resp. P_C) consists of partitions whose Young diagrams have even rows (resp. columns). For partitions σ and μ with at most k parts, we define the following sums of the Littlewood–Richardson coefficients:

$$C^\sigma_\mu := \sum_{\nu\in P_R, l(\nu)\leq k} c^\sigma_{\mu\nu} \quad \text{and} \quad D^\sigma_\mu := \sum_{\nu\in P_C, l(\nu)\leq k} c^\sigma_{\mu\nu}. \tag{5.3}$$

To compute the K-type multiplicities of $L(\mu)$ using Proposition 4.6, we first need to know the Dirac cohomology $H_D(L(\mu))$. This is known explicitly, since by Theorem 3.10 it can be read off from the corresponding $\mathfrak{p}^+$-cohomology which is determined by [E]. (Here we are exchanging roles of $\mathfrak{p}^+$ and $\mathfrak{p}^-$, which also exchanges roles of lowest and highest weight modules.) We can also compute $H_D(L(\mu))$ directly in some cases; see for example [HPP]. The other ingredient of Proposition 4.6 is knowing the K-type multiplicities for generalized Verma modules $N(\mu)$. Since $N(\mu) = S(\mathfrak{p}^+) \otimes F^\mu$ as a K-module, one needs to understand the K-module structure of $S(\mathfrak{p}^+)$. In general, this was first done by Schmid [S]. The following completely explicit special cases can be found in [GW, §5.2.5] and [GW, §5.2.6]. Let $\mathbb{C}^k$ denote the standard module for $GL_k(\mathbb{C})$, and let P_R, P_C be as in (5.2).

Lemma 5.4. **(i)** *For the Hermitian symmetric pair* $(\mathfrak{sp}(2k,\mathbb{R}), \mathfrak{u}(k))$, $\mathfrak{p}^+ = S^2(\mathbb{C}^k)$ *as a module for* $\mathfrak{k} = \mathfrak{gl}(k,\mathbb{C})$, *and the* $\mathfrak{k}$*-module* $S(\mathfrak{p}^+)$ *decomposes as*

$$S(\mathfrak{p}^+) = \bigoplus_{\nu \in P_R, l(\nu) \leq k} F^\nu.$$

(ii) *For the Hermitian symmetric pair* $(\mathfrak{so}^*(2k), \mathfrak{u}(k))$, $\mathfrak{p}^+ = \bigwedge^2(\mathbb{C}^k)$ *as a module for* $\mathfrak{k} = \mathfrak{gl}(k,\mathbb{C})$, *and the* $\mathfrak{k}$*-module* $S(\mathfrak{p}^+)$ *decomposes as*

$$S(\mathfrak{p}^+) = \bigoplus_{\nu \in P_C, l(\nu) \leq k} F^\nu.$$

To get the K-multiplicities for $N(\mu)$ we tensor the above formulas with F^μ and recall the sums of the Littlewood–Richardson coefficients C^σ_μ, D^σ_μ defined above in (5.3). Consequently, we obtain the following proposition.

Proposition 5.5. **(i)** *For the Hermitian symmetric pair* $(\mathfrak{sp}(2k,\mathbb{R}), \mathfrak{u}(k))$, *the multiplicity of the* $\mathfrak{gl}(k,\mathbb{C})$*-module* F^σ *in* $N(\mu)$ *is* C^σ_μ, *the sum of the Littlewood–Richardson coefficients defined in (5.3).*

(ii) *For the Hermitian symmetric pair* $(\mathfrak{so}^*(2k), \mathfrak{u}(k))$, *the multiplicity of the* $\mathfrak{gl}(k,\mathbb{C})$*-module* F^σ *in* $N(\mu)$ *is* D^σ_μ, *the sum of the Littlewood–Richardson coefficients defined in (5.3).*

Theorem 5.6. *Assume that* $H^+_D(L(\mu)) = \sum_\xi F^\xi$ *and* $H^-_D(L(\mu)) = \sum_\eta F^\eta$. *Then*

(i) *For the Hermitian symmetric pair* $(\mathfrak{sp}(2k,\mathbb{R}), \mathfrak{u}(k))$,

$$[L(\mu), F^\sigma] = \sum_\xi C^\sigma_{\xi+\rho_n} - \sum_\eta C^\sigma_{\eta+\rho_n}.$$

(ii) *For the Hermitian symmetric pair* $(\mathfrak{so}^*(2k), \mathfrak{u}(k))$,

$$[L(\mu), F^\sigma] = \sum_\xi D^\sigma_{\xi+\rho_n} - \sum_\eta D^\sigma_{\eta+\rho_n}.$$

Here σ and μ are regarded as k-tuples with the sums running over the same sets as above, and C^σ_ξ and D^σ_η are defined in (5.3).

Proof. Let $N(\mu)$ be a lowest weight generalized Verma module as above, with unique irreducible quotient $L(\mu)$. Assume that $H_D^+(L(\mu)) = \sum_\xi F^\xi$ and $H_D^-(L(\mu)) = \sum_\eta F^\eta$, where ξ and η run over some finite sets of dominant $\mathfrak{k}$-weights. Then

$$\operatorname{ch} L(\mu) = \sum_\xi \operatorname{ch} N(\xi + \rho_n) - \sum_\eta \operatorname{ch} N(\eta + \rho_n),$$

with ξ and η running over the same sets as above. Then the theorem follows from Proposition 5.5. □

Remark 5.7. The Dirac cohomology $H_D(L(\mu))$ appearing in the above theorem can be calculated very explicitly in many cases. For example, the Proposition 3.13 gives the result for the case $L(\mu)$ is finite dimensional. The main result in [HPP] deals with the case when μ is one-dimensional. The formula for the general case is Theorem 6.16 in [HX].

6 Branching Rules for $O_n \hookrightarrow GL_n$ and $Sp_{2n} \hookrightarrow GL_{2n}$

In this final section we recall the generalized Littlewood Restriction Formulae obtained in [HPZ]. These formulae are branching rules for $O_n \hookrightarrow GL_n$ and $Sp_{2n} \hookrightarrow GL_{2n}$. It was important for the proof in [HPZ] to use Howe's dual pair correspondence (see [H1] and [H2]) and the see-saw dual pairs introduced by Kudla [Ku] (see also in [HK]).

We first recall that D. E. Littlewood [L1] gave a formula in 1940 for the decomposition of some representations of GL_k restricted to O_k, or Sp_k for k even. We use the same notation as in Sect. 5; namely, the representations of classical groups are parametrized by partitions. The Littlewood Restriction Formulae can be described as follows (see [L1] and [L2]).

(i) **(Littlewood Restriction Formula for $O_k(\mathbb{C}) \subset GL_k(\mathbb{C})$).** For partitions σ and μ with at most $\frac{k}{2}$ parts,

$$\dim \operatorname{Hom}_{O_k(\mathbb{C})}(E^\mu, F^\sigma) = C^\sigma_\mu.$$

(ii) **(Littlewood Restriction Formula for $Sp_k(\mathbb{C}) \subset GL_k(\mathbb{C})$, k even).** For partitions σ and μ with at most $\frac{k}{2}$ parts,

$$\dim \operatorname{Hom}_{Sp_k(\mathbb{C})}(V^\mu, F^\sigma) = D^\sigma_\mu.$$

Here σ and μ are regarded as k-tuples, C_ξ^σ and D_η^σ are defined in (5.3). We note the definition of C_ξ^σ and D_η^σ in (5.3) extends to general k-tuples σ and μ, which are regarded as highest weights of irreducible $\mathfrak{gl}(k, \mathbb{C})$-modules.

Inspired by the work of Enright and Willenbring [EW], a generalized Littlewood Restriction Formulae in terms of Dirac cohomology was obtained in [HPZ]. It was proved that those formulae are equivalent to the formulae that were proved in [EW] in terms of nilpotent Lie algebra cohomology.

Recall that for any n-tuple σ, it is defined in [HPZ]

$$\sigma^\diamond = \sigma + \underbrace{(\frac{k}{2}, \dots, \frac{k}{2})}_{n} \quad \text{and} \quad \sigma^{-\diamond} = \sigma - \underbrace{(\frac{k}{2}, \dots, \frac{k}{2})}_{n}. \tag{6.1}$$

We recall that $L(\mu)$ is the lowest weight $(\mathfrak{g}, K)$-module of $Sp(2n, \mathbb{R})$ or $SO^*(2n, \mathbb{R})$ defined and studied in Sects. 2 and 4 and we specialize here to the case $n = k$. These branching formulae involve some twist (partitions are shifted by $\pm\diamond$). This twist was due to Proposition 4.1 of [HPZ] which is proved by the Howe dual pair correspondence. The classical case of Littlewood Restriction Theorem has no such twist (see Remark 4.7 in [HPZ]). To understand the proof and the twist we recall the proposition needed in the proof.

Proposition 6.2. ([HPZ] Proposition 4.1)

(i) *(For generalization of Littlewood Restriction Formula for $O_k(\mathbb{C}) \subset GL_k(\mathbb{C})$). Let σ and μ be partitions with at most k parts. Then*

$$[E^\mu, F^\sigma] = \dim \mathrm{Hom}_{O_k(\mathbb{C})}(E^\mu, F^\sigma) = \dim \mathrm{Hom}_{\mathfrak{u}(k)}(F^{\sigma^\diamond}, L(\mu^\diamond)).$$

(ii) *(For generalization of Littlewood Restriction Formula for $Sp_k(\mathbb{C}) \subset GL_k(\mathbb{C})$, k even). For partitions σ and μ with $l(\sigma) \leq k$ and $l(\mu) \leq \frac{k}{2}$,*

$$[V^\mu, F^\sigma] = \dim \mathrm{Hom}_{Sp_k(\mathbb{C})}(V^\mu, F^\sigma) = \dim \mathrm{Hom}_{\mathfrak{u}(k)}(F^{\sigma^\diamond}, L(\mu^\diamond)).$$

Then Theorem 5.6 and the above proposition give the Generalized Littlewood Restriction Formulae in [HPZ].

Theorem 6.3. ([HPZ] Theorem 4.9)
Assume that $H_D^+(L(\mu^\diamond)) = \sum_\xi F^\xi$ and $H_D^-(L(\mu^\diamond)) = \sum_\eta F^\eta$.

(i) *Let σ and μ be partitions with at most k parts. Assume that the sum of the first two columns of the Young diagram of μ is at most k. Then*

$$[E^\mu, F^\sigma] = \dim \mathrm{Hom}_{O_k(\mathbb{C})}(E^\mu, F^\sigma) = \sum_\xi C^\sigma_{\xi^{-\diamond}+\rho_n} - \sum_\eta C^\sigma_{\eta^{-\diamond}+\rho_n}.$$

(ii) *Let k be even. Let σ and μ be partitions with $l(\sigma) \leq k$ and $l(\mu) \leq \frac{k}{2}$. Then*

$$[V^\mu, F^\sigma] = \dim \mathrm{Hom}_{Sp_k(\mathbb{C})}(V^\mu, F^\sigma) = \sum_\xi D^\sigma_{\xi^{-\diamond}+\rho_n} - \sum_\eta D^\sigma_{\eta^{-\diamond}+\rho_n}.$$

Here σ and μ are regarded as k-tuples with the sums running over the same sets as above, and C^σ_ξ and D^σ_η are defined in (5.3).

Recall from the previous section that $\rho_n = (\frac{k+1}{2}, \ldots, \frac{k+1}{2})$ for $\mathfrak{g}_0 = \mathfrak{sp}(2k, \mathbb{R})$ and $\rho_n = (\frac{k-1}{2}, \ldots, \frac{k-1}{2})$ for $\mathfrak{g}_0 = \mathfrak{so}^*(2k)$. It follows that $\xi^{-\diamond} + \rho_n = \xi + (\frac{1}{2}, \cdots, \frac{1}{2})$ in case $\mathfrak{g}_0 = \mathfrak{sp}(2k, \mathbb{R})$ and $\xi^{-\diamond} + \rho_n = \xi - (\frac{1}{2}, \cdots, \frac{1}{2})$ in case $\mathfrak{g}_0 = \mathfrak{so}^*(2k)$. Then we have the following corollary.

Corollary 6.4. *Assume that $H_D^+(L(\mu^\diamond)) = \sum_\xi F^\xi$ and $H_D^-(L(\mu^\diamond)) = \sum_\eta F^\eta$.*

(i) *Let σ and μ be partitions with at most k parts. Assume that the sum of the first two columns of the Young diagram of μ is at most k. Then*

$$[E^\mu, F^\sigma] = \sum_\xi C^\sigma_{\xi+(\frac{1}{2},\cdots,\frac{1}{2})} - \sum_\eta C^\sigma_{\eta+(\frac{1}{2},\cdots,\frac{1}{2})}.$$

(ii) *Let k be even. Let σ and μ be partitions with $l(\sigma) \leq k$ and $l(\mu) \leq \frac{k}{2}$. Then*

$$[V^\mu, F^\sigma] = \sum_\xi D^\sigma_{\xi-(\frac{1}{2},\cdots,\frac{1}{2})} - \sum_\eta D^\sigma_{\eta-(\frac{1}{2},\cdots,\frac{1}{2})}.$$

Here σ and μ are regarded as k-tuples with the sums running over the same sets as above, and C^σ_ξ and D^σ_η are defined in (5.3).

References

[BGG] J. Bernstein, I. Gelfand, S. Gelfand, *Category of $\mathfrak{g}$-modules*, Funct. Anal. and Appl. **10** (1976), 87–92.

[DH] C.-P. Dong, J.-S. Huang, *Jacquet modules and Dirac cohomology*, Adv. Math. **226** (2011), 2911–2934.

[E] T. Enright, *Analogues of Kostant's $\mathfrak{u}$-cohomology formulas for unitary highest weight modules*, J. Reine Angew. Math. **392** (1988), 27–36.

[EHW] T. Enright, R. Howe, N. Wallach, *A classification of unitary highest weight modules*, in *Representation theory of reductive groups*, Park City, Utah, 1982, Birkhäuser, Boston, 1983, 97–143.

[EW] T. Enright, J. Willenbring, *Hilbert series, Howe duality and branching rules for classical groups*, Ann. Math. **159** (2004), no.1, 337–375.

[GW] R. Goodman, N. Wallach, *Representations and Invariants of the Classical Groups*, Encyc. Math. and its Appl. **68**, Cambridge Univ. Press, Cambridge, 1998.

[HS] H. Hecht, W. Schmid, *A proof of Blattner's conjecture*, Invent. Math. **31** (1975), no. 2, 129–154.

[H1] R. Howe, *Remarks on classical invariant theory*, Trans. Amer. Math. Soc. **313** (1989), no.2, 539–570.

[H2] R. Howe, *Perspectives on invariant theory: Schur duality, multiplicity-free actions and beyond*, The Schur lectures (1992), Israel Math. Conf. Proc., Bar-Ilan Univ., Ramat Gan, 1995, 1–182.

[HK] R. Howe, H. Kraft, *Principal covariants, multiplicity free actions and the K-types of holomorphic discrete series*, in Geometry and representation theory of real and p-adic Lie groups (Córdoba 1995), J. Tirao, D. Vogan, J.A. Wolf (eds), Progr. Math. 158, Birkhäuser, Boston, 1997, 209–242.

[HL] R. Howe, S. Lee, *Why should the Littlewood–Richardson rule be true*, Bulletin of Amer. Math. Soc. **49** (2012), 187–236.

[HTW] R. Howe, E.-C. Tan, P. Willenbring, *Stable branching rules for classical symmetric pairs*, Trans. Amer. Math. Soc. **357** (2004), no.4, 1601–1626.

[HKP] J.-S. Huang, Y.-F. Kang, P. Pandžić, *Dirac cohomology of some Harish-Chandra modules*, Transform. Groups **14** (2009), no. 1, 163–173.

[HP1] J.-S Huang, P. Pandžić, *Dirac cohomology, unitary representations and a proof of a conjecture of Vogan*, J. Amer. Math. Soc. **15** (2002), 185–202.

[HP2] J.-S Huang, P. Pandžić, *Dirac operator in Representation Theory*, Math. Theory Appl., Birkhäuser, 2006.

[HPR] J.-S Huang, P. Pandžić, D. Renard, *Dirac operators and Lie algebra cohomology*, Represent. Theory **10** (2006), 299–313.

[HPP] J.-S. Huang, P. Pandžić, V. Protsak, *Dirac cohomology of Wallach representations*, Pacific J. Math. **250** (2011), no. 1, 163–190.

[HPZ] J.-S. Huang, P. Pandžić, F.-H. Zhu, *Dirac cohomology, K-Characters and branching laws*, American Jour. Math. **135** (2013), 1253–1269.

[HX] J.-S. Huang, W. Xiao, *Dirac cohomology of highest weight modules*, Selecta Math. New Series **18** (2012), 803–824.

[Hum] J. Humphreys, *Representations of Semisimple Lie Algebras in the BGG Category $\mathcal{O}$*. GSM. **94**, Amer. Math. Soc., 2008.

[Kn] A. Knapp, *Lie Groups Beyond An Introduction*, Progress in Math. Vol. **140**, Birkhäuser, 2002.

[K1] B. Kostant, *A cubic Dirac operator and the emergence of Euler number multiplets of representations for equal rank subgroups*, Duke Math. J. **100** (1999), 447–501.

[K2] B. Kostant, *Dirac cohomology for the cubic Dirac operator, Studies in Memory of Issai Schur*, Progress in Math. Vol. **210** (2003), 69–93.

[Ku] S. Kudla, *Seesaw dual reductive pairs*, Automorphic forms of several variables (Katata, 1983), Progr. Math. 46, Birkhäuser, Boston, 1984, pp. 244–268.

[L1] D. E. Littlewood, *The Theory of Group Characters and Matrix Representations of Groups*, Oxford Univ. Press, New York, 1940.

[L2] D. E. Littlewood, *On invariant theory under restricted groups*, Philos. Trans. Roy. Soc. London, Ser. A, **239** (1944), 387–417.

[P] R. Parthasarathy, *Dirac operator and the discrete series*, Ann. Math. **96** (1972), 1–30.

[S] W. Schmid, *Die Randwerte holomorpher Funktionen auf hermitesch symmetrischen Räumen*, Invent. Math. **9** (1969/1970), 61–80.

[V] D. Vogan Jr., *Dirac operators and unitary representations*, 3 talks at MIT Lie groups seminar, Fall 1997.

Cleft Extensions and Quotients of Twisted Quantum Doubles

Geoffrey Mason and Siu-Hung Ng

Abstract Given a pair of finite groups F, G and a normalized 3-cocycle ω of G, where F acts on G as automorphisms, we consider quasi-Hopf algebras defined as a cleft extension $\Bbbk_{\omega}^{G}\#_c \Bbbk F$ where c denotes some suitable cohomological data. When $F \to \overline{F} := F/A$ is a quotient of F by a central subgroup A acting trivially on G, we give necessary and sufficient conditions for the existence of a surjection of quasi-Hopf algebras and cleft extensions of the type $\Bbbk_{\omega}^{G}\#_c \Bbbk F \to \Bbbk_{\omega}^{G}\#_{\overline{c}} \Bbbk\overline{F}$. Our construction is particularly natural when $F = G$ acts on G by conjugation, and $\Bbbk_{\omega}^{G}\#_c\Bbbk G$ is a twisted quantum double $D^{\omega}(G)$. In this case, we give necessary and sufficient conditions that Rep($\Bbbk_{\omega}^{G}\#_{\overline{c}} \Bbbk\overline{G}$) is a modular tensor category.

Key words Twisted quantum double • Quasi Hopf algebra • Modular tensor category

Mathematics Subject Classification (2010): 16T10, 18D10.

1 Introduction

Given finite groups F, G with a right action of F on G as *automorphisms*, one can form the *cross product* $\Bbbk^{G}\#\Bbbk F$, which is naturally a Hopf algebra and a *trivial cleft extension*. Moreover, given a normalized 3-cocycle ω of G and suitable

G. Mason (✉)
University of California, Santa Cruz, CA 95064, USA
e-mail: gem@ucsc.edu

S.-H. Ng
Louisiana State University, Baton Rouge, LA 70803, USA
e-mail: rng@math.lsu.edu

G. Mason et al. (eds.), *Developments and Retrospectives in Lie Theory: Algebraic Methods*, Developments in Mathematics 38,
DOI 10.1007/978-3-319-09804-3_11

cohomological data c, this construction can be 'twisted' to yield a quasi-Hopf algebra $\Bbbk_\omega^G \#_c \Bbbk G$. (Details are deferred to Sect. 2.) For a surjection of groups $\pi : F \to \overline{F}$ such that $\ker \pi$ acts trivially on G, we consider the possibility of constructing another quasi-Hopf algebra $\Bbbk_\omega^G \#_{\overline{c}} \Bbbk \overline{F}$ (for suitable data $\overline{c}$) for which there is a 'natural' surjection of quasi-Hopf algebras $f : \Bbbk_\omega^G \#_c \Bbbk F \to \Bbbk_\omega^G \#_{\overline{c}} \Bbbk \overline{F}$. In general such a construction is not possible. The main result of the present paper (Theorem 3.6) gives *necessary and sufficient* conditions for the existence of $\Bbbk_\omega^G \#_{\overline{c}} \Bbbk \overline{F}$ and f in the important case when $\ker \pi$ is contained in the *center* $Z(F)$ of F. The conditions involve rather subtle cohomological conditions on $\ker \pi$; when they are satisfied we obtain interesting new quasi-Hopf algebras.

A special case of this construction applies to the twisted quantum double $D^\omega(G)$ [2], where $F = G$ acts on G by conjugation and the condition that $\ker \pi$ acts trivially on G is *equivalent* to the centrality of $\ker \pi$. In this case, we obtain quotients $\Bbbk_\omega^G \#_{\overline{c}} \Bbbk \overline{G}$ of the twisted quantum double whenever the relevant cohomological conditions hold. Related objects were considered in [5], and in the case that $\pm I \in G \subseteq SU_2(\mathbb{C})$ the fusion rules were investigated. In fact, we can prove that the modular data of each of the orbifold conformal field theories $V_{\widehat{\mathfrak{sl}}_2}^{\overline{G}}$, where $\widehat{\mathfrak{sl}}_2$ is the level 1 affine Kac–Moody Lie algebra of type $\mathfrak{sl}_2$ and $\overline{G} = G/\pm I$, are reproduced by the modular data of $\Bbbk_\omega^G \#_{\overline{c}} \Bbbk \overline{G}$ for suitable choices of cohomological data ω and $\overline{c}$. This result will be appear elsewhere.

The paper is organized as follows. In Sect. 2 we introduce a category associated to a fixed quasi-Hopf algebra $\Bbbk_\omega^G$ whose objects are the cleft extensions we are interested in. In Sect. 3 we focus on central extensions and establish the main existence result (Theorem 3.6). In Sects. 4 and 5 we consider the special case of twisted quantum doubles. The main result here (Theorem 5.5) gives necessary and sufficient conditions for $\mathrm{Rep}(\Bbbk_\omega^G \#_{\overline{c}} \Bbbk \overline{G})$ to be a modular tensor category.

2 Quasi-Hopf Algebras and Cleft Extensions

A quasi-Hopf algebra is a tuple $(H, \Delta, \epsilon, \phi, \alpha, \beta, S)$ consisting of a quasi-bialgebra $(H, \Delta, \epsilon, \phi)$ together with an antipode S and distinguished elements $\alpha, \beta \in H$ which together satisfy various consistency conditions. See, for example, [1, 6, 10]. A Hopf algebra is a quasi-Hopf algebra with $\alpha = \beta = 1$ and trivial Drinfel'd associator $\phi = 1 \otimes 1 \otimes 1$. As long as α is invertible, $(H, \Delta, \epsilon, \phi, 1, \beta\alpha^{-1}, S_\alpha)$ is also a quasi-Hopf algebra for some antipode S_α ([1]). All of the examples of quasi-Hopf algebras in this paper, constructed from data associated to a group, will satisfy the condition $\alpha = 1$.

Suppose that G is a finite group, $\Bbbk$ a field, and $\omega \in Z^3(G, \Bbbk^\times)$ a normalized (multiplicative) 3-cocycle. There are several well-known quasi-Hopf algebras associated to this data. The group algebra $\Bbbk G$ is a Hopf algebra, whence it is a quasi-Hopf algebra too. The dual group algebra is also a quasi-Hopf algebra $\Bbbk^G_\omega$ when equipped with the Drinfel'd associator

$$\phi = \sum_{a,b,c\in G} \omega(a,b,c)^{-1} e_a \otimes e_b \otimes e_c, \tag{1}$$

where $\{e_a \mid a \in G\}$ is the basis of $\Bbbk^G$ dual to the basis of group elements $\{a \mid a \in G\}$ in $\Bbbk G$. Here, $\beta = \sum_{a\in G} \omega(a, a^{-1}, a)e_a$ and $S(a) = a^{-1}$ for $a \in G$. In particular, $\Bbbk^G = \Bbbk^G_1$ is the usual dual Hopf algebra of $\Bbbk G$.

We are particularly concerned with *cleft extensions* determined by a pair of finite groups F, G. We assume that there is a right action $\triangleleft$ of F on G as *automorphisms* of G. The right F-action induces a natural left $\Bbbk F$-action on $\Bbbk^G$, making $\Bbbk^G$ a left $\Bbbk F$-module algebra. If we consider $\Bbbk F$ as a trivial $\Bbbk^G$-comodule (i.e., G acts trivially on $\Bbbk F$), then $(\Bbbk F, \Bbbk^G)$ is a *Singer pair*. Throughout this paper, only these special kinds of Singer pairs will be considered.

A *cleft object* of $\Bbbk^G_\omega$ (or simply G) consists of a triple $c = (F, \gamma, \theta)$ where $c_0 = F$ is a group with a right action $\triangleleft$ on G as automorphisms, and $c_1 = \gamma \in C^2(G, (\Bbbk^F)^\times)$, $c_2 = \theta \in C^2(F, (\Bbbk^G)^\times)$ are normalized 2-cochains. They are required to satisfy the following conditions:

$$\theta_{g\triangleleft x}(y,z)\theta_g(x,yz) = \theta_g(xy,z)\theta_g(x,y), \tag{2}$$

$$\gamma_x(gh,k)\gamma_x(g,h)\omega(g\triangleleft x, h\triangleleft x, k\triangleleft x) = \gamma_x(h,k)\gamma_x(g,hk)\omega(g,h,k), \tag{3}$$

$$\frac{\gamma_{xy}(g,h)}{\gamma_x(g,h)\gamma_y(g\triangleleft x, h\triangleleft x)} = \frac{\theta_g(x,y)\theta_h(x,y)}{\theta_{gh}(x,y)}, \tag{4}$$

where $\theta_g(x,y) := \theta(x,y)(g)$, $\gamma_x(g,h) := \gamma(g,h)(x)$ for $x, y \in F$ and $g, h \in G$.

Associated to a cleft object c of G is a quasi-Hopf algebra

$$H = \Bbbk^G_\omega \#_c \Bbbk F \tag{5}$$

with underlying linear space $\Bbbk^G \otimes \Bbbk F$; the ingredients necessary to define the quasi-Hopf algebra structure are as follows:

$$e_g x \cdot e_h y = \delta_{g\triangleleft x,h}\, \theta_g(x,y)\, e_g xy, \quad 1_H = \sum_{g\in G} e_g,$$

$$\Delta(e_g x) = \sum_{ab=g} \gamma_x(a,b) e_a x \otimes e_b x, \quad \varepsilon(e_g x) = \delta_{g,1}\,,$$

$$S(e_g x) = \theta_{g^{-1}}(x, x^{-1})^{-1}\gamma_x(g, g^{-1})^{-1} e_{g^{-1}\triangleleft x} x^{-1},$$

$$\alpha = 1_H, \quad \beta = \sum_{g\in G} \omega(g, g^{-1}, g) e_g,$$

where $e_g x \equiv e_g \otimes x$ and $e_g \equiv e_g \otimes 1_F$. The Drinfel'd associator ϕ is again given by (1). This quasi-Hopf algebra is also called the *cleft extension* of kF by $\Bbbk^G_\omega$ (cf. [8]). The proof that (5) is indeed a quasi-Hopf algebra when equipped with these structures is rather routine, and is similar to that of the *twisted quantum double* $D^\omega(G)$, which is the case when $F = G$ and the action on G is conjugation ([2, 6]). We shall return to this example in due course. Note that these cleft extensions admit the canonical morphisms of quasi-Hopf algebras

$$\Bbbk^G_\omega \xrightarrow{i} \Bbbk^G_\omega \#_c \Bbbk F \xrightarrow{p} \Bbbk F \tag{6}$$

where

$$i(e_g) = e_g, \quad p(e_g x) = \delta_{g,1} x.$$

Introduce the category Cleft($\Bbbk^G_\omega$) whose objects are the cleft objects of $\Bbbk^G_\omega$; a morphism from $c = (F, \gamma, \theta)$ to $c' = (F', \overline{\gamma}, \overline{\theta})$ is a pair (f_1, f_2) of quasi-bialgebra homomorphisms satisfying that

(i) f_2 preserves the actions on G, i.e. $g \triangleleft x = g \triangleleft f_2(x)$, and
(ii) The diagram

$$\begin{array}{ccccc}
\Bbbk^G_\omega & \xrightarrow{i} & \Bbbk^G_\omega \#_c \Bbbk F & \xrightarrow{p} & \Bbbk F \\
\downarrow{\scriptstyle \mathrm{id}} & & \downarrow{\scriptstyle f_1} & & \downarrow{\scriptstyle f_2} \\
\Bbbk^G_\omega & \xrightarrow{i} & \Bbbk^G_\omega \#_{c'} \Bbbk F' & \xrightarrow{p'} & \Bbbk F'
\end{array}$$

commutes.

It is worth noting that Cleft($\Bbbk^G_\omega$) is essentially the category of cleft extensions of group algebras by $\Bbbk^G_\omega$.

Remark 2.1. The quasi-Hopf algebra $\Bbbk^G_\omega \#_c \Bbbk F$ also admits a natural F-grading which makes it an F-graded algebra. This F-graded structure can be described in terms of the $\Bbbk F$-comodule via the structure map $\rho_c = (\mathrm{id} \otimes p)\Delta$. A morphism $(f_1, f_2) : c \to c'$ in Cleft($\Bbbk^G_\omega$) induces the right $\Bbbk F'$-comodule structure $\rho'_c = (\mathrm{id} \otimes f_2)\rho_c$ on $\Bbbk^G_\omega \#_c \Bbbk F$, and $f_1 : \Bbbk^G_\omega \#_c \Bbbk F \to \Bbbk^G_\omega \#_{c'} \Bbbk F'$ is then a right $\Bbbk F'$-comodule map. In the language of group-grading, f_2 induces an F'-grading on $\Bbbk^G_\omega \#_c \Bbbk F$ and f_1 is an F'-graded linear map. Since f_1 is an algebra map and preserves F'-grading, $f_1(e_g x) = \chi_x(g) e_g \overline{x}$ for some scalar $\chi_x(g)$, where $\overline{x} = f_2(x) \in F'$ for $x \in F$.

Remark 2.2. In general, a quasi-bialgebra homomorphism between two quasi-Hopf algebras is *not* a quasi-Hopf algebra homomorphism. However, if (f_1, f_2) is a morphism in Cleft($\Bbbk^G_\omega$), then both f_1 and f_2 are quasi-Hopf algebra homomorphisms.

We leave this observation as an exercise to readers (cf. (13) and (14) in the proof of Theorem 3.6 below).

In Cleft($\Bbbk_\omega^G$), there is a trivial object $\underline{1}$ in which the group F is trivial and θ, γ are both identically 1. This cleft object is indeed the trivial cleft extension of $\Bbbk_\omega^G$: $\Bbbk_\omega^G \xrightarrow{\text{id}} \Bbbk_\omega^G \xrightarrow{\epsilon} \Bbbk$. It is straightforward to check that $\underline{1}$ is an initial object of Cleft($\Bbbk_\omega^G$).

Suppose we are given a cleft object $c = (F, \gamma, \theta)$ and a quotient map $\pi_{\bar{F}} : F \to \overline{F}$ of F which preserves their actions on G. We ask the following question: is there a cleft object $\overline{c} = (\overline{F}, \overline{\gamma}, \overline{\theta})$ of $\Bbbk_\omega^G$ and a quasi-bialgebra homomorphism $\pi : \Bbbk_\omega^G \#_c \Bbbk F \to \Bbbk_\omega^G \#_{\overline{c}} \Bbbk \overline{F}$ such that $(\pi, \pi_{\bar{F}}) : c \to \overline{c}$ is a morphism of Cleft($\Bbbk_\omega^G$)? Equivalently, the diagram

$$
\begin{array}{ccccc}
\Bbbk_\omega^G & \xrightarrow{i} & \Bbbk_\omega^G \#_c \Bbbk F & \xrightarrow{p} & \Bbbk F \\
\downarrow{\scriptstyle \text{id}} & & \downarrow{\scriptstyle \pi} & & \downarrow{\scriptstyle \pi_{\bar{F}}} \\
\Bbbk_\omega^G & \xrightarrow{i} & \Bbbk_\omega^G \#_{\overline{c}} \Bbbk \overline{F} & \xrightarrow{\overline{p}} & \Bbbk \overline{F}
\end{array}
\tag{7}
$$

commutes. Generally, one can expect the answer to this question to be 'no'. In the following section, we will provide a complete answer in an important special case.

3 Central Quotients

Throughout this section we assume $\Bbbk$ is a field of *any* characteristic, $c = (F, \gamma, \theta)$ an object of Cleft($\Bbbk_\omega^G$) with the associated quasi-Hopf algebra monomorphism $i : \Bbbk_\omega^G \to \Bbbk_\omega^G \#_c \Bbbk F$ and epimorphism $p : \Bbbk_\omega^G \#_c \Bbbk F \to \Bbbk F$. We use the same notation as before, and write $H = \Bbbk_\omega^G \#_c \Bbbk F$.

We now suppose that $A \subseteq Z(F)$ is a *central* subgroup of F such that the restriction of the F-action $\triangleleft$ on G to A is *trivial*. Then the quotient group $\overline{F} = F/A$ inherits the right action, giving rise to an induced Singer pair $(\Bbbk\overline{F}, \Bbbk^G)$. With this setup, we will answer the question raised in the previous section about the existence of the diagram (7). To explain the answer, we need some preparations.

Definition 3.1. (i) $0 \neq u \in H$ is *group-like* if $\Delta(u) = u \otimes u$. The sets of group-like elements and central group-like elements of H are denoted by $\Gamma(H)$ and $\Gamma_0(H)$ respectively.

(ii) $x \in F$ is called *γ-trivial* if $\gamma_x \in B^2(G, \Bbbk^\times)$ is a 2-coboundary. The set of γ-trivial elements is denoted by F^γ.

(iii) $a \in F$ is *c-central* if there is $t_a \in C^1(G, \Bbbk^\times)$ such that

$$\sum_{g \in G} t_a(g) e_g a \in \Gamma_0(H) . \tag{8}$$

The set of c-central elements is denoted by $Z_c(F)$.

Let $\hat{G} = \mathrm{Hom}(G, \Bbbk^\times)$ be the group of linear characters of G. The following lemma concerning the sets F^γ, $\Gamma(H)$ and $\hat{G}$ is similar to an observation in [9].

Lemma 3.2. *The following statements concerning F^γ and $\Gamma(H)$ hold.*

(i) F^γ is a subgroup of F, $\Gamma(H)$ is a subgroup of the group of units in H, and $p(\Gamma(H)) = F^\gamma$. Moreover, for $x \in F^\gamma$ and $t_x \in C^1(G, \Bbbk^\times)$,

$$\sum_{g \in G} t_x(g) e_g x \in \Gamma(H) \text{ if, and only if, } \delta t_x = \gamma_x .$$

(ii) The sequence of groups

$$1 \to \hat{G} \xrightarrow{i} \Gamma(H) \xrightarrow{p} F^\gamma \to 1 \tag{9}$$

is exact. The 2-cocycle $\beta \in Z^2(F^\gamma, \hat{G})$ associated with the section $x \mapsto \sum_{g \in G} t_x(g) e_g x$ of p in (9) is given by

$$\beta(x, y)(g) = \frac{t_x(g) t_y(g \triangleleft x)}{t_{xy}(g)} \theta_g(x, y) \quad (x, y \in F^\gamma, g \in G). \tag{10}$$

Proof. The proofs of (i) and (ii) are similar to Lemma 3.3 in [9]. □

Remark 3.3. Equation (9) is a *central extension* if F acts trivially on $\hat{G}$, but in general it is *not* a central extension.

Remark 3.4. If $a \in Z_c(F)$, then a central group-like element $\sum_{g \in G} t_a(g) e_g a \in \Gamma_0(H)$ will be mapped to the central element a in $\Bbbk F$ under p. Therefore, by Lemma 3.2, we always have $Z_c(F) \subseteq Z(F) \cap F^\gamma$. By direct computation, the condition (8) for $a \in Z_c(F)$ is equivalent to the conditions:

$$\delta t_a = \gamma_a, \; t_a(g) \theta_g(a, y) = t_a(g \triangleleft y) \theta_g(y, a) \text{ and } g \triangleleft a = g \quad (g \in G, y \in F).$$

In particular, $\theta_g(a, b) = \theta_g(b, a)$ for all $a, b \in Z_c(F)$.

By Lemma 3.2, we can parameterize the elements $u = u(\chi, x) \in \Gamma(H)$ by $(\chi, x) \in \hat{G} \times F^\gamma$. More precisely, for a fixed family of 1-cochains $\{t_x\}_{x \in F^\gamma}$ satisfying $\delta t_x = \gamma_x$, every element $u \in \Gamma(H)$ is uniquely determined by a pair $(\chi, x) \in \hat{G} \times F^\gamma$ given by

$$u = u(\chi, x) = \sum_{g \in G} \chi(g) t_x(g) e_g x .$$

Note that a choice of such a family of 1-cochains $\{t_x\}_{x \in F^\gamma}$ satisfying $\delta t_x = \gamma_x$ is equivalent to a section of p in (9). With this convention we have $i(\chi) = u(\chi, 1)$ and $p(u(\chi, x)) = x$ for all $\chi \in \hat{G}$ and $x \in F^\gamma$.

Lemma 3.5. *The set $Z_c(F)$ of c-central elements is a subgroup of $Z(F)$, and it acts trivially on G. Moreover, $\Gamma_0(H)$ is a central extension of $Z_c(F)$ by $\hat{G}^F$ via the exact sequence:*

$$1 \to \hat{G}^F \xrightarrow{i} \Gamma_0(H) \xrightarrow{p} Z_c(F) \to 1\,, \tag{11}$$

where $\hat{G}^F$ is the group of F-invariant linear characters of G.

If we choose t_x such that $u(1, x) \in \Gamma_0(H)$ whenever $x \in Z_c(F)$, then the formula (10) for $\beta(x, y)$ defines a 2-cocycle for the exact sequence (11).

Proof. By Lemma 3.2 and the preceding paragraph, $u(\chi, x) \in \Gamma_0(H)$ for some $\chi \in \hat{G}$ if, and only if, $x \in Z_c(F)$. In particular, $p(\Gamma_0(H)) = Z_c(F)$. It follows from Remark 3.4 that $Z_c(F)$ is a subgroup of $F^\gamma \cap Z(F)$ and $Z_c(F)$ acts trivially on G. By Remark 3.4 again, $u(\chi, 1) \in \Gamma_0(H)$ is equivalent to

$$\chi(g)t_1(g)\theta_g(1, y) = \chi(g \triangleleft y)t_1(g \triangleleft y)\theta_g(y, 1) \text{ for all } g \in G,\ y \in F.$$

In particular, $\hat{G}^F = \ker p|_{\Gamma_0(H)}$, and this establishes the exact sequence (11). If t_x is chosen such that $u(1, x) \in \Gamma_0(H)$ whenever $x \in Z_c(F)$, the second statement follows immediately from Lemma 3.2 (ii) and the commutative diagram:

$$\begin{array}{ccccccccc}
1 & \longrightarrow & \hat{G} & \xrightarrow{i} & \Gamma(H) & \xrightarrow{p} & F^\gamma & \longrightarrow & 1 \\
 & & \uparrow{\scriptstyle incl} & & \uparrow{\scriptstyle incl} & & \uparrow{\scriptstyle incl} & & \\
1 & \longrightarrow & \hat{G}^F & \xrightarrow{i} & \Gamma_0(H) & \xrightarrow{p} & Z_c(F) & \longrightarrow & 1
\end{array}\quad .$$

□

Theorem 3.6. *Let the notation be as before, with $A \subseteq Z(F)$ a subgroup acting trivially on G, and with the right action of $\overline{F} = F/A$ on G inherited from that of F. Then the following statements are equivalent:*

(i) There exist a cleft object $\overline{c} = (\overline{F}, \overline{\gamma}, \overline{\theta})$ of $\Bbbk_\omega^G$ and a quasi-bialgebra map $\pi : \Bbbk_\omega^G \#_c \Bbbk F \to \Bbbk_\omega^G \#_{\overline{c}} \Bbbk\overline{F}$ such that the diagram

$$\begin{array}{ccccc}
\Bbbk_\omega^G & \xrightarrow{i} & \Bbbk_\omega^G \#_c \Bbbk F & \xrightarrow{p} & \Bbbk F \\
\downarrow{\scriptstyle \mathrm{id}} & & \downarrow{\scriptstyle \pi} & & \downarrow{\scriptstyle \pi_{\overline{F}}} \\
\Bbbk_\omega^G & \xrightarrow{i} & \Bbbk_\omega^G \#_{\overline{c}} \Bbbk\overline{F} & \xrightarrow{p'} & \Bbbk\overline{F}
\end{array} \tag{12}$$

commutes.

(ii) $A \subseteq Z_c(F)$ and the subextension

$$1 \to \hat{G}^F \xrightarrow{i} p|_{\Gamma_0(H)}^{-1}(A) \xrightarrow{p} A \to 1$$

of (11) splits.

(iii) $A \subseteq Z_c(F)$ *and there exist* $\{t_a\}_{a \in A}$ *in* $C^1(G, \Bbbk^\times)$ *and* $\{\tau_g\}_{g \in G}$ *in* $C^1(A, \Bbbk^\times)$ *such that* $\delta t_a = \gamma_a$, $\delta \tau_g = \theta_g|_A$ *and*

$$s_a(g) = t_a(g)\tau_g(a)$$

defines a F*-invariant linear character on* G *for all* $a \in A$.

Proof. ((i) ⇒ (ii)) Suppose there exist a cleft object $\overline{c} = (\overline{F}, \overline{\gamma}, \overline{\theta})$ of $\Bbbk^G_\omega$ and a quasi-bialgebra map $\pi : \Bbbk^G_\omega \#_c \Bbbk F \to \Bbbk^G_\omega \#_{\overline{c}} \Bbbk \overline{F}$ such that the diagram (12) commutes. Then $\pi(e_g) = e_g$ for all $g \in G$. Since π is an algebra map, $\pi(e_g x) = \sum_{\overline{y} \in F} \chi_x(g, \overline{y}) e_g \overline{y}$ for some scalars $\chi_x(g, \overline{y})$. Here, we simply write $\overline{y}$ for $\pi_{\overline{F}}(y)$.

By Remark 2.1, π is a $\overline{F}$-graded linear map and so we have $\pi(e_g x) = \chi_x(g, \overline{x}) e_g \overline{x}$. Therefore, we simply denote $\chi_x(g)$ for $\chi_x(g, \overline{x})$. In particular, $\chi_1 = 1$ and $\chi_x(1) = 1$ by the commutativity of (12). Moreover, we find

$$\gamma_x(g, h)\chi_x(g)\chi_x(h) = \overline{\gamma}_{\overline{x}}(g, h)\chi_x(gh), \tag{13}$$

$$\overline{\theta}_g(\overline{x}, \overline{y})\chi_x(g)\chi_y(g \triangleleft x) = \theta_g(x, y)\chi_{xy}(g) \tag{14}$$

for all $x, y \in F$ and $g, h \in G$. An immediate consequence of these equations is that $\chi_x \in C^1(G, \Bbbk^\times)$ for $x \in F$.

For $a \in A$, $\overline{\theta}_g(\overline{a}, \overline{y}) = \overline{\gamma}_{\overline{a}}(g, h) = 1$. Then, (13) and (14) imply

$$\gamma_a = \delta \chi_a^{-1}, \quad 1 = \frac{\chi_{ay}(g)}{\chi_a(g)\chi_y(g)}\theta_g(a, y) = \frac{\chi_{ya}(g)}{\chi_y(g)\chi_a(g \triangleleft y)}\theta_g(y, a) \tag{15}$$

for all $a \in A$, $g \in G$ and $y \in F$. These equalities in turn yield

$$\sum_{g \in G} \chi_a^{-1}(g) e_g a \in \Gamma_0(H)$$

for all $a \in A$. Therefore $A \subseteq Z_c(F)$.

In particular, $A \subseteq F^\gamma$. If we choose $t_a = \chi_a^{-1}$ for all $a \in A$, then the restriction of the 2-cocycle β, given in (10), on A is constant function 1. Therefore, the subextension

$$1 \to \hat{G}^F \xrightarrow{i} p|^{-1}_{\Gamma_0(H)}(A) \xrightarrow{p} A \to 1$$

of (11) splits.

((ii) ⇒ (i) and (iii)) Assume $A \subseteq Z_c(F)$ and the restriction of β on A is a coboundary. By Remark 3.4, we let $t_a \in C^1(G, \Bbbk^\times)$ such that $\delta t_a = \gamma_a$ and

$$t_a(g)\theta_g(a, y) = t_a(g \triangleleft y)\theta_g(y, a) \tag{16}$$

for all $a \in A$, $y \in F$ and $g \in G$. In particular,

$$\sum_{g \in G} t_a(g) e_g a \in \Gamma_0(H)$$

for all $a \in A$. By Lemma 3.5, $\beta(a, b) \in \hat{G}^F$ for all $a, b \in A$. Suppose $\nu = \{\nu_a \mid a \in A\}$ is a family in $\hat{G}^F$ such that $\beta(a, b) = \nu_a \nu_b \nu_{ab}^{-1}$ for all $a, b \in A$.

Let $\overline{r} : \overline{F} \to F$ be a section of $\pi_{\overline{F}}$ such that $\overline{r}(\overline{1}) = 1$. For $x \in F$, we set $r(x) = \overline{r}(\overline{x})$ and

$$\chi_x(g) = \frac{\nu_a(g)}{t_a(g)\theta_g(a, r(x))} \tag{17}$$

for all $g \in G$, where $a = xr(x)^{-1}$. It is easy to see that $\chi_1 = 1$ and χ_x is a normalized 1-cochain of G. Note that for $b \in A$, $\theta_g(a, b) = \theta_g(b, a)$, so we have

$$\frac{\chi_{bx}(g)}{\chi_b(g)\chi_x(g)} = \frac{\nu_{ab}(g)}{t_{ab}(g)\theta_g(ab, r(x))} \frac{t_b(g)}{\nu_b(g)} \frac{t_a(g)\theta_g(a, r(x))}{\nu_a(g)} = \theta_g(b, x)^{-1}, \tag{18}$$

$$\chi_b(g \triangleleft x)\theta_g(b, x) = \chi_b(g)\theta_g(x, b) \quad \text{and} \quad \delta\chi_b^{-1} = \gamma_b\,. \tag{19}$$

Let $\tau_g(a) = \chi_a(g)$ for all $a \in A$ and $g \in G$. Equation (18) implies that $\delta\tau_g = \theta_g|_A$ and

$$\nu_a(g) = t_a(g)\tau_g(a)\,,$$

and this proves (iii).

Define the maps $\overline{\gamma} \in C^2(G, (\Bbbk^{\overline{F}})^\times)$ and $\overline{\theta} \in C^2(\overline{F}, (\Bbbk^G)^\times)$ as follows:

$$\overline{\gamma}_{\overline{x}}(g, h) = \frac{\chi_x(g)\chi_x(h)}{\chi_x(gh)}\gamma_x(g, h)\,, \tag{20}$$

$$\overline{\theta}_g(\overline{x}, \overline{y}) = \frac{\chi_{xy}(g)}{\chi_x(g)\chi_y(g \triangleleft x)}\theta_g(x, y)\,. \tag{21}$$

We need to show that these functions are well defined. Let $b \in A$, $x, y \in F$ and $g, h \in G$. By (4), (18) and (19), we find

$$\frac{\chi_{bx}(g)\chi_{bx}(h)}{\chi_{bx}(gh)}\gamma_{bx}(g, h) = \frac{\chi_x(g)\chi_x(h)}{\chi_x(gh)}\gamma_x(g, h)\,,$$

and this proves $\overline{\gamma}$ is well defined. To show that $\overline{\theta}$ is also well defined, it suffices to prove

$$\frac{\chi_{bxy}(g)}{\chi_{bx}(g)\chi_y(g \triangleleft bx)}\theta_g(bx, y) = \frac{\chi_{xy}(g)}{\chi_x(g)\chi_y(g \triangleleft x)}\theta_g(x, y) = \frac{\chi_{xby}(g)}{\chi_x(g)\chi_{by}(g \triangleleft x)}\theta_g(x, by)$$

for all $b \in A$, $x, y \in F$ and $g, h \in G$. However, the first equality follows from (18) and (2), while the second equality is a consequence of (2), (18) and (19).

It is straightforward to verify that $\overline{c} = (\overline{F}, \overline{\gamma}, \overline{\theta})$ defines cleft object of $\Bbbk^G_\omega$ and $\pi : \Bbbk^G_\omega \#_c \Bbbk F \to \Bbbk^G_\omega \#_{\overline{c}} \Bbbk \overline{F}$, $e_g x \mapsto \chi_x(g) e_g \overline{x}$ defines a quasi-bialgebra homomorphism which makes the diagram (12) commute. We leave routine details to the reader.

((iii) $\Rightarrow$ (ii)) Since $s_a(g) = t_a(g)\tau_g(a)$ defines a F-invariant linear character of G for each a, then $\nu(a) = s_a$ defines a 1-cochain in $C^1(A, \hat{G}^F)$ and

$$\delta\nu = \beta|_A$$

where β is the 2-cocycle given in (10). In particular, $\beta|_A$ is a coboundary. □

Remark 3.7. Suppose we are given $A \subseteq Z_c(A)$ satisfying condition (ii) of the preceding theorem, and $\{t_a\}_{a\in A}$ a fixed family of cochains in $C^1(G, \Bbbk^\times)$ such that $\sum_{g\in G} t_a(g) e_g a \in \Gamma_0(H)$ for $a \in A$. Then the set $\mathscr{S}(A)$ of group homomorphism sections of $p : p^{-1}(A) \to A$ is in one-to-one correspondence with $\mathscr{B}(A) = \{\nu \in C^1(A, \hat{G}^F) \mid \delta\nu = \beta \text{ on } A\}$. For $\nu \in \mathscr{B}(A)$, it is easy to see that

$$\tilde{p}_\nu(a) = \sum_{g\in G} \frac{t_a(g)}{\nu(a)(g)} e_g a \quad (a \in A)$$

defines a group homomorphism in $\mathscr{S}(A)$. Conversely, if $\tilde{p}' \in \mathscr{S}(A)$, then there exists a group homomorphism $f : A \to \hat{G}^F$ such that $i(f(a))\tilde{p}'(a) = \tilde{p}(a)$ for all $a \in A$. In particular, if $\tilde{p}'(a) = \sum_{g\in G} t'_a(g) e_g a$ for $a \in A$, then

$$t'_a = \frac{t_a}{\nu(a) f(a)}$$

and $\nu' = \nu f \in \mathscr{S}(A)$. Therefore, $\tilde{p}' = \tilde{p}_{\nu'}$.

The cleft object $\overline{c} = (F/A, \overline{g}, \overline{\theta})$ and morphism π constructed in the proof of Theorem 3.6 are *not* unique. The definition of $\chi_x(g)$ is determined by the choice of the section $\overline{r} : \overline{F} \to F$ of $\pi_{\overline{F}}$ and $\nu \in \mathscr{B}(A)$. If $\nu' \in \mathscr{B}(A)$, then $\nu' = \nu f$ for some group homomorphism $f : A \to \hat{G}^F$. Thus, the corresponding

$$\chi'_x(g) = f(xr(x)^{-1})(g)\chi_x(g)\,.$$

This implies $\overline{c}' = (F/A, \overline{\gamma}', \overline{\theta}')$ where $\overline{\gamma}' = \overline{\gamma}$ but

$$\overline{\theta}'_g(\overline{x}, \overline{y}) = \frac{\overline{\theta}_g(\overline{x}, \overline{y})}{f(r(x)r(y)r(xy)^{-1})(g)}\,.$$

Therefore, $\overline{c}$ as well as π can be altered by the choice of any group homomorphism $f : A \to \hat{G}^F$ for a given section $\overline{r} : \overline{F} \to F$ of $\pi_{\overline{F}}$.

4 Cleft Objects for the Twisted Quantum Double $D^{\omega}(G)$

Consider the right action of a finite group $F = G$ on itself by conjugation with $\omega \in Z^3(G, \mathbb{k}^{\times})$ a normalized 3-cocycle. We will write $x^g = g^{-1}xg$. There is a *natural* cleft object $c_\omega = (G, \gamma, \theta)$ of $\mathbb{k}^G_\omega$ given by

$$\gamma_g(x, y) = \frac{\omega(x, y, g)\omega(g, x^g, y^g)}{\omega(x, g, y^g)}, \quad \theta_g(x, y) = \frac{\omega(g, x, y)\omega(x, y, g^{xy})}{\omega(x, g^x, y)}. \tag{22}$$

Note that $\gamma_z = \theta_z$ for any $z \in Z(G)$. The associated quasi-Hopf algebra $D^{\omega}_{\mathbb{k}}(G) = \mathbb{k}^G_\omega \#_{c_\omega} \mathbb{k}G$ of this natural cleft object c_ω is the *twisted quantum double* of G [2]. From now on, we simply abbreviate $D^{\omega}_{\mathbb{k}}(G)$ as $D^{\omega}(G)$ when $\mathbb{k}$ is the field of complex numbers $\mathbb{C}$.

For the cleft object c_ω, we can characterize the c_ω-central elements in the following result (cf. Lemma 3.5).

Proposition 4.1. *The c_ω-center $Z_{c_\omega}(G)$ is given by*

$$Z_{c_\omega}(G) = Z(G) \cap G^{\gamma}.$$

The group $\Gamma_0(D^{\omega}(G))$ of central group-like elements of $D^{\omega}(G)$ is the middle term of the short exact sequence

$$1 \to \hat{G} \xrightarrow{i} \Gamma_0(D^{\omega}(G)) \xrightarrow{p} Z(G) \cap G^{\gamma} \to 1\,.$$

In addition, if $H^2(G, \mathbb{k}^{\times})$ is trivial, then $Z(G) = Z_{c_\omega}(G)$.

Proof. The inclusion $Z_{c_\omega}(G) \subseteq Z(G) \cap G^{\gamma}$ follows directly from Remark 3.4. Suppose $z \in Z(G) \cap G^{\gamma}$ and choose $t_z \in C^1(G, \mathbb{k}^{\times})$ so that $\delta t_z = \gamma_z$. Since $z \in Z(G)$, $\theta_z = \gamma_z$ and so $\theta_z = \delta t_z$. This implies

$$\frac{\theta_z(y, g^y)}{\theta_z(g, y)} = \frac{t_z(g^y)}{t_z(g)} \quad (g, y \in G).$$

It follows directly from the definition (22) of θ that

$$\frac{\theta_g(z, y)}{\theta_g(y, z)} = \frac{\theta_z(y, g^y)}{\theta_z(g, y)}\,.$$

Thus we have

$$t_z(g)\theta_g(z, y) = t_z(g^y)\theta_g(y, z) \quad (g, y \in G).$$

It follows from Remark 3.4 that $z \in Z_{c_\omega}(G)$. Since $\hat{G} = \hat{G}^G$, the exact sequence follows from Lemma 3.5.

Finally, if $H^2(G, \Bbbk^\times)$ is trivial and $z \in Z(G)$, then $\gamma_z \in B^2(G, \Bbbk^\times)$ and therefore $z \in G^\gamma$. The equality $Z(G) = Z(G) \cap G^\gamma = Z_{c_\omega}(G)$ follows. □

Definition 4.2. In light of Theorem 3.6, for the canonical cleft object $c_\omega = (G, \gamma, \theta)$ of $\Bbbk_\omega^G$, a subgroup $A \subseteq Z(G)$ is called *ω-admissible* if A satisfies one of the conditions in Theorem 3.6. The quasi-Hopf algebra $\Bbbk_\omega^G \#_{\overline{c}_\omega} \Bbbk\overline{G}$ of an associated cleft object $\overline{c}_\omega = (\overline{G} = G/A, \overline{\gamma}, \overline{\theta})$ is simply denoted by $D^\omega_{r,\tilde{p}}(G, A)$. It depends on the choice of a section r of $\pi_{\overline{G}} : G \to \overline{G}$ and a group homomorphism section $\tilde{p} : A \to \Gamma_0(D^\omega(G))$ of $p : p^{-1}(A) \to A$ (cf. Remark 3.7). We drop the subscripts $r, \tilde{p}$ if there is no ambiguity.

Remark 4.3. The quasi-Hopf algebra $D^\omega(G, N)$ constructed in [5], where $N \trianglelefteq G$ and ω is an inflation of a 3-cocycle of G/N, is a completely different construction from the one presented with the same notation in the preceding definition. Both are attempts to generalized the twisted quantum double construction by taking subgroups into account.

Example 4.4. Let Q be the quaternion group of order 8 and $A = Z(Q)$. Since $H^2(Q, \mathbb{C}^\times) = 1$, A is c_ω-central for all $\omega \in Z^3(Q, \mathbb{C}^\times)$. Since $\hat{Q} \cong \mathbb{Z}_2 \times \mathbb{Z}_2$, the associated 2-cocycle β of the extension

$$1 \to \hat{Q} \to \Gamma_0(D^\omega(Q)) \to Z(Q) \to 1$$

has order 1 or 2. Thus, if ω is a square of another 3-cocycle, $\beta = 1$ and so A is ω-admissible. In fact, A is ω-admissible for all 3-cocycles of Q but the proof is a bit more complicated.

5 Simple Currents and ω-Admissible Subgroups

For simplicity, we will mainly work over the base field $\mathbb{C}$ for the remaining discussion. Again, we assume that G is a finite group and $\omega \in Z^3(G, \mathbb{C}^\times)$ a normalized 3-cocycle. An isomorphism class of a 1-dimensional $D^\omega(G)$-module is also called a *simple current* of $D^\omega(G)$. The set $\mathsf{SC}(G, \omega)$ of all simple currents of $D^\omega(G)$ forms a finite group with respect to tensor product of $D^\omega(G)$-modules. The inverse of a simple current V is the left dual $D^\omega(G)$-module V^*. $\mathsf{SC}(G, \omega)$ is also called the group of invertible objects of $\mathrm{Rep}(D^\omega(G))$ in some articles. Since the category $\mathrm{Rep}(D^\omega(G))$ of finite-dimensional $D^\omega(G)$-modules is a braided monoidal category, $\mathsf{SC}(G, \omega)$ is *abelian*.

Recall that each simple module $V(K, t)$ of $D^\omega(G)$ is characterized by a conjugacy class K of G and an irreducible character t of the twisted group algebra $\mathbb{C}^{\theta_{g_K}}(C_G(g_K))$, where g_K is a fixed element of K and $C_G(g_K)$ is the centralizer of g_K in G. The degree of the module $V(K, t)$ is equal to $|K|t(1)$ (cf. [2,7]).

Suppose $V(K,t)$ is 1-dimensional. Then $K = \{z\}$ for some $z \in Z(G)$ and t is a 1-dimensional character of $\mathbb{C}^{\theta_z}(G)$. Thus, for $g, h \in G$, we have

$$\theta_z(g,h)t(\widetilde{gh}) = t(\tilde{g})t(\tilde{h}), \tag{23}$$

where $\tilde{g}$ denotes g regarded as an element of $\mathbb{C}^{\theta_z}(G)$. Defining $t(g) = t(\tilde{g})$ for $g \in G$, we see that $\theta_z = \gamma_z = \delta t$ is a 2-coboundary of G. Hence $z \in G^\gamma \cap Z(G)$. By Proposition 4.1, $z \in Z_{C_\omega}(G)$. Conversely, if $z \in Z_{C_\omega}(G)$, then there exists $t_z \in C^1(G, \mathbb{C}^\times)$ such that $\delta t_z = \gamma_z$. Then $V(z, t_z)$ is a 1-dimensional $D^\omega(G)$-module. Thus we have proved

Lemma 5.1. *Let K be a conjugacy class of G, g_K a fixed element of K and t an irreducible character of $\mathbb{C}^{\theta_{g_K}}(C_G(g_K))$. Then $V(K,t)$ is a simple current of $D^\omega(G)$ if, and only if, $K = \{z\}$ for some $z \in Z_{C_\omega}(G)$ and $\delta t = \theta_z$.* □

For simplicity, we denote the simple current $V(\{z\}, t)$ by $V(z,t)$. By [2] or [7] the character $\xi_{z,t}$ of $V(z,t)$ is given by

$$\xi_{z,t}(e_g x) = \delta_{g,z} t(x)\,. \tag{24}$$

Fix a family of normalized 1-cochains $\{t_z\}_{z \in Z_{C_\omega}(G)}$ such that $\delta t_z = \gamma_z$. Then for any simple current $V(z,t)$ of $D^\omega(G)$, t is a normalized 1-cochain of G satisfying $\delta t = \theta_z$. Thus, $t = t_z \chi$ for some $\chi \in \hat{G}$. Therefore,

$$\mathsf{SC}(G,\omega) = \{V(z, t_z\chi) \mid z \in Z_{C_\omega}(G) \text{ and } \chi \in \hat{G}\}\,.$$

Suppose $V(z', t_{z'}\chi')$ is another simple current of $D^\omega(G)$. Note that

$$\gamma_x(z,z') = \theta_x(z,z') \text{ and } \gamma_z(x,y) = \theta_z(x,y) \tag{25}$$

for all $z, z' \in Z(G)$ and $x, y \in G$. By considering the action of $e_g x$, we find

$$V(z, t_z\chi) \otimes V(z', t_{z'}\chi') = V(zz', t_{zz'}\beta(z,z')\chi\chi') \tag{26}$$

where β is given by (10). Therefore, we have an exact sequence

$$1 \longrightarrow \hat{G} \xrightarrow{\;i\;} \mathsf{SC}(G,\omega) \xrightarrow{\;p\;} Z_{C_\omega}(G) \longrightarrow 1$$

of abelian groups, where $i : \chi \mapsto V(1,\chi)$ and $p : V(z, t_z\chi) \mapsto z$. With the same fixed family $\{t_z\}_{z \in Z_{C_\omega}(G)}$ of 1-cochains, $u(\chi, z) = \sum_{g \in G} t_z(g) e_g z$ ($z \in Z_{C_\omega}(G)$, $\chi \in \hat{G}$) are all the central group-like elements of $D^\omega(G)$. By Lemma 3.5, the 2-cocycle associated with the extension

$$1 \longrightarrow \hat{G} \xrightarrow{\ i\ } \Gamma_0(D^{\omega}(G)) \xrightarrow{\ p\ } Z_{c_\omega}(G) \longrightarrow 1$$

is also β, and so we have proved

Proposition 5.2. *Fix a family $\{t_z\}_{z\in Z_{c_\omega}(G)}$ in $C^1(G,\mathbb{C}^\times)$ such that $\delta t_z = \theta_z$. Then the map $\zeta : \Gamma_0(D^\omega(G)) \to \mathsf{SC}(G,\omega)$, $u(\chi,z) \mapsto V(z,t_z\chi)$ for $\chi \in \hat{G}$ and $z \in Z_{c_\omega}(G)$, defines an isomorphism of the following extensions:*

$$\begin{array}{ccccccccc} 1 & \longrightarrow & \hat{G} & \xrightarrow{\ i\ } & \mathsf{SC}(G,\omega) & \xrightarrow{\ p\ } & Z_{c_\omega}(G) & \longrightarrow & 1 \\ & & \uparrow{\scriptstyle \mathrm{id}} & & \uparrow{\scriptstyle \zeta} & & \uparrow{\scriptstyle \mathrm{id}} & & \\ 1 & \longrightarrow & \hat{G} & \xrightarrow{\ i\ } & \Gamma_0(D^\omega(G)) & \xrightarrow{\ p\ } & Z_{c_\omega}(G) & \longrightarrow & 1. \end{array}$$

□

Remark 5.3. The preceding proposition implies that these extensions depend only on the cohomology class of ω. In fact, if ω and ω' are cohomologous 3-cocycles of G, then $Z_{c_\omega}(G) = Z_{c_{\omega'}}(G)$ but $\Gamma(D^\omega(G))$ and $\Gamma(D^{\omega'}(G))$ are not necessarily isomorphic.

In view of Proposition 5.2, we will identify the group of simple currents $\mathsf{SC}(G,\omega)$ with the group $\Gamma_0(D^\omega(G))$ of central group-like elements of $D^\omega(G)$ under the map ζ. In particular, we simply write the simple current $V(z,t_z\chi)$ as $u(\chi,z)$.

The associativity constraint ϕ and the braiding c of $\mathrm{Rep}(D^\omega(G))$ define an Eilenberg–MacLane 3-cocycle $(\tilde{\phi},d)$ of $\mathsf{SC}(G,\omega)$ ([3,4]) given by

$$\begin{aligned} &\tilde{\phi}^{-1}(u(\chi_1,z_1),u(\chi_2,z_2),u(\chi_3,z_3)) \\ &:= \Big((u(\chi_1,z_1)\otimes u(\chi_2,z_2))\otimes u(\chi_3,z_3) \xrightarrow{\phi\cdot} u(\chi_1,z_1)\otimes u(\chi_2,z_2)\otimes u(\chi_3,z_3)\Big) \end{aligned} \tag{27}$$

and

$$\begin{aligned} &d(u(\chi_1,z_1)|u(\chi_2,z_2)) := c_{u(\chi_1,z_1),u(\chi_2,z_2)} \\ &= \Big(u(\chi_1,z_1)\otimes u(\chi_2,z_2) \xrightarrow{R\cdot} u(\chi_1,z_1)\otimes u(\chi_2,z_2) \xrightarrow{flip} u(\chi_2,z_2)\otimes u(\chi_1,z_1)\Big), \end{aligned} \tag{28}$$

where $R = \sum_{g,h\in G} e_g \otimes e_h g$ is the universal R-matrix of $D^\omega(G)$. By (24), one can compute directly that

$$\tilde{\phi}(u(\chi_1, z_1), u(\chi_2, z_2), u(\chi_3, z_3)) = \omega(z_1, z_2, z_3)\,, \tag{29}$$

$$d(u(\chi_1, z_1)|u(\chi_2, z_2)) = \chi_2(z_1)t_{z_2}(z_1)\,. \tag{30}$$

The double braiding on $u(\chi_1, z_1) \otimes u(\chi_2, z_2)$ is then the scalar

$$d(u(\chi_1, z_1)|u(\chi_2, z_2)) \cdot d(u(\chi_2, z_2)|u(\chi_1, z_1)),$$

which defines a symmetric bicharacter $(\cdot|\cdot)$ on $\mathsf{SC}(G, \omega)$. Using (24) to compute directly, we obtain

$$(u(\chi_1, z_1)|u(\chi_2, z_2)) = \chi_1(z_2)\chi_2(z_1)t_{z_2}(z_1)t_{z_1}(z_2)$$

for all $u(\chi_1, z_1), u(\chi_2, z_2) \in \mathsf{SC}(G, \omega)$. In general, $\mathsf{SC}(G, \omega)$ is degenerate relative to this symmetric bicharacter $(\cdot|\cdot)$. However, there could be nondegenerate subgroups of $\mathsf{SC}(G, \omega)$.

Remark 5.4. It follows from [11, Cor 7.11] or [12, Cor. 2.16] that a subgroup $A \subseteq \mathsf{SC}(G, \omega)$ is nondegenerate if, and only if, the full subcategory $\mathscr{A}$ of $\mathrm{Rep}(D^\omega(G))$ generated by A is a modular tensor category.

We now assume A is an ω-admissible subgroup of G. Let ν be a normalized cochain in $C^1(A, \hat{G})$ such that $\beta(a, b) = \nu(a)\nu(b)\nu(ab)^{-1}$ for all $a, b \in A$. Therefore, by Remark 3.7, the assignment $\tilde{p}_\nu : a \mapsto u(\nu(a)^{-1}, a)$ defines a group monomorphism from A to $\mathsf{SC}(G, \omega)$ which is also a section of $p : p^{-1}(A) \to A$. Hence A admits a bicharacter $(\cdot|\cdot)_\nu$ via the restriction of $(\cdot|\cdot)$ to $\tilde{p}_\nu(A)$. In particular,

$$(a|b)_\nu = (\tilde{p}_\nu(a)|\tilde{p}_\nu(b)) = \frac{t_b(a)t_a(b)}{\nu(b)(a)\nu(a)(b)}\,. \tag{31}$$

Obviously, $(\cdot|\cdot)_\nu$ is nondegenerate if, and only if, $\tilde{p}_\nu(A)$ is a nondegenerate subgroup of $\mathsf{SC}(G, \omega)$. On the other hand, ν also defines the quasi-Hopf algebra $D^\omega(G, A)$ and a surjective quasi-Hopf algebra homomorphism $\pi_\nu : D^\omega(G) \to D^\omega(G, A)$. In particular, $\mathrm{Rep}(D^\omega(G, A))$ is a tensor (full) subcategory of $\mathrm{Rep}(D^\omega(G))$, so it inherits the braiding c of $\mathrm{Rep}(D^\omega(G))$. We can now state the main theorem in this section.

Theorem 5.5. *Let A be an ω-admissible subgroup of G, ν a normalized cochain in $C^1(A, \hat{G})$, and $\tilde{p}_\nu : A \to \mathsf{SC}(G, \omega)$ the associated group monomorphism. Then*

$$c_{\tilde{p}_\nu(a), V} \circ c_{V, \tilde{p}_\nu(a)} = \mathrm{id}_{V \otimes \tilde{p}_\nu(a)}$$

for all $a \in A$ and irreducible $V \in \mathrm{Rep}(D^\omega(G, A))$. Moreover, $\mathrm{Rep}(D^\omega(G, A))$ is a modular tensor category if, and only if, the bicharacter $(\cdot|\cdot)_\nu$ on A is nondegenerate.

Proof. Since a braiding $c_{U,V} : U \otimes V \to V \otimes U$ is a natural isomorphism and the regular representation U of $D^\omega(G, A)$ has every irreducible $V \in \mathrm{Rep}(D^\omega(G, A))$ as a summand, it suffices to show that

$$c_{\tilde{p}_\nu(a),U} \circ c_{U,\tilde{p}_\nu(a)} = \mathrm{id}_{U\otimes\tilde{p}_\nu(a)}$$

for all $a \in A$. Let $\overline{c}_\omega = (G/A = \overline{G}, \overline{\theta}, \overline{\gamma})$ be an associated cleft object of $\mathbb{C}^G_\omega$ and $\pi_\nu : D^\omega(G) \to D^\omega(G, A)$ an epimorphism of quasi-Hopf algebras constructed in the proof of Theorem 3.6 using ν. In particular, $\pi_\nu(e_g x) = \chi_x(g)\, e_g \overline{x}$ for all $g, x \in G$ where $\overline{x}$ denotes the coset xA and the scalar $\chi_x(g)$ is given by (17).

Let $\mathbb{1}_{\tilde{p}_\nu(a)}$ denote a basis element of $\tilde{p}_\nu(a) = V(a, t_a \nu(a)^{-1})$. Then, by (24),

$$e_g x \cdot \mathbb{1}_{\tilde{p}_\nu(a)} = \delta_{g,a} \frac{t_a(x)}{\nu(a)(x)} \mathbb{1}_{\tilde{p}_\nu(a)} .$$

Note that we can take $U = D^\omega(G, A)$ as a $D^\omega(G)$-module via π_ν, and so

$$e_g x \cdot e_h \overline{y} = \pi_\nu(e_g x) e_h \overline{y} = \delta_{g^x,h}\, \chi_x(g)\, \overline{\theta}_g(\overline{x}, \overline{y}) e_g \overline{xy}$$

for all $g, h, x, y \in G$. Since the R-matrix of $D^\omega(G)$ is given by $R = \sum_{g,h\in G} e_g \otimes e_h g$, we have

$$\begin{aligned} c_{\tilde{p}_\nu(a),U} \circ c_{U,\tilde{p}_\nu(a)}(e_g\overline{y} \otimes \mathbb{1}_{\tilde{p}_\nu(a)}) &= R^{21} R \cdot (e_g\overline{y} \otimes \mathbb{1}_{\tilde{p}_\nu(a)}) \\ &= \frac{t_a(g)}{\nu(a)(g)} R^{21} \cdot (e_g\overline{y} \otimes \mathbb{1}_{\tilde{p}_\nu(a)}) \\ &= \frac{t_a(g)}{\nu(a)(g)} \chi_a(g)\overline{\theta}_g(\overline{a}, \overline{y})\, e_g\overline{y} \otimes \mathbb{1}_{\tilde{p}_\nu(a)} \\ &= e_g\overline{y} \otimes \mathbb{1}_{\tilde{p}_\nu(a)} \end{aligned}$$

for all $a \in A$. This proves the first assertion.

Let $\mathscr{A}$ be the full subcategory of $\mathscr{C} = \mathrm{Rep}(D^\omega(G))$ generated by $\tilde{p}_\nu(A)$. The first assertion of the theorem implies that $\mathrm{Rep}(D^\omega(G, A))$ is a full subcategory of the centralizer $C_\mathscr{C}(\mathscr{A})$ of $\mathscr{A}$ in $\mathscr{C}$. Since $\dim\mathscr{A} = |A|$ and $\mathrm{Rep}(D^\omega(G))$ is a modular tensor category, by [12, Thm. 3.2],

$$\dim C_\mathscr{C}(\mathscr{A}) = \dim D^\omega(G)/\dim\mathscr{A} = |G|^2/|A| = \dim D^\omega(G, A).$$

Therefore

$$C_\mathscr{C}(\mathscr{A}) = \mathrm{Rep}(D^\omega(G, A)) \text{ and } C_\mathscr{C}(\mathrm{Rep}(D^\omega(G, A))) = \mathscr{A} .$$

By Remark 5.4, $\mathscr{A}$ is a modular category if, and only if, $\tilde{p}_\nu(A)$ is nondegenerate subgroup of $\mathsf{SC}(G, \omega)$; this is equivalent to the assertion that the bicharacter $(\cdot|\cdot)_\nu$ on A is nondegenerate. It follows from [12, Thm. 3.2 and Cor. 3.5] that $\mathscr{A}$ is modular if, and only if, $C_\mathscr{C}(\mathscr{A})$ is modular. This proves the second assertion. □

The choice of cochain $\nu \in C^1(A, \hat{G})$ in the preceding theorem determines an embedding $\tilde{p}_\nu$ of A into $\mathsf{SC}(G, \omega)$. Therefore, the degeneracy of $\tilde{p}_\nu(A)$ in $\mathsf{SC}(G, \omega)$ depends on the choice of ν. However, the degeneracy of $\tilde{p}_\nu(A)$ can also be independent of the choice of ν in some situations. Important examples of this are contained in the next result.

Lemma 5.6. *If A is an ω-admissible subgroup of G such that $A \cong \mathbb{Z}_2$ or $A \leq [G, G]$. Then the bicharacter $(\cdot|\cdot)_\nu$ on A is independent of the choice of ν.*

Proof. Suppose $\nu' \in C^1(A, \hat{G})$ is another cochain satisfying the condition of Theorem 5.5. Then there is a group homomorphism $f : A \to \hat{G}$ such that $\nu'(a)(b) = f(a)(b)\nu(a)(b)$. Thus the associated bicharacter $(\cdot|\cdot)_{\nu'}$ is given by

$$(a|b)_{\nu'} = f(a)(b)^{-1}\nu(a)(b)^{-1}f(b)(a)^{-1}\nu(b)(a)^{-1}t_a(b)t_b(a)$$
$$= f(a)(b)^{-1}f(b)(a)^{-1}(a|b)_\nu. \quad (32)$$

If $A \subseteq G'$, then $f(a)(b) = f(b)(a) = 1$ for all $a, b \in A$, whence $(a|b)_\nu = (a|b)_{\nu'}$.

On the other hand, if A is a group of order 2 generated by z, then $f(z)(z)^2 = 1$, so that

$$(z, z)_{\nu'} = f(z)(z)^2(z|z)_\nu = (z|z)_\nu \,.$$

□

Acknowledgements Research of the first author was partially supported by NSA and NSF. The second author was supported by NSF DMS1303253.

References

[1] V. Drinfel'd, *Quasi-Hopf algebras*, Leningrad Math. J. **1** (1990), 1419–1457.
[2] R. Dijkgraaf, V. Pasquier and P. Roche, Quasi-Hopf algebras, group cohomology and orbifold models, in *Integrable Systems and Quantum groups*, World Scientific Publishing, NJ, 75–98.
[3] S. Eilenberg and S. MacLane, *Cohomology theory of Abelian groups and homotopy theory. I*, Proc. Nat. Acad. Sci. U. S. A. **36** (1950), 443–447.
[4] S. Eilenberg and S. MacLane, *Cohomology theory of Abelian groups and homotopy theory. II*, Proc. Nat. Acad. Sci. U. S. A. **36** (1950), 657–663.
[5] C. Goff and G. Mason, *Generalized twisted quantum doubles and the McKay correspondence*, J. Algebra **324** (2010), no. 11, 3007–3016.
[6] C. Kassel, *Quantum Groups*, Springer, New York, 1995.
[7] Y. Kashina, G. Mason and S. Montgomery, *Computing the Schur indicator for abelian extensions of Hopf algebras*, J. Algebra **251** (2002), no. 2, 888–913.
[8] A. Masuoka, *Hopf algebra extensions and cohomology*, New directions in Hopf algebras, Math. Sci. Res. Inst. Publ., Vol. **43**, Cambridge Univ. Press, Cambridge, 2002, pp. 167–209.
[9] G. Mason and S.-H. Ng, *Group cohomology and gauge equivalence of some twisted quantum doubles*, Trans. Amer. Math. Soc. **353** (2001), no . 9, 3465–3509.

[10] G. Mason and S.-H. Ng, *Central invariants and Frobenius-Schur indicators for semisimple quasi-Hopf algebras*, Adv. Math. **190** (2005), no. 1, 161–195.
[11] M. Müger, *From subfactors to categories and topology. II. The quantum double of tensor categories and subfactors*, J. Pure Appl. Algebra **180** (2003), no. 1–2, 159–219.
[12] M. Müger, *On the structure of modular categories*, Proc. London Math. Soc. (3) **87** (2003), no. 2, 291–308.

On the Structure of $\mathbb{N}$-Graded Vertex Operator Algebras

Geoffrey Mason* **and Gaywalee Yamskulna**†

Abstract We consider the algebraic structure of $\mathbb{N}$-graded vertex operator algebras with conformal grading $V = \oplus_{n\geq 0} V_n$ and $\dim V_0 \geq 1$. We prove several results along the lines that the vertex operators $Y(a, z)$ for a in a Levi factor of the Leibniz algebra V_1 generate an affine Kac–Moody subVOA. If V arises as a shift of a self-dual VOA of CFT-type, we show that V_0 has a "de Rham structure" with many of the properties of the de Rham cohomology of a complex connected manifold equipped with Poincaré duality.

Key words Vertex operator algebra • Lie algebra • Leibniz algebra.

Mathematics Subject Classification (2010): 17B65, 17B69.

1 Introduction

The purpose of this paper is the study of the algebraic structure of $\mathbb{N}$-*graded vertex operator algebras* (VOAs). A VOA $V = (V, Y, \mathbf{1}, \omega)$ is called $\mathbb{N}$-graded if it has no nonzero states of negative conformal weight, so that its conformal grading takes the form

$$V = \oplus_{n=0}^{\infty} V_n. \tag{1}$$

*Supported by the NSF

†Partially supported by a grant from the Simons Foundation (# 207862)

G. Mason (✉)
Department of Mathematics, University of California, Santa Cruz, CA 95064, USA
e-mail: gem@ucsc.edu

G. Yamskulna
Department of Mathematical Sciences, Illinois State University, Normal, IL 61790, USA
e-mail: gyamsku@ilstu.edu

G. Mason et al. (eds.), *Developments and Retrospectives in Lie Theory: Algebraic Methods*, Developments in Mathematics 38,
DOI 10.1007/978-3-319-09804-3_12

The VOAs in this class which have been most closely investigated hitherto are those of *CFT-type*, where one assumes that $V_0 = \mathbb{C}\mathbf{1}$ is spanned by the vacuum vector. (It is well known that a VOA of CFT-type is necessarily $\mathbb{N}$-graded.) Our main interest here is in the contrary case, when $\dim V_0 \geq 2$.

There are several available methods of constructing $\mathbb{N}$-graded vertex algebras. One that particularly motivates the present paper arises from the cohomology of the chiral de Rham complex of a complex manifold M, due to Malikov, Schechtman and Vaintrob [MS1,MS2,MSV]. In this construction V_0 (which is always a commutative algebra with respect to the -1th operation $ab := a(-1)b$) is identified with the de Rham cohomology $H^*(M)$. One can also consider algebraic structures defined on $V_0 \oplus V_1$ or closely related spaces, variously called 1-*truncated conformal algebras, vertex A-algebroids, and Lie A-algebroids* [GMS, Br, LY], and construct $\mathbb{N}$-graded vertex algebras from a 1-truncated conformal algebra much as one constructs affine VOAs from a simple Lie algebra. A third method involves *shifted* VOAs [DM3]. Here, beginning with a VOA $V = (V, Y, \mathbf{1}, \omega)$, one replaces ω by a second conformal vector $\omega_h := \omega + h(-2)\mathbf{1}$ $(h \in V_1)$ so that $V^h := (V, Y, \mathbf{1}, \omega_h)$ is a new VOA with the same Fock space, vacuum and set of fields as V. We call V^h a *shifted VOA*. For propitious choices of V and h (lattice theories were used in [DM3]) one can construct lots of shifted VOAs that are $\mathbb{N}$-graded. In particular, if V is rational, then V^h is necessarily also rational, and in this way one obtains $\mathbb{N}$-graded rational VOAs that are not of CFT-type.

Beyond these construction techniques, the literature devoted to the study of $\mathbb{N}$-graded VOAs per se is sparse. There are good reasons for this. For a VOA of CFT-type the weight 1 space V_1 carries the structure of a Lie algebra L with respect to the bracket $[ab] = a(0)b$ $(a, b \in V_1)$, and the modes of the corresponding vertex operators $Y(a, z)$ close on an affinization $\widehat{L}$ of L. For a general VOA, $\mathbb{N}$-graded or not, this no longer pertains. Rather, V_1 satisfies the weaker property of being a *left Leibniz algebra* (a sort of Lie algebra for which skew-symmetry fails), but one can still ask the question:

$$\text{what is the nature of the algebra spanned by the vertex operators } Y(a, z) \text{ for } a \in V_1? \tag{2}$$

Next we give an overview of the contents of this paper. Section 2 is concerned with question (2) for an *arbitrary* VOA. After reviewing general facts about Leibniz algebras and their relation to VOAs, we consider the annihilator $F \subseteq V_1$ of the *Leibniz kernel* of V_1. F is itself a Leibniz algebra, and we show (Theorem 1) that the vertex operators $Y(a, z)$ for a belonging to a fixed Levi subalgebra $S \subseteq F$ close on an affine algebra $U \subseteq V$. Moreover, all such Levi factors F are conjugate in Aut(V), so that U is an invariant of V. (Finite-dimensional Leibniz algebras have a Levi decomposition in the style of Lie algebras, and the semisimple part is a true Lie algebra.) This result generalizes the 'classical' case of VOAs of CFT-type discussed above, to which it reduces if $\dim V_0 = 1$, and provides a partial answer to (2). We do not know if, more generally, the same result holds if we replace S by a Levi factor of V_1.

From Sect. 3 on we consider simple ℕ-graded VOAs that are also *self-dual* in the sense that they admit a nonzero invariant bilinear form $(\, , \,) : V \times V \to \mathbb{C}$ (cf. [L]). By results in [DM2] this implies that V_0 carries the structure of a *local, commutative, symmetric algebra*, and in particular it has a unique minimal ideal $\mathbb{C}t$. This result is fundamental for everything that follows. It permits us to introduce a second bilinear form $\langle \, , \rangle : V_1 \times V_1 \to \mathbf{C}$ on V_1, defined in terms of $(\, , \,)$ and t, and we try to determine its radical. Section 3 covers background results, and in Sect. 4 we show (Proposition 2) exactly how $\mathrm{Rad}\langle \, , \, \rangle$ is related to the annihilator of the endomorphism $t(-1)$ acting on V_1. In all cases known to us we have

$$\mathrm{Rad}\langle \, , \, \rangle = \mathrm{Ann}_{V_1}(t(-1)), \tag{3}$$

and it is of interest to know if this is always true.

In Sects. 5 and 6 we consider shifted VOAs, more precisely we consider the set-up in which we have a *self-dual* VOA $(W, Y, \mathbf{1}, \omega')$ of CFT-type together with an element $h \in W_1$ such that the shifted theory $W^h = (W, Y, \mathbf{1}, \omega'_h)$ as previously defined is a self-dual, ℕ-graded VOA V. As we mentioned, triples (W, h, V) of this type are readily constructed, and they have very interesting properties. The main result of Sect. 5 is Theorem 2, which, roughly speaking, asserts that V_0 looks just like the de Rham cohomology of a complex manifold equipped with Poincaré duality. More precisely, we prove that the eigenvalues of $h(0)$ acting on V_0 are nonnegative integers; the maximal eigenvalue is ν, say, and the ν-eigenspace is 1-dimensional and spanned by t; and the restriction of the nonzero invariant bilinear form on V to V_0 induces a perfect pairing between the λ- and $(\nu - \lambda)$-eigenspaces. One may compare this result with the constructions of Malikov et al. in the chiral de Rham complex, where the same conclusions arise directly from the identification of the lowest weight space with $H^*(M)$ for a complex manifold M. There is, of course, no a priori complex manifold associated to the shifted triple (W, h, V), but one can ask whether, at least in some instances, the cohomology of the chiral de Rham complex arises from the shifted construction?

In Sect. 8 we present several examples that illustrate the theory described in the previous paragraph. In particular, we take for W the affine Kac–Moody theory $L_{\widehat{sl_2}}(k, 0)$ of positive integral level k and show that it has a canonical shift to a self-dual, ℕ-graded VOA $V = W^H$ ($2H$ is semisimple and part of a Chevalley basis for sl_2). It turns out that the algebra structure on V_0 is naturally identified with $H^*(\mathbb{CP}^k)$. We also look at shifts of lattice theories W_L, where the precise structure of V_0 depends on L.

Keeping the notation of the previous paragraph, in Sect. 6 we use the results of Sect. 5 to prove that the shifted VOA V indeed satisfies (3). Moreover, if the Lie algebra W_1 on the weight 1 space of the CFT-type VOA W is *reductive*, we prove that Rad$\langle\ ,\ \rangle$ is the *nilpotent radical* of the Leibniz algebra V_1, i.e., the smallest ideal in V_1 such that the quotient is a reductive Lie algebra. It was precisely for the purpose of proving such a result that the form $\langle\ ,\ \rangle$ was introduced. It is known [DM1] that W_1 is indeed reductive if W is *regular* (rational and C_2-cofinite), so for VOAs obtained as a shift of such a W, we get a precise description of the nilpotent radical, generalizing the corrresponding result of [DM3].

In Sect. 7 we study simple, self-dual $\mathbb{N}$-graded VOAs that are C_2-cofinite. After reviewing rationality and C_2-cofiniteness of vertex operator algebras, we prove (Theorem 4) that in this case Rad$\langle\ ,\ \rangle$ lie between the nilpotent radical of V_1 and the solvable radical of V_1. In particular, the restriction of $\langle\ ,\ \rangle$ to a Levi factor $S \subseteq V_1$ is nondegenerate; furthermore, the vertex operators $Y(a,z)$ $(a \in S)$ close on a tensor product of WZW models, i.e., simple affine algebras $L_{\hat{\mathfrak{g}}}(k,0)$ of positive integral level k. Thus we obtain a partial answer to (2) which extends results in [DM4], where the result was proved for CFT-type VOAs.

2 Leibniz Algebras and Vertex Operator Algebras

In this section, we assume that V is any simple vertex operator algebra

$$V = \oplus_{n \geq n_0} V_n,$$

with *no* restriction on the nature of the conformal grading.

A *left Leibniz algebra* is a $\mathbb{C}$-linear space L equipped with a bilinear product, or bracket, [ ] satisfying

$$[a[bc]] = [[ab]c] + [b[ac]], \quad (a, b, c \in V).$$

Thus $[a*]$ is a left derivation of the algebra L, and L is a Lie algebra if, and only if, the bracket is also skew-symmetric. We refer to [MY] for facts about Leibniz algebras that we use below.

Lemma 1. *V is a $\mathbb{Z}$-graded left Leibniz algebra with respect to the 0th operation $[ab] := a(0)b$. Indeed, there is a triangular decomposition*

$$V = \left\{\oplus_{n \leq 0} V_n\right\} \oplus V_1 \oplus \left\{\oplus_{n \geq 2} V_n\right\} \tag{4}$$

into left Leibniz subalgebras. Moreover, $\oplus_{n \leq 0} V_n$ is nilpotent.

Proof. Recall the commutator formula

$$[u(p), v(q)]w = \sum_{i=0}^{\infty} \binom{p}{i} (u(i)v)(p+q-i)w. \tag{5}$$

Upon taking $p = q = 0$, (5) specializes to

$$u(0)v(0)w - v(0)u(0)w = (u(0)v)(0)w,$$

which is the identity required to make V a left Leibniz algebra. The remaining assertions are consequences of

$$u(0)(V_n) \subseteq V_{n+k-1} \quad (u \in V_k).$$

□

Remark 1. A *right* Leibniz algebra L has a bracket with respect to which L acts as right derivations. Generally, a left Leibniz algebra is *not* a right Leibniz algebra, and in particular a vertex operator algebra is generally not a right Leibniz algebra.

It is known (e.g., [B, MY]) that a finite-dimensional left Leibniz algebra has a *Levi decomposition*. In particular, this applies to the middle summand V_1 in (4). Thus there is a decomposition

$$V_1 = S \oplus B, \tag{6}$$

where S is a semisimple *Lie* subalgebra and B is the solvable radical of V_1. As in the case of a Lie algebra, we call S a *Levi subalgebra*. Unlike Lie algebras, Levi factors are generally *not* conjugate to each other by exponential automorphisms, i.e., Malcev's theorem does *not* extend to Leibniz algebras [MY].

This circumstance leads to several interesting questions in VOA theory. In particular, what is the nature of the subalgebra of V generated by a Levi subalgebra $S \subseteq V_1$? Essentially, we want a description of the Lie algebra of operators generated by the modes $a(n)$ $(a \in S, n \in \mathbb{Z})$. In the case when V is of CFT-type (i.e., $V_0 = \mathbb{C}\mathbf{1}$ is spanned by the vacuum), it is a fundamental fact that these modes generate an affine algebra. Moreover, all Levi subfactors of V_1 are conjugate in $\mathrm{Aut}(V)$ (cf. [M]), so that the affine algebra is an invariant of V. It would be interesting to know if these facts continue to hold for arbitrary vertex operator algebras. We shall deal here with a special case.

To describe our result, introduce the *Leibniz kernel* defined by

$$N := \langle a(0)a \mid a \in V_1\rangle = \langle a(0)b + b(0)a \mid a, b \in V_1\rangle \quad \text{(linear span)}.$$

N is the smallest 2-sided ideal of V_1 such that V_1/N is a Lie algebra. The annihilator of the Leibniz kernel is

$$F := \mathrm{Ann}_{V_1}(N) = \{a \in V_1 \mid a(0)N = 0\}.$$

This is a 2-sided ideal of V_1, in particular it is a Leibniz subalgebra and itself contains Levi factors. We will prove

Theorem 1. *Let V be a simple vertex operator algebra, with N and F as above. Then the following hold:*

(a) $\mathrm{Aut}(V)$ *acts transitively on the Levi subalgebras of F.*
(b) Let $S \subseteq F$ be a Levi subalgebra of F. Then $u(1)v \in \mathbb{C}\mathbf{1}$ $(u, v \in S)$, and the vertex operators $Y(u, z)$ $(u \in S)$ close on an affine algebra, i.e.,

$$[u(m), v(n)] = (u(0)v)(m+n) + m\alpha(u, v)\delta_{m+n,0}Id_V,$$

where $u(1)v = \alpha(u, v)\mathbf{1}$.

We prove the theorem in a sequence of lemmas. Fix a Levi subalgebra $S \subseteq F$, and set

$$W := \oplus_{n \le 0} V_n.$$

Lemma 2. *W is a* trivial *left S-module, i.e., $u(0)w = 0$ $(u \in S, w \in W)$.*

Proof. We have to show that each homogeneous space V_n $(n \le 0)$, is a trivial left S-module. Because $L(-1) : V_n \to V_{n+1}$ is an *injective* V_1-equivariant map for $n \ne 0$, it suffices to show that V_0 is a trivial S-module.

Consider $L(-1) : V_0 \to V_1$, and set $N' := L(-1)V_0$. Because $[L(-1), u(0)] = 0$, $L(-1)$ is V_1-equivariant. By skew-symmetry we have $u(0)u = 1/2L(-1)u(1)u$. This shows that $N \subseteq N'$. Now because $(L(-1)v)(0) = 0$ $(v \in V)$ then in particular $N'(0)V_1 = 0$. Therefore, $S(0)N' = \langle u(0)v + v(0)u \mid u \in S, v \in N'\rangle \subseteq N$. But S is semisimple and it annihilates N. It follows that S annihilates N'.

Because V is simple, its center $Z(V) = \mathrm{ker}L(-1)$ coincides with $\mathbb{C}\mathbf{1}$. By Weyl's theorem of complete reducibility, there is an S-invariant decomposition

$$V_0 = \mathbb{C}\mathbf{1} \oplus J,$$

and restriction of $L(-1)$ is an S-isomorphism $J \xrightarrow{\cong} N'$. Because S annihilates N', it must annihilate J. It therefore also annihilates V_0, as we see from the previous display. This completes the proof of the lemma. □

Lemma 3. *We have*

$$u(k)w = 0 \quad (u \in S, w \in W, k \ge 0). \tag{7}$$

Proof. Because S is semisimple, we may, and shall, assume without loss that u is a commutator $u = a(0)b$ $(a, b \in S)$. Then

$$(a(0)b)(k)w = a(0)b(k)w - b(k)a(0)w = 0.$$

The last equality holds thanks to Lemma 2, and because $b(k)w \in W$ for $k \geq 0$. The lemma is proved. □

Lemma 4. *We have*

$$[u(m), w(n)] = 0 \ \ (u \in S, w \in W; m, n \in \mathbb{Z}). \tag{8}$$

Proof. First notice that by Lemma 2,

$$[u(0), w(n)] = (u(0)w)(n) = 0. \tag{9}$$

Once again, it is suffices to assume that $u = a(0)b$ $(a, b \in F)$. In this case we obtain, using several applications of (9), that

$$\begin{aligned}
[u(m), w(n)] &= [(a(0)b)(m), w(n)] \\
&= [[a(0), b(m)], w(n)] \\
&= [a(0), [b(m), w(n)]] - [b(m), [a(0), w(n)]] \\
&= [a(0), (b(0)w)(m+n) + m(b(1)w)(m+n-1)] \\
&= 0.
\end{aligned}$$

This completes the proof of the lemma. □

Consider the Lie algebra L of operators on V defined by

$$L := \langle u(m), w(n) \mid u \in S, w \in W; m, n \in \mathbb{Z} \rangle.$$

If $w, x \in W$, then

$$[w(m), x(n)] = \sum_{i \geq 0} \binom{m}{i} (w(i)x)(m+n-i),$$

and $w(i)x$ has weight *less* than that of w and x whenever $w, x \in W$ are homogeneous and $i \geq 0$. This shows that the operators $w(m)$ $(w \in W, m \in \mathbb{Z})$ span a nilpotent ideal of L, call it P. Let L_0 be the Lie subalgebra generated by $u(m)$ $(u \in S_0, m \in \mathbb{Z})$. By Lemma 4, L_0 is also an ideal of L; indeed

$$L = P + L_0, \ [P, L_0] = 0.$$

Next, for $u, v \in S$ we have

$$[u(m), v(n)] = (u(0)v)(m+n) + \sum_{i \geq 1} \binom{m}{i} (u(i)v)(m+n-i). \tag{10}$$

So if $w \in S$, then by Lemma 4 once more,

$$[w(0), [u(m), v(n)]] = [w(0), (u(0)v)(m+n)] = (w(0)(u(0)v))(m+n).$$

This shows that L_0 coincides with its derived subalgebra. Furthermore, the short exact sequence

$$0 \to P \cap L_0 \to L_0 \to L_0/(P \cap L_0) \to 0$$

shows that L_0 is a perfect central extension of the loop algebra $\widehat{L}(S_0) \cong L_0/(P \cap L_0)$. Because $H^2(\widehat{L}(G))$ is 1-dimensional for a finite-dimensional simple Lie algebra G, we can conclude that $\dim(P \cap L_0)$ is *finite*.

Taking $m = 1$ in (10), it follows that

$$(u(1)v)(n) \in P \cap L_0 \ (n \in \mathbb{Z}). \tag{11}$$

Now if $u(1)v \notin Z(V)$, then all of the modes $(u(1)v)(n), n < 0$, are nonzero, and indeed linearly independent. This follows from the creation formula

$$\sum_{n \leq -1} (u(1)v)(n)\mathbf{1}z^{-n-1} = e^{zL(-1)}u(1)v.$$

Because $P \cap L_0$ is finite-dimensional and contains all of these modes, this is not possible. We deduce that in fact $u(1)v \in Z(V) = \mathbb{C}$, say $u(1)v = \alpha(u,v)\mathbf{1}, \alpha(u,v) \in \mathbb{C}$.

Taking $m = 2, 3, \ldots$ in (10), we argue in the same way that $u(i)v \in Z(V)$ for $i \geq 2$. Since $Z(V) \subseteq V_0$, this means that $u(i)v = 0$ for $i \geq 2$. Therefore, (10) now reads

$$[u(m), v(n)] = (u(0)v)(m+n) + m\alpha(u,v)\delta_{m+n,0}Id, \tag{12}$$

where $u(1)v = \alpha(u,v)\mathbf{1}$. This completes the proof of part (b) of the Theorem.

It remains to show that $\mathrm{Aut}(V)$ acts transitively on the set of Levi subalgebras of F.

Lemma 5. *[FF] consists of* primary states, *i.e.*, $L(k)[FF] = 0$ $(k \geq 1)$.

Proof. It suffices to show that $L(k)a(0)b = 0$ for $a, b \in F$ and $k \geq 1$. Since $L(k)b \in W$ then $a(0)L(k)b = 0$ by Lemma 2. Using induction on k, we then have

$$\begin{aligned} L(k)a(0)b &= [L(k), a(0)]b \\ &= (L(-1)a)(k+1)b + (k+1)(L(0)a)(k)b + (L(k)a)(0)b \end{aligned}$$

$$= (L(k)a)(0)b$$
$$= \sum_{i\geq 0}(-1)^{i+1}/i!L(-1)^i b(i)L(k)a = 0,$$

where we used skew-symmetry for the fourth equality, and Lemma 3 for the last equality. The lemma is proved. □

Finally, by [MY], Theorem 3.1, if S_1, S_2 are a pair of Levi subalgebras of F, then we can find $x \in [FF]$ such that $e^{x(0)}(S_1) = S_2$. Because x is a primary state, it is well known that $e^{x(0)}$ is an automorphism of V. This completes the proof of Theorem 1. □

3 ℕ-Graded Vertex Operator Algebras

In this section, we assume that V is a *simple, self-dual, ℕ-graded* vertex operator algebra. We are mainly interested in the case that $\dim V_0 \geq 2$. There is a lot of structure available to us in this situation, and in this section we review some of the details, and at the same time introduce some salient notation.

The self-duality of V means that there is a *nonzero* bilinear form

$$(\ ,\ ) : V \times V \to \mathbb{C}$$

that is *invariant* in the sense that

$$(Y(u,z)v, w) = \left(v, Y(e^{zL(1)}(-z^{-2})^{L(0)}u, z^{-1})w\right) \quad (u, v, w \in V). \tag{13}$$

$(\ ,\ )$ is necessarily *symmetric* [FHL], and because V is simple, it is then *nondegenerate*. The simplicity of V also implies (Schur's Lemma) that $(\ ,\ )$ is *unique* up to scalars. By results of Li [L], there is an isomorphism between the space of invariant bilinear forms and $V_0/L(1)V_1$. Therefore, $L(1)V_1$ has codimension 1 in V_0. For now, we fix a nonzero form $(\ ,\ )$, but do not choose any particular normalization.

If $u \in V_k$ is *quasiprimary* (i.e., $L(1)u = 0$), then (13) is equivalent to

$$(u(n)v, w) = (-1)^k(v, u(2k-n-2)w) \quad (n \in \mathbb{Z}). \tag{14}$$

In particular, taking u to be the conformal vector $\omega \in V_2$, which is always quasiprimary, and $n = 1$ or 2 yields

$$(L(0)v, w) = (v, L(0)w), \tag{15}$$
$$(L(1)v, w) = (v, L(-1)w). \tag{16}$$

We write $P \perp Q$ for the direct sum of subspaces $P, Q \subseteq V$ that are orthogonal with respect to $(\, , \,)$. Thus $(V_n, V_m) = 0$ for $n \neq m$ by (15), so that

$$V = \perp_{n \geq 0} V_n.$$

In particular, the restriction of $(\, , \,)$ to each V_n is nondegenerate. We adopt the following notational convention for $U \subseteq V_n$:

$$U^{\perp} := \{a \in V_n \mid (a, U) = 0\}.$$

The *center* of V is defined to be $Z(V) := \ker L(-1)$. Because V is simple, we have $Z(V) = \mathbb{C}\mathbf{1}$ (cf. [LL, DM2]). Then from (16) we find that

$$(L(1)V_1)^{\perp} = \mathbb{C}\mathbf{1}. \tag{17}$$

V_0 carries the structure of a *commutative associative algebra* with respect to the operation $a(-1)b$ $(a, b \in V_0)$. Since all elements in V_0 are quasiprimary, we can apply (14) with $u, v, w \in V_0$ to obtain

$$(u(-1)v, w) = (v, u(-1)w). \tag{18}$$

Thus $(\, , \,)$ is a nondegenerate, symmetric, invariant bilinear form on V_0, whence V_0 is a commutative *symmetric algebra*, or *Frobenius algebra*.

What is particularly important for us is that because V is simple, V_0 is a *local algebra*, i.e., the Jacobson radical $J := J(V_0)$ is the unique maximal ideal of V_0, and every element of $V_0 \setminus J$ is a unit. This follows from results of Dong–Mason ([DM2], Theorem 2 and Remark 3).

For a symmetric algebra, the map $I \to I^{\perp}$ is an inclusion-reversing duality on the set of ideals. In particular, because V_0 is a local algebra, it has a *unique minimal* (nonzero) ideal, call it T, and T is 1-dimensional. Indeed,

$$T = J^{\perp} = \mathrm{Ann}_{V_0}(J) = \mathbb{C}t, \tag{19}$$

for some fixed, but arbitrary, nonzero element $t \in T$. We have

$$T \oplus L(1)V_1 = V_0.$$

This is a consequence of the nondegeneracy of $(\, , \,)$ on V_0, which entails that $L(1)V_1$ contains no nonzero ideals of V_0. In particular, (17) implies that

$$(t, \mathbf{1}) \neq 0. \tag{20}$$

We will change some of the notation from the previous section by setting $N := L(-1)V_0$ (it was denoted N' before). In the proof of Lemma 2 we showed that N contains the Leibniz kernel of V_1. In particular, V_1/N is a Lie algebra. We write

$$N_0/N = \mathrm{Nil}(V_1/N),\ N_1/N = \mathrm{Nil}p(V_1/N), B/N = \mathrm{solv}(V_1/N), \qquad (21)$$

the *nil radical*, *nilpotent radical*, and *solvable radical* respectively of V_1/N. N_0/N is the largest nilpotent ideal in V_1/N, N_1/N is the intersection of the annihilators of simple V_1/N-modules, and B/N the largest solvable ideal in V_1/N. It is well known that $N_1 \subseteq N_0 \subseteq B$. Moreover, $N_1/N = [V_1/N, V_1/N] \cap B/N$, V_1/N_1 is a *reductive* Lie algebra, and N_1 is the smallest ideal in V_1 with this property. Note that N_0 and B are also the largest nilpotent, and solvable ideals respectively in the left Leibniz algebra V_1.

Each of the homogeneous spaces V_n is a left V_1-module with respect to the 0th bracket. Because $u(0) = 0$ for $u \in N$, it follows that V_n is also a left module over the Lie algebra V_1/N. Since $V_0 = \mathbb{C}\mathbf{1} \oplus J$, $L(-1)$ induces an *isomorphism* of V_1-modules

$$L(-1) : J \overset{\cong}{\to} N. \qquad (22)$$

Remark 2. Most of the structure we have been discussing concerns the 1-truncated conformal algebra $V_0 \oplus V_1$ [Br, GMS, LY], and many of our results can be couched in this language.

4 The Bilinear Form ⟨ , ⟩

We keep the notation of the previous section; in particular $t \in V_0$ spans the unique minimal ideal of V_0. We introduce the bilinear form $\langle\ ,\ \rangle : V_1 \otimes V_1 \to \mathbb{C}$, defined as follows:

$$\langle u, v\rangle := (u(1)v, t), \quad (u, v \in V_1). \qquad (23)$$

We are interested in the *radical* of $\langle\ ,\ \rangle$, defined as

$$\mathrm{rad}\langle\ ,\ \rangle := \{u \in V_1 \mid \langle u, V_1\rangle = 0\}.$$

We will see that $\langle\ ,\ \rangle$ is a symmetric, invariant bilinear form on the Leibniz algebra V_1. The main result of this section (Proposition 2) determines the radical in terms of certain other subspaces that we introduce in due course. In order to study $\langle\ ,\ \rangle$ and its radical, we need some preliminary results.

Lemma 6. *We have $u(0)J \subseteq J$ and $u(0)T \subseteq T$ for $u \in V_1$. Moreover, the left annihilator*

$$M := \{u \in V_1 \mid u(0)T = 0\}$$

is a 2-sided ideal of V_1 of codimension 1, and $M = (L(-1)T)^{\perp}$.

Proof. Any derivation of a finite-dimensional commutative algebra B leaves invariant both the Jacobson radical $J(B)$ and its annihilator. In the case that the derivation is $u(0)$, $u \in V_1$, acting on V_0, this says that the left action of $u(0)$ leaves both J and T invariant (using (19) for the second assertion). This proves the first two statements of the lemma.

For $u \in V_1$ we have

$$(L(-1)t, u) = (t(-2)\mathbf{1}, u) = (\mathbf{1}, t(0)u) = -(\mathbf{1}, u(0)t). \tag{24}$$

Now because T is the unique minimal ideal in V_0 then $T \subseteq J$ and hence $\dim L(-1)T = 1$ by (22). Then (24) and (20) show that $(L(-1)T)^{\perp} = M$ has codimension exactly 1 in V_1.

Finally, using the commutator formula $[u(0), v(0)] = (u(0)v)(0)$ applied with one of $u, v \in M$ and the other in V_1, we see that $(u(0)v)(0)T = 0$ in either case. Thus $u(0)v \in M$, whence M is a 2-sided ideal in V_1. This completes the proof of the lemma. □

Lemma 7. *We have*

$$t(-2)J = 0.$$

Proof. Let $a \in J, u \in V_1$. Then

$$(t(-2)a, u) = (a, t(0)u) = -(a, u(0)t) = 0.$$

The last equality follows from $u(0)t \in T$ (Lemma 6) and $T = J^{\perp}$ (19). We deduce that $t(-2)J \subseteq V_1^{\perp} = 0$, and the lemma follows. □

Proposition 1. *$\langle\ ,\ \rangle$ is a symmetric bilinear form that is* invariant *in the sense that*

$$\langle v(0)u, w\rangle = \langle v, u(0)w\rangle \quad (u, v, w \in V_1).$$

Moreover $N \subseteq \mathrm{rad}\langle\ ,\ \rangle$.

Proof. By skew-symmetry we have $u(1)v = v(1)u$ for $u, v \in V_1$, so the symmetry of $\langle\ ,\ \rangle$ follows immediately from the definition (23). If $u \in N$, then $u = L(-1)a$ for some $a \in J$ by (22), and we have

$$\begin{aligned}\langle u, v\rangle &= ((L(-1)a)(1)v, t) = -(a(0)v, t)\\ &= -(v, a(-2)t) = (v, t(-2)a - L(-1)t(-1)a) = 0.\end{aligned}$$

Here, we used $t(-2)a = 0$ (Lemma 7) and $t(-1)a \in t(-1)J = 0$ to obtain the last equality. This proves the assertion that $N \subseteq \mathrm{rad}\langle\ ,\ \rangle$.

As for the invariance, we have

$$\begin{aligned}\langle u(0)v, w\rangle &= ((u(0)v)(1)w, t) = (u(0)v(1)w - v(1)u(0)w, t).\\ &= (u(0)v(1)w, t) - \langle v, u(0)w\rangle.\end{aligned}$$

Now $V_0 = \mathbb{C}\mathbf{1} \oplus J$, $u(0)\mathbf{1} = 0$, and $u(0)J \subseteq J = T^{\perp}$. Therefore, $(u(0)v(1)w, t) = 0$, whence we obtain $\langle u(0)v, w\rangle = -\langle v, u(0)w\rangle$ from the previous display. Now because $N \subseteq \mathrm{rad}\langle\ ,\ \rangle$ we see that

$$\langle v(0)u, w\rangle = -\langle u(0)v - L(-1)u(1)v, w\rangle = \langle v, u(0)w\rangle,$$

as required. This completes the proof of the proposition. □

Lemma 8. *We have*

$$\langle u, v\rangle = -(v, u(-1)t), \quad u, v \in V_1. \tag{25}$$

In particular,

$$\mathrm{rad}\langle\ ,\ \rangle = \{u \in V_1 \mid u(-1)t = 0\}.$$

Proof. The first statement implies the second, so it suffices to establish (25). To this end, we apply (13) with $u, v \in V_1$, $w = t$ to find that

$$\langle u, v\rangle = (u(1)v, t) = -(v, u(-1)t) - (v, (L(1)u)(-2)t).$$

On the other hand, $L(1)u \in V_0 = \mathbb{C}\mathbf{1} \oplus J$, so that $(L(1)u)(-2)t = a(-2)t = -t(-2)a + L(-1)t(-1)a$ for some $a \in J$. Since $t(-1)a = t(-2)a = 0$ (the latter equality thanks to Lemma 7), the final term of the previous display vanishes, and what remains is (25). The lemma is proved. □

We introduce

$$\begin{aligned}P &:= \{u \in V_1 \mid \langle u, M\rangle = 0\},\\ \mathrm{Ann}_{V_1}(t(-1)) &:= \{u \in V_1 \mid t(-1)u = 0\}.\end{aligned}$$

Lemma 9. *We have*

$$P = \{u \in V_1 \mid t(-1)u \in L(-1)T\},$$

and this is a 2-sided ideal of V_1.

Proof. Let $m \in M, u \in V_1$. By (25) we have

$$\langle u, m\rangle = -(m, u(-1)t).$$

But by Lemma 6 we have $M^\perp = L(-1)T$. Hence, the last display implies that $P = \{u \in V_1 \mid u(-1)t \in L(-1)T\}$. Furthermore, we have $u(-1)t = t(-1)u - L(-1)t(0)u = t(-1)u + L(-1)u(0)t \in t(-1)u + L(-1)T$ by Lemma 6 once more. Thus $u(-1)t \in L(-1)T$ if, and only if, $t(-1)u \in L(-1)T$. The first assertion of the lemma follows.

Because $N \subseteq \text{rad}\langle\ ,\ \rangle$ thanks to Proposition 1, then certainly $N \subseteq P$. So in order to show that P is a 2-sided ideal in V_1, it suffices to show that it is a right ideal. To see this, let $a \in P, n \in M, u \in V_1$. By Lemma 6 and Proposition 1 we find that

$$\langle a(0)u, n\rangle = \langle a, u(0)n\rangle \in \langle a, M\rangle = 0.$$

This completes the proof of the lemma. □

Lemma 10. *We have* $M \cap \text{Ann}_{V_1}(t(-1)) = M \cap \text{rad}\langle\ ,\ \rangle$.

Proof. If $u \in M$, then $t(-1)u = u(-1)t - L(-1)u(0)t = u(-1)t$. Hence for $u \in M$, we have $u \in \text{Ann}_{V_1}(t(-1)) \Leftrightarrow u(-1)t = 0 \Leftrightarrow u \in \text{rad}\langle\ ,\ \rangle$, where we used Lemma 8 for the last equivalence. The lemma follows. □

Lemma 11. *At least one of the containments* $\text{rad}\langle\ ,\ \rangle \subseteq M$, $\text{Ann}_{V_1}(t(-1)) \subseteq M$ *holds.*

Proof. Suppose that we can find $v \in \text{rad}\langle\ ,\ \rangle \setminus M$. Then $v(-1)t = 0$ by Lemma 8, and $v(0)t = \lambda t$ for a scalar $\lambda \neq 0$. Rescaling v, we may, and shall, take $\lambda = 1$. Then

$$\begin{aligned} 0 = v(-1)t &= t(-1)v - L(-1)t(0)v \\ &= t(-1)v + L(-1)v(0)t = t(-1)v + L(-1)t. \end{aligned}$$

Then for $u \in \text{Ann}_{V_1}(t(-1))$ we have

$$(L(-1)t, u) = -(t(-1)v, u) = -(v, t(-1)u) = 0,$$

which shows that $\text{Ann}_{V_1}(t(-1)) \subseteq (L(-1)t)^\perp = M$ (using Lemma 6). This completes the proof of the Lemma. □

The next result almost pins down the radical of $\langle\ ,\ \rangle$.

Proposition 2. *Exactly one of the following holds:*

$$(i)\ \mathrm{Ann}_{V_1}(t(-1)) = \mathrm{rad}\langle\ ,\ \rangle \subset P;$$

$$(ii)\ \mathrm{Ann}_{V_1}(t(-1)) \subset \mathrm{rad}\langle\ ,\ \rangle = P;$$

$$(iii)\ \mathrm{rad}\langle\ ,\ \rangle \subset \mathrm{Ann}_{V_1}(t(-1)) = P.$$

In each case, the containment $\subset$ *is one in which the smaller subspace has* codimension one *in the larger subspace.*

Proof. First note from Lemma 9 that $\mathrm{Ann}_{V_1}(t(-1)) \subseteq P$; indeed, since $\dim L(-1)T = 1$ then the codimension is at most 1. Also, it is clear from the definition of P that $\mathrm{rad}\langle\ ,\ \rangle \subseteq P$.

Suppose first that the containment $\mathrm{Ann}_{V_1}(t(-1)) \subset P$ is *proper*. Then we can choose $v \in P\backslash \mathrm{Ann}_{V_1}t(-1)$ such that $t(-1)v = L(-1)t \neq 0$. If $u \in \mathrm{Ann}_{V_1}(t(-1))$, we then obtain

$$(L(-1)t, u) = (t(-1)v, u) = (v, t(-1)u) = 0,$$

whence $u \in (L(-1)t)^{\perp} = M$ by Lemma 6. This shows that $\mathrm{Ann}_{V_1}(t(-1)) \subseteq M$. By Lemma 10 it follows that $\mathrm{Ann}_{V_1}(t(-1)) = M \cap \mathrm{rad}\langle\ ,\ \rangle$. Now if also $\mathrm{rad}\langle\ ,\rangle \subseteq M$, then Case 1 of the theorem holds. On the other hand, if $\mathrm{rad}\langle\ ,\rangle \nsubseteq M$, then we have $\mathrm{Ann}_{V_1}(t(-1)) \subset \mathrm{rad}\langle\ ,\rangle \subseteq P$ and the containment is proper; since $\mathrm{Ann}_{V_1}(t(-1))$ has codimension at most 1 in P then we are in Case 2 of the theorem.

It remains to consider the case that $\mathrm{Ann}_{V_1}(t(-1)) = P \supseteq \mathrm{rad}\langle\ ,\rangle$. Suppose the latter containment is proper. Because M has codimension 1 in V_1, it follows from Lemma 10 that $\mathrm{rad}\langle\ ,\ \rangle$ has codimension exactly 1 in $\mathrm{Ann}_{V_1}(t(-1))$, whence Case 3 of the theorem holds. The only remaining possibility is that $\mathrm{Ann}_{V_1}(t(-1)) = P = \mathrm{rad}\langle\ ,\rangle$, and we have to show that this cannot occur. By Lemma 11 we must have $\mathrm{rad}\langle\ ,\ \rangle \subseteq M$, so that $M/\mathrm{rad}\langle\ ,\ \rangle$ is a subspace of codimension 1 in the nondegenerate space $V_1/\mathrm{rad}\langle\ ,\ \rangle$ (with respect to $\langle\ ,\ \rangle$). But then the space orthogonal to $M/\mathrm{rad}\langle\ ,\ \rangle$, that is $P/\mathrm{rad}\langle\ ,\ \rangle$, is 1-dimensional. This contradiction completes the proof of the Theorem. □

Remark 3. In all cases that we know of, it is (i) of Proposition 2 that holds. This circumstance leads us to raise the question, whether this is always the case? We shall later see several rather general situations where this is so. At the same time, we will see how $\mathrm{rad}\langle\ ,\ \rangle$ is related to the Leibniz algebra structure of V_1.

5 The de Rham Structure of Shifted Vertex Operator Algebras

In the next few sections we consider ℕ-graded vertex operator algebras that are *shifts* of vertex operator algebras of CFT-type [DM3].

Let us first recall the idea of a *shifted* vertex operator algebra [DM3]. Suppose that $W = (W, Y, \mathbf{1}, \omega')$ is an $\mathbb{N}$-graded vertex operator algebra of central charge c' and $Y(\omega', z) =: \sum_n L'(n)z^{-n-2}$. It is easy to see that for any $h \in W_1$, the state $\omega'_h := \omega' + L'(-1)h$ is also a Virasoro vector, i.e., the modes of ω'_h satisfy the relations of a Virasoro algebra of some central charge c'_h (generally different from c'). (The proof of Theorem 3.1 in [DM3] works in the slightly more general context that we are using here.) Now consider the quadruple

$$W^h := (W, Y, \mathbf{1}, \omega'_h), \tag{26}$$

which is generally *not* a vertex operator algebra. If it *is*, we call it a *shifted* vertex operator algebra.

We emphasize that in this situation, W and W^h share the *same* underlying Fock space, the *same* set of vertex operators, and the *same* vacuum vector. Only the Virasoro vectors differ, although this has a dramatic effect because it means that W and W^h have quite different conformal gradings, so that the two vertex operator algebras seem quite different.

Now let $V = (V, Y, \mathbf{1}, \omega)$ be a simple, self-dual $\mathbb{N}$-graded vertex operator algebra as in the previous two sections. The assumption of this section is that there is a self-dual VOA W of CFT-type such that $W^h = V$. That is, V arises as a shift of a vertex operator algebra of CFT-type as described above. Thus $h \in W_1$ and

$$(W, Y, \mathbf{1}, \omega'_h) = (V, Y, \mathbf{1}, \omega).$$

(Note that by definition, W has CFT-type if $W_0 = \mathbb{C}\mathbf{1}$. In this case, W is necessarily $\mathbb{N}$-graded by [DM3], Lemma 5.2.) Although the two vertex operator algebras share the same Fock space, it is convenient to distinguish between them, and we shall do so in what follows. We sometimes refer to (W, h, V) as a *shifted triple*. Examples are constructed in [DM3], and it is evident from those calculations that there are large numbers of shifted triples.

There are a number of consequences of the circumstance that (W, h, V) is a shifted triple. We next discuss some that we will need. Because $\omega = \omega'_h = \omega' + L'(-1)h$ then

$$L(n) = (\omega' + L(-1)h)(n+1) = L'(n) - (n+1)h(n), \tag{27}$$

in particular $L(0) = L'(0) - h(0)$. Because $h \in W_1$, we also have $[L'(0), h(0)] = 0$. Then because $L'(0)$ is semisimple with integral eigenvalues, the same is true of $h(0)$. Set

$$W_{m,n} := \{w \in W \mid L'(0)w = mw, h(0)w = nw\}.$$

Hence,

$$V_n = \oplus_{m \geq 0} W_{m,m-n},$$

and in particular

$$V_0 = \mathbb{C}\mathbf{1} \oplus_{m \geq 1} W_{m,m}, \tag{28}$$

$$V_1 = \oplus_{m \geq 1} W_{m,m-1}. \tag{29}$$

(28) follows because W is of CFT-type, so that $W_{0,0} = \mathbb{C}\mathbf{1}$ and $W_{m,n} = 0$ for $n < 0$.

We have $L(0)h = L'(0)h - h(0)h$. Because W is of CFT-type then W_1 is a Lie algebra with respect to the 0th bracket, and in particular $h(0)h = 0$. Therefore, $L(0)h = h$, that is $h \in V_1$. Thus $h(0)$ induces a derivation in its action on the commutative algebra V_0. The decomposition (28) is one of $h(0)$-eigenspaces, and it confers on V_0 a structure that looks very much like the de Rham cohomology of a (connected) complex manifold equipped with its Poincaré duality. This is what we mean by the *de Rham structure of* V_0. Specifically, we have

Theorem 2. *Set* $A = V_0$ *and* $A^\lambda := W_{\lambda,\lambda}$, *the* λ-*eigenspace for the action of* $h(0)$ *on* A. *Then the following hold;*

(*i*) $A = \oplus_\lambda A^\lambda$, *and if* $A^\lambda \neq 0$, *then* λ *is a* nonnegative *integer.*

(*ii*) $A^0 = \mathbb{C}\mathbf{1}$.

(*iii*) *Let* $h(1)h = (\nu/2)\mathbf{1}$. *Then* $A^\nu = T = \mathbb{C}t$.

(*iv*) $A^\lambda(-1)A^\mu \subseteq A^{\lambda+\mu}$.

(*v*) $A^\lambda \perp A^\mu = 0$ *if* $\lambda + \mu \neq \nu$.

(*vi*) *If* $\lambda + \mu = \nu$, *the bilinear form* (,) *induces a perfect pairing*

$$A^\lambda \times A^\mu \to \mathbb{C}.$$

(Here, (,) *is the invariant bilinear form on* V, *and* T *the unique minimal ideal of* V_0, *as in Sects. 4 and 5.)*

Proof. (i) and (ii) are just restatements of the decomposition (28).

Next we prove (iv). Indeed, because $h(0)$ is a derivation of the algebra A, if $a \in A^\lambda$, $b \in A^\mu$, then $h(0)a(-1)b = [h(0), a(-1)]b + a(-1)h(0)b = (h(0)a)(-1)b + \mu a(-1)b = (\lambda + \mu)a(-1)b$. Part (iv) follows.

Next we note that because W is of CFT-type then certainly $h(1)h = (\nu/2)\mathbf{1}$ for some scalar ν. Now let a, b be as in the previous paragraph. Then

$$\begin{aligned}
\lambda(a, b) &= (h(0)a, b) = \mathrm{Res}_z(Y(h, z)a, b) \\
&= \mathrm{Res}_z(a, Y(e^{zL(1)}(-z^{-2})^{L(0)}h, z^{-1})b) \ \text{(by (13))} \\
&= -\mathrm{Res}_z z^{-2}(a, Y(h + zL(1)h, z^{-1})b) \\
&= -(a, h(0)b) - (a, (L(1)h)(-1)b) \\
&= -\mu(a, b) - (a, L'(1)h - 2h(1)h)(-1)b) \\
&= -\mu(a, b) + 2(a, (h(1)h)(-1)b) \\
&= -\mu(a, b) + \nu(a, b).
\end{aligned}$$

Here we used the assumption that W is self-dual and of CFT-type to conclude that $L'(1)W_1 = 0$, and in particular $L'(1)h = 0$. Thus we have obtained

$$(\lambda + \mu - \nu)(a, b) = 0. \tag{30}$$

If $\lambda + \mu \neq \nu$, then we must have $(a, b) = 0$ for all choices of a, b, and this is exactly what (v) says. Because the bilinear form $(\ ,\ )$ is nondegenerate, it follows that it must induce a perfect pairing between A^λ and A^μ whenever $\lambda + \mu = \nu$. So (vi) holds.

Finally, taking $\lambda = 0$, we know that $A^0 = \mathbb{C}\mathbf{1}$ by (ii). Thus A^0 pairs with A^ν and $\dim A^\nu = 1$. Because $(\mathbf{1}, t) \neq 0$ by (20), and T is an eigenspace for $h(0)$ (Lemma 6), we see that $A^\nu = T$. This proves (iii), and completes the proof of the theorem. □

6 The Bilinear Form in the Shifted Case

We return to the issue, introduced in Sect. 5, of the nature of the radical of the bilinear form $\langle\ ,\ \rangle$ for an $\mathbb{N}$-graded vertex operator algebra V, assuming now that V is a shift of a simple, self-dual vertex operator algebra $W = (W, Y, \mathbf{1}, \omega')$ of CFT-type as in Sect. 5. We will also assume that $\dim V_0 \geq 2$.

We continue to use the notations of Sects. 3–5. We shall see that the question raised in Remark 3 has an affirmative answer in this case, and that $\mathrm{rad}\langle\ ,\ \rangle/N$ is *exactly* the nilpotent radical N_1/N of the Lie algebra V_1/N when W_1 is reductive. The precise result is as follows.

Theorem 3. *We have*

$$\oplus_{m\geq 2} W_{m,m-1} \subseteq \mathrm{Ann}_{V_1}(t(-1)) = \mathrm{rad}\langle\ ,\ \rangle. \tag{31}$$

Moreover, if W_1 is reductive, *then*

$$N_1 = \oplus_{m\geq 2} W_{m,m-1} = \mathrm{Ann}_{V_1}(t(-1)) = \mathrm{rad}\langle\ ,\ \rangle. \tag{32}$$

Recall that V_1 is a Leibniz algebra, $N = L(-1)V_0$ is a 2-sided ideal in V_1, V_1/N is a Lie algebra, and N_1/N is the *nilpotent radical* of V_1/N (21). Because W is a VOA of CFT-type then W_1 is a Lie algebra with bracket $a(0)b$ $(a, b \in W_1)$, and $W_{1,0} = C_{W_1}(h)$ is the *centralizer* of h in W_1.

Lemma 12. $h \in P$.

Proof. We have to show that $\langle h, M\rangle = (h(1)M, t) = 0$. Since M is an ideal in V_1 then M is the direct sum of its $h(0)$-eigenspaces. Let $M^p = \{m \in M \mid h(0)m = pm\}$. Now $h(0)h(1)m = h(1)h(0)m = ph(1)m$ $(m \in M^p)$, showing that $h(1)M^p \subseteq A^p$. If $p \neq 0$, then $(A^p, t) = 0$ by Theorem 2, so that $(h(1)M^p, t) = 0$ in this case.

It remains to establish that $(h(1)M^0, t) = 0$. To see this, first note that because W is self-dual then $L'(1)W_1 = 0$. Since $L'(1) = L(1)+2h(1)$ then $h(1)W_1 = L(1)W_1$, and in particular $h(1)M^0 = L(1)M^0$ (because $M^0 \subseteq V_1^0 = W_{1,0} \subseteq W_1$). Therefore, $(h(1)M^0, t) = (L(1)M^0, t) = (M^0, L(-1)t) = 0$, where the last equality holds by Lemma 6. The lemma is proved. □

Lemma 13. $h \notin \mathrm{Ann}_{V_1}(t(-1)) \cup \mathrm{rad}\langle\, ,\, \rangle$.

Proof. First recall that $h(1)h = \nu/2\mathbf{1}$. Then we have

$$\langle h, h\rangle = (h(1)h, t) = \nu/2(\mathbf{1}, t) \neq 0.$$

Here, $\nu \neq 0$ thanks to Theorem 2 and because we are assuming that $\dim V_0 \geq 2$. Because $h \in V_1$, this shows that $\langle h, V_1\rangle \neq 0$, so that $h \notin \mathrm{rad}\langle\, ,\, \rangle$.

Next, using (27) we have $L(1)h = L'(1)h - 2h(1)h$. Because W is assumed to be self-dual then $L'(1)h = 0$, so that $L(1)h = -2h(1)h = -\nu\mathbf{1}$. Now

$$(t(-1)h, h) = (h(-1)t - L(-1)h(0)t, h).$$

Therefore,

$$(L(-1)h(0)t, h) = (h(0)t, L(1)h) = -\nu^2(t, \mathbf{1}).$$

Also,

$$(h(-1)t, h) = (t, -h(1)h - (L(1)h)(0)h) = -\nu/2(t, \mathbf{1}).$$

Therefore,

$$(t(-1)h, h) = (\nu^2 - \nu/2)(t, \mathbf{1}).$$

Because ν is a positive integer, the last displayed expression is nonzero. Therefore, $t(-1)h \neq 0$, i.e., $h \notin \mathrm{Ann}_{V_1}(t(-1))$. This completes the proof of the lemma. □

We turn to the proof of Theorem 3. First note that by combining Lemmas 12 and 13 together with Proposition 2, we see that cases (ii) and (iii) of Proposition 2 *cannot* hold. Therefore, case (i) must hold, that is

$$\mathrm{rad}\langle\, ,\, \rangle = \mathrm{Ann}_{V_1}(t(-1)).$$

From (28) it is clear that, up to scalars, $\mathbf{1}$ is the only state in V_0 annihilated by $h(0)$. It then follow from Lemma 6 that $J = \oplus_{m\geq 1} W_{m,m}$. In particular, $(W_{m,m}, t) = 0$ $(m \geq 1)$ by (19).

Now let $u \in W_{m,m-1}$, $v \in W_{k,k-1}$ with $m \geq 1, k \geq 2$. Then $u(1)v \in V_0$ and $L'(0)u(1)v = (m+k-2)u(1)v$. Therefore, $u(1)v \in W_{m+k-2,m+k-2}$, and because $m+k-2 \geq 1$ it follows that

$$\langle u, v \rangle = (u(1)v, t) = 0.$$

Because this holds for all $u \in W_{m,m-1}$ and all $m \geq 1$, we conclude that $v \in \mathrm{rad}\langle\, , \rangle$. This proves that $\oplus_{m\geq 2} W_{m,m-1} \subseteq \mathrm{rad}\langle\, ,\, \rangle$. Now (31) follows immediately.

Now suppose that W_1 is reductive. Because W is self-dual and of CFT-type, it has (up to scalars) a unique nonzero invariant bilinear form. Let us denote it by $((\, ,\,))$. In particular, we have

$$((u, v))\mathbf{1} = u(1)v \quad (u, v \in W_1).$$

Because V is simple and V, W have the same set of fields, then W is also simple. In particular, $((\, ,\,))$ must be nondegenerate. Now if L is a (finite-dimensional, complex) reductive Lie algebra equipped with a nondegenerate, symmetric invariant bilinear form, then the restriction of the form to the centralizer of any semisimple element in L is also nondegenerate. In the present situation, this tells us that the restriction of $((\, ,\,))$ to $C_{W_1}(h)$ is nondegenerate.

On the other hand, we have

$$\langle u, v \rangle = (u(1)v, t) = ((u, v))(\mathbf{1}, t)$$

and $(\mathbf{1}, t) \neq 0$ by (20). This shows that the restrictions of $\langle\, ,\, \rangle$ and $((\, ,\,))$ to $C_{W_1}(h)$ are *equivalent* bilinear forms. Since the latter is nondegenerate, so is the former. Therefore, $\mathrm{rad}\langle\, ,\, \rangle \cap C_{W_1}(h) = 0$. Now the second and third equalities of (32) follow from (31) and the decomposition $V_1 = C_{W_1}(h) \oplus \oplus_{m\geq 2} W_{m,m-1}$.

To complete the proof of the theorem it suffices to prove the next result.

Lemma 14. *We have*

$$N_1 = \mathrm{Nilp(V1)}(C_{W_1}(h)) \oplus_{m\geq 2} W_{m,m-1}. \tag{33}$$

In particular, if W_1 is a reductive *Lie algebra, then*

$$N_1 = \oplus_{m\geq 2} W_{m,m-1}. \tag{34}$$

Proof. Let $u \in W_{k,k-1}$, $v \in W_{m,m-1}$ with $k \geq 1, m \geq 2$. Then $v(0)u \in W_{m+k-1} \cap V_1 \subseteq W_{m+k-1,m+k-2}$ and $m+k-1 > k$. This shows that $\oplus_{m\geq 2} W_{m,m-1}$ is an ideal in V_1. Moreover, because there is a maximum integer r for which $W_{r,r-1} \neq 0$, it follows that $v(0)^l V_1 = 0$ for large enough l, so that the left adjoint action of $v(0)$ on V_1 is *nilpotent*. This shows that $\oplus_{m\geq 2} W_{m,m-1}$ is a nilpotent ideal. Because $h(0)$ acts on $W_{m,m-1}$ as multiplication by $m-1$, then $W_{m,m-1} = [h, W_{m,m-1}]$ for $m \geq 2$, whence in fact $\oplus_{m\geq 2} W_{m,m-1} \subseteq N_1$. Then (33) follows immediately.

Finally, if W_1 is reductive, the centralizer of any semisimple element in W_1 is also reductive. In particular, this applies to $C_{W_1}(h)$ since $h(0)$ is indeed semisimple, so (34) follows from (33). This completes the proof of the lemma, and hence also that of Theorem 3. □

The VOA W is *strongly regular* if it is self-dual and CFT-type as well as both *C_2-cofinite* and *rational*. For a general discussion of such VOAs see, for example, [M]. It is known ([M] and [DM1], Theorem 1.1) that in this case W_1 is necessarily reductive. Consequently, we deduce from Theorem 3 that the following holds.

Corollary 1. *Suppose that W is a strongly regular VOA, and that V is a self-dual, $\mathbb{N}$-graded VOA obtained as a shift of W. Then* $\mathrm{rad}\langle\,,\,\rangle = \mathrm{Nilp}(V_1)$. □

Remark 4. The corollary applies, for example, to the shifted theories $V = L_{\hat{sl_2}}(k,0)^H$ discussed in Sect. 8 below. In this case, one can directly compute the relevant quantities.

7 The C_2-Cofinite Case

In this section we are mainly concerned with simple VOAs V that are self-dual and $\mathbb{N}$-graded as before, but that are also *rational*, or *C_2-cofinite*, or both. Recall [DLM1, DLM2, Z] that V is rational if every admissible (or $\mathbb{N}$-gradable) V-module is completely reducible; C_2-cofinite if the span of the states $u(-2)v$ $(u, v \in V)$ has finite codimension in V; regular if it is *both* rational and C_2-cofinite; and *strongly regular* if it is both regular and self-dual (as discussed in Sect. 3). It is known [DLM1, DLM2, Z] that both rationality and C_2-cofiniteness imply that V-Mod has only finitely many simple objects.

To motivate the main results of this section, we recall some results about vertex operator algebras V with $V_0 = \mathbb{C}\mathbf{1}$. In this case, V_1 is a Lie algebra, and if V is strongly regular, then V_1 is *reductive* ([DM1], Theorem 1.1). It is also known ([DM4], Theorem 3.1) that if V is C_2-cofinite, but not necessarily rational, and $S \subseteq V_1$ is a Levi factor, then the vertex operator subalgebra U of V generated by S satisfies

$$U \cong L_{\hat{\mathfrak{g}}_1}(k_1, 0) \oplus \ldots \oplus L_{\hat{\mathfrak{g}}_r}(k_r, 0), \tag{35}$$

i.e., a direct sum of simple affine Kac–Moody Lie algebras $L_{\hat{\mathfrak{g}}_j}(k_j, 0)$ of positive integral level k_j.

We want to know to what extent these results generalize to the more general case when $\dim V_0 > 1$. With $N = L(-1)V_0$ as before, we have seen that V_1/N is a Lie algebra. Now $V_0 = \mathbb{C}\mathbf{1}$ precisely when $N = 0$, but the natural guess that V_1/N is reductive if V is rational and C_2-cofinite is generally false. Thus we need to understand the nilpotent radical N_1/N of this Lie algebra. That is where the bilinear form $\langle\ ,\ \rangle$ comes in. These questions are naturally related to the issue, already addressed in Sect. 2, of the structure of the subalgebra of V generated by a Levi subalgebra of V_1. The main result is

Theorem 4. *Let V be a simple, self-dual, $\mathbb{N}$-graded vertex operator algebra that is C_2-cofinite, and let $V_1 = B \oplus S$ with Levi factor S and solvable radical B. Then the following hold.*

1. *$N_1 \subseteq \mathrm{rad}\langle\ ,\ \rangle \subseteq B$, and the restriction of $\langle\ ,\ \rangle$ to S is nondegenerate. In particular, $a(1)b \in \mathbb{C}\mathbf{1}$ for all $a, b \in S$.*
2. *If U is the vertex operator algebra generated by S, then U satisfies (35).*

We start with

Proposition 3. *Let $V = \oplus_{n=0}^{\infty} V_n$ be an $\mathbb{N}$-graded vertex operator algebra such that $dim\ V_0 > 1$. Let $X = \{x^i\}_{i\in I} \cup \{y^j\}_{j\in J}$ be a set of homogeneous elements in V which are representatives of a basis of $V/C_2(V)$. Here x^i are vectors whose weights are greater than or equal to 1 and y^j are vectors whose weights are zero. Then V is spanned by elements of the form*

$$x^{i_1}(-n_1)\ldots x^{i_s}(-n_s)y^{j_1}(-m_1)\ldots y^{j_k}(-m_k)\mathbf{1}$$

where $n_1 > n_2 > \ldots. > n_s > 0$ and $m_1 \geq m_2 \geq \ldots \geq m_k > 0$.

Proof. The result follows by modifying the proof of Proposition 8 in [GN]. □

Notice that for a Lie algebra $W \subset V_1$, we have $u(0)v = -v(0)u$ for $u, v \in W$. Hence, $L(-1)u(1)v = 0$ and $u(1)v \in \mathbb{C}\mathbf{1}$. Moreover, we have $\langle\ ,\ \rangle\mathbf{1} = (u(1)v, t)\mathbf{1} = u(1)v$ for $u, v \in W$.

Let $\mathfrak{g}$ be a finite-dimensional simple Lie algebra, let $\mathfrak{h} \subset \mathfrak{g}$ be a Cartan subalgebra, and let Ψ be the associated root system with simple roots Δ. Also, we set $(\cdot,\cdot)$ to be the nondegenerate symmetric invariant bilinear form on $\mathfrak{g}$ normalized so that the longest positive root $\theta \in \Psi$ satisfies $(\theta, \theta) = 2$. The corresponding affine Kac–Moody Lie algebra $\hat{\mathfrak{g}}$ is defined as

$$\hat{\mathfrak{g}} = \mathfrak{g} \otimes \mathbb{C}[t, t^{-1}] \oplus \mathbb{C}K,$$

where K is central and the bracket is defined for $u, v \in \mathfrak{g}$, $m, n \in \mathbb{Z}$ as

$$[u(m), v(n)] = [u, v](m+n) + m\delta_{m+n,0}(u, v)K \quad (u(m) = u \otimes t^m).$$

Let $W \subset V_1$ be a Lie algebra such that $\mathfrak{g} \subset W$ and $\langle\ ,\ \rangle$ is nondegenerate on W. If $\langle\ ,\ \rangle$ is nondegenerate on $\mathfrak{g}$, then the map

$$\hat{\mathfrak{g}} \to \mathrm{End}(V);\ u(m) \mapsto u(m),\ \ u \in \mathfrak{g}, m \in \mathbb{Z},$$

is a representation of the affine Kac–Moody algebra $\hat{g}$ of level k where

$$\langle\ ,\ \rangle = u(1)v = k(u,v) \text{ for } u, v \in \mathfrak{g}.$$

Theorem 5. *Let $W \subset V_1$ be a Lie algebra such that $\langle\ ,\ \rangle$ is nondegenerate on W. If $\mathfrak{g}$ is a simple Lie subalgebra of W and U is the vertex operator subalgebra of V generated by $\mathfrak{g}$, then $\langle\ ,\ \rangle$ is nondegenerate on $\mathfrak{g}$, and U is isomorphic to $L_{\hat{\mathfrak{g}}}(k,0)$. Furthermore, k is a positive integer and V is an integrable $\hat{\mathfrak{g}}$-module.*

Proof. We will follow the proof in Theorem 3.1 of [DM4]. First, assume that $\mathfrak{g} = sl_2(\mathbb{C})$ with standard basis $\alpha, x_\alpha, x_{-\alpha}$. Hence, $(\alpha,\alpha) = 2$. Since each homogeneous subspace of V is a completely reducible $\mathfrak{g}$-module, then V is also a completely reducible $\mathfrak{g}$-module. A nonzero element $v \in V$ is called a weight vector for $\mathfrak{g}$ of $\mathfrak{g}$-weight λ ($\lambda \in \mathbb{C}\alpha$) if $\alpha(0)v = (\alpha,\lambda)v$. Here, $\lambda \in \frac{1}{2}\mathbb{Z}\alpha$.

We now make use of Proposition 3. Let $X = \{x^i\}_{i\in I} \cup \{y^j\}_{j\in J}$ be a set of homogeneous weight vectors in V which are representatives of a basis of $V/C_2(V)$. The x^i are vectors whose weights are greater than or equal to 1 and y^j are vectors whose weights are zero. Since X is finite, there is a nonnegative elements $\lambda_0 = m\alpha \in \frac{1}{2}\mathbb{Z}\alpha$ such that the weight of each element in X is bounded above by λ_0. For any integer $t \geq 0$, we have

$$\begin{aligned}\oplus_{n\leq t(t+1)/2} V_n \subseteq \mathrm{Span}_{\mathbb{C}}\{&x^{i_1}(-n_1)\ldots x^{i_s}(-n_s)y^{j_1}(-m_1)\ldots y^{j_r}(-m_r)\mathbf{1} \mid \\ &n_1 > n_2 > \ldots > n_s > 0, m_1 \geq m_2 \geq \ldots \geq m_r > 0, \\ &0 \leq s, r \leq t\}. \end{aligned} \tag{36}$$

Furthermore, if $n \leq \frac{t(t+1)}{2}$, then a $\mathfrak{g}$-weight vector in V_n has $\mathfrak{g}$-weight less than or equal to $2t\lambda_0 = 2tm\alpha$.

Let l be an integer such that $l + 1 > 4m$ and we let

$$u = (x_\alpha)(-1)^{l(l+1)/2}\mathbf{1}.$$

We claim that $u = 0$. Assume $u \neq 0$. By (36), we can conclude that the $\mathfrak{g}$-weight of u is at most $2lm\alpha$. This contradicts the direct calculation which shows that the $\mathfrak{g}$-weight of u is $\frac{l(l+1)}{2}\alpha$. Hence, $u = 0$. This implies that U is integrable. Furthermore, we have V is integrable, k is a positive integer and $\langle\ ,\ \rangle$ is nondegenerate.

This proves the theorem for $\mathfrak{g} = sl_2$. The general case follows easily from this (cf. [DM4]). □

Lemma 15. *Let S be a Levi subalgebra of V_1. Then $\langle\ ,\ \rangle$ is nondegenerate on S and* $\mathrm{Rad}\langle\ ,\ \rangle \cap S = \{0\}$.

Proof. Clearly, for $u, v \in S$, we have $u(1)v \in \mathbb{C}\mathbf{1}$. Let $f : S \times S \to \mathbb{C}\mathbf{1}$ be a map defined by $f((u,v)) = u(1)v$. Since $u(1)v = v(1)u$ and

$$(w(0)u)(1)v = -(u(0)w)(1)v = -(u(0)w(1)v - w(1)u(0)v) = w(1)u(0)v$$

for $u, v, w \in S$, we can conclude that f is a symmetric invariant bilinear form on S. For convenience, we set $X = \mathrm{Rad}(f)$. Since S is semisimple and X is a S-module, these imply that $S = X \oplus W$ for some S-module W. Note that W and X are semisimple and $S \cap \mathrm{Rad}\langle\ ,\ \rangle \subseteq X$.

For $u, v \in X$, we have $u(1)v = 0$. Hence, the vertex operators $Y(u, z)$, $u \in X$, generate representation of the loop algebra in the sense that

$$[u(m), v(n)] = (u(0)v)(m+n), \quad \text{for } u, v \in X.$$

Following the proof of Theorem 3.1 in [DM4], we can show that the representation on V is integrable and the vertex operator subalgebra U generated by a simple component of X is the corresponding simple vertex operator algebra $L(k, 0)$ and $k = 0$. However, the maximal submodule of the Verma module $V(0, 0)$, whose quotient is $L(0, 0)$, has co-dimension one. This is not possible if $X \neq \{0\}$. Consequently, we have $X = 0$ and $S \cap \mathrm{Rad}\langle\ ,\ \rangle = \{0\}$. Hence $\langle\ ,\ \rangle$ is nondegenerate on S. □

Theorem 4 follows from these results.

8 Examples of Shifted Vertex Operator Algebras

To illustrate previous results, in this section we consider some particular classes of shifted vertex operator algebras.

8.1 Shifted $\widehat{sl_2}$

We will show that the simple vertex operator algebra (WZW model) $L_{\widehat{sl_2}}(k, 0)$ corresponding to affine sl_2 at (positive integral) level k has a canonical shift to an $\mathbb{N}$-graded vertex operator algebra $L_{\hat{sl_2}}(k, 0)^H$, and that the resulting de Rham structure on V_0 is that of complex projective space $\mathbb{CP}^k$. The precise result is the following.

Theorem 6. *Let e, f, h be Chevalley generators of sl_2, and set $H = h/2$. Then the following hold:*

(*a*) *$L_{\hat{sl_2}}(k, 0)^H$ is a simple, $\mathbb{N}$-graded, self-dual vertex operator algebra.*
(*b*) *The algebra structure on the zero weight space of $L_{\hat{sl_2}}(k, 0)^H$ is isomorphic to $\mathbb{C}[x]/\langle x^{k+1}\rangle$, where $x = e$.*

Proof. Let $W = L_{\hat{sl}_2}(k, 0)$. It is spanned by states $v_{IJK} := e_I f_J h_K \mathbf{1}$, where we write $e_I = e(-l_1)\ldots e(-l_r)$, $f_J = f(-m_1)\ldots f(-m_s)$, $h_K = h(-n_1)\ldots h(-n_t)$ for $l_i, m_i, n_i > 0$. Note that $v_{IJK} \in W_n$, where $n = \sum l_i + \sum m_i + \sum n_i$

Recall from (27) that $L_H(0) = L(0) - H(0)$. We have

$$[H(0), e(n)] = [H, e](n) = e(n),$$
$$[H(0), f(n)] = [H, f](n) = -f(n),$$
$$[H(0), h(n)] = [H, h](n) = 0.$$

Then $H(0)v_{IJK} = (r - s)v_{IJK}$, so that

$$L_H(0)v_{IJK} = \left(\sum(l_i - 1) + \sum(m_j + 1) + \sum n_k\right) v_{IJK}. \tag{37}$$

It is well known (e.g., [DL], Propositions 13.16 and 13.17) that $Y(e, z)^{k+1} = 0$. Thus we may take r to be no greater than k. It follows from (37) that the eigenvalues of $L_H(0)$ are integral and bounded below by 0, and that the eigenspaces are finite-dimensional. Therefore, $V := W^H$ is indeed an ℕ-graded vertex operator algebra. Because W is simple then so too is V, since they share the same fields.

Next we show that V is self-dual, which amounts to the assertion that $L_H(1)V_1$ is *properly* contained in V_0. Observe from (37) that the states $e(-1)^p\mathbf{1}$ $(0 \leq p \leq k)$ span in V_0, while V_1 is spanned by the states $\{h(-1)e(-1)^i\mathbf{1}, e(-2)e(-1)^i\mathbf{1} \mid 0 \leq i \leq k-1\}$. We will show that $e(-1)^k\mathbf{1}$ does *not* lie in the image of $L_H(1)$.

For $g \in sl_2$, $m \geq 1$, we have

$$[L(1), g(-m)] = mg(1 - m).$$

Since $L(1)e = 0$ and

$$L(1)e(-1)^{j+1}\mathbf{1} = e(-1)L(1)e^j(-1)\mathbf{1} + e(0)e^j(-1)\mathbf{1} = e(-1)L(1)e^j(-1)\mathbf{1}$$

for $j \geq 0$, we can conclude by induction that

$$L(1)e(-1)^i\mathbf{1} = 0 \text{ for all } i \geq 1.$$

Similarly, because $H(1)e = 0$ and

$$H(1)e(-1)^{j+1}\mathbf{1} = e(-1)H(1)e(-1)^j\mathbf{1} + e(0)e(-1)^j\mathbf{1} = e(-1)H(1)e(-1)^j\mathbf{1},$$

then

$$H(1)e(-1)^i\mathbf{1} = 0 \text{ for all } i \geq 0.$$

We can conclude that for $0 \le i \le k-1$,

$$\begin{aligned} L_H(1)h(-1)e(-1)^i\mathbf{1} &= (L(1) - 2H(1))h(-1)e(-1)^i\mathbf{1} \\ &= 2ie(-1)^i\mathbf{1} - 2ke(-1)^i\mathbf{1} \\ &= 2(i-k)e(-1)^i\mathbf{1}, \end{aligned}$$

while

$$L_H(1)e(-2)e(-1)^j\mathbf{1} = (L(1) - 2H(1))e(-2)e(-1)^j\mathbf{1} = 0.$$

Our assertion that $e(-1)^k\mathbf{1} \notin imL_H(1)$ follows from these calculations. This establishes part (a) of the theorem.

Finally, if we set $x := e(-1)\mathbf{1} = e$, then by induction $e(-1)^i\mathbf{1} = x.x^{i-1} = x^i$, so the algebra structure on V_0 is isomorphic $\mathbb{C}[x]/x^{k+1}$ and part (b) holds. This completes the proof of the theorem. □

Remark 5. Suitably normalized, the invariant bilinear form on V_0 satisfies $(x^p, x^q) = \delta_{p+q,k}$ (cf. Theorem 2). V_0 can be identified with the de Rham cohomology of $\mathbb{CP}^k$ (x has degree 2) equipped with the pairing arising from Poincaré duality.

8.2 Shifted Lattice Theories

Let L be a positive-definite even lattice of rank d with inner product $(\ ,\ ) : L \times L \to \mathbb{Z}$. Let $H = \mathbb{C} \otimes L$ be the corresponding complex linear space equipped with the $\mathbb{C}$-linear extension of $(\ ,\ )$. The dual lattice of L is

$$L^\circ = \{ f \in \mathbb{R} \otimes L \mid (f, \alpha) \in \mathbb{Z} \text{ all } \alpha \in L\}.$$

Let $(M(1), Y, \mathbf{1}, \omega_L)$ be the free bosonic vertex operator algebra based on H and let $(V_L, Y, \mathbf{1}, \omega_L)$ be the corresponding lattice vertex operator algebra. Both vertex operator algebras have central charge d, and the Fock space of V_L is

$$V_L = M(1) \otimes \mathbb{C}[L],$$

where $\mathbb{C}[L]$ is the group algebra of L.

For a state $h \in H \subset (V_L)_1$, we set $\omega_h = \omega_L + h(-2)\mathbf{1}$, with $V_{L,h} = (V_L, Y, \mathbf{1}, \omega_h)$.

Lemma 16. *([DM3]). Suppose that $h \in L^0$. Then $V_{L,h}$ is a vertex operator algebra, and it is self-dual if, and only if, $2h \in L$.*

For the rest of this section, we assume that $0 \neq h \in L^\circ$ and $2h \in L$, so that $V_{L,h}$ is a self-dual, simple vertex operator algebra. Set

$$Y(\omega_h, z) = \sum_{n\in\mathbb{Z}} L_h(n)z^{-n-2}.$$

Then

$$L_h(0)(u \otimes e^\alpha) = (n + \frac{1}{2}(\alpha, \alpha) - (h, \alpha))u \otimes e^\alpha \ (u \in M(1)_n).$$

It follows that $V_{L,h}$ is $\mathbb{N}$-graded if, and only if, the following condition holds:

$$(2h, \alpha) \leq (\alpha, \alpha) \ (\alpha \in L). \tag{38}$$

From now on, we *assume that (38) is satisfied.* It is equivalent to the condition $(\alpha-h, \alpha-h) \geq (-h, -h)$, i.e., $-h$ has the least (squared) length among all elements in the coset $L - h$. Set

$$A := \{\alpha \in L \mid (\alpha, \alpha) = (2h, \alpha)\}.$$

Note that $0, 2h \in A$. We have

$$(V_{L,h})_0 = \mathrm{Span}_{\mathbb{C}}\{e^\alpha \mid \alpha \in A\}.$$

We want to understand the commutative algebra structure of $(V_{L,h})_1$, defined by the -1th product $a(-1)b$. The identity element is $\mathbf{1} = e^0$.

First note that if $\alpha, \beta \in A$, then $(-h, -h) \leq (\alpha + \beta - h, \alpha + \beta - h) = (h, h) + 2(\alpha, \beta)$ shows that $(\alpha, \beta) \geq 0$. Moreover, $(\alpha, \beta) = 0$ if, and only if, $\alpha + \beta \in A$. We employ standard notation for vertex operators in the lattice theory V_L [LL]. Then

$$\begin{aligned} e^\alpha(-1)e^\beta &= Res_z z^{-1}E^-(-\alpha, z)E^+(-\alpha, z)e_\alpha z^\alpha \cdot e^\beta \\ &= \epsilon(\alpha, \beta)Res_z z^{(\alpha,\beta)-1}E^-(-\alpha, z)E^+(-\alpha, z)e^{\alpha+\beta}, \end{aligned}$$

where

$$E^-(-\alpha, z)E^+(-\alpha, z) = exp\left\{-\sum_{n>0}\frac{\alpha(-n)}{n}z^n\right\} exp\left\{\sum_{n>0}\frac{\alpha(n)}{n}z^{-n}\right\}.$$

It follows that

$$e^\alpha(-1)e^\beta = \begin{cases} \epsilon(\alpha, \beta)e^{\alpha+\beta} & \text{if } \alpha + \beta \in A \\ 0 & \text{otherwise.} \end{cases} \tag{39}$$

If $0 \neq \alpha \in A$, then $(2h, \alpha) = (\alpha, \alpha) \neq 0$. Thus $2h + \alpha \notin A$, and the last calculation shows that $e^\alpha(-1)e^{2h} = 0$. It follows that e^{2h} spans the unique minimal ideal $T \subseteq (V_{L,h})_1$

Recall [LL] $\epsilon : L \times L \to \{\pm 1\}$ is a (bilinear) 2-cocycle satisfying $\epsilon(\alpha,\beta)\epsilon(\beta,\alpha) = (-1)^{(\alpha,\beta)}$. Thus we have proved

Lemma 17. *There are signs $\epsilon(\alpha,\beta) = \pm 1$ such that multiplication in $(V_{L,h})_1$ is given by (39). The minimal ideal T is spanned by e^{2h}.* □

References

[B] D. Barnes, On Levi's Theorem for Leibniz algebras, *Bull. Aust. Math. Soc.* **86(2)** (2012), 184–185.

[Br] P. Bressler, Vertex algebroids I, arXiv: math. AG/0202185.

[DL] C. Dong and J. Lepowsky, *Generalized Vertex Algebras and Relative Vertex Operators*, Progress in Mathematics Vol. **112**, Birkhäuser, Boston, 1993.

[DLM1] C. Dong, H. Li and G. Mason, Regularity of vertex operator algebras, Adv. Math. **132** (1997) No. 1, 148–166.

[DLM2] C. Dong, H. Li and G. Mason, Twisted representations of vertex operator algebras, *Math. Ann.* **310** (1998) No. 3, 571–600.

[DM1] C. Dong and G. Mason, Rational vertex operator algebras and the effective central charge, *Int, Math. Res. Not.* **56** (2004), 2989–3008.

[DM2] C. Dong and G. Mason, Local and semilocal vertex operator algebras, *J. Algebra* **280** (2004), 350–366.

[DM3] C. Dong and G. Mason, Shifted vertex operator algebras, *Math. Proc. Cambridge Phil. Soc.* **141** (2006), 67–80.

[DM4] C. Dong and G. Mason, Integrability of C_2-cofinite Vertex Operator Algebra, *IMRN*, Article ID 80468 (2006), 1–15.

[FHL] I. Frenkel, Y.-Z. Huang and J. Lepowsky, *On Axiomatic Approaches to Vertex Operator Algebras and Modules*, Mem. A.M.S. Vol. **104** No. 4, 1993.

[GN] M. Gaberdiel and A. Neitzke, Rationality, quasirationality and finite W-algebras, *Comm. Math. Phys.* **238** (2008), 305–331.

[GMS] V. Gorbounov, F. Malikov and V. Schechtman, Gerbes of chiral differential operators II, Vertex algebroids, *Invent. Math.* **155** (2004), 605–680.

[L] H. Li, Symmetric invariant bilinear forms on vertex operator algebras, *J. Pure and Appl. Math.* **96** (1994), no.1, 279–297.

[LL] J. Lepowsky and H.-S. Li, *Introduction to Vertex Operator Algebras and Their Representations*, Progress in Math. Vol. **227**, Birkhäuser, Boston, 2003.

[LY] H. Li and G. Yamskulna, On certain vertex algebras and their modules associated with vertex algebroids, *J. Alg* **283** (2005), 367–398.

[M] G. Mason, Lattice subalgebras of vertex operator algebras, to appear in Proceedings of the Heidelberg Conference, Springer, arXiv: 1110.0544.

[MS1] F. Malikov and V. Schechtman, Chiral Poincaré duality, Math. Res. Lett. **6** (1999), 533–546, in *Differential Topology, Infinite-Dimensional Lie Algebras, and Applications*, Amer. Math. Soc.Transl. Ser. 2 **194** (1999), 149–188.

[MS2] F. Malikov and V. Schechtman, Chiral de Rham complex II, in *Differential Topology, Infinite-Dimensional Lie Algebras, and Applications*, Amer. Math. Soc.Transl. Ser. 2 **194** (1999), 149–188.

[MSV] F. Malikov, V. Schechtman and A. Vaintrob, Chiral de Rham complex, *Comm. Math. Phys.* **204** (1999), 439–473.

[MY] G. Mason and G. Yamskulna, Leibniz algebras and Lie algebras, *SIGMA* **9** (2013), no 63, 10 pages.

[Z] Y. Zhu, Modular invariance of characters of vertex operator algebras, *J. Amer. Math. Soc.* **9** (1996) No. 1, 237–302.

Variations on a Casselman–Osborne Theme

Dragan Miličić

Abstract We discuss two classical results in homological algebra of modules over an enveloping algebra – lemmas of Casselman–Osborne and Wigner. They have a common theme: they are statements about derived functors. While the statements for the functors itself are obvious, the statements for derived functors are not and the published proofs were completely different from each other. First we give simple, pedestrian arguments for both results based on the same principle. Then we give a natural generalization of these results in the setting of derived categories.

Key words Derived functors • Centers of categories • Casselman–Osborne lemma • Wigner lemma

Mathematics Subject Classification (2010): Primary 16E35. Secondary 22E47.

Introduction

This paper is inspired by two classical results in homological algebra of modules over an enveloping algebra—lemmas of Casselman–Osborne and Wigner. They have a common theme: they are statements about derived functors. While the statements for the functors themselves are obvious, the statements for derived functors are not and the published proofs were completely different from each other.

In the first section we give simple, pedestrian arguments for both results based on the same principle. They suggest a common generalization which is the topic of this paper.

D. Miličić (✉)
Department of Mathematics, University of Utah, Salt Lake City, UT 84112, USA
e-mail: milicic@math.utah.edu

G. Mason et al. (eds.), *Developments and Retrospectives in Lie Theory: Algebraic Methods*, Developments in Mathematics 38,
DOI 10.1007/978-3-319-09804-3_13

In the second section we discuss some straightforward properties of centers of abelian categories and their derived categories. In the third section, we consider a class of functors and prove a simple result about their derived functors which generalizes the first two results.

The original arguments were considerably more complicated and based on different ideas [1, 3] and [5].

1 Classical Approach

1.1 Wigner's Lemma

Let $\mathfrak{g}$ be a complex Lie algebra, $\mathscr{U}(\mathfrak{g})$ its enveloping algebra and $\mathscr{Z}(\mathfrak{g})$ the center of $\mathscr{U}(\mathfrak{g})$. Denote by $\mathscr{M}(\mathscr{U}(\mathfrak{g}))$ the category of $\mathscr{U}(\mathfrak{g})$-modules.

Let $\chi : \mathscr{Z}(\mathfrak{g}) \longrightarrow \mathbb{C}$ be an algebra morphism of $\mathscr{Z}(\mathfrak{g})$ into the field of complex numbers. We say that a module V in $\mathscr{M}(\mathfrak{g})$ has an *infinitesimal character* χ if

$$z \cdot v = \chi(z)v \text{ for any } z \in \mathscr{Z}(\mathfrak{g}) \text{ and any } v \in V.$$

Theorem 1.1. *Let U and V be two objects in $\mathscr{M}(\mathscr{U}(\mathfrak{g}))$ with infinitesimal characters χ_U and χ_V. Then $\chi_U \neq \chi_V$ implies* $\operatorname{Ext}^p_{\mathscr{U}(\mathfrak{g})}(U, V) = 0$ *for all $p \in \mathbb{Z}_+$.*

Proof. Clearly, the center $\mathscr{Z}(\mathfrak{g})$ of $\mathscr{U}(\mathfrak{g})$ acts naturally on $\operatorname{Hom}_{\mathscr{U}(\mathfrak{g})}(U, V)$ for any two $\mathscr{U}(\mathfrak{g})$-modules U and V, by

$$z(T) = z \cdot T = T \cdot z \text{ for any } z \in \mathscr{Z}(\mathfrak{g}) \text{ and any } T \in \operatorname{Hom}_{\mathscr{U}(\mathfrak{g})}(U, V),$$

i.e., we can view it as a bifunctor from the category of $\mathscr{U}(\mathfrak{g})$-modules into the category of $\mathscr{Z}(\mathfrak{g})$-modules. Hence, its derived functors $\operatorname{Ext}^*_{\mathscr{U}(\mathfrak{g})}$ are bifunctors from the category of $\mathscr{U}(\mathfrak{g})$-modules into the category of $\mathscr{Z}(\mathfrak{g})$-modules.

Fix now a $\mathscr{U}(\mathfrak{g})$-module U with infinitesimal character χ_U. Consider the functor $F = \operatorname{Hom}_{\mathscr{U}(\mathfrak{g})}(U, -)$ from the category $\mathscr{M}(\mathscr{U}(\mathfrak{g}))$ into the category of $\mathscr{Z}(\mathfrak{g})$-modules. Since the infinitesimal character of U is χ_U, any element of $z \in \mathscr{Z}(\mathfrak{g})$ acts on $F(V) = \operatorname{Hom}_{\mathscr{U}(\mathfrak{g})}(U, V)$ as multiplication by $\chi_U(z)$ for any object V in $\mathscr{M}(\mathscr{U}(\mathfrak{g}))$.

Fix now a $\mathscr{U}(\mathfrak{g})$-module V with infinitesimal character χ_V. Let

$$0 \longrightarrow V \longrightarrow I^0 \longrightarrow I^1 \longrightarrow \dots \longrightarrow I^n \longrightarrow \dots$$

be an injective resolution of V. Let $z \in \ker \chi_V$. Then we have the commutative diagram

$$\begin{array}{ccccccccccc} 0 & \longrightarrow & I^0 & \longrightarrow & I^1 & \longrightarrow & \dots & \longrightarrow & I^n & \longrightarrow & \dots \\ & & \downarrow z & & \downarrow z & & & & \downarrow z & & \\ 0 & \longrightarrow & I^0 & \longrightarrow & I^1 & \longrightarrow & \dots & \longrightarrow & I^n & \longrightarrow & \dots \end{array}.$$

We can interpret this as a morphism $\phi^{\cdot} : I^{\cdot} \longrightarrow I^{\cdot}$ of complexes. Clearly, since $H^0(I^{\cdot}) = V$, we have $H^0(\phi^{\cdot}) = 0$. Therefore, $\phi^{\cdot}$ is homotopic to 0. By applying the functor F to this diagram we get

$$\begin{array}{ccccccccccc} 0 & \longrightarrow & F(I^0) & \longrightarrow & F(I^1) & \longrightarrow & \dots & \longrightarrow & F(I^n) & \longrightarrow & \dots \\ & & \downarrow F(z) & & \downarrow F(z) & & & & \downarrow F(z) & & \\ 0 & \longrightarrow & F(I^0) & \longrightarrow & F(I^1) & \longrightarrow & \dots & \longrightarrow & F(I^n) & \longrightarrow & \dots \end{array},$$

i.e., a morphism $F(\phi^{\cdot}) : F(I^{\cdot}) \longrightarrow F(I^{\cdot})$ of complexes. Since $\phi^{\cdot}$ is homotopic to 0, $F(\phi^{\cdot})$ is also homotopic to 0. This implies that all $H^p(\phi^{\cdot}) : H^p(I^{\cdot}) \longrightarrow H^p(I^{\cdot})$, $p \in \mathbb{Z}$, are equal to 0. Since $H^p(I^{\cdot}) = R^pF(V) = \operatorname{Ext}^p_{\mathscr{U}(\mathfrak{g})}(U, V)$, we see that $\operatorname{Ext}^p_{\mathscr{U}(\mathfrak{g})}(U, V)$ are annihilated by z.

On the other hand, by the first remark in the proof, z must act on $\operatorname{Ext}^p_{\mathscr{U}(\mathfrak{g})}(U, V)$ as multiplication by $\chi_U(z)$.

Since $\chi_U \neq \chi_V$, there exists $z \in \ker \chi_V$ such that $\chi_U(z) \neq 0$. This implies that $\operatorname{Ext}^p_{\mathscr{U}(\mathfrak{g})}(U, V)$ must be zero for all $p \in \mathbb{Z}_+$. □

1.2 *Casselman–Osborne Lemma*

Now we assume that $\mathfrak{g}$ is a complex semisimple Lie algebra. Let $\mathfrak{h}$ be a Cartan subalgebra of $\mathfrak{g}$, R the root system of $(\mathfrak{g}, \mathfrak{h})$ and R^+ a set of positive roots. Let $\mathfrak{n}$ be the nilpotent Lie algebra spanned by root subspaces of positive roots. Let $\gamma : \mathscr{Z}(\mathfrak{g}) \longrightarrow \mathscr{U}(\mathfrak{h})$ be the Harish-Chandra homomorphism, i.e., the algebra morphisms such that $z - \gamma(z) \in \mathfrak{n}\,\mathscr{U}(\mathfrak{g})$ [2, Ch. VIII, §6, no. 4].

Let V be a $\mathscr{U}(\mathfrak{g})$-module. Since $\mathfrak{h}$ normalizes $\mathfrak{n}$, the quotient $V/\mathfrak{n}V = \mathbb{C} \otimes_{\mathscr{U}(\mathfrak{n})} V$ has a natural structure of $\mathscr{U}(\mathfrak{h})$-module. Also, $\mathscr{Z}(\mathfrak{g})$ acts naturally on $V/\mathfrak{n}V$, and this action is given by the composition of γ and the $\mathscr{U}(\mathfrak{h})$-action.

We can consider $F(V) = V/\mathfrak{n}V$ as a right exact functor F from the category of $\mathscr{U}(\mathfrak{g})$-modules into the category of $\mathscr{U}(\mathfrak{h})$-modules. Let $\mathrm{For}_{\mathfrak{g}}$ denote the forgetful functor from the category of $\mathscr{U}(\mathfrak{g})$-modules into the category of $\mathscr{U}(\mathfrak{n})$-modules. Let $\mathrm{For}_{\mathfrak{h}}$ denote the forgetful functor from the category of $\mathscr{U}(\mathfrak{h})$-modules into the category of linear spaces. Then we have the following commutative diagram

$$\begin{array}{ccc} \mathcal{M}(\mathcal{U}(\mathfrak{g})) & \xrightarrow{F} & \mathcal{M}(\mathcal{U}(\mathfrak{h})) \\ {\scriptstyle \mathrm{For}_{\mathfrak{g}}}\big\downarrow & & \big\downarrow{\scriptstyle \mathrm{For}_{\mathfrak{h}}} \\ \mathcal{M}(\mathcal{U}(\mathfrak{n})) & \xrightarrow[H_0(\mathfrak{n},-)]{} & \mathcal{M}(\mathbb{C}) \end{array} .$$

By the Poincaré–Birkhoff–Witt theorem, a free $\mathcal{U}(\mathfrak{g})$-module is a free $\mathcal{U}(\mathfrak{n})$-module, hence we can use free left resolutions in $\mathcal{M}(\mathcal{U}(\mathfrak{g}))$ to calculate Lie algebra homology $H_p(\mathfrak{n}, -)$ of $\mathcal{U}(\mathfrak{g})$-modules, i.e., we get the commutative diagram

$$\begin{array}{ccc} \mathcal{M}(\mathcal{U}(\mathfrak{g})) & \xrightarrow{L_pF} & \mathcal{M}(\mathcal{U}(\mathfrak{h})) \\ {\scriptstyle \mathrm{For}_{\mathfrak{g}}}\big\downarrow & & \big\downarrow{\scriptstyle \mathrm{For}_{\mathfrak{h}}} \\ \mathcal{M}(\mathcal{U}(\mathfrak{n})) & \xrightarrow[H_p(\mathfrak{n},-)]{} & \mathcal{M}(\mathbb{C}) \end{array} ,$$

for any $p \in \mathbb{Z}_+$. Therefore, Lie algebra homology groups $H_p(\mathfrak{n}, -)$ of $\mathcal{U}(\mathfrak{g})$-modules have the structure of $\mathcal{U}(\mathfrak{h})$-modules.

Theorem 1.2. *Let V be an object in $\mathcal{M}(\mathcal{U}(\mathfrak{g}))$. Let $z \in \mathcal{Z}(\mathfrak{g})$ be an element which annihilates V. Then $\gamma(z)$ annihilates $H_p(\mathfrak{n}, V)$, $p \in \mathbb{Z}_+$.*

Proof. Let $z \in \mathcal{Z}(\mathfrak{g})$. Let

$$\dots \longrightarrow P_n \longrightarrow \dots \longrightarrow P_1 \longrightarrow P_0 \longrightarrow V \longrightarrow 0$$

be a projective resolution of V in $\mathcal{M}(\mathcal{U}(\mathfrak{g}))$. Multiplication by z gives the following commutative diagram:

$$\begin{array}{ccccccccc} \dots \longrightarrow & P_n & \longrightarrow \dots \longrightarrow & P_1 & \longrightarrow & P_0 & \longrightarrow & 0 \\ & {\scriptstyle z}\big\downarrow & & \big\downarrow{\scriptstyle z} & & \big\downarrow{\scriptstyle z} & & \\ \dots \longrightarrow & P_n & \longrightarrow \dots \longrightarrow & P_1 & \longrightarrow & P_0 & \longrightarrow & 0 \end{array}$$

We can interpret this diagram as a morphism $\psi. : P. \longrightarrow P.$ of complexes of $\mathcal{U}(\mathfrak{h})$-modules. Applying the functor F we get the diagram

$$\begin{array}{ccccccccc} \cdots \longrightarrow & F(P_n) & \longrightarrow \cdots \longrightarrow & F(P_1) & \longrightarrow & F(P_0) & \longrightarrow 0 \\ & \downarrow F(z) & & \downarrow F(z) & & \downarrow F(z) & \\ \cdots \longrightarrow & F(P_n) & \longrightarrow \cdots \longrightarrow & F(P_1) & \longrightarrow & F(P_0) & \longrightarrow 0 \end{array}$$

representing $F(\psi.)$, where $F(z)$ is the multiplication by $\gamma(z)$.

Now, assume that $z \in \mathscr{Z}(\mathfrak{g})$ annihilates V. Then we have $H^0(\psi.) = 0$. It follows that $\psi.$ is homotopic to 0. This in turn implies that $F(\psi.)$ is homotopic to 0. Hence, the multiplication by $\gamma(z)$ on $F(P.)$ is homotopic to zero. Therefore, the multiplication by $\gamma(z)$ annihilates the cohomology groups of the complex $F(P.)$, i.e., $\gamma(z) \cdot H_p(\mathfrak{n}, V) = 0$ for $p \in \mathbb{Z}_+$. □

2 Centers of Derived Categories

2.1 Center of an Additive Category

Let $\mathscr{A}$ be an additive category. This implies that for any object V in $\mathscr{A}$, all its endomorphisms form a ring $\operatorname{End}(V)$ with identity id_V.

An endomorphism z of the identity functor on $\mathscr{A}$ is an assignment to each object U in $\mathscr{A}$ of an endomorphism z_U of U such that for any two objects U and V in $\mathscr{A}$ and any morphism $\varphi : U \longrightarrow V$ we have $z_V \circ \varphi = \varphi \circ z_U$.

Lemma 2.1. *Let z be an endomorphism of the identity functor on $\mathscr{A}$ and V an object in $\mathscr{A}$. Then z_V is in the center of the ring* $\operatorname{End}(V)$.

Proof. Let $e : V \longrightarrow V$ be an endomorphism of V. Then, $z_V \circ e = e \circ z_V$, i.e., z_V commutes with e. This implies that z_V is in the center of $\operatorname{End}(V)$. □

All endomorphisms of the identity functor on $\mathscr{A}$ form a commutative ring with identity which is called the *center* $Z(\mathscr{A})$ of $\mathscr{A}$.

Let $\mathscr{B}$ be the full additive subcategory of $\mathscr{A}$. Then, by restriction, any element of the center of $\mathscr{A}$ determines an element of the center of $\mathscr{B}$. Clearly, the induced map $r : Z(\mathscr{A}) \longrightarrow Z(\mathscr{B})$ is a ring homomorphism. If the inclusion functor $\mathscr{B} \longrightarrow \mathscr{A}$ is an equivalence of categories, the morphism of centers is an isomorphism.

Let U and V be two objects in $\mathscr{A}$. Then the center $Z(\mathscr{A})$ acts naturally on $\operatorname{Hom}(U, V)$ by

$$z(\varphi) = z_V \circ \varphi = \varphi \circ z_U$$

for $z \in Z(\mathscr{A})$. Therefore, $\operatorname{Hom}(U, V)$ has a natural structure of a $Z(\mathscr{A})$-module. Clearly, in this way $\operatorname{Hom}(-,-)$ becomes a bifunctor from $\mathscr{A}^\circ \times \mathscr{A}$ into the category of $Z(\mathscr{A})$-modules.[1]

Assume that $\mathscr{C}$ is a triangulated category and T its translation functor. Let z be an element of the center of $\mathscr{C}$. Let U and V be two objects in $\mathscr{C}$ and $\varphi : U \longrightarrow V$ a morphism. Then $T^{-1}(\varphi) : T^{-1}(U) \longrightarrow T^{-1}(V)$ is a morphism and we have

$$z_{T^{-1}(V)} \circ T^{-1}(\varphi) = T^{-1}(\varphi) \circ z_{T^{-1}(U)}.$$

By applying T to this equality we get

$$T(z_{T^{-1}(V)}) \circ \varphi = \varphi \circ T(z_{T^{-1}(U)}).$$

Since $\varphi : U \longrightarrow V$ is arbitrary, we conclude that the assignment $U \longmapsto T(z_{T^{-1}(U)})$ is an element of the center of $\mathscr{A}$, which we denote by $T(z)$. It follows that T induces an automorphism of the center $Z(\mathscr{C})$ of $\mathscr{C}$. The elements of the center $Z(\mathscr{C})$ fixed by this automorphisms form a subring with identity which we call the *t-center* of $\mathscr{C}$ and denote by $Z_0(\mathscr{C})$.

Let

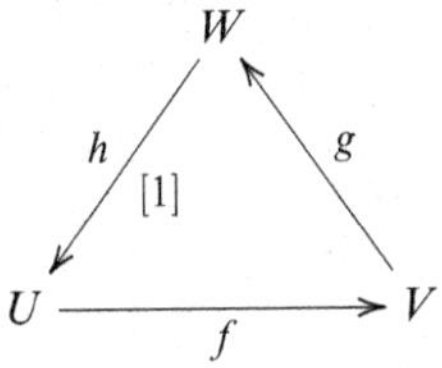

be a distinguished triangle in $\mathscr{C}$ and z an element of the t-center $Z_0(\mathscr{C})$ of $\mathscr{C}$. Clearly, since z is in the center, we have the commutative diagram

$$\begin{array}{ccccc} U & \xrightarrow{f} & V & \xrightarrow{g} & W \\ {\scriptstyle z_U}\downarrow & & \downarrow{\scriptstyle z_V} & & \downarrow{\scriptstyle z_W} \\ U & \xrightarrow[f]{} & V & \xrightarrow[g]{} & W \end{array}.$$

Moreover, since z is in the t-center of $\mathscr{C}$, we have $T(z_U) = z_{T(U)}$ and the diagram

[1] $\mathscr{A}^\circ$ is the category opposite to $\mathscr{A}$.

$$\begin{array}{ccc} W & \xrightarrow{h} & T(U) \\ {\scriptstyle z_W}\downarrow & & \downarrow{\scriptstyle T(z_U)} \\ W & \xrightarrow[h]{} & T(U) \end{array}$$

commutes. Therefore,

$$\begin{array}{ccccccc} U & \xrightarrow{f} & V & \xrightarrow{g} & W & \xrightarrow{h} & T(U) \\ {\scriptstyle z_U}\downarrow & & \downarrow{\scriptstyle z_V} & & \downarrow{\scriptstyle z_W} & & \downarrow{\scriptstyle T(z_U)} \\ U & \xrightarrow[f]{} & V & \xrightarrow[g]{} & W & \xrightarrow[h]{} & T(U) \end{array}$$

is an endomorphism of the above distinguished triangle. It follows that the elements of the t-center induce endomorphisms of distinguished triangles in $\mathscr{C}$.

Let X be another object of $\mathscr{C}$. The above remark implies that the distinguished triangle determines long exact sequences

$$\cdots \to \mathrm{Hom}(X, U) \to \mathrm{Hom}(X, V) \to \mathrm{Hom}(X, W) \to \mathrm{Hom}(X, T(U)) \to \dots$$

and

$$\cdots \to \mathrm{Hom}(T(U), X) \to \mathrm{Hom}(W, X) \to \mathrm{Hom}(V, X) \to \mathrm{Hom}(U, X) \to \dots$$

of $Z_0(\mathscr{C})$-modules.

2.2 Center of a Derived Category

Let $C^*(\mathscr{A})$ (where $*$ is b, $+$, $-$ or nothing, respectively) be the category of (bounded, bounded from below, bounded from above or unbounded) complexes of objects of $\mathscr{A}$. Then $C^*(\mathscr{A})$ is also an additive category.

Let z be an element of the center of $\mathscr{A}$. If

$$\dots \longrightarrow V^0 \longrightarrow V^1 \longrightarrow \dots \longrightarrow V^n \longrightarrow \dots$$

is an object in $C^*(\mathscr{A})$, we get the commutative diagram

$$
\begin{array}{ccccccccccc}
\dots & \longrightarrow & V^0 & \longrightarrow & V^1 & \longrightarrow & \dots & \longrightarrow & V^n & \longrightarrow & \dots \\
 & & \Big\downarrow{\scriptstyle z_{V^0}} & & \Big\downarrow{\scriptstyle z_{V^1}} & & & & \Big\downarrow{\scriptstyle z_{V^n}} & & \\
\dots & \longrightarrow & V^0 & \longrightarrow & V^1 & \longrightarrow & \dots & \longrightarrow & V^n & \longrightarrow & \dots
\end{array}
$$

which we can interpret as an endomorphism $z_{V^\cdot}$ of $V^\cdot$.

Let $\varphi^\cdot : U^\cdot \longrightarrow V^\cdot$ be a morphism in $C^*(\mathscr{A})$. Then $z_{V^p} \circ \varphi^p = \varphi^p \circ z_{U^p}$ for any $p \in \mathbb{Z}$, i.e., $z_{V^\cdot} \circ \varphi^\cdot = \varphi^\cdot \circ z_{U^\cdot}$. Therefore, the assignment $V^\cdot \longmapsto z_{V^\cdot}$ defines an element $C^*(z)$ of the center of $C^*(\mathscr{A})$. Moreover, we have the following trivial observation.

Lemma 2.2. *The map $z \longmapsto C^*(z)$ defines a homomorphism of the center $Z(\mathscr{A})$ of $\mathscr{A}$ into the center $Z(C^*(\mathscr{A}))$ of $C^*(\mathscr{A})$.*

Let $K^*(\mathscr{A})$ be the corresponding homotopic category of complexes. Let $[z_{V^\cdot}]$ be the homotopy class of endomorphism $z_{V^\cdot}$ of $V^\cdot$ in $C^*(\mathscr{A})$. Then it defines an endomorphism of $V^\cdot$ in $K^*(\mathscr{A})$. Clearly, the assignment $V^\cdot \longmapsto [z_{V^\cdot}]$ is an endomorphism $K^*(z)$ of the identity functor in $K^*(\mathscr{A})$. Moreover, the category $K^*(\mathscr{A})$ is triangulated and the translation functor is given by $T(U^\cdot)^p = U^{p+1}$ for any $p \in \mathbb{Z}$ for any object $U^\cdot$ in $K^*(\mathscr{A})$. If a morphism $\varphi : U^\cdot \longrightarrow V^\cdot$ is the homotopy class of a morphism of complexes $f^\cdot : U^\cdot \longrightarrow V^\cdot$, the morphism $T(\varphi) : T(U^\cdot) \longrightarrow T(V^\cdot)$ is the homotopy class of the morphism of complexes given by $f^{p+1} : T(U^\cdot)^p \longrightarrow T(V^\cdot)^p$ for $p \in \mathbb{Z}$. This immediately implies that $T([z_{U^\cdot}]) = [z_{T(U^\cdot)}]$ for any element z of the center of $\mathscr{A}$. It follows that $K^*(z)$ is in the t-center $Z_0(K^*(\mathscr{A}))$ of $K^*(\mathscr{A})$.

Therefore, we have the following observation.

Lemma 2.3. *The map $z \longmapsto K^*(z)$ defines a homomorphism of the center $Z(\mathscr{A})$ of $\mathscr{A}$ into the t-center $Z_0(K^*(\mathscr{A}))$ of $K^*(\mathscr{A})$.*

Finally, assume that $\mathscr{A}$ is an abelian category and let $D^*(\mathscr{A})$ be the corresponding derived category of $\mathscr{A}$, i.e., the localization of $K^*(\mathscr{A})$ with respect to all quasiisomorphisms. Clearly, for any $z \in Z(\mathscr{A})$, $[z_{V^\cdot}]$ determines an endomorphism $[[z_{V^\cdot}]]$ of $V^\cdot$ in $D^*(\mathscr{A})$.

Let $U^\cdot$ and $V^\cdot$ be two complexes in $D^*(\mathscr{A})$ and $\varphi : U^\cdot \longrightarrow V^\cdot$ a morphism of $U^\cdot$ into $V^\cdot$ in $D^*(\mathscr{A})$. We can represent φ by a roof (see, for example [4]):

$$
\begin{array}{ccccc}
 & & W^\cdot & & \\
 & {\scriptstyle s}\swarrow\,{\scriptstyle \sim} & & \searrow{\scriptstyle f} & \\
U^\cdot & & & & V^\cdot
\end{array}
$$

where $s : U^\cdot \longrightarrow W^\cdot$ is a quasiisomorphism and $f : W^\cdot \longrightarrow V^\cdot$ is a morphism in $K^*(\mathscr{A})$. On the other hand, $[[z_{U^\cdot}]]$ and $[[z_{V^\cdot}]]$ are represented by roofs

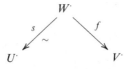

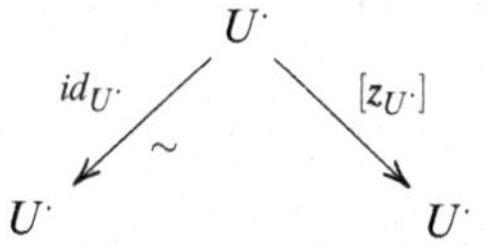

and

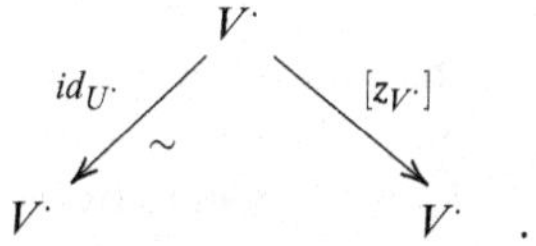

To calculate the composition $[[z_{V^\cdot}]] \circ \varphi$ we consider the composition diagram

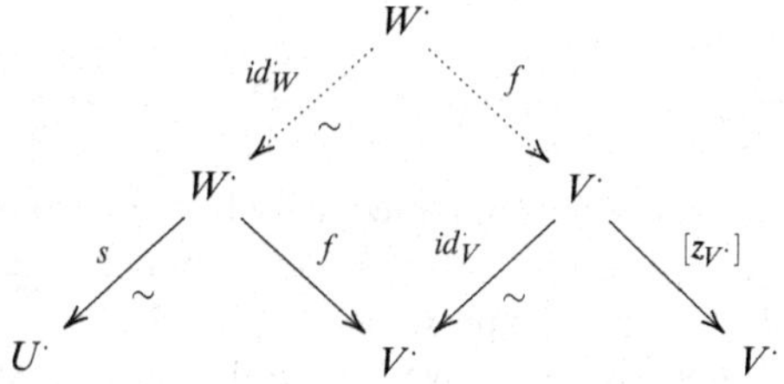

which obviously commutes. This implies that the composition is represented by the roof

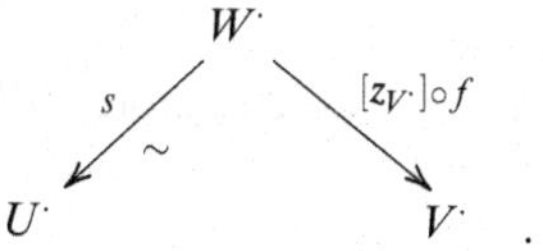

Analogously, to calculate $\varphi \circ [[z_{U^\cdot}]]$ we consider the composition diagram

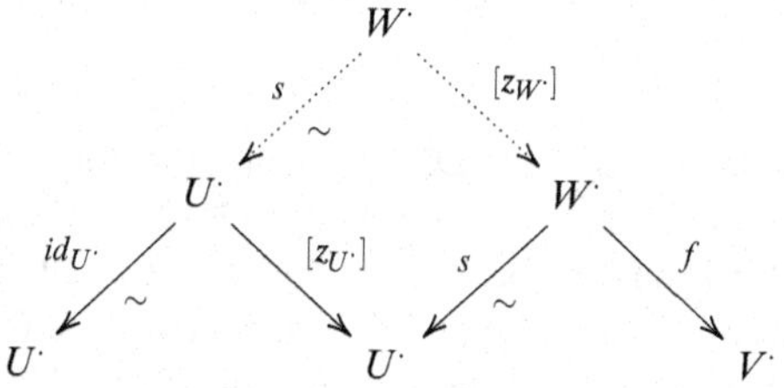

which commutes since $K^*(z)$ is in the center of $K^*(\mathscr{A})$. This implies that the composition is represented by the roof

$$\begin{array}{ccc} & W^{\cdot} & \\ {\scriptstyle s}\swarrow{\scriptstyle\sim} & & \searrow{\scriptstyle f\circ[z_{W^{\cdot}}]} \\ U^{\cdot} & & V^{\cdot} \end{array} .$$

Since $f \circ [z_{W^{\cdot}}] = [z_{V^{\cdot}}] \circ f$, these two roofs are identical and $[[z_{V^{\cdot}}]] \circ \varphi = \varphi \circ [[z_{V^{\cdot}}]]$. Hence, the assignment $V^{\cdot} \longmapsto [[z_{V^{\cdot}}]]$ defines an element of the t-center $Z_0(D^*(\mathscr{A}))$ of $D^*(\mathscr{A})$ which we denote by $D^*(z)$. Moreover, we have the following result.

Lemma 2.4. *The map $z \longmapsto D^*(z)$ defines an injective morphism of the center $Z(\mathscr{A})$ of $\mathscr{A}$ into the t-center $Z_0(D^*(\mathscr{A}))$ of $D^*(\mathscr{A})$.*

For any $z \in Z(\mathscr{A})$, we have

$$H^p([[z_{V^{\cdot}}]]) = z_{H^p(V^{\cdot})} \text{ for any } V^{\cdot} \text{ in } D^*(\mathscr{A}) \text{ and any } p \in \mathbb{Z}.$$

Proof. The second statement follows immediately from the construction.

To prove injectivity, assume that $D^*(z) = 0$ for some $z \in Z(\mathscr{A})$. For an object V in $\mathscr{A}$, denote by $D(V)^{\cdot}$ the complex such that $D(V)^0 = V$ and $D(V)^p = 0$ for $p \neq 0$. By our assumption, we have $[[z_{D(V)^{\cdot}}]] = 0$. This implies that $z_V = H^0([[z_{D(V)^{\cdot}}]]) = 0$. Therefore, $z_V = 0$ for any V in $\mathscr{A}$, i.e., $z = 0$. □

Let z be an element of the t-center $Z_0(D^*(\mathscr{A}))$ of $D^*(\mathscr{A})$. Then

$$H^{p+1}(z_{U^{\cdot}}) = H^p(T(z_{U^{\cdot}})) = H^p(z_{T(U^{\cdot})})$$

for any object $U^{\cdot}$ in $D^*(\mathscr{A})$ and $p \in \mathbb{Z}$. Therefore, $H^0(z_{U^{\cdot}}) = 0$ for all objects $U^{\cdot}$ in $D^*(\mathscr{A})$ is equivalent to $H^p(z_{U^{\cdot}}) = 0$ for all objects $U^{\cdot}$ in $D^*(\mathscr{A})$ and $p \in \mathbb{Z}$. In particular,

$$\begin{aligned} I_0(D^*(\mathscr{A})) &= \{z \in Z_0(D^*(\mathscr{A})) \mid H^p(z_{U^{\cdot}}) = 0 \text{ for all } U^{\cdot} \text{ in } D^*(\mathscr{A}) \text{ and } p \in \mathbb{Z}\} \\ &= \{z \in Z_0(D^*(\mathscr{A})) \mid H^0(z_{U^{\cdot}}) = 0 \text{ for all } U^{\cdot} \text{ in } D^*(\mathscr{A})\} \end{aligned}$$

is an ideal in $Z_0(D^*(\mathscr{A}))$.

On the other hand, let $D : \mathscr{A} \longrightarrow D^*(\mathscr{A})$ be the functor which attaches to each object V in $\mathscr{A}$ the complex $D(V)^{\cdot}$, such that $D(V)^0 = V$ and $D(V)^p = 0$ for all $p \neq 0$. This functor is an isomorphism of $\mathscr{A}$ onto the full additive subcategory of $D^*(\mathscr{A})$ consisting of all complexes $U^{\cdot}$ such that $U^p = 0$ for all $p \neq 0$ [4]. Therefore, we have a natural homomorphism r of $Z(D^*(\mathscr{A}))$ into $Z(\mathscr{A})$ which attaches to an element z of the center of $D^*(\mathscr{A})$ the element of the center of $\mathscr{A}$

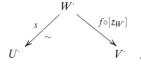

given by $V \longmapsto H^0(z_{D(V)})$ for any V in $\mathscr{A}$. In particular, we have a natural homomorphism $r : Z_0(D^*(\mathscr{A})) \longrightarrow Z(\mathscr{A})$.

From Lemma 2.4, we see that

$$r(D^*(z))_V = H^0(D^*(z)_{D(V)}) = H^0([[z_{D(V)}]]) = z_V$$

for any z in the center of $\mathscr{A}$ and any V in $\mathscr{A}$. Therefore, we have the following result.

Proposition 2.5. *The natural homomorphism $r : Z_0(D^*(\mathscr{A})) \longrightarrow Z(\mathscr{A})$ is a left inverse of the homomorphism $D^* : Z(\mathscr{A}) \longrightarrow Z_0(D^*(\mathscr{A}))$. In particular, it is surjective.*

Its kernel is the ideal $I_0(D^*(\mathscr{A}))$.

The situation is particularly nice for bounded derived categories.[2]

Proposition 2.6. *The natural homomorphism $r : Z_0(D^b(\mathscr{A})) \longrightarrow Z(\mathscr{A})$ is an isomorphism.*

Proof. We have to prove that $I_0(D^b(\mathscr{A})) = 0$. Let z be an element of $I_0(D^b(\mathscr{A}))$.

Clearly, for any object V in $\mathscr{A}$, we have $z_{D(V)} = 0$. Moreover, since z is in the t-center, $z_{T^p(D(V))} = 0$ for any $p \in \mathbb{Z}$.

For any object $U^\cdot$ in $D^b(\mathscr{A})$ we put

$$\ell(U^\cdot) = \operatorname{Card}\{p \in \mathbb{Z} \mid H^p(U^\cdot) \neq 0\},$$

and call $\ell(U^\cdot)$ the *cohomological length* of $U^\cdot$.

Now we want to prove that $z_{U^\cdot} = 0$ for all $U^\cdot$ in $D^b(\mathscr{A})$. The proof is by induction in the cohomological length $\ell(U^\cdot)$. If $\ell(U^\cdot) = 0$, $U^\cdot = 0$ and $z_{U^\cdot} = 0$. If $\ell(U^\cdot) = 1$, there exists $p \in \mathbb{Z}$ such that $H^q(U^\cdot) = 0$ for all $q \neq p$. In this case, $U^\cdot$ is isomorphic to the complex which is zero in all degrees $q \neq p$ and in degree p is equal to $H^p(U^\cdot)$, i.e., to $T^{-p}(D(H^p(U^\cdot)))$. Hence, by the above remark, $z_{U^\cdot} = 0$.

Assume now that $\ell(U^\cdot) > 1$. Let $\tau_{\leq p}$ and $\tau_{\geq p}$ be the usual truncation functors [4]. Then, for any $p \in \mathbb{Z}$, we have the truncation distinguished triangle

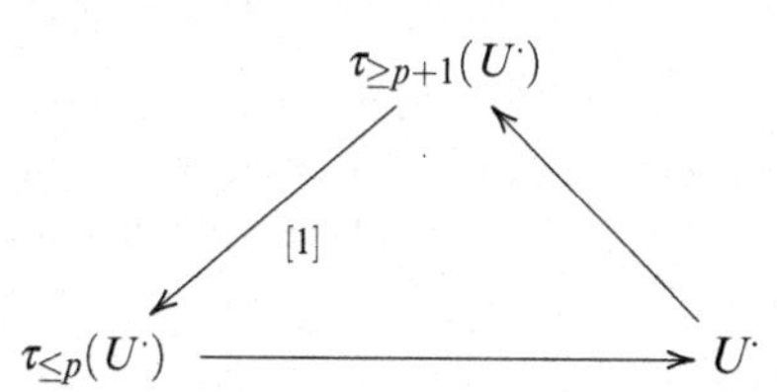

[2]I do not know any example where this result fails in unbounded case.

and by choosing a right $p \in \mathbb{Z}$, we have $\ell(\tau_{\leq p}(U^\cdot)) < \ell(U^\cdot)$ and $\ell(\tau_{\geq p+1}(U^\cdot)) < \ell(U^\cdot)$. Therefore, by the induction assumption, there exists $p \in \mathbb{Z}$ such that $z_{\tau_{\leq p}(U^\cdot)} = 0$ and $z_{\tau_{\geq p+1}(U^\cdot)} = 0$. As we remarked before, this distinguished triangle leads to the long exact sequence

$$\cdots \to \operatorname{Hom}(U^\cdot, \tau_{\leq p}(U^\cdot)) \to \operatorname{Hom}(U^\cdot, U^\cdot) \to \operatorname{Hom}(U^\cdot, \tau_{\geq p+1}(U^\cdot)) \to \dots$$

of $Z_0(D^b(\mathscr{A}))$-modules. By our construction, z annihilates the first and third module. Therefore, it must annihilate $\operatorname{Hom}(U^\cdot, U^\cdot)$ too. This implies that

$$0 = z(\mathrm{id}_{U^\cdot}) = \mathrm{id}_{U^\cdot} \circ z_{U^\cdot} = z_{U^\cdot}.$$

□

3 Centers and Derived Functors

3.1 Homogeneous Functors

Let $\mathscr{A}$ and $\mathscr{B}$ be two abelian categories. Let R be a commutative ring with identity and $\alpha : R \longrightarrow Z(\mathscr{A})$ and $\beta : R \longrightarrow Z(\mathscr{B})$ ring morphisms of rings with identity.

By Lemma 2.4, α and β define ring morphisms $\boldsymbol{\alpha} = D^* \circ \alpha : R \longrightarrow Z_0(D^*(\mathscr{A}))$ and $\boldsymbol{\beta} = D^* \circ \beta : R \longrightarrow Z_0(D^*(\mathscr{B}))$

Let $F : \mathscr{A} \longrightarrow \mathscr{B}$ be an additive functor. We say that F is *R-homogeneous* if for any $r \in R$ we have

$$\beta(r)_{F(V)} = F(\alpha(r)_V) \text{ for any object } V \text{ in } \mathscr{A}.$$

Assume now that F is left exact. Assume that there exists a subcategory $\mathscr{R}$ of $\mathscr{A}$ right adapted to F [4, ch. III, §6, no. 3].[3] Then F has the right derived functor $RF : D^+(\mathscr{A}) \longrightarrow D^+(\mathscr{B})$.

Theorem 3.1. *The functor $RF : D^+(\mathscr{A}) \longrightarrow D^+(\mathscr{B})$ is R-homogeneous.*

Proof. Let $V^\cdot$ be a complex in $D^+(\mathscr{A})$. Since $\mathscr{R}$ is right adapted to F, there exists a bounded from below complex $R^\cdot$ consisting of objects in $\mathscr{R}$ and a quasiisomorphism $q : V^\cdot \longrightarrow R^\cdot$. Let z be an element of the center of $\mathscr{A}$. Then we have the commutative diagram

[3] I would prefer a proof of the next theorem which doesn't use the construction of the derived functor, but its universal property. Unfortunately, I do not know such argument.

$$\begin{array}{ccc} V^{\cdot} & \xrightarrow{q} & R^{\cdot} \\ {\scriptstyle [[z_{V^{\cdot}}]]}\downarrow & & \downarrow{\scriptstyle [[z_{R^{\cdot}}]]} \\ V^{\cdot} & \xrightarrow[q]{} & R^{\cdot} \end{array} .$$

By applying the functor RF to it, we get the diagram

$$\begin{array}{ccc} RF(V^{\cdot}) & \xrightarrow{RF(q)} & F(R^{\cdot}) \\ {\scriptstyle RF([[z_{V^{\cdot}}]])}\downarrow & & \downarrow{\scriptstyle [[F(z)_{F(R^{\cdot})}]]} \\ RF(V^{\cdot}) & \xrightarrow[RF(q)]{} & F(R^{\cdot}) \end{array} .$$

If $r \in R$, $\alpha(r)$ is in the center of $\mathscr{A}$ and the above diagram implies that

$$\begin{array}{ccc} RF(V^{\cdot}) & \xrightarrow{RF(q)} & F(R^{\cdot}) \\ {\scriptstyle [[\beta(r)_{RF(V^{\cdot})}]]}\downarrow & & \downarrow{\scriptstyle [[\beta(r)_{F(R^{\cdot})}]]} \\ RF(V^{\cdot}) & \xrightarrow[RF(q)]{} & F(R^{\cdot}) \end{array} .$$

is commutative. Moreover, $\beta(r)$ is in the center of $\mathscr{B}$, hence we also have

$$\begin{array}{ccc} RF(V^{\cdot}) & \xrightarrow{RF(q)} & F(R^{\cdot}) \\ {\scriptstyle [[\beta(r)_{RF(V^{\cdot})}]]}\downarrow & & \downarrow{\scriptstyle [[\beta(r)_{F(R^{\cdot})}]]} \\ RF(V^{\cdot}) & \xrightarrow[RF(q)]{} & F(R^{\cdot}) \end{array} .$$

Hence, we conclude that $RF([[\alpha(r)_{V^{\cdot}}]]) = [[\beta(r)_{RF(V^{\cdot})}]]$, i.e., RF is R-homogeneous. □

Let V be an object in $\mathscr{A}$. Then

$$\boldsymbol{\beta}(r)_{RF(D(V))} = RF(\boldsymbol{\alpha}(r)_{D(V)}) \text{ for any } r \in R.$$

By taking cohomology, we get

$$\beta(r)_{R^pF(V)} = H^p(\boldsymbol{\beta}(r)_{RF(D(V))}) = R^pF(\alpha(r)_V) \text{ for any } r \in R \text{ and } p \in \mathbb{Z}_+.$$

Therefore, we have the following consequence.

Corollary 3.2. *The functors $R^pF : \mathscr{A} \longrightarrow \mathscr{B}$ are R-homogeneous.*

We leave to the reader the formulation and proofs of the analogous results for a right exact functor F and its left derived functor $LF : D^-(\mathscr{A}) \longrightarrow D^-(\mathscr{B})$.

3.2 Special Cases

Now we are going to illustrate how Theorems 1.1 and 1.2 follow from the above discussion.

First, we prove a well-known result about the center of the category of modules. This is not necessary for our applications, but puts the constructions in a proper perspective.

Let A be a ring with identity and Z its center. Let $\mathscr{M}(A)$ be the category of A-modules. Any element z in Z determines an endomorphism z_U of an A-module U. Clearly, the assignment $U \longmapsto z_U$ defines an element of the center $Z(\mathscr{M}(A))$ of $\mathscr{M}(A)$. Therefore, we have a natural homomorphism $i : Z \longrightarrow Z(\mathscr{M}(A))$ of rings.

Lemma 3.3. *The morphism $i : Z \longrightarrow Z(\mathscr{M}(A))$ is an isomorphism.*

Proof. If we consider A as an A-module for the left multiplication, we see that $i(z)_A$ is the multiplication by z for any $z \in Z$. Therefore, $i(z)_A(1) = z$ and $i : Z \longrightarrow Z(\mathscr{M}(A))$ is injective.

Let ζ be an element of the center of $\mathscr{A}$. Then ζ_A is an endomorphism of A considered as A-module for left multiplication. Let $z = \zeta_A(1)$. Then

$$\zeta_A(a) = a\zeta_A(1) = az$$

for any $a \in A$. Moreover, any $b \in A$ defines an endomorphism φ_b of A given by $\varphi_b(a) = ab$ for all $a \in A$. Since we must have $\zeta_A \circ \varphi_b = \varphi_b \circ \zeta_A$, it follows that

$$bz = (\zeta_A \circ \varphi_b)(1) = (\varphi_b \circ \zeta_A)(1) = zb.$$

Since $b \in A$ is arbitrary, z must be in the center Z of A.

Let M be an arbitrary A-module and $m \in M$. Then m determines a module morphism $\psi_m : A \longrightarrow M$ given by $\psi_m(a) = am$ for any $a \in A$. Therefore,

$$\zeta_M(m) = (\zeta_M \circ \psi_m)(1) = (\psi_m \circ \zeta_A)(1) = zm = i(z)_M m.$$

Hence $\zeta = i(z)$, and i is surjective. □

Now we return to the notation from the first section. By Lemma 3.3, the center of the category $\mathscr{M}(\mathscr{U}(\mathfrak{g}))$ is isomorphic to $\mathscr{Z}(\mathfrak{g})$.

First we discuss Theorem 1.1. The functor $F = \operatorname{Hom}_{\mathscr{U}(\mathfrak{g})}(U, -)$ is a functor from the category $\mathscr{U}(\mathfrak{g})$ into the category of $\mathscr{Z}(\mathfrak{g})$-modules. If we define α as the natural morphism of $\mathscr{Z}(\mathfrak{g})$ into the center of $\mathscr{M}(\mathscr{U}(\mathfrak{g}))$ and β as multiplication by $\chi_U(z)$, F is clearly $\mathscr{Z}(\mathfrak{g})$-homogeneous. This implies that the functors R^pF are

$\mathscr{Z}(\mathfrak{g})$-homogeneous. Hence, for any V in $\mathscr{M}(\mathscr{U}(\mathfrak{g}))$ we have $R^pF(z_V) = \chi_U(z)$ for all $p \in \mathbb{Z}_+$. In, particular, if $z \in \ker \chi_V$ we have

$$0 = R^pF(0) = R^pF(z_V) = \chi_U(z).$$

This clearly contradicts $\chi_U \neq \chi_V$ if $\operatorname{Ext}^p_{\mathscr{U}(\mathfrak{g})}(U, V) \neq 0$ for some $p \in \mathbb{Z}$.

No we discuss Theorem 1.2. The functor $F = H_0(\mathfrak{n}, -)$ is a functor from the category $\mathscr{U}(\mathfrak{g})$ into the category of $\mathscr{Z}(\mathfrak{g})$-modules. If we define α as the natural morphism of $\mathscr{Z}(\mathfrak{g})$ into the center of $\mathscr{M}(\mathscr{U}(\mathfrak{g}))$ and β as the composition of the Harish-Chandra homomorphism with the natural morphism of $\mathscr{U}(\mathfrak{h})$ into the center of $\mathscr{M}(\mathscr{U}(\mathfrak{h}))$, F is clearly $\mathscr{Z}(\mathfrak{g})$-homogeneous. This implies that the functors L_pF are $\mathscr{Z}(\mathfrak{g})$-homogeneous. Hence for any V in $\mathscr{M}(\mathscr{U}(\mathfrak{g}))$, we have $L_pF(z_V) = \gamma(z)_{L_pF(V)}$ for all $p \in \mathbb{Z}_+$. In, particular, if z annihilates V, $\gamma(z)$ annihilates $L_pF(V) = H_p(\mathfrak{n}, V)$ for all $p \in \mathbb{Z}$.

References

1. A. Borel and N. Wallach, *Continuous cohomology, discrete subgroups, and representations of reductive groups*, second ed., Mathematical Surveys and Monographs, Vol. 67, American Mathematical Society, Providence, RI, 2000.
2. Nicolas Bourbaki, *Groupes et algèbres de Lie*, Hermann, 1968.
3. William Casselman and M. Scott Osborne, *The* $\mathfrak{n}$*-cohomology of representations with an infinitesimal character*, Compositio Math. **31** (1975), no. 2, 219–227.
4. Sergei I. Gelfand and Yuri I. Manin, *Methods of homological algebra*, Springer-Verlag, Berlin, 1996.
5. Anthony W. Knapp and David A. Vogan, Jr., *Cohomological induction and unitary representations*, Princeton Mathematical Series, Vol. 45, Princeton University Press, Princeton, NJ, 1995.

Tensor Representations of Mackey Lie Algebras and Their Dense Subalgebras

Ivan Penkov and Vera Serganova

Abstract In this article we review the main results of the earlier papers [PStyr, PS] and [DPS], and establish related new results in considerably greater generality. We introduce a class of infinite-dimensional Lie algebras $\mathfrak{g}^M$, which we call Mackey Lie algebras, and define monoidal categories $\mathbb{T}_{\mathfrak{g}^M}$ of tensor $\mathfrak{g}^M$-modules. We also consider dense subalgebras $\mathfrak{a} \subset \mathfrak{g}^M$ and corresponding categories $\mathbb{T}_{\mathfrak{a}}$. The locally finite Lie algebras $\mathfrak{sl}(V, W), \mathfrak{o}(V), \mathfrak{sp}(V)$ are dense subalgebras of respective Mackey Lie algebras. Our main result is that if $\mathfrak{g}^M$ is a Mackey Lie algebra and $\mathfrak{a} \subset \mathfrak{g}^M$ is a dense subalgebra, then the monoidal category $\mathbb{T}_{\mathfrak{a}}$ is equivalent to $\mathbb{T}_{\mathfrak{sl}(\infty)}$ or $\mathbb{T}_{\mathfrak{o}(\infty)}$; the latter monoidal categories have been studied in detail in [DPS]. A possible choice of $\mathfrak{a}$ is the well-known Lie algebra of generalized Jacobi matrices.

Key words Finitary Lie algebra • Mackey Lie algebra • Linear system • Tensor representation • Socle filtration

Mathematics Subject Classification (2010): Primary 17B10, 17B65. Secondary 18D10.

Support by the DFG Priority Program for both authors.

Vera Serganova acknowledges support from NSF via grant 1303301.

I. Penkov (✉)
Jacobs University Bremen, Campus Ring 1,
28759, Bremen, Germany
e-mail: i.penkov@jacobs-university.de

V. Serganova
Department of Mathematics, University of California, Berkeley, CA 94720-3840, USA
e-mail: serganov@math.berkeley.edu

G. Mason et al. (eds.), *Developments and Retrospectives in Lie Theory: Algebraic Methods*, Developments in Mathematics 38,
DOI 10.1007/978-3-319-09804-3_14

Introduction

This paper combines a review of some results on locally finite Lie algebras, mostly from [PStyr, PS] and [DPS], with new results about categories of representations of a class of (not locally finite) infinite-dimensional Lie algebras which we call Mackey Lie algebras. Locally finite Lie algebras (i.e., Lie algebras in which any finite set of elements generates a finite-dimensional Lie subalgebra) and their representations have been gaining the attention of researchers in the past 20 years. An incomplete list of references on this topic is: [Ba1, BB, BS, DiP1, DiP3, DPS, DPSn, DPW, DaPW, N, Na, NP, NS, O, PS, PStyr, PZ]. In particular, in [PStyr, PS] and [DPS] integrable representations of the three classical locally finite Lie algebras $\mathfrak{g} = \mathfrak{sl}(\infty), \mathfrak{o}(\infty), \mathfrak{sp}(\infty)$ have been studied from various points of view. An important step in the development of the representation theory of these Lie algebras has been the introduction of the category of tensor modules $\mathbb{T}_{\mathfrak{g}}$ in [DPS].

In the present article we shift the focus to understanding a natural generality in which the category $\mathbb{T}_{\mathfrak{g}}$ is defined. In particular, we consider the finitary locally simple Lie algebras $\mathfrak{g} = \mathfrak{sl}(V, W), \mathfrak{o}(V), \mathfrak{sp}(V)$, where V is an arbitrary vector space (not necessarily of countable dimension), and either a nondegenerate pairing $V \times W \to \mathbb{C}$ is given, or V is equipped with a nondegenerate symmetric, or antisymmetric form. In Sects. 1–5 we reproduce the most important results from [PStyr] and [DPS] in this greater generality. In fact, we study five different categories of integrable modules, see Sect. 3.6, but pay maximum attention to the category $\mathbb{T}_{\mathfrak{g}}$. The central new result in this part of the paper is Theorem 5.5, claiming that the category $\mathbb{T}_{\mathfrak{g}}$ for $\mathfrak{g} = \mathfrak{sl}(V, W), \mathfrak{o}(V), \mathfrak{sp}(V)$ is canonically equivalent, as a monoidal category, to the respective category $\mathbb{T}_{\mathfrak{sl}(\infty)}$, $\mathbb{T}_{\mathfrak{o}(\infty)}$ or $\mathbb{T}_{\mathfrak{sp}(\infty)}$. It is shown in [DPS] that each of the latter categories is Koszul and that $\mathbb{T}_{\mathfrak{sl}(\infty)}$ is self-dual Koszul, while $\mathbb{T}_{\mathfrak{o}(\infty)}$ and $\mathbb{T}_{\mathfrak{sp}(\infty)}$ are not self-dual but are equivalent.

In the second part of the paper, starting with Sect. 6, we explore several new ideas. The first one is that given a nondegenerate pairing $V \times W \to \mathbb{C}$ between two vector spaces, or a nondegenerate symmetric or antisymmetric form on a vector space V, there is a canonical, in general not locally finite, Lie algebra attached to this datum. Indeed, fix a pairing $V \times W \to \mathbb{C}$. Then the Mackey Lie algebra $\mathfrak{gl}^M(V, W)$ is the Lie algebra of all endomorphisms of V whose duals keep W stable (this definition is given in a more precise form at the beginning of Sect. 6). Similarly, if V is equipped with a nondegenerate form, the respective Lie algebra $\mathfrak{o}^M(V)$ or $\mathfrak{sp}^M(V)$ is the Lie algebra of all endomorphisms of V for which the form is invariant.

The Lie algebras $\mathfrak{gl}^M(V, W), \mathfrak{o}^M(V), \mathfrak{sp}^M(V)$ are not simple as they have obvious ideals: these are respectively $\mathfrak{gl}(V, W) \oplus \mathbb{C}\mathrm{Id}$, $\mathfrak{o}(\infty)$, and $\mathfrak{sp}(\infty)$. However, we prove that, if both V and W are countable dimensional, the quotients $\mathfrak{gl}^M(V, W)/(\mathfrak{gl}(V, W) \oplus \mathbb{C}\mathrm{Id})$, $\mathfrak{o}^M(V)/\mathfrak{o}(V)$, $\mathfrak{sp}^M(V)/\mathfrak{sp}(V)$ are simple Lie algebras. This result is an algebraic analogue of the simplicity of the Calkin algebra in functional analysis.

Despite the fact that the Lie algebras $\mathfrak{gl}^M(V, W), \mathfrak{o}^M(V), \mathfrak{sp}^M(V)$ are completely natural objects, the representation theory of these Lie algebras has not yet been explored. We are undertaking the first step of such an exploration by introducing the categories of tensor modules $\mathbb{T}_{\mathfrak{g}^M}$ for $\mathfrak{g}^M = \mathfrak{gl}^M(V, W), \mathfrak{o}^M(V), \mathfrak{sp}^M(V)$. Our main result about these categories is Theorem 7.10 which implies that $\mathbb{T}_{\mathfrak{gl}^M(V,W)}$ is equivalent to $\mathbb{T}_{\mathfrak{sl}(\infty)}$, and $\mathbb{T}_{\mathfrak{o}^M(V)}$ and $\mathbb{T}_{\mathfrak{sp}^M(V)}$ are equivalent respectively to $\mathbb{T}_{\mathfrak{o}(\infty)}$ and $\mathbb{T}_{\mathfrak{sp}(\infty)}$.

A further idea is to consider dense subalgebras $\mathfrak{a}$ of the Lie algebras $\mathfrak{g}^M$ (see the definition in Sect. 7). We show that if $\mathfrak{a} \subset \mathfrak{g}$ is a dense subalgebra, the category $\mathbb{T}_{\mathfrak{a}}$, whose objects are tensor modules of $\mathfrak{g}$ considered as $\mathfrak{a}$-modules, is canonically equivalent to $\mathbb{T}_{\mathfrak{g}^M}$, and hence to one of the categories $\mathbb{T}_{\mathfrak{sl}(\infty)}$ or $\mathbb{T}_{\mathfrak{o}(\infty)}$. It is interesting that this result applies to the Lie algebra of generalized Jacobi matrices (infinite matrices with "finitely many nonzero diagonals") which has been studied for over 30 years, see for instance [FT].

In short, the main point of this paper is that the categories of tensor modules $\mathbb{T}_{\mathfrak{sl}(\infty)}, \mathbb{T}_{\mathfrak{o}(\infty)}, \mathbb{T}_{\mathfrak{sp}(\infty)}$ introduced in [DPS] are in some sense universal, being naturally equivalent to the respective categories of tensor representations of a large class of, possibly not locally finite, infinite-dimensional Lie algebras.

1 Preliminaries

The ground field is $\mathbb{C}$. By M^* we denote the dual space of a vector space M, i.e., $M^* = \mathrm{Hom}_{\mathbb{C}}(M, \mathbb{C})$. S_n stands for the symmetric group on n letters. The sign $\subset$ denotes not necessarily strict inclusion. By definition, a *natural representation* (or a *natural module*) of a classical simple finite-dimensional Lie algebra is a simple nontrivial finite-dimensional representation of minimal dimension.

In this paper $\mathfrak{g}$ denotes a *locally simple locally finite* Lie algebra, i.e., an infinite-dimensional Lie algebra $\mathfrak{g}$ obtained as the direct limit $\varinjlim \mathfrak{g}_\alpha$ of a directed system of embeddings (i.e., injective homomorphisms) $\mathfrak{g}_\alpha \hookrightarrow \mathfrak{g}_\beta$ of finite-dimensional simple Lie algebras parametrized by a directed set of indices. It is clear that any such $\mathfrak{g}$ is a simple Lie algebra. If $\mathfrak{g}$ is countable dimensional, then the above directed set can always be chosen as $\mathbb{Z}_{\geq 1}$, and the corresponding directed system can be chosen as a chain

$$\mathfrak{g}_1 \hookrightarrow \mathfrak{g}_2 \hookrightarrow \ldots \hookrightarrow \mathfrak{g}_i \hookrightarrow \mathfrak{g}_{i+1} \hookrightarrow \ldots . \tag{1}$$

In this case we write $\mathfrak{g} = \varinjlim \mathfrak{g}_i$. Moreover, if $\mathfrak{g}_i = \mathfrak{sl}(i+1)$, then up to isomorphism there is only one such Lie algebra which we denote by $\mathfrak{sl}(\infty)$. Similarly, if $\mathfrak{g}_i = \mathfrak{o}(i)$ or $\mathfrak{g}_i = \mathfrak{sp}(2i)$, up to isomorphism one obtains only two Lie algebras: $\mathfrak{o}(\infty)$ and $\mathfrak{sp}(\infty)$. The Lie algebras $\mathfrak{sl}(\infty)$, $\mathfrak{o}(\infty)$, $\mathfrak{sp}(\infty)$ are often referred to as the *finitary locally simple Lie algebras* [Ba1, Ba2, BS], or as the *classical locally simple Lie algebras* [PS].

A more general (and very interesting) class of locally finite locally simple Lie algebras are the diagonal locally finite Lie algebras introduced by Y. Bahturin and H. Strade in [BhS]. We recall that an injective homomorphism $\mathfrak{g}_1 \hookrightarrow \mathfrak{g}_2$ of simple classical Lie algebras of the same type $\mathfrak{sl}$, $\mathfrak{o}$, $\mathfrak{sp}$, is *diagonal* if the pull-back $V_{\mathfrak{g}_2\downarrow\mathfrak{g}_1}$ of a natural representation $V_{\mathfrak{g}_2}$ of $\mathfrak{g}_2$ to $\mathfrak{g}_1$ is isomorphic to a direct sum of copies of a natural representation $V_{\mathfrak{g}_1}$, of its dual $V^*_{\mathfrak{g}_1}$, and of the trivial 1-dimensional representation. In this paper, by a *diagonal Lie algebra* $\mathfrak{g}$ we mean an infinite-dimensional Lie algebra obtained as the limit of a directed system of diagonal homomorphisms of classical simple Lie algebras $\mathfrak{g}_\alpha$. We say that a diagonal Lie algebra *is of type* $\mathfrak{sl}$ (respectively, $\mathfrak{o}$ or $\mathfrak{sp}$) if all $\mathfrak{g}_\alpha$ can be chosen to have type $\mathfrak{sl}$ (respectively, $\mathfrak{o}$ or $\mathfrak{sp}$).

Countable-dimensional diagonal Lie algebras have been classified up to isomorphism by A. Baranov and A. Zhilinskii [BaZh]. S. Markouski [Ma] has determined when there is an embedding $\mathfrak{g} \hookrightarrow \mathfrak{g}'$ for given countable-dimensional diagonal Lie algebras $\mathfrak{g}$ and $\mathfrak{g}'$. If both $\mathfrak{g}$ and $\mathfrak{g}'$ are classical locally simple Lie algebras, then an embedding $\mathfrak{g} \hookrightarrow \mathfrak{g}'$ always exists, and such embeddings have been studied in detail in [DiP2].

Let V and W be two infinite-dimensional vector spaces with a nondegenerate pairing $V \times W \to \mathbb{C}$. G. Mackey calls such a pair V, W a *linear system* and was the first to study linear systems in depth [M]. The tensor product $V \otimes W$ is an associative algebra (without identity), and we denote the corresponding Lie algebra by $\mathfrak{gl}(V, W)$. The pairing $V \times W \to \mathbb{C}$ induces a homomorphism of Lie algebras $\mathrm{tr} : \mathfrak{gl}(V, W) \to \mathbb{C}$. The kernel of this homomorphism is denoted by $\mathfrak{sl}(V, W)$. The Lie algebra $\mathfrak{sl}(V, W)$ is a locally simple locally finite Lie algebra. A corresponding directed system is given by $\{\mathfrak{sl}(V_f, W_f)\}$, where V_f and W_f run over all finite-dimensional subspaces $V_f \subset V$, $W_f \subset W$ such that the restriction of the pairing $V \times W \to \mathbb{C}$ to $V_f \times W_f$ is nondegenerate. If V and W are countable dimensional, then $\mathfrak{sl}(V, W)$ is isomorphic to $\mathfrak{sl}(\infty)$. In what follows we call a pair of finite-dimensional subspaces $V_f \subset V$, $W_f \subset W$ a *finite-dimensional nondegenerate pair* if the restriction of the pairing $V \times W \to \mathbb{C}$ to $V_f \times W_f$ is nondegenerate. We can also define $\mathfrak{gl}(V, W)$ as a Lie algebra of finite rank linear operators in $V \oplus W$ preserving V, W and the pairing $V \times W \to \mathbb{C}$.

There is an obvious notion of *isomorphism of linear systems*: given two linear systems $V \times W \to \mathbb{C}$ and $V \times W' \to \mathbb{C}$, an isomorphism of these linear systems is a pair of isomorphisms of vector spaces $\varphi : V \to W$, $\psi : W \to W'$ or $\varphi : V \to W'$, $\psi : W \to V'$, commuting with the respective pairings. If V and W are countable dimensional then, as shown by G. Mackey [Ma], there exists a basis $\{v_1, v_2, \dots\}$ of V such that $V_* = \mathrm{span}\{v_1^*, v_2^*, \dots\}$, where $\{v_1^*, v_2^*, \dots\}$ is the set of linear functionals dual to $\{v_1, v_2, \dots\}$, i.e., $v_i^*(v_j) = \delta_{ij}$. Consequently, up to isomorphism, there exists only one linear system $V \times W \to \mathbb{C}$ such that V and W are countable dimensional. The choice of a basis of V as above identifies $\mathfrak{gl}(V, W)$ with the Lie algebra $\mathfrak{gl}(\infty)$ consisting of infinite matrices $X = (x_{ij})_{i\geq 1, j\geq 1}$ with finitely many nonzero entries. The Lie algebra $\mathfrak{sl}(V, W)$ is identified with $\mathfrak{sl}(\infty)$ realized as the Lie algebra of traceless matrices $X = (x_{ij})_{i\geq 1, j\geq 1}$ with finitely many nonzero entries.

Now let V be a vector space endowed with a nondegenerate symmetric (respectively, antisymmetric) form $(\cdot,\cdot)$. Then $\Lambda^2 V$ (respectively, $S^2 V$) has a Lie algebra structure, defined by

$$[v_1\wedge v_2, w_1\wedge w_2] = -(v_1,w_1)v_2\wedge w_2+(v_2,w_1)v_1\wedge w_2+(v_1,w_2)v_2\wedge w_1-(v_2,w_2)v_1\wedge w_1$$

(respectively, by

$$[v_1v_2, w_1w_2] = (v_1,w_1)v_2w_2 + (v_2,w_1)v_1w_2 + (v_1,w_2)v_2w_1 + (v_2,w_2)v_1w_1).$$

We denote the Lie algebra $\Lambda^2 V$ by $\mathfrak{o}(V)$, and the Lie algebra $S^2 V$ by $\mathfrak{sp}(V)$. Let $V_f \subset V$ be an n-dimensional subspace such that the restriction of the form on V_f is nondegenerate. Then $\mathfrak{o}(V_f) \subset \mathfrak{o}(V)$ (respectively, $\mathfrak{sp}(V_f) \subset \mathfrak{sp}(V)$) is a simple subalgebra isomorphic to $\mathfrak{o}(n)$ (respectively, $\mathfrak{sp}(n)$). Therefore, $\mathfrak{o}(V)$ (respectively, $\mathfrak{sp}(V)$) is the direct limit of all its subalgebras $\mathfrak{o}(V_f)$ (respectively, $\mathfrak{sp}(V_f)$). This shows that both $\mathfrak{o}(V)$ and $\mathfrak{sp}(V)$ are locally simple locally finite Lie algebras. We can also identify $\mathfrak{o}(V)$ (respectively, $\mathfrak{sp}(V)$) with the Lie subalgebra of all finite rank operators in V under which the form $(\cdot,\cdot)$ is invariant.

If V is countable dimensional, there always is a basis $\{v_i, w_j\}_{i,j\in\mathbb{Z}}$ of V such that span$\{v_i\}_{i\in\mathbb{Z}}$ and span$\{w_j\}_{j\in\mathbb{Z}}$ are isotropic spaces and $(v_i, w_j) = 0$ for $i \neq j$, $(v_i, w_i) = 1$. Therefore, in this case $\mathfrak{o}(V) \simeq \mathfrak{o}(\infty)$ and $\mathfrak{sp}(V) \simeq \mathfrak{sp}(\infty)$.

Note that if V is not finite or countable dimensional, then V may have several inequivalent nondegenerate symmetric forms. Indeed, let for instance $V := W \oplus W^*$ for some countable-dimensional space W. Extend the pairing between W and W^* to a nondegenerate symmetric form $(\cdot,\cdot)$ on V for which W and W^* are both isotropic. It is clear that W is a maximal isotropic subspace of V. On the other hand, choose a basis $\mathbf{b}$ in V and let $(\cdot,\cdot)'$ be the symmetric form on V for which $\mathbf{b}$ is an orthonormal basis. Then V does not have countable-dimensional maximal isotropic subspaces for the form $(\cdot,\cdot)'$. Hence the forms $(\cdot,\cdot)$ and $(\cdot,\cdot)'$ are not equivalent.

Proposition 1.1. *(a) Two Lie algebras $\mathfrak{sl}(V, W)$ and $\mathfrak{sl}(V', W')$ are isomorphic if and only if the linear systems $V \times W \to \mathbb{C}$ and $V' \times W' \to \mathbb{C}$ are isomorphic.*

(b) Two Lie algebras $\mathfrak{o}(V)$ and $\mathfrak{o}(V')$ (respectively, $\mathfrak{sp}(V)$ and $\mathfrak{sp}(V')$) are isomorphic if and only if there is an isomorphism of vector spaces $V \simeq V'$ transferring the form defining $\mathfrak{o}(V)$ (respectively $\mathfrak{sp}(V)$) into the form defining $\mathfrak{o}(V')$ (respectively, $\mathfrak{sp}(V')$).

We first prove a lemma.

Lemma 1.2 (cf. Proposition 2.3 in [DiP2]).

(a) Let $\mathfrak{g}_1 \subset \mathfrak{g}_3$ be an inclusion of classical finite-dimensional simple Lie algebras such that a natural $\mathfrak{g}_3$-module restricts to $\mathfrak{g}_1$ as the direct sum of a natural $\mathfrak{g}_1$-module and a trivial $\mathfrak{g}_1$-module. If $\mathfrak{g}_2$ is an intermediate classical simple subalgebra, $\mathfrak{g}_1 \subseteq \mathfrak{g}_2 \subseteq \mathfrak{g}_3$, then a natural $\mathfrak{g}_3$-module restricts to $\mathfrak{g}_2$ as the direct sum of a natural $\mathfrak{g}_2$-module and a trivial module.

(b) *Assume* $\mathrm{rk}\mathfrak{g}_1 > 4$. *If* $\mathfrak{g}_1 \simeq \mathfrak{sl}(i)$, *then* $\mathfrak{g}_2$ *is isomorphic to* $\mathfrak{sl}(k)$ *for some* $k \geq i$. *If* $\mathfrak{g}_3 \simeq \mathfrak{o}(j)$ *(respectively,* $\mathfrak{sp}(2j)$*), then* $\mathfrak{g}_2$ *is isomorphic to* $\mathfrak{o}(k)$ *(respectively,* $\mathfrak{sp}(2k)$*) for some* $k \leq j$.

Proof. Let V_3 be a natural $\mathfrak{g}_3$-module. We have a decomposition of $\mathfrak{g}_1$-modules, $V_3 = V_1 \oplus W$, where V_1 is a natural $\mathfrak{g}_1$-module and W is a trivial $\mathfrak{g}_1$-module. Let $V' \subset V_3$ be the minimal $\mathfrak{g}_2$-submodule containing V_1. Then $V_3 = V' \oplus W'$, where W' is a complementary $\mathfrak{g}_2$-submodule. Since $\mathfrak{g}_1$ acts trivially on W' and $\mathfrak{g}_2$ is simple, we obtain that W' is a trivial $\mathfrak{g}_2$-module and V' is a simple $\mathfrak{g}_2$-module.

We now prove that V' is a natural $\mathfrak{g}_2$-module. Recall that for an arbitrary nontrivial module M over a simple Lie algebra $\mathfrak{k}$ the symmetric form $B_M(X, Y) = \mathrm{tr}_M(XY)$ for $X, Y \in \mathfrak{k}$ is nondegenerate. Moreover, $B_M = t_M B$, where B is the Killing form. If M is a simple $\mathfrak{k}$-module with highest weight λ, then

$$t_M = \frac{\dim M}{\dim \mathfrak{k}}(\lambda + 2\rho, \lambda),$$

where ρ is the half-sum of positive roots and $(\cdot, \cdot)$ is the form on the weight lattice of $\mathfrak{k}$ induced by B. It is easy to check that a natural module is a simple module with minimal t_M. Let V_2 be a natural $\mathfrak{g}_2$-module. Note that the restriction of $B_{V'}$ on $\mathfrak{g}_1$ equals B_{V_1} and the restriction of B_{V_2} on $\mathfrak{g}_1$ equals tB_{V_1} for some $t \geq 1$. On the other hand, $t = \frac{t_{V_2}}{t_{V'}}$. Since t_{V_2} is minimal, we have $t = 1$ and $t_{V_2} = t_{V'}$. Hence, V' is a natural module, i.e., (a) is proved.

To prove (b), note that a classical simple Lie algebra of rank greater than 4 admits, up to isomorphism, two (mutually dual) natural representations when it is of type $\mathfrak{sl}$, and one natural representation when it is of type $\mathfrak{o}$ or $\mathfrak{sp}$. Moreover, in the orthogonal (respectively, symplectic) case the natural module admits an invariant symmetric (respectively, skew-symmetric) bilinear form.

Now, assume $\mathfrak{g}_1 \simeq \mathfrak{sl}(i)$. We claim that $\mathfrak{g}_2 \simeq \mathfrak{sl}(k)$ for some $i \leq k \leq j$. Indeed, if $\mathfrak{g}_2$ is not isomorphic to $\mathfrak{sl}(k)$, then V' is self-dual. Therefore its restriction to $\mathfrak{g}_1$ is self-dual, and we obtain a contradiction as V_1 is not a self-dual $\mathfrak{sl}(i)$-module for $i \geq 3$.

Finally, assume $\mathfrak{g}_3 \simeq \mathfrak{o}(j)$ (respectively, $\mathfrak{sp}(2j)$). Then $V' \oplus W'$, and hence V', admits an invariant symmetric (respectively, skew-symmetric) form. Therefore $\mathfrak{g}_2 \simeq \mathfrak{o}(k)$ (respectively, $\mathfrak{sp}(2k)$). □

Corollary 1.3 (cf. [DiP2, Corollary 2.4]). *Let* $\mathfrak{g} = \mathfrak{sl}(V, W)$ *and* $\mathfrak{g} = \varinjlim \mathfrak{g}_\alpha$ *for some directed system* $\{\mathfrak{g}_\alpha\}$ *of simple finite-dimensional Lie subalgebras* $\mathfrak{g}_\alpha \subset \mathfrak{g}$. *Then there exists a subsystem* $\{\mathfrak{g}_{\alpha'}\}$ *such that* $\mathfrak{g} = \varinjlim \mathfrak{g}_{\alpha'}$ *and, for every* α', $\mathfrak{g}_{\alpha'} = \mathfrak{sl}(V_{\alpha'}, W_{\alpha'})$ *for some finite-dimensional nondegenerate pair* $V_{\alpha'} \subset V$, $W_{\alpha'} \subset W$. *Similarly, if* $\mathfrak{g} = \mathfrak{o}(V)$ *(respectively,* $\mathfrak{sp}(V)$*), then there exists a subsystem* $\{\mathfrak{g}_{\alpha'}\}$ *such that* $\mathfrak{g} = \varinjlim \mathfrak{g}_{\alpha'}$ *and, for every* α', $\mathfrak{g}_{\alpha'} = \mathfrak{o}(V_{\alpha'})$ *(respectively,* $\mathfrak{sp}(V_{\alpha'})$*) for some finite-dimensional nondegenerate* $V_{\alpha'} \subset V$.

Proof. Let $\mathfrak{g} = \mathfrak{sl}(V, W)$. One fixes a Lie subalgebra $\mathfrak{sl}(V_f, W_f) \subset \mathfrak{g}$ where $V_f \subset V$, $W_f \subset W$ is a finite-dimensional nondegenerate pair, and considers the directed subsystem $\{\mathfrak{g}_{\alpha'}\}$ of all $\mathfrak{g}_{\alpha'}$ such that $\mathfrak{sl}(V_f, W_f) \subset \mathfrak{g}_{\alpha'}$. There exists another finite-dimensional nondegenerate pair V'_f, W'_f such that $\mathfrak{sl}(V_f, W_f) \subset \mathfrak{g}_{\alpha'} \subset \mathfrak{sl}(V'_f, W'_f)$. Then, by Lemma 1.2, $\mathfrak{g}_{\alpha'} = \mathfrak{sl}(V_{\alpha'}, W_{\alpha'})$ for appropriate $V_{\alpha'} \subset V$, $W_{\alpha'} \subset W$. The cases $\mathfrak{g} = \mathfrak{o}(V), \mathfrak{sp}(V)$ are similar. □

Proof of Proposition 1.1. We consider the case $\mathfrak{g} = \mathfrak{sl}(V, W)$ and leave the remaining cases to the reader. Let $\mathfrak{g} = \mathfrak{sl}(V, W)$ be isomorphic to $\mathfrak{sl}(V', W')$. Then $\mathfrak{g} = \varinjlim \mathfrak{sl}(V_f, W_f)$ over all finite-dimensional nondegenerate pairs $V_f \subset V, W_f \subset W$, and at the same time $\mathfrak{g} = \varinjlim \mathfrak{sl}(V'_f, W'_f)$ over all finite-dimensional nondegenerate pairs $V'_f \subset V', W'_f \subset W'$. By Corollary 1.3 and Lemma 1.2, for each $V_f \subset V, W_f \subset W$ one can find $V'_f \subset V', W'_f \subset W'$ and an embedding of Lie algebras $\mathfrak{sl}(V_f, W_f) \subset \mathfrak{sl}(V'_f, W'_f)$ as in Lemma 1.2. That implies the existence of embeddings $V_f \hookrightarrow V'_f, W_f \hookrightarrow W'_f$ or $V_f \hookrightarrow W'_f, W_f \hookrightarrow V'_f$ preserving the pairing. After a twist by transposition we may assume that $V_f \hookrightarrow V'_f, W_f \hookrightarrow W'_f$. Therefore we have embeddings $V = \varinjlim V_f \hookrightarrow V', W = \varinjlim W_f \hookrightarrow W'$ preserving the pairing. On the other hand, both maps are surjective since $\mathfrak{sl}(V', W') = \varinjlim \mathfrak{sl}(V_f, W_f)$. Therefore the linear systems $V \times W \to \mathbb{C}$ and $V' \times W' \to \mathbb{C}$ are isomorphic. □

Assume next that $\mathfrak{g}$ is an arbitrary locally finite locally simple Lie algebra. If we can choose a Cartan subalgebra $\mathfrak{h}_\alpha \subset \mathfrak{g}_\alpha$ such that $\mathfrak{h}_\alpha \hookrightarrow \mathfrak{h}_\beta$ for any embedding $\mathfrak{g}_\alpha \hookrightarrow \mathfrak{g}_\beta$, then $\mathfrak{h} := \varinjlim \mathfrak{h}_\alpha$ is called a *local Cartan subalgebra.*

In general, a local Cartan subalgebra may not exist. For example, the following proposition implies that the Lie algebra $\mathfrak{g} = \mathfrak{sl}(V, V^*)$ does not have a local Cartan subalgebra.

Proposition 1.4. *Let* $\mathfrak{g} = \mathfrak{sl}(V, W)$. *Then a local Cartan subalgebra of* $\mathfrak{g}$ *exists if and only if* V *admits a basis* $\{v_\gamma\}$ *such that* $W = \operatorname{span}\{v^*_\gamma\}$, *where* $v^*_{\tilde{\gamma}}(v_\gamma) = \delta_{\tilde{\gamma}\gamma}$. *In this case, every local Cartan subalgebra of* $\mathfrak{g}$ *is of the form* $\operatorname{span}\left\{v_\gamma \otimes v^*_\gamma - v_{\tilde{\gamma}} \otimes v^*_{\tilde{\gamma}}\right\}_{\gamma,\tilde{\gamma}}$ *for a basis* $\{v_\gamma\}$ *as above.*

Proof. By Corollary 1.3 we may assume

$$\mathfrak{g} = \mathfrak{sl}(V, W) = \varinjlim \mathfrak{g}_\alpha = \varinjlim \mathfrak{sl}(V_\alpha, W_\alpha),$$

where $V_\alpha \subset V$, $W_\alpha \subset W$ are certain nondegenerate finite-dimensional pairs, and that $\mathfrak{h} = \varinjlim \mathfrak{h}_\alpha$ where $\mathfrak{h}_\alpha$ is a Cartan subalgebra of $\mathfrak{g}_\alpha$. Note that for any α we have $\mathfrak{h}_\alpha \cdot V_\alpha = V_\alpha$ and $\mathfrak{h}_\alpha \cdot W_\alpha = W_\alpha$. Since $\mathfrak{h}$ is abelian, we have $\mathfrak{h} \cdot V_\alpha = V_\alpha$ and $\mathfrak{h} \cdot W_\alpha = W_\alpha$. Therefore V and W are semisimple $\mathfrak{h}$-modules. This means that V is the direct sum of nontrivial one-dimensional $\mathfrak{h}$-submodules V_γ, i.e., $V = \bigoplus_\gamma V_\gamma$; similarly, $W = \bigoplus_{\gamma'} W_{\gamma'}$. Since however, for any α, the spaces V_α and W_α are dual to each other, γ' and γ run over the same set of indices and $W_\gamma(V_{\tilde{\gamma}}) \neq 0$ precisely for $\gamma = \tilde{\gamma}$. This yields a basis v_γ as required: v_γ can be chosen as any nonzero vector in V_γ and v^*_γ is the unique vector in W_γ with $v^*_\gamma(v_\gamma) = 1$. Finally,

$\mathfrak{h} = \operatorname{span}\left\{v_\gamma \otimes v_\gamma^* - v_{\tilde{\gamma}} \otimes v_{\tilde{\gamma}}^*\right\}$ as, clearly, $\mathfrak{h} \cap \mathfrak{g}_\alpha = \operatorname{span}\left\{v_\gamma \otimes v_\gamma^* - v_{\tilde{\gamma}} \otimes v_{\tilde{\gamma}}^*\right\}$ for $v_\gamma, v_{\tilde{\gamma}} \in V_\alpha$.

In the other direction, given a basis v_γ of V such that $\left\{v_\gamma^*\right\}$ is a basis of W, it is clear that $\mathfrak{g} = \varinjlim \mathfrak{sl}\left(\operatorname{span}\left\{v_\gamma\right\}_{\gamma\in A}, \operatorname{span}\left\{v_\gamma^*\right\}_{\gamma\in A}\right)$ for all finite sets of indices A, and that $\mathfrak{h} = \varinjlim \left(\mathfrak{h} \cap \operatorname{span}\left\{v_\gamma \otimes v_\gamma^* - v_{\tilde{\gamma}} \otimes v_{\tilde{\gamma}}^*\right\}_{\gamma,\tilde{\gamma}\in A}\right)$. □

In [DPSn] (and also in earlier work, see the references in [DPSn]) *Cartan subalgebras* are defined as maximal toral subalgebras of $\mathfrak{g}$ (i.e., as subalgebras each vector in which is ad-semisimple). *Splitting* Cartan subalgebras are Cartan subalgebras for which the adjoint representation is semisimple. It is shown in [PStr] that a countable dimensional locally finite, locally simple Lie algebra $\mathfrak{g}$ admits a splitting Cartan subalgebra if and only if $\mathfrak{g} \simeq \mathfrak{sl}(\infty), \mathfrak{o}(\infty), \mathfrak{sp}(\infty)$. Proposition 1.4 determines when Lie algebras of the form $\mathfrak{g} = \mathfrak{sl}(V, W), \mathfrak{o}(V), \mathfrak{sp}(V)$ admit local Cartan subalgebras and implies that the notions of local Cartan subalgebra and of splitting Cartan subalgebra coincide for these Lie algebras.

In what follows, we denote by V, V_* a pair of infinite-dimensional spaces (of not necessarily countable dimension) arising from a linear system $V \times V_* \to \mathbb{C}$ for which there is a basis $\{v_\gamma\}$ of V such that $V_* = \operatorname{span}(\{v_\gamma^*\})$ where $v_{\tilde{\gamma}}^*(v_\gamma) = \delta_{\tilde{\gamma}\gamma}$.

2 The Category $\mathrm{Int}_\mathfrak{g}$

Let $\mathfrak{g}$ be an arbitrary locally simple locally finite Lie algebra. An *integrable* $\mathfrak{g}$-module is a $\mathfrak{g}$-module M which is locally finite as a module over any finite-dimensional subalgebra $\mathfrak{g}'$ of $\mathfrak{g}$. In other words, $\dim U(\mathfrak{g}') \cdot m < \infty \quad \forall\, m \in M$. We denote the category of integrable $\mathfrak{g}$-modules by $\mathrm{Int}_\mathfrak{g}$: $\mathrm{Int}_\mathfrak{g}$ is a full subcategory of the category $\mathfrak{g}$-mod of all $\mathfrak{g}$-modules. It is clear that $\mathrm{Int}_\mathfrak{g}$ is an abelian category and a monoidal category with respect to usual tensor product. Note that the adjoint representation of $\mathfrak{g}$ is an object of $\mathrm{Int}_\mathfrak{g}$.

The *functor of* $\mathfrak{g}$*-integrable vectors*

$$\Gamma_\mathfrak{g} \;:\; \mathfrak{g} - \mathrm{mod} \rightsquigarrow \mathrm{Int}_\mathfrak{g},$$
$$\Gamma_\mathfrak{g}(M) := \left\{m \in M \mid \dim U(\mathfrak{g}') \cdot m < \infty \;\forall \text{ finite-dim. subalgebras } \mathfrak{g}' \subset \mathfrak{g}\right\}$$

is a well-defined left-exact functor. This follows from the fact that the functor of $\mathfrak{g}'$-finite vectors $\Gamma_{\mathfrak{g}'}$ is well defined for any finite-dimensional subalgebra $\mathfrak{g}' \subset \mathfrak{g}$, see for instance [Z], and that $\mathfrak{g}$ equals the direct limit of its finite-dimensional subalgebras.

Theorem 2.1. *(a) Let M be an object of* $\mathrm{Int}_\mathfrak{g}$*. Then $\Gamma_\mathfrak{g}(M^*)$ is an injective object of* $\mathrm{Int}_\mathfrak{g}$.

(b) $\mathrm{Int}_{\mathfrak{g}}$ *has enough injectives. More precisely, for any object* M *of* $\mathrm{Int}_{\mathfrak{g}}$ *there is a canonical injective homomorphism of* $\mathfrak{g}$*-modules*

$$M \to \Gamma_{\mathfrak{g}}(\Gamma_{\mathfrak{g}}(M^*)^*).$$

Proof. In [PS], see Proposition 3.2 and Corollary 3.3, the proof is given under the assumption that $\mathfrak{g}$ is countable dimensional. The reader can check that this assumption is inessential. □

3 Five Subcategories of $\mathrm{Int}_{\mathfrak{g}}$

3.1 *The Category* $\mathrm{Int}_{\mathfrak{g}}^{\mathrm{alg}}$

We start by defining the full subcategory $\mathrm{Int}_{\mathfrak{g}}^{\mathrm{alg}} \subset \mathrm{Int}_{\mathfrak{g}}$. Its objects are integrable $\mathfrak{g}$-modules M such that for any simple finite-dimensional subalgebra $\mathfrak{g}' \subset \mathfrak{g}$, the restriction of M to $\mathfrak{g}'$ is a direct sum of finitely many $\mathfrak{g}'$-isotypic components. Clearly, if $\dim M = \infty$, at least one of these isotypic components must be infinite dimensional. If $\mathfrak{g}$ is diagonal, the adjoint representation of $\mathfrak{g}$ is easily seen to be an object of $\mathrm{Int}_{\mathfrak{g}}^{\mathrm{alg}}$.

The following proposition provides equivalent definitions of $\mathrm{Int}_{\mathfrak{g}}^{\mathrm{alg}}$.

Proposition 3.1. *(a)* $M \in \mathrm{Int}_{\mathfrak{g}}^{\mathrm{alg}}$ *iff* M *and* M^* *are integrable.*

(b) An integrable $\mathfrak{g}$*-module* M *is an object of* $\mathrm{Int}_{\mathfrak{g}}^{\mathrm{alg}}$ *iff for any* $X \in \mathfrak{g}$ *there exists a nonzero polynomial* $p(t) \in \mathbb{C}[t]$ *such that* $p(X) \cdot M = 0$.

Proof. (a) In the countable-dimensional case the statement is proven in [PS, Lemma 4.1]. In general, let $\mathfrak{g}' \subset \mathfrak{g}$ be a finite-dimensional simple subalgebra and let $M = \oplus_\alpha M_\alpha$ be the decomposition of M into $\mathfrak{g}'$-isotypic components. Then it is straightforward to check that $M^* = \prod_\alpha M_\alpha^*$ is an integrable $\mathfrak{g}'$-module iff the direct product is finite. This proves (a), since a $\mathfrak{g}$-module is integrable iff it is $\mathfrak{g}'$-integrable for all finite-dimensional Lie subalgebras $\mathfrak{g}' \subset \mathfrak{g}$.

(b) Let $M \in \mathrm{Int}_{\mathfrak{g}}^{\mathrm{alg}}$. Any $X \in \mathfrak{g}$ lies in some finite-dimensional Lie subalgebra $\mathfrak{g}' \subset \mathfrak{g}$. For each $\mathfrak{g}'$-isotypic component M_i of M there exists $p_i(t)$ such that $p_i(X) \cdot M_i = 0$. Since there are finitely many $\mathfrak{g}'$-isotypic components, we can set $p(t) = \prod_i p_i(t)$. Then $p(X) \cdot M = 0$.

On the other hand, if $M \notin \mathrm{Int}_{\mathfrak{g}}^{\mathrm{alg}}$, then there are infinitely many isotypic components for some finite-dimensional simple $\mathfrak{g}' \subset \mathfrak{g}$. That implies the existence of a semisimple $X \in \mathfrak{g}'$ which has infinitely many eigenvalues in M. Therefore $p(X) \cdot M \neq 0$ for any $0 \neq p(t) \in \mathbb{C}[t]$. □

It is obvious that $\mathrm{Int}_{\mathfrak{g}}^{\mathrm{alg}}$ is an abelian monoidal subcategory of $\mathfrak{g}$-mod. It is also closed under dualization.

Proposition 3.2. $\mathrm{Int}_{\mathfrak{g}}^{\mathrm{alg}}$ *contains a nontrivial module iff* $\mathfrak{g}$ *is diagonal.*

Proof. Again, for a countable dimensional $\mathfrak{g}$ the statement is proven in [PS] (see Proposition 4.3). In fact, we prove in [PS] that if $\mathfrak{g} = \varinjlim \mathfrak{g}_i$ has a non-trivial integrable module such that M^* is also integrable, then the embedding $\mathfrak{g}_i \hookrightarrow \mathfrak{g}_{i+1}$ is diagonal for all sufficiently large i.

To give a general proof, it remains to show that if $\mathfrak{g}$ is not diagonal, then $\mathrm{Int}_{\mathfrak{g}}^{\mathrm{alg}}$ contains no nontrivial modules. Assume that $\mathfrak{g} = \varinjlim \mathfrak{g}_\alpha$ is not diagonal. Fix a simple finite-dimensional Lie algebra $\mathfrak{g}_{\alpha_1}$ and a simple $\mathfrak{g}$-module $M \in \mathrm{Int}_{\mathfrak{g}}^{\mathrm{alg}}$ such that $M_{\downarrow \mathfrak{g}_{\alpha_1}}$ is nontrivial. We claim that one can find a chain of proper embeddings of simple finite-dimensional Lie algebras

$$\mathfrak{g}_{\alpha_1} \hookrightarrow \mathfrak{g}_{\alpha_2} \hookrightarrow \cdots \hookrightarrow \mathfrak{g}_{\alpha_i} \hookrightarrow \mathfrak{g}_{\alpha_{i+1}} \hookrightarrow \cdots$$

such that the embeddings $\mathfrak{g}_{\alpha_i} \hookrightarrow \mathfrak{g}_{\alpha_{i+1}}$ are not diagonal. Indeed, otherwise there will exist β_0 so that the embedding $\mathfrak{g}_{\beta_0} \hookrightarrow \mathfrak{g}_\alpha$ is diagonal for all $\alpha > \beta_0$. Then, since $\mathfrak{g} = \varinjlim_{\alpha>\beta_0} \mathfrak{g}_\alpha$, $\mathfrak{g}$ is diagonal. This shows that the existence of β_0 is contradictory. Now Proposition 4.3 in [PS] implies that $M_{\downarrow \varinjlim \mathfrak{g}_{\alpha_i}}$ is a trivial module, which shows that the assumption that $M_{\downarrow \mathfrak{g}_{\alpha_1}}$ is nontrivial is false. □

Let $\mathfrak{g} = \mathfrak{sl}(V, W)$ (respectively, $\mathfrak{g} = \mathfrak{o}(V), \mathfrak{sp}(V)$). Then the tensor products $T^{m,n} := V^{\otimes m} \otimes W^{\otimes n}$ (respectively, $T^m := V^{\otimes m}$) and their subquotients are objects of $\mathrm{Int}_{\mathfrak{g}}^{\mathrm{alg}}$.

Here is a less trivial example of a simple object of $\mathrm{Int}_{\mathfrak{g}}^{\mathrm{alg}}$ for $\mathfrak{gl} = \mathfrak{sl}(V, V_*)$ where V is a countable-dimensional vector space. Let $\mathfrak{g} = \varinjlim \mathfrak{g}_i$ where $\mathfrak{g}_i = \mathfrak{sl}(V_i)$, $\dim V_i = i + 1$, and $\varinjlim V_i = V$. Define $\Lambda^{[\frac{\infty}{2}]}V$ as the direct limit $\varinjlim \Lambda^{[\frac{i}{2}]}(V_i)$ for $i \geq 2$. Then $\Lambda^{[\frac{\infty}{2}]}V$ is a simple object of $\mathrm{Int}_{\mathfrak{g}}^{\mathrm{alg}}$ and is not isomorphic to a subquotient of a tensor product of the form $T^{m,n}$.

Given a $\mathfrak{g}$-module $M \in \mathrm{Int}_{\mathfrak{g}}^{\mathrm{alg}}$, where $\mathfrak{g} = \varinjlim \mathfrak{g}_\alpha$, for each α we can assign to $\mathfrak{g}_\alpha$ the finite set of isomorphism classes of simple finite-dimensional $\mathfrak{g}_\alpha$-modules which occur in the restriction $M_{\downarrow \mathfrak{g}_\alpha}$. A. Zhilinskii has defined a *coherent local system of finite-dimensional representations* of $\mathfrak{g} = \varinjlim \mathfrak{g}_\alpha$ as a function of α with values in the set of isomorphism classes of finite-dimensional $\mathfrak{g}_\alpha$-modules, with the following compatibility condition: if $\beta < \alpha$, then the representations assigned to β are obtained by restriction from the representations assigned to α. Thus, every $M \in \mathrm{Int}_{\mathfrak{g}}^{\mathrm{alg}}$ determines a coherent local system of *finite type*, i.e., a local system containing finitely many isomorphism classes for any α.

Zhilinskii has classified all coherent local systems under the condition that $\mathfrak{g}$ is countable dimensional [Zh1, Zh2] (see also [PP] for an application of Zhilinskii's result). In particular, he has proved that proper coherent local systems, i.e., coherent local systems different from the ones assigning the trivial 1-dimensional module to all α, or all finite-dimensional $\mathfrak{g}_\alpha$-modules to α, exist only if $\mathfrak{g}$ is diagonal. This leads to another proof of Proposition 3.2.

The category $\mathrm{Int}_{\mathfrak{g}}^{\mathrm{alg}}$ has enough injectives: this follows immediately from Proposition 3.1 (a) and Theorem 2.1. We know of no classification of simple modules in $\mathrm{Int}_{\mathfrak{g}}^{\mathrm{alg}}$.

3.2 *The Category* $\mathrm{Int}_{\mathfrak{g},\mathfrak{h}}^{\mathrm{wt}}$

Given a local Cartan subalgebra $\mathfrak{h} \subset \mathfrak{g}$, we define $\mathrm{Int}_{\mathfrak{g},\mathfrak{h}}^{\mathrm{wt}}$ as the full subcategory of $\mathrm{Int}_{\mathfrak{g}}$ consisting of $\mathfrak{h}$-*semisimple* integrable $\mathfrak{g}$-modules, i.e., integrable $\mathfrak{g}$-modules M admitting an $\mathfrak{h}$-weight decomposition

$$M = \oplus_{\lambda \in \mathfrak{h}^*} M^\lambda \tag{2}$$

where

$$M^\lambda := \{m \in M \mid h \cdot m = \lambda(h)m \ \ \forall h \in \mathfrak{h}\} .$$

If $\mathfrak{g} = \mathfrak{sl}(V, W), \mathfrak{o}(V), \mathfrak{sp}(V)$ for countable-dimensional V, W, then V (and W in case $\mathfrak{g} = \mathfrak{sl}(V, W)$) is a simple object of $\mathrm{Int}_{\mathfrak{g},\mathfrak{h}}^{\mathrm{wt}}$ for any $\mathfrak{h}$. Moreover, if $\mathfrak{g}$ is a countable-dimensional locally simple Lie algebra, it is proved in [PStr] that the adjoint representation of $\mathfrak{g}$ is an object of $\mathrm{Int}_{\mathfrak{g},\mathfrak{h}}^{\mathrm{wt}}$ iff $\mathfrak{g} \simeq \mathfrak{sl}(\infty), \mathfrak{o}(\infty), \mathfrak{sp}(\infty)$. The simple modules of $\mathrm{Int}_{\mathfrak{g},\mathfrak{h}}^{\mathrm{wt}}$ for $\mathfrak{g} = \mathfrak{sl}(\infty), \mathfrak{o}(\infty), \mathfrak{sp}(\infty)$ have been studied in [DiP1], however there is no classification of such modules.

Assume that $\mathfrak{g}$ is a locally simple diagonal countable-dimensional Lie algebra. Without loss of generality, assume that $\mathfrak{g} = \varinjlim \mathfrak{g}_i$, where all $\mathfrak{g}_i$ are of the same type A, B, C, or D. The very definition of $\mathfrak{g}$ implies that there is a well-defined chain

$$V_{\mathfrak{g}_1} \overset{\kappa_1}{\hookrightarrow} V_{\mathfrak{g}_2} \overset{\kappa_2}{\hookrightarrow} \ldots \hookrightarrow V_{\mathfrak{g}_i} \overset{\kappa_i}{\hookrightarrow} V_{\mathfrak{g}_{i+1}} \hookrightarrow \ldots \tag{3}$$

of embeddings of natural $\mathfrak{g}_i$-modules, and we call its direct limit V a *natural representation* of $\mathfrak{g}$. Moreover, a fixed natural representation V is a simple object of $\mathrm{Int}_{\mathfrak{g},\mathfrak{h}}^{\mathrm{wt}}$ for some local Cartan subalgebra $\mathfrak{h}$. To see this, we use induction to define a local Cartan subalgebra $\mathfrak{h} \subset \mathfrak{g}$ so that $V \in \mathrm{Int}_{\mathfrak{g},\mathfrak{h}}^{\mathrm{wt}}$. Given $\mathfrak{h}_i \subset \mathfrak{g}_i$ and an $\mathfrak{h}_i$-eigenbasis $\mathbf{b}_i$ of V_i, let $\mathfrak{h}_{i+1}$ be a Cartan subalgebra of $\mathfrak{g}_{i+1}$ whose eigenbasis $\mathbf{b}_{i+1}$ of V_{i+1} contains $\mathbf{b}_i$. The assumption that $\mathfrak{g}_i$ and $\mathfrak{g}_{i+1}$ are of the same type A, B, C or D (in the sense of the classification of simple Lie algebras [Bou]) implies that $\mathfrak{h}_{i+1}$ exists as required. Moreover, $\mathfrak{h} := \varinjlim \mathfrak{h}_i$ is a well-defined local Cartan subalgebra of $\mathfrak{g}$ and $V \in \mathrm{Int}_{\mathfrak{g},\mathfrak{h}}^{\mathrm{wt}}$.

Assume next that $\mathfrak{g}$ is a locally simple Lie algebra which admits a local Cartan subalgebra $\mathfrak{h}$ such that the adjoint representation belongs to $\mathrm{Int}_{\mathfrak{g},\mathfrak{h}}^{\mathrm{wt}}$. This certainly holds for $\mathfrak{g} = \mathfrak{sl}(\infty), \mathfrak{o}(\infty), \mathfrak{sp}(\infty)$, but also for instance for $\mathfrak{g} = \mathfrak{sl}(V, V_*)$ where V is an arbitrary vector space. In this case we can define a left exact functor $\Gamma_{\mathfrak{h}}^{\mathrm{wt}} : \mathrm{Int}_{\mathfrak{g}} \rightsquigarrow \mathrm{Int}_{\mathfrak{g},\mathfrak{h}}^{\mathrm{wt}}$ by setting

$$\Gamma_{\mathfrak{h}}^{\mathrm{wt}}(M) := \oplus_{\lambda \in \mathfrak{h}^*} M^{\lambda},$$

where M^{λ} is given by (3). It is easy to see that $\Gamma_{\mathfrak{h}}^{\mathrm{wt}}$ is right adjoint to the inclusion functor $\mathrm{Int}_{\mathfrak{g},\mathfrak{h}}^{\mathrm{wt}} \rightsquigarrow \mathrm{Int}_{\mathfrak{g}}$. Hence $\Gamma_{\mathfrak{h}}^{\mathrm{wt}}$ maps injectives to injectives, and therefore $\mathrm{Int}_{\mathfrak{g},\mathfrak{h}}^{\mathrm{wt}}$ has enough injectives. We do not know whether $\mathrm{Int}_{\mathfrak{g},\mathfrak{h}}^{\mathrm{wt}}$ has enough injectives in the case when the adjoint representation is not an object of $\mathrm{Int}_{\mathfrak{g},\mathfrak{h}}^{\mathrm{wt}}$.

We conjecture that for nondiagonal Lie algebras $\mathfrak{g}$, the category $\mathrm{Int}_{\mathfrak{g},\mathfrak{h}}^{\mathrm{wt}}$ consists of trivial modules only.

3.3 *The Category* $\mathbf{Int}_{\mathfrak{g},\mathfrak{h}}^{\mathbf{fin}}$

By $\mathrm{Int}_{\mathfrak{g},\mathfrak{h}}^{\mathrm{fin}}$ we denote the full subcategory of $\mathrm{Int}_{\mathfrak{g},\mathfrak{h}}^{\mathrm{wt}}$ consisting of integrable $\mathfrak{g}$-modules satisfying $\dim M^{\lambda} < \infty \ \forall \lambda \in \mathfrak{h}^*$.

Note that for $\mathfrak{g} = \mathfrak{sl}(V, V_*)$ (respectively, for $\mathfrak{g} = \mathfrak{o}(\infty), \mathfrak{sp}(\infty)$) the tensor products $T^{m,0} = V^{\otimes m}$ and $T^{0,n} = W^{\otimes n}$ (respectively, $T^m = V^{\otimes m}$) are objects of $\mathrm{Int}_{\mathfrak{g},\mathfrak{h}}^{\mathrm{fin}}$ for every local Cartan subalgebra $\mathfrak{g}$. However, the adjoint representation is not in $\mathrm{Int}_{\mathfrak{g},\mathfrak{h}}^{\mathrm{fin}}$ for any $\mathfrak{h}$.

If $\mathfrak{g}$ is countable dimensional diagonal then, as shown above, for each natural representation V there is a local Cartan subalgebra $\mathfrak{h}$ so that V (and more generally $V^{\otimes m}$) is an object of $\mathrm{Int}_{\mathfrak{g},\mathfrak{h}}^{\mathrm{wt}}$. In fact, $V^{\otimes m} \in \mathrm{Int}_{\mathfrak{g},\mathfrak{h}}^{\mathrm{fin}}$ for any $m \geq 0$.

Here is a more interesting example of a simple module in $\mathrm{Int}_{\mathfrak{g},\mathfrak{h}}^{\mathrm{fin}}$ for $\mathfrak{g} = \mathfrak{sl}(V, V_*)$, where V is a countable-dimensional vector space. Fix a chain of embeddings

$$\mathfrak{g}_1 \hookrightarrow \mathfrak{g}_2 \hookrightarrow \cdots \hookrightarrow \mathfrak{g}_i \hookrightarrow \mathfrak{g}_{i+1} \hookrightarrow \cdots$$

so that $\mathfrak{g} = \mathfrak{sl}(V_i)$ for $\dim V_i = i + 1$, $V = \varinjlim V_i$, $\mathfrak{g} = \varinjlim \mathfrak{g}_i$. Note that there is a canonical injection of $\mathfrak{g}_i$-modules $S^{i+1}(V_i) \hookrightarrow S^{i+2}(V_{i+1})$, and set $\Delta := \varinjlim S^{i+1}(V_i)$. Then one can check that Δ is a multiplicity free $\mathfrak{h}$-module, where $\mathfrak{h}$ is such that $\mathfrak{h}_i := \mathfrak{h} \cap \mathfrak{g}_i$ is a Cartan subalgebra of $\mathfrak{g}_i$.

The following result is proved in [PS].

Proposition 3.3. *Let* $\mathfrak{g} = \mathfrak{sl}(\infty), \mathfrak{o}(\infty), \mathfrak{sp}(\infty)$. *Then the category* $\mathrm{Int}_{\mathfrak{g},\mathfrak{h}}^{\mathrm{fin}}$ *is semi-simple.*

This result should be considered an extension of Weyl's semisimplicity theorem to the case of direct limit Lie algebras. It is an interesting question whether the category $\mathrm{Int}^{\mathrm{fin}}_{\mathfrak{g},\mathfrak{h}}$ is semisimple whenever it is well defined.

3.4 *The Category* $\widetilde{\mathbf{Tens}}_{\mathfrak{g}}$

Let M be a $\mathfrak{g}$-module. Recall that the *socle* $\mathrm{soc}\, M = \mathrm{soc}^1 M$ of M is the unique maximal semisimple submodule of M, and

$$\mathrm{soc}^k M := \pi^{-1}(\mathrm{soc}(M/\mathrm{soc}^{k-1} M))$$

for $k \geq 2$, where $\pi : M \to M/\mathrm{soc}^k M$ is the natural projection. The ascending chain

$$0 \subset \mathrm{soc}\, M = \mathrm{soc}^1 M \subset \mathrm{soc}^2 M \subset \cdots \subset \mathrm{soc}^k M \subset \ldots$$

is by definition the *socle filtration* of M. The $\mathfrak{g}$-module M has *finite Loewy length* if it has a finite and exhaustive socle filtration, i.e.,

$$M = \mathrm{soc}^l M$$

for some l.

By definition, $\widetilde{\mathrm{Tens}}_{\mathfrak{g}}$ is the full subcategory of $\mathrm{Int}_{\mathfrak{g}}$ whose objects are integrable $\mathfrak{g}$-modules with the property that both M and $\Gamma_{\mathfrak{g}}(M^*)$ have finite Loewy length.

The category $\widetilde{\mathrm{Tens}}_{\mathfrak{g}}$ is studied in detail in [PS] for $\mathfrak{g} = \mathfrak{sl}(\infty), \mathfrak{o}(\infty), \mathfrak{sp}(\infty)$, where it is shown in particular that $\Gamma_{\mathfrak{g}}(M^*) = M^*$ for any object M of $\widetilde{\mathrm{Tens}}_{\mathfrak{g}}$. A major result of [PS] is that, up to isomorphism, the simple objects of $\widetilde{\mathrm{Tens}}_{\mathfrak{g}}$ are precisely the simple subquotients of the tensor algebra $T(V \oplus V_*)$ for $\mathfrak{g} = \mathfrak{sl}(V, V_*) \simeq \mathfrak{sl}(\infty)$, and of the tensor algebra $T(V)$ for $\mathfrak{g} = \mathfrak{o}(V) \simeq \mathfrak{o}(\infty)$ or $\mathfrak{g} = \mathfrak{sp}(V) \simeq \mathfrak{sp}(\infty)$. These simple modules are discussed in more detail in Sect. 4 below. Note that the objects of $\widetilde{\mathrm{Tens}}_{\mathfrak{g}}$ have in general infinite length and are not objects of $\mathrm{Int}^{\mathrm{wt}}_{\mathfrak{g},\mathfrak{h}}$ for any $\mathfrak{h}$. An example of infinite length module in $\widetilde{\mathrm{Tens}}_{\mathfrak{g}}$ for $\mathfrak{g} = \mathfrak{sl}(V, V_*) \simeq \mathfrak{sl}(\infty)$ is V^*: there is a nonsplitting exact sequence of $\mathfrak{g}$-modules

$$0 \to V_* = \mathrm{soc}\, V^* \to V^* \to V^*/V_* \to 0$$

and V^*/V_* is a trivial module of uncountable dimension.

For $\mathfrak{g} = \mathfrak{sl}(\infty), \mathfrak{o}(\infty), \mathfrak{sp}(\infty)$, the category $\widetilde{\mathrm{Tens}}_{\mathfrak{g}}$ has enough injectives [PS, Corollary 6.7a)].

3.5 The Category $\mathbb{T}_{\mathfrak{g}}$

The fifth subcategory we would like to introduce in this section is the category of tensor modules $\mathbb{T}_{\mathfrak{g}}$. We define this category only for $\mathfrak{g} = \mathfrak{sl}(V, W), \mathfrak{o}(V), \mathfrak{sp}(V)$, and discuss it in detail in Sect. 5.

We call a subalgebra $\mathfrak{k} \subset \mathfrak{sl}(W, V)$ a *finite-corank subalgebra* if it contains the subalgebra $\mathfrak{sl}(W_0^{\perp}, V_0^{\perp})$ for some finite-dimensional nondegenerate pair $V_0 \subset V, W_0 \subset W$. Similarly, we call $\mathfrak{k} \subset \mathfrak{o}(V)$ (respectively, $\mathfrak{sp}(V)$) a *finite corank subalgebra* if it contains $\mathfrak{o}(V^{\perp})$ (respectively, $\mathfrak{sp}(V_0^{\perp})$) for some finite-dimensional $V_0 \subset V$ such that the restriction of the form on V_0 is nondegenerate.

We say that a $\mathfrak{g}$-module L *satisfies the large annihilator condition* if the annihilator in $\mathfrak{g}$ of any $l \in L$ contains a finite-corank subalgera. It follows immediately from definition that if L_1 and L_2 satisfy the large annihilator condition, then the same holds also for $L_1 \oplus L_2$ and $L_1 \otimes L_2$.

By $\mathbb{T}_{\mathfrak{g}}$ we denote the category of finite length integrable $\mathfrak{g}$-modules which satisfy the large annihilator condition. By definition, $\mathbb{T}_{\mathfrak{g}}$ is a full subcategory of $\mathrm{Int}_{\mathfrak{g}}$. It is clear that $\mathbb{T}_{\mathfrak{g}}$ is a monoidal category with respect to usual tensor product $\otimes$.

3.6 Inclusion Pattern

The following diagram summarizes the inclusion pattern for the five subcategories of $\mathrm{Int}_{\mathfrak{g}}$ introduced above:

$$
\begin{array}{ccc}
 & \widetilde{\mathrm{Tens}}_{\mathfrak{g}} \subset \mathrm{Int}_{\mathfrak{g}}^{\mathrm{alg}} & \\
\subset & & \cap \\
\mathbb{T}_{\mathfrak{g}} & & \mathrm{Int}_{\mathfrak{g}} \\
\cap & & \subset \\
 & \mathrm{Int}_{\mathfrak{g},\mathfrak{h}}^{\mathrm{wt}} & \\
 & \cup & \\
 & \mathrm{Int}_{\mathfrak{g},\mathfrak{h}}^{\mathrm{fin}} &
\end{array}
$$

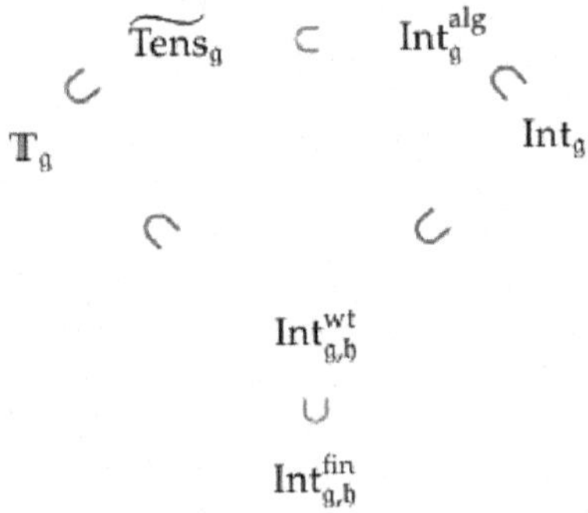

Note that all categories except $\mathbb{T}_{\mathfrak{g}}$ are defined for any locally simple Lie algebra $\mathfrak{g}$, while $\mathbb{T}_{\mathfrak{g}}$ is defined only for $\mathfrak{g} = \mathfrak{sl}(V, W), \mathfrak{o}(V), \mathfrak{sp}(V)$. Moreover, under the latter assumption all inclusions are strict. We support this claim by a list of examples and leave it to the reader to complete the proof.

Examples. Let $\mathfrak{g} = \mathfrak{sl}(V, V_*), \mathfrak{o}(V), \mathfrak{sp}(V)$, where V is countable dimensional. The simple objects of $\mathbb{T}_{\mathfrak{g}}$ and $\widetilde{\mathrm{Tens}}_{\mathfrak{g}}$ are the same, however $V^* \in \widetilde{\mathrm{Tens}}_{\mathfrak{g}}$ while $V^* \notin$

$\mathbb{T}_{\mathfrak{g}}$. Moreover, $V^* \notin \mathrm{Int}^{\mathrm{wt}}_{\mathfrak{g},\mathfrak{h}}$ for any local Cartan subalgebra $\mathfrak{h}$. The module Δ from Sect. 3.3 is an object of $\mathrm{Int}^{\mathrm{fin}}_{\mathfrak{g},\mathfrak{h}}$ but not an object of $\mathrm{Int}^{\mathrm{alg}}_{\mathfrak{g}}$. The adjoint representation is an object of $\mathrm{Int}^{\mathrm{wt}}_{\mathfrak{g},\mathfrak{h}}$ but not of $\mathrm{Int}^{\mathrm{fin}}_{\mathfrak{g},\mathfrak{h}}$.

4 Mixed Tensors

In this section $\mathfrak{g} = \mathfrak{sl}(V, W), \mathfrak{o}(V), \mathfrak{sp}(V)$. By definition, V is a $\mathfrak{g}$-module. For $\mathfrak{g} = \mathfrak{sl}(V, W)$, W is also a $\mathfrak{g}$-module.

Consider the tensor algebra $T(V)$ of V. Then, as it is easy to see, finite-dimensional Schur duality implies that

$$T(V) = \bigoplus_{\lambda} \mathbb{C}_\lambda \otimes V_\lambda, \tag{4}$$

where λ runs over all Young diagrams (i.e., over all partitions of all integers $m \in \mathbb{Z}_{\geq 0}$), $\mathbb{C}_\lambda$ denotes the irreducible $S_{|\lambda|}$-module (where $|\lambda|$ is the degree of λ) corresponding to λ, and V_λ is the image of the Schur projector corresponding to λ. For $\mathfrak{g} = \mathfrak{sl}(V, W)$, V_λ is an irreducible $\mathfrak{g}$-module as it is isomorphic to the direct limit $\varinjlim (V_f)_\lambda$ of the directed system $\{(V_f)_\lambda\}$ of irreducible $\mathfrak{sl}(V_f, W_f)$-modules for sufficiently large nondegenerate finite-dimensional pairs $V_f \subset V, W_f \subset W$. For $\mathfrak{g} = \mathfrak{o}(V), \mathfrak{sp}(V)$, V_λ is in general a reducible $\mathfrak{g}$-module.

Similarly, for $\mathfrak{g} = \mathfrak{sl}(V, W)$,

$$T(W) = \bigoplus_{\lambda} \mathbb{C}_\lambda \otimes W_\lambda.$$

Let $\mathfrak{g} = \mathfrak{sl}(V, W)$. Recall that $T^{m,n} = V^{\otimes m} \otimes W^{\otimes n}$. Then

$$T^{m,n} = \bigoplus_{|\lambda|=n,\ |\mu|=m} \mathbb{C}_\lambda \otimes \mathbb{C}_\mu \otimes V_\lambda \otimes W_\mu.$$

Note that, as a $\mathfrak{g}$-module $T(V, W) := \bigoplus_{m,n \geq 0} T^{m,n}$ is not completely reducible. This follows simply from the observation that the exact sequence

$$0 \to \mathfrak{g} \to V \otimes W \to \mathbb{C} \to 0$$

does not split as $V \otimes W$ has no trivial submodule. In [PStyr] the structure of $T(V, W)$ has been studied in detail for countable-dimensional V and W.

For each ordered set $I = \{i_1, \dots, i_k, j_1, \dots, j_k\}$, where $i_1, \dots, i_k \in \{1, \dots, m\}$, $j_1, \dots, j_k \in \{1, \dots, n\}, k \leq \min\{m, n\}$, there is a well-defined surjective morphism of $\mathfrak{g}$-modules

$$\varphi_I : T^{m,n} \longrightarrow T^{m-k,n-k}$$

such that

$$\varphi_I(v_1 \otimes \cdots \otimes v_m \otimes w_1 \otimes \cdots \otimes w_n) = \prod_s \varphi(v_{i_s} \otimes w_{j_s})(\otimes_{i \neq i_s} v_i) \otimes (\otimes_{j \neq j_s} w_j)$$

for $s = 1, \ldots, k$, where $\varphi : V \otimes W \to \mathbb{C}$ is the linear operator induced by the pairing $V \times W \to \mathbb{C}$.

We now define a filtration of $T^{m,n}$ by setting

$$F_0^{m,n} := 0,\ F_k^{m,n} := \cap_I \ker \varphi_I \text{ for } k = 1, \ldots, \min\{m,n\},\ F_{\min\{m,n\}+1}^{m,n} := T^{m,n}, \tag{5}$$

where I runs over all ordered sets $\{i_1, \ldots, i_k, j_1, \ldots, j_k\}$ as above.

Let $|\lambda| = m$, $|\mu| = n$. We set

$$V_{\lambda,\mu} := F_1^{m,n} \cap (V_\lambda \otimes W_\mu).$$

Note that, for sufficiently large finite-dimensional nondegenerate pairs $V_f \subset V$, $W_f \subset W$, the $\mathfrak{sl}(V_f, W_f)$-module $T(V_f, W_f) \cap V_{\lambda,\mu}$ is simple. Therefore $V_{\lambda,\mu}$ is a simple $\mathfrak{sl}(V, W)$-module.

Theorem 4.1. *$\{F_k^{m,n}\}_{0 \leq k \leq \min\{m,n\}+1}$ is the socle filtration of $T^{m,n}$ as a $\mathfrak{sl}(V, W)$-module.*

Proof. In [PStyr] this theorem is proven in the countable-dimensional case. Here we give a proof for arbitrary V and W.

Recall that if M is a $\mathfrak{g}$-module, $M^{\mathfrak{g}}$ stands for the space of $\mathfrak{g}$-invariants in M.

Lemma 4.2. *Let $\mathfrak{g} = \mathfrak{sl}(V, W)$ (respectively, $\mathfrak{o}(V)$ or $\mathfrak{sp}(V)$). Then $(T^{m,n})^{\mathfrak{g}} = 0$ for $m + n > 0$ (respectively, $(T^m)^{\mathfrak{g}} = 0$ for $m > 0$).*

Proof. We prove the statement for $\mathfrak{g} = \mathfrak{sl}(V, W)$ and $m > 0$. The other cases are similar. Let $u \in T^{m,n} = V^{\otimes m} \otimes W^{\otimes n}$, $u \neq 0$. Then $u \in V_f^{\otimes m} \otimes W_f^{\otimes n}$ for some finite-dimensional nondegenerate pair $V_f \subset V$, $W_f \subset W$. Choose bases in V_f and W_f and write

$$u = \sum_{i=1}^{t} c_i v_1^i \otimes \cdots \otimes v_m^i \otimes w_1^i \otimes \cdots \otimes w_n^i,$$

where all v_j^i and w_j^i are basis vectors respectively of V_f and W_f. Pick $w \in W$ such that $\mathrm{tr}(v_1^1 \otimes w) = 1$ and $\mathrm{tr}(v_j^i \otimes w) = 0$ for all $v_j^i \neq v_1^1$. Let $v \in V \backslash V_f$ and $w \in W_f^{\perp}$. Then

$$(v \otimes w) \cdot u = \sum_{i=1}^{t} \sum_{j=1}^{m} c_i \mathrm{tr}(v_j^i \otimes w) v_1^i \otimes \cdots \otimes v_{j-1}^i \otimes v \otimes v_{j+1}^i \otimes \cdots \otimes v_m^i \otimes w_1^i \otimes \cdots \otimes w_n^i.$$

Our choice of v and w ensures that at least one term in the right-hand side is not zero and there is no repetition in the tensor monomials appearing with nonzero coefficients. That implies $(v \otimes w) \cdot u \neq 0$. Hence $u \notin (V^{\otimes m} \otimes W^{\otimes n})^{\mathfrak{g}}$. □

Lemma 4.3. *Let* $\mathfrak{g} = \mathfrak{sl}(V, W)$. *If* $\mathrm{Hom}_{\mathfrak{g}}(V_{\lambda,\mu}, T^{m,n}) \neq 0$, *then* $|\lambda| = m$, $|\mu| = n$.

Proof. Choose a finite-dimensional nondegenerate pair $V_f \subset V$, $W_f \subset W$ such that $\dim V_f \geq \max\{m, n, |\lambda|, |\mu|\}$. Then $(V_f)_{\lambda,\mu} := T(V_f, W_f) \cap V_{\lambda,\mu}$ is annihilated by the finite corank subalgebra $\mathfrak{k} = \mathfrak{sl}(W_f^{\perp}, V_f^{\perp})$ of $\mathfrak{g}$. Let $\mathfrak{l} = \mathfrak{sl}(V_f, W_f) \oplus \mathfrak{k}$. Then

$$\begin{aligned}\mathrm{Hom}_{\mathfrak{l}}((V_f)_{\lambda,\mu}, T^{m,n}) &= \mathrm{Hom}_{\mathfrak{sl}(V_f,W_f)}((V_f)_{\lambda,\mu}, (T^{m,n})^{\mathfrak{k}}) \\ &= \mathrm{Hom}_{\mathfrak{sl}(V_f,W_f)}((V_f)_{\lambda,\mu}, V_f^{\otimes m} \otimes W_f^{\otimes n}).\end{aligned}$$

Therefore a homomorphism $\varphi \in \mathrm{Hom}_{\mathfrak{g}}(V_{\lambda,\mu}, T^{m,n})$ has a well-defined restriction $\varphi_f \in \mathrm{Hom}_{\mathfrak{sl}(V_f,W_f)}((V_f)_{\lambda,\mu}, V_f^{\otimes m} \otimes W_f^{\otimes n})$. According to finite-dimensional representation theory, $\varphi_f \neq 0$ implies that φ_f is a composition

$$(V_f)_{\lambda,\mu} \to (V_f^{\otimes|\lambda|} \otimes W_f^{\otimes|\mu|}) \otimes (V_f^{\otimes(m-|\lambda|)} \otimes W_f^{\otimes(n-|\mu|)})^{\mathfrak{sl}(V_f,W_f)} \to V_f^{\otimes m} \otimes W_f^{\otimes n}.$$

Since φ is the inverse limit of φ_f, φ is a composition

$$V_{\lambda,\mu} \to T^{|\lambda|,|\mu|} \otimes (T^{m-|\lambda|,n-|\mu|})^{\mathfrak{sl}(V,W)} \to T^{m,n}.$$

However, by Lemma 4.2, $(T^{m-|\lambda|,n-|\mu|})^{\mathfrak{sl}(V,W)} \neq 0$ only if $|\lambda| = m$, $|\mu| = n$. □

Note that Lemma 4.3 implies

$$\mathrm{soc} T^{m,n} = \mathrm{soc}^1 T^{m,n} = F_1^{m,n}. \tag{6}$$

Consider now the exact sequence

$$0 \to F_{k-1}^{m,n} \to T^{m,n} \to \bigoplus_I T^{m-k+1,n-k+1}, \tag{7}$$

where I runs over the same set as in (5). It follows from (6) that (7) induces an exact sequence

$$0 \to F_{k-1}^{m,n} \to F_k^{m,n} \to \bigoplus_I F_1^{m-k+1,n-k+1}.$$

Therefore induction on k yields $\mathrm{soc}^k T^{m,n} = F_k^{m,n}$. Theorem 4.1 is proved. □

As a corollary we obtain that the $\mathfrak{sl}(V, W)$-module $V_\lambda \otimes W_\mu$ is indecomposable since its socle $V_{\lambda,\mu}$ is simple. Further one shows that any simple subquotient of $T(V, W)$ is isomorphic to $V_{\lambda,\mu}$ for an appropriate pair of partitions λ, μ. The k-th layer of the socle filtration of $V_\lambda \otimes W_\mu$, i.e., the quotient $\mathrm{soc}^k(V_\lambda \otimes W_\mu)/\mathrm{soc}^{k-1}(V_\lambda \otimes W_\mu)$, can have only simple constituents isomorphic to $V_{\lambda',\mu'}$ where λ' is obtained from λ by removing $k-1$ boxes and μ' is obtained from μ by removing $k-1$ boxes. An explicit formula for the multiplicity of $V_{\lambda'\mu'}$ in $\mathrm{soc}^k(V_\lambda \otimes W_\mu)/\mathrm{soc}^{k-1}(V_\lambda \otimes W_\mu)$ is given in [PStyr].

Next, consider the associative algebra $\mathcal{A}_{\mathfrak{sl}(V,W)} \subset \mathrm{End}_{\mathfrak{sl}(V,W)}(T(V,W))$ generated by all contractions $\varphi_{i,j}$ and by the direct sum of group algebras $\bigoplus_{m,n\geq 0} \mathbb{C}[S_m \times S_n]$. It is clear that $\mathcal{A}_{\mathfrak{sl}(V,W)}$ does not depend on the choice of the linear system $V \times W \to \mathbb{C}$. In what follows we use the notation $\mathcal{A}_{\mathfrak{sl}}$. One can equip $\mathcal{A}_{\mathfrak{sl}}$ with a $\mathbb{Z}_{\geq 0}$-grading $\mathcal{A}_{\mathfrak{sl}} = \bigoplus_{q\geq 0}(\mathcal{A}_{\mathfrak{sl}})_q$ by setting $(\mathcal{A}_{\mathfrak{sl}})_q := \bigoplus_{m,n\geq 0} \mathrm{Hom}_{\mathfrak{sl}(V,W)}(T^{m,n}, T^{m-q,n-q}) \cap \mathcal{A}_{\mathfrak{sl}}$. If we set $T^{\leq r}(V,W) := \bigoplus_{m+n\leq r} T^{m,n}$ and denote by $\mathcal{A}^{(r)}_{\mathfrak{sl}}$ the intersection of $\mathcal{A}_{\mathfrak{sl}}$ with $\mathrm{End}_{\mathfrak{sl}(V,W)}(T^{\leq r}(V,W))$, then, obviously, $\mathcal{A}_{\mathfrak{sl}} = \varinjlim \mathcal{A}^{(r)}_{\mathfrak{sl}}$.

The following statement is a central result in [DPS].

Proposition 4.4. *(a) If V is countable dimensional, then*

$$(\mathcal{A}_{\mathfrak{sl}})_q = \bigoplus_{m,n\geq 0} \mathrm{Hom}_{\mathfrak{sl}(V,V_*)}(T^{m,n}, T^{m-q,n-q}).$$

(b) $\mathcal{A}^{(r)}_{\mathfrak{sl}}$ is a Koszul self-dual ring for any $r \geq 0$.

Now let $\mathfrak{g} = \mathfrak{o}(V)$ (respectively, $\mathfrak{sp}(V)$). Recall that $T^m = V^{\otimes m}$. Assume $m \geq 2$. For a pair of indices $1 \leq i < j \leq m$ we have a contraction map $\varphi_{i,j} \in \mathrm{Hom}_{\mathfrak{g}}(V^{\otimes m}, V^{\otimes m-2})$. If V is countable dimensional, the socle filtration of $T(V)$ considered as a $\mathfrak{g}$-module is described in [PStyr]. Recall the decomposition (4). Each V_λ is an indecomposable $\mathfrak{g}$-module with simple socle which we denote by $V_{\lambda,\mathfrak{g}}$. Moreover,

$$\mathrm{soc}^k\, V_\lambda = \mathrm{soc}^k\, (V_\lambda \cap V^{\otimes|\lambda|}) = V_\lambda \cap (\cap_{I_1,\dots,I_k} \ker(\varphi_{I_1,\dots,I_k} : V^{\otimes|\lambda|} \to V^{\otimes|\lambda|-2k})),$$

where $I_1, \dots, I_k$ run over all sets of k distinct pairs of indices $1, \dots, |\lambda|$ and $\varphi_{I_1,\dots,I_k} = \varphi_{I_1} \circ \dots \circ \varphi_{I_k}$.

Next, let $\mathcal{A}_{\mathfrak{g}} \subset \mathrm{End}_{\mathfrak{g}}(T(V))$ be the graded subalgebra of $\mathrm{End}_{\mathfrak{g}}(T(V))$ generated by $\bigoplus_{m\geq 0} \mathbb{C}[S_m]$ and the contractions $\varphi_{i,j}$. We define a $\mathbb{Z}_{\geq 0}$- grading $\mathcal{A}_{\mathfrak{g}} = \bigoplus_{q\geq 0}(\mathcal{A}_{\mathfrak{g}})_q$ by setting

$$(\mathcal{A}_{\mathfrak{g}})_q := \bigoplus_{m\geq 0} \mathrm{Hom}_{\mathfrak{g}}(T^m, T^{m-2q}) \cap \mathcal{A}_{\mathfrak{g}}.$$

If we set $T^{\leq r}(V) := \bigoplus_{m\leq r} T^m$ and denote by $\mathcal{A}^{(r)}_{\mathfrak{g}}$ the intersection of $\mathcal{A}_{\mathfrak{g}}$ with $\mathrm{End}_{\mathfrak{g}}(T^{\leq r}(V))$, then $\mathcal{A}_{\mathfrak{g}} = \varinjlim \mathcal{A}^{(r)}_{\mathfrak{g}}$. It is clear that the algebra $\mathcal{A}_{\mathfrak{g}}$ can depend only on the symmetry type of the form on V but not on V and the form itself. This justifies the notations $\mathcal{A}_{\mathfrak{o}}$ and $\mathcal{A}_{\mathfrak{sp}}$.

Proposition 4.5 ([DPS]).

(a) $\mathcal{A}_{\mathfrak{o}}^{(r)} \simeq \mathcal{A}_{\mathfrak{sp}}^{(r)}$ *for each* $r \geq 0$, *and* $\mathcal{A}_{\mathfrak{o}} \simeq \mathcal{A}_{\mathfrak{sp}}$.
(b) *If* V *is countable dimensional, then* $(\mathcal{A}_{\mathfrak{o}})_q = \bigoplus_{m\geq 0} \mathrm{Hom}_{\mathfrak{o}(V)}(T^m, T^{m-2q})$, $(\mathcal{A}_{\mathfrak{sp}})_q = \bigoplus_{m\geq 0} \mathrm{Hom}_{\mathfrak{sp}(V)}(T^m, T^{m-2q})$.
(c) $\mathcal{A}_{\mathfrak{o}}^{(r)} \simeq \mathcal{A}_{\mathfrak{sp}}^{(r)}$ *is a Koszul ring for any* $r \geq 0$.

In each of the three cases $\mathfrak{g} = \mathfrak{sl}(\infty), \mathfrak{o}(\infty), \mathfrak{sp}(\infty)$ we call the modules $V_{\lambda,\mu}$, respectively $V_{\lambda,\mathfrak{g}}$, the *simple tensor modules* of $\mathfrak{g}$.

5 The Category $\mathbb{T}_{\mathfrak{g}}$

5.1 The Countable-Dimensional Case

In this subsection we assume that $\mathfrak{g} = \mathfrak{sl}(V, V_*), \mathfrak{o}(V)$ or $\mathfrak{sp}(V)$ for a countable-dimensional space V. The category $\mathbb{T}_{\mathfrak{g}}$ has been studied in [DPS], and here we review some key results.

Denote by $\tilde{G}$ the group of automorphisms of V under which V_* is stable for $\mathfrak{g} = \mathfrak{sl}(V, V_*)$, and the group of automorphisms of V which keep fixed the form on V which defines $\mathfrak{g}$. The group $\tilde{G}$ is a subgroup of $\mathrm{Aut}\mathfrak{g}$ and therefore acts naturally on isomorphism classes of $\mathfrak{g}$-modules: to each $\mathfrak{g}$-module M one assigns the twisted $\mathfrak{g}$-module $M_{\tilde{g}}$ for $\tilde{g} \in \tilde{G}$. A $\mathfrak{g}$-module M is $\tilde{G}$*-invariant* if $M \simeq M_{\tilde{g}}$ for all $\tilde{g} \in \tilde{G}$.

Furthermore, define a $\mathfrak{g}$-module M to be an *absolute weight module* if the decomposition (2) holds for any local Cartan subalgebra of $\mathfrak{g}$, i.e., if M is a weight module for any local Cartan subalgebra $\mathfrak{h}$ of $\mathfrak{g}$. In [DPS] we have given five equivalent characterizations of the objects of $\mathbb{T}_{\mathfrak{g}}$.

Theorem 5.1 ([DPS]). *The following conditions on a* $\mathfrak{g}$*-module* M *of finite length are equivalent:*

(i) M *is an object of* $\mathbb{T}_{\mathfrak{g}}$;
(ii) M *is a weight module for some local Cartan subalgebra* $\mathfrak{h} \subset \mathfrak{g}$ *and* M *is* $\tilde{G}$*-invariant;*
(iii) M *is a subquotient of* $T(V \oplus V_*)$ *for* $\mathfrak{g} = \mathfrak{sl}(V, V_*)$ *(respectively, of* $T(V)$ *for* $\mathfrak{g} = \mathfrak{o}(V), \mathfrak{sp}(V)$*);*
(iv) M *is a submodule of* $T(V \oplus V_*)$ *for* $\mathfrak{g} = \mathfrak{sl}(V, V_*)$ *(respectively, of* $T(V)$ *for* $\mathfrak{g} = \mathfrak{o}(V), \mathfrak{sp}(V)$*);*
(v) M *is an absolute weight module.*

Furthermore, the following two theorems are crucial for understanding the structure of $\mathbb{T}_{\mathfrak{g}}$.

Theorem 5.2 ([PS, DPS]). *The simple objects in the categories* $\widetilde{\mathrm{Tens}}_{\mathfrak{g}}$ *and* $\mathbb{T}_{\mathfrak{g}}$ *coincide and are all of the form* $V_{\lambda,\mu}$ *for* $\mathfrak{g} = \mathfrak{sl}(V, V_*)$, *or respectively* $V_{\lambda,\mathfrak{g}}$ *for* $\mathfrak{g} = \mathfrak{o}(V), \mathfrak{sp}(V)$.

Theorem 5.3 ([DPS]).

(a) $\mathbb{T}_{\mathfrak{g}}$ *has enough injectives. If* $\mathfrak{g} = \mathfrak{sl}(V, V_*)$, *then* $V_\lambda \otimes (V_*)_\mu$ *is an injective hull of* $V_{\lambda,\mu}$. *If* $\mathfrak{g} = \mathfrak{o}(V)$ *or* $\mathfrak{sp}(V)$, *then* V_λ *is an injective hull of* $V_{\lambda,\mathfrak{g}}$.

(b) $\mathbb{T}_{\mathfrak{g}}$ *is anti-equivalent to the category of locally unitary finite-dimensional* $\mathcal{A}_{\mathfrak{g}}$*−modules.*

Theorem 5.3 means that the category $\mathbb{T}_{\mathfrak{g}}$ is "Koszul" in the sense that it is anti-equivalent to a module category over the infinite-dimensional Koszul algebra $\mathcal{A}_{\mathfrak{g}}$.

Corollary 5.4. $\mathbb{T}_{\mathfrak{o}(\infty)}$ *and* $\mathbb{T}_{\mathfrak{sp}(\infty)}$ *are equivalent abelian categories.*

In fact, the stronger result that $\mathbb{T}_{\mathfrak{o}(\infty)}$ and $\mathbb{T}_{\mathfrak{sp}(\infty)}$ are equivalent as monoidal categories also holds, see [SS] and [S].

5.2 The General Case

In this subsection we prove the following result.

Theorem 5.5. *Let* $\mathfrak{g} = \mathfrak{sl}(V, W), \mathfrak{o}(V), \mathfrak{sp}(V)$. *Then, as a monoidal category,* $\mathbb{T}_{\mathfrak{g}}$ *is equivalent to* $\mathbb{T}_{\mathfrak{sl}(\infty)}$ *or* $\mathbb{T}_{\mathfrak{o}(\infty)}$.

The proof of Theorem 5.5 is accomplished by proving several lemmas and corollaries.

Lemma 5.6. *(a) Let* $\mathfrak{g} = \mathfrak{sl}(V, W)$ *and* $C_{m,n} := \mathrm{Hom}_{\mathfrak{g}}(T^{m,n}, \mathbb{C})$. *If* $m \neq n$, *then* $C_{m,n} = 0$, *and if* $m = n$, *then* $C_{m,m}$ *is spanned by* τ_π *for all* $\pi \in S_m$, *where*

$$\tau_\pi(v_1 \otimes \cdots \otimes v_m \otimes w_1 \otimes \cdots \otimes w_m) = \prod_{i=1}^{m} \mathrm{tr}(v_i \otimes w_{\pi(i)}).$$

(b) Let $\mathfrak{g} = \mathfrak{o}(V)$ *or* $\mathfrak{sp}(V)$. *Then* $\mathrm{Hom}_{\mathfrak{g}}(T^{2m+1}, \mathbb{C}) = 0$ *and* $\mathrm{Hom}_{\mathfrak{g}}(T^{2m}, \mathbb{C})$ *is spanned by* σ_π *for all* $\pi \in S_m$, *where*

$$\sigma_\pi(v_1 \otimes \cdots \otimes v_{2m}) = \prod_{i=1}^{m} (v_i, v_{m+\pi(i)}).$$

Proof. In the finite-dimensional case the same statement is the fundamental theorem of invariant theory. Since $T^{m,n}$ for $\mathfrak{g} = \mathfrak{sl}(V, W)$ (respectively, T^m for $\mathfrak{g} = \mathfrak{o}(V), \mathfrak{sp}(V)$) is a direct limit of finite-dimensional representations of the same type, the statement follows from the fundamental theorem of invariant theory. □

Let L be a $\mathfrak{g}$-module and let $\mathfrak{g}'$ denote a subalgebra of $\mathfrak{g}$ of the form $\mathfrak{sl}(V', W')$ (respectively, $\mathfrak{o}(V'), \mathfrak{sp}(V')$) for some nondegenerate pair $V' \subset V, W' \subset W$

(respectively, nondegenerate subspace $V' \subset V$). Let (V'_f, W'_f) be a finite-dimensional nondegenerate pair satisfying $V'_f \subset V'$, $W'_f \subset W'$ (respectively, $V'_f \subset V'$) and let $\mathfrak{k}' = \mathfrak{sl}((W'_f)^\perp, (V'_f)^\perp) \subset \mathfrak{g}$ (respectively $\mathfrak{k}' = \mathfrak{o}((V'_f)^\perp), \mathfrak{sp}(V'_f)^\perp)$). Then $L^{\mathfrak{k}'}$ is an $\mathfrak{sl}(W'_f, V'_f)$-module (respectively, an $\mathfrak{o}(V'_f)-$ or $\mathfrak{sp}(V'_f)$-module), and moreover if we let $\mathfrak{k}'$ vary, the corresponding $\mathfrak{sl}(V'_f, W'_f)$-modules (respectively, $\mathfrak{o}(V'_f)-$ or $\mathfrak{sp}(V'_f)$-modules) form a directed system whose direct limit

$$\Gamma^{ann}_{\mathfrak{g}'}(L) = \varinjlim L^{\mathfrak{k}'}$$

is a $\mathfrak{g}'$-module. Note that $\Gamma^{ann}_{\mathfrak{g}'}(L)$ may simply be defined as the union $\bigcup_{\mathfrak{k}'} L^{\mathfrak{k}'}$ of subspaces $L^{\mathfrak{k}'} \subset L$.

It is easy to check that $\Gamma^{ann}_{\mathfrak{g}'}$ is a well-defined functor from the category $\mathfrak{g}-$mod to its subcategory of $\mathfrak{g}'-$mod consisting of modules satisfying the large annihilator condition. In particular, $\Gamma^{ann}_{\mathfrak{g}}$ is a well-defined functor from $\mathfrak{g}-$mod to the category of $\mathfrak{g}$-modules satisfying the large annihilator condition, and the restriction of $\Gamma^{ann}_{\mathfrak{g}}$ to $\mathbb{T}_{\mathfrak{g}}$ is the identity functor.

In the case when $\mathfrak{g}'$ is finite dimensional the functor $\Gamma^{ann}_{\mathfrak{g}'}$ and its right derived functors are studied in detail in [SSW].

Lemma 5.7. *(a) Let* $\mathfrak{g} = \mathfrak{sl}(V, W)$*; then*

$$\Gamma^{ann}_{\mathfrak{g}}((T^{m,n})^*) \simeq \bigoplus_{k \geq 0} b_k T^{n-k,m-k}$$

where $b_k = \binom{m}{k}\binom{n}{k}k!$.
(b) Let $\mathfrak{g} = \mathfrak{o}(V)$ *or* $\mathfrak{sp}(V)$*, then*

$$\Gamma^{ann}_{\mathfrak{g}}((T^m)^*) \simeq \bigoplus_{k \geq 0} c_k T^{m-2k}$$

where $c_k = \binom{m}{2k}k!$.

Proof. We prove (a) and leave (b) to the reader. Choose a finite-dimensional nondegenerate pair $V_f \subset V, W_f \subset W$, and let $\mathfrak{k} = \mathfrak{sl}(W_f^\perp, V_f^\perp)$. There is an isomorphism of $\mathfrak{k}$-modules

$$(T^{m,n})^* = (V^{\otimes m} \otimes W^{\otimes n})^* \simeq \bigoplus_{k\geq 0, l \geq 0} d_{k,l}(W_f^{\otimes m-k} \otimes V_f^{\otimes n-l}) \otimes ((V_f^\perp)^{\otimes k} \otimes (W_f^\perp)^{\otimes l})^* \tag{8}$$

where $d_{k,l} = \binom{m}{k}\binom{n}{l}$.

Using (8) and Lemma 5.6 (a) applied to $\mathfrak{k}$ in place of $\mathfrak{g}$, we compute that

$$\left((T^{m,n})^*\right)^{\mathfrak{k}} \simeq \bigoplus_{k\geq 0} b_k(W_f^{\otimes m-k} \otimes V_f^{\otimes n-k}).$$

Now the statement follows by taking the direct limit of $\mathfrak{k}$-invariants over all nondegenerate finite-dimensional pairs $V_f \subset V$, $W_f \subset W$. □

Corollary 5.8. *$T^{m,n}$ is an injective object of $\mathbb{T}_{\mathfrak{sl}(V,W)}$, and T^m is an injective object of $\mathbb{T}_{\mathfrak{g}}$ for $\mathfrak{g} = \mathfrak{o}(V), \mathfrak{sp}(V)$.*

Proof. We consider only the case $\mathfrak{g} = \mathfrak{sl}(V, W)$. Recall (Theorem 2.1) that if M is an integrable module such that M^* is integrable, then M^* is injective in $\mathrm{Int}_{\mathfrak{g}}$. In particular, $(T^{m,n})^*$ is injective in $\mathrm{Int}_{\mathfrak{g}}$. Next, note that $\Gamma_{\mathfrak{g}}^{ann}$ is right adjoint to the inclusion functor $\mathbb{T}_{\mathfrak{g}} \rightsquigarrow \mathrm{Int}_{\mathfrak{g}}$, i.e., for any $L \in \mathbb{T}_{\mathfrak{g}}$ and any $Y \in \mathrm{Int}_{\mathfrak{g}}$, we have

$$\mathrm{Hom}_{\mathfrak{g}}(L, Y) = \mathrm{Hom}_{\mathfrak{g}}(L, \Gamma_{\mathfrak{g}}^{ann}(Y)).$$

Hence, $\Gamma_{\mathfrak{g}}^{ann}$ transforms injectives in $\mathrm{Int}_{\mathfrak{g}}$ to injectives in $\mathbb{T}_{\mathfrak{g}}$. This implies that $\Gamma_{\mathfrak{g}}^{ann}((T^{n,m})^*)$ is injective in $\mathbb{T}_{\mathfrak{g}}$. By Lemma 5.7, $T^{m,n}$ is a direct summand in $\Gamma_{\mathfrak{g}}^{ann}((T^{n,m})^*)$, and the statement follows. □

Next we impose the condition that our fixed subalgebra $\mathfrak{g}' \subset \mathfrak{g}$ is countable dimensional. In the rest of the paper we set $\mathfrak{g}_c := \mathfrak{g}'$. More precisely, we choose strictly increasing chains of finite-dimensional subspaces

$$V_1 \subset V_2 \subset \ldots \subset V_i \subset V_{i+1} \subset \ldots, \quad W_1 \subset W_2 \subset \ldots \subset W_i \subset W_{i+1} \subset \ldots$$

and set $\mathfrak{g}_c = \mathfrak{sl}(V_c, W_c)$ where $V_c := \varinjlim V_i$, $W_c := \varinjlim W_i$. It is clear that $V_c \times W_c \to \mathbb{C}$ is a countable-dimensional linear system, hence $\mathfrak{g}_c \simeq \mathfrak{sl}(\infty)$. If $\mathfrak{g} = \mathfrak{o}(V), \mathfrak{sp}(V)$, choose a strictly increasing chain of nondegenerate finite-dimensional subspaces $V_1 \subset V_2 \subset \ldots \subset V_i \subset V_{i+1} \subset \ldots$ and set $V_c := \varinjlim V_i$, $\mathfrak{g}_c = \mathfrak{o}(V_c), \mathfrak{sp}(V_c)$.

By Φ we denote the restriction of $\Gamma_{\mathfrak{g}_c}^{ann}$ to $\mathbb{T}_{\mathfrak{g}}$. Note that for any $L \in \mathbb{T}_{\mathfrak{g}}$, $\Phi(L)$ is a $\mathfrak{g}_c$-submodule of L.

Lemma 5.9. *Let $L, L' \in \mathbb{T}_{\mathfrak{g}}$.*

(a) $\Phi(L)$ generates L.
(b) The homomorphism $\Phi(L, L') : \mathrm{Hom}_{\mathfrak{g}}(L, L') \to \mathrm{Hom}_{\mathfrak{g}_c}(\Phi(L), \Phi(L'))$ is injective.

Proof. Again we consider only the case $\mathfrak{g} = \mathfrak{sl}(V, W)$ since the other cases are similar. Let $SL(V, W)$ denote the direct limit group $\varinjlim SL(V_f, W_f)$ for all nondegenerate finite-dimensional pairs $V_f \subset V$, $W_f \subset W$, where $SL(V_f, W_f) \simeq SL(\dim V_f)$.

(a) Since L has finite length and satisfies the large annihilator condition, there is a finite-dimensional nondegenerate pair $V_f \subset V, W_f \subset W$ and a finite-dimensional $\mathfrak{gl}(V_f, W_f)$-submodule $L_f \subset L$ annihilated by $\mathfrak{sl}((W_f)^\perp, (V_f)^\perp)$ such that L is generated by L_f over $\mathfrak{g}$. Choose i so that $\dim V_f < \dim V_i$. Then there exists $g \in SL(V, W)$ such that $g(V_f) \subset V_i, g(W_f) \subset W_i$. Note that $g = \exp x$ for some $x \in \mathfrak{sl}(V, W)$. By the integrability of L as a $\mathfrak{g}$-module, the action of g is well defined on L, and $g(L_f)$ also generates L over $\mathfrak{g}$. On the other hand, by construction $g(L_f)$ is annihilated by $g\mathfrak{sl}((W_f)^\perp, (V_f)^\perp)g^{-1}$. Observe that

$$\mathfrak{sl}((W_i)^\perp, (V_i)^\perp) \subset \mathfrak{sl}(g(W_f)^\perp, g(V_f)^\perp) = g\mathfrak{sl}((W_f)^\perp, (V_f)^\perp)g^{-1}.$$

Hence $g(L_f) \subset \Phi(L)$. The statement follows.

(b) Follows immediately from (a). □

Lemma 5.10. *(a)* $\Phi(T^{m,n}) = V_c^{\otimes m} \otimes W_c^{\otimes n}$ *for* $\mathfrak{g} = \mathfrak{sl}(V, W)$, *and* $\Phi(T^m) = V_c^{\otimes m}$ *for* $\mathfrak{g} = \mathfrak{o}(V), \mathfrak{sp}(V)$;

(b) The homomorphisms

$$\Phi(T^{m,n}, T^{k,l}) : \mathrm{Hom}_{\mathfrak{g}}(T^{m,n}, T^{k,l}) \to \mathrm{Hom}_{\mathfrak{g}_c}(V_c^{\otimes m} \otimes W_c^{\otimes n}, V_c^{\otimes k} \otimes W_c^{\otimes l})$$

for $\mathfrak{g} = \mathfrak{sl}(V, W)$, *and*

$$\Phi(T^m, T^k) : \mathrm{Hom}_{\mathfrak{g}}(T^m, T^k) \to \mathrm{Hom}_{\mathfrak{g}_c}(V_c^{\otimes k}, V_c^{\otimes k})$$

for $\mathfrak{g} = \mathfrak{o}(V)$ *or* $\mathfrak{sp}(V)$, *are isomorphisms.*

(c) Let $X \subset \bigoplus_i V_c^{\otimes m_i} \otimes W_c^{\otimes n_i}$, *(respectively,* $X \subset \bigoplus_i V_c^{m_i}$ *for* $\mathfrak{g} = \mathfrak{o}(V), \mathfrak{sp}(V)$*) be a* $\mathfrak{g}_c$*-submodule. Then* $\Phi(U(\mathfrak{g}) \cdot X) = X$.

(d) If $X \subset V_c^{\otimes m} \otimes W_c^{\otimes n}$ *(respectively,* $X \subset \bigoplus_i V_c^{m_i}$ *for* $\mathfrak{g} = \mathfrak{o}(V), \mathfrak{sp}(V)$*) is a simple submodule, then* $U(\mathfrak{g}) \cdot X$ *is a simple* $\mathfrak{g}$*-module.*

Proof. (a) follows easily from the observation that

$$(T^{m,n})^{\mathfrak{k}} = V_i^{\otimes m} \otimes W_i^{\otimes n}$$

for any finite corank subalgebra $\mathfrak{k} = \mathfrak{sl}(W_i^\perp, V_i^\perp)$. This observation is a straightforward consequence of Lemma 4.2.

To prove (b), note that the injectivity of the homomorphisms $\Phi(T^{m,n}, T^{k,l})$ follows from (a) and Lemma 5.9 (b). To prove surjectivity, we observe that $\mathrm{Hom}_{\mathfrak{g}_c}(V_c^{\otimes m} \otimes W_c^{\otimes n}, V_c^{\otimes k} \otimes W_c^{\otimes l})$ is generated by permutations and contractions according to Proposition 4.5 (b). Both are defined in $\mathrm{Hom}_{\mathfrak{g}}(T^{m,n}, T^{k,l})$ by the same formulae. Therefore the homomorphisms $\Phi(T^{m,n}, T^{k,l})$ are surjective.

We now prove (c). Note that $X = \ker\alpha$ for some $\alpha \in \mathrm{Hom}_{\mathfrak{g}_c}(\bigoplus_i V_c^{\otimes m_i} \otimes W_c^{\otimes n_i}, \bigoplus_j V_c^{\otimes m_j} \otimes W_c^{\otimes n_j})$. Using (b) we have $U(\mathfrak{g}) \cdot X \subset \ker\Phi^{-1}(\alpha)$. Hence, $\Phi(U(\mathfrak{g}) \cdot X) \subset \ker\alpha = X$. Since the inclusion $X \subset \Phi(U(\mathfrak{g}) \cdot X)$ is obvious, the statement follows.

To prove (d), suppose $U(\mathfrak{g}) \cdot X$ is not simple, i.e., there is an exact sequence

$$0 \to L \to U(\mathfrak{g}) \cdot X \to L' \to 0$$

for some nonzero L, L'. By the exactness of Φ and by (c), we have an exact sequence

$$0 \to \Phi(L) \to X \to \Phi(L') \to 0.$$

By Lemma 5.9 (a), $\Phi(L)$ and $\Phi(L')$ are both nonzero. This contradicts the assumption that X is simple. □

Lemma 5.11. *For $\mathfrak{g} = \mathfrak{sl}(V, W)$ (respectively, for $\mathfrak{g} = \mathfrak{o}(V), \mathfrak{sp}(V)$) any simple object in the category $\mathbb{T}_{\mathfrak{g}}$ is isomorphic to a submodule in $T^{m,n}$ for suitable m and n (respectively, in T^m for a suitable m).*

Proof. We assume that $\mathfrak{g} = \mathfrak{sl}(V, W)$ and leave the other cases to the reader. Let L be a simple module in $\mathbb{T}_{\mathfrak{g}}$. By Lemma 5.9 (a), $\Phi(L) \neq 0$. Let $L_i = L^{\mathfrak{sl}(W_i^{\perp}, V_i^{\perp})} \neq 0$ for some i, and let $L' \subset L_i$ be a simple $\mathfrak{sl}(V_i, W_i)$-submodule. Consider the $\mathbb{Z}$-grading $\mathfrak{g} = \mathfrak{g}^{-1} \oplus \mathfrak{g}^0 \oplus \mathfrak{g}^1$ where $\mathfrak{g}^0 = \mathfrak{gl}(V_i, W_i) \oplus \mathfrak{sl}(W_i^{\perp}, V_i^{\perp})$, $\mathfrak{g}^1 = V_i \otimes V_i^{\perp}$, $\mathfrak{g}^{-1} = W_i^{\perp} \otimes W_i$. There exists a finite-dimensional subspace $W' \subset V_i^{\perp}$, such that $S(V_i \otimes W')$ generates $S(\mathfrak{g}^1)$ as a module over $\mathfrak{sl}(W_i^{\perp}, V_i^{\perp})$. By the integrability of L, $(V_i \otimes W')^q \cdot L' = 0$ for sufficiently large $q \in \mathbb{Z}_{\geq 0}$, and thus $(\mathfrak{g}^1)^q \cdot L' = 0$. Hence, there is a nonzero vector $l \in L_i \subset L$ annihilated by $\mathfrak{g}^1$, and consequently there is a simple $\mathfrak{g}^0$-submodule $L'' \subset L$ annihilated by $\mathfrak{g}^1$. Therefore L is isomorphic to a quotient of the parabolically induced module $U(\mathfrak{g}) \otimes_{U(\mathfrak{g}^0 \oplus \mathfrak{g}^1)} L''$. The latter module is a direct limit of parabolically induced modules for finite-dimensional subalgebras of $\mathfrak{g}$. Hence it has a unique integrable quotient, and this quotient is isomorphic to L. On the other hand, L'' is a simple $\mathfrak{g}_0$-submodule of $T^{m,n}$ for some m and n. Thus, by Frobenius reciprocity, a quotient of $U(\mathfrak{g}) \otimes_{U(\mathfrak{g}^0 \oplus \mathfrak{g}^1)} L''$ is isomorphic to a submodule of $T^{m,n}$. Since $T^{m,n}$ is integrable, this quotient is isomorphic to L. □

Corollary 5.12. *(a) If $\mathfrak{g} = \mathfrak{sl}(V, W)$, then $\mathcal{A}_{\mathfrak{sl}} = \bigoplus_{m,n,q} \mathrm{Hom}_{\mathfrak{g}}(T^{m,n}, T^{m-q,n-q})$. If $\mathfrak{g} = \mathfrak{o}(V), \mathfrak{sp}(V)$, then $\mathcal{A}_{\mathfrak{g}} = \bigoplus_{m,q} \mathrm{Hom}\,(T^m, T^{m-2q})$. Furthermore,*

$$\mathcal{A}_{\mathfrak{sl}} = \varinjlim \mathrm{End}_{\mathfrak{g}}\Big(\bigoplus_{m+n \leq r} T^{m,n}\Big),$$

and for $\mathfrak{g} = \mathfrak{o}(V), \mathfrak{sp}(V)$

$$\mathcal{A}_{\mathfrak{o}} = \varinjlim \operatorname{End}_{\mathfrak{g}}(\bigoplus_{m \leq r} T^m).$$

(b) *Up to isomorphism, the objects of* $\mathbb{T}_{\mathfrak{g}}$ *are precisely all finite length submodules of* $T(V, W)^{\oplus k}$ *for* $\mathfrak{g} = \mathfrak{sl}(V, W)$*, and of* $T(V)^{\oplus k}$ *for* $\mathfrak{g} = \mathfrak{o}(V), \mathfrak{sp}(V)$*. Equivalently, up to isomorphism, the objects of* $\mathbb{T}_{\mathfrak{g}}$ *are the finite length subquotients of* $T(V, W)^{\oplus k}$ *for* $\mathfrak{g} = \mathfrak{sl}(V, W)$*, and of* $T(V)^{\oplus k}$ *for* $\mathfrak{g} = \mathfrak{o}(V), \mathfrak{sp}(V)$.

Proof. Claim (a) is a consequence of Lemma 5.10. Claim (b) follows from Lemma 5.11 and Corollary 5.8. □

Lemma 5.13. *For any* $L \in \mathbb{T}_{\mathfrak{g}}$, $\Phi(L) \in \mathbb{T}_{\mathfrak{g}_c}$. *Moreover, the functor* $\Phi : \mathbb{T}_{\mathfrak{g}} \to \mathbb{T}_{\mathfrak{g}_c}$ *is fully faithful and essentially surjective.*

Proof. By Corollary 5.12 (b), L is isomorphic to a submodule in a direct sum of finitely many copies of $T(V, W)$. Then $\Phi(L)$ is isomorphic to a submodule in a direct sum of finitely many copies of $T(V_c, W_c)$. That implies the first assertion. The fact that Φ is faithful follows from Lemma 5.9 (b).

To prove that Φ is full, consider $L, L' \in \mathbb{T}_{\mathfrak{g}}$ and let $I(L), I(L')$ denote respective injective hulls in $\mathbb{T}_{\mathfrak{g}}$. Then

$$\operatorname{Hom}_{\mathfrak{g}}(L, L') \subset \operatorname{Hom}_{\mathfrak{g}}(I(L), I(L'))$$

and

$$\operatorname{Hom}_{\mathfrak{g}_c}(\Phi(L), \Phi(L')) \subset \operatorname{Hom}_{\mathfrak{g}_c}(\Phi(I(L)), \Phi(I(L'))).$$

By Corollary 5.12 (a), the homomorphism

$$\Phi(I(L), I(L')) : \operatorname{Hom}_{\mathfrak{g}}(I(L), I(L')) \to \operatorname{Hom}_{\mathfrak{g}_c}(\Phi(I(L)), \Phi(I(L')))$$

is surjective. Therefore for any $\varphi \in \operatorname{Hom}_{\mathfrak{g}_c}(\Phi(L), \Phi(L'))$ there exists $\psi \in \operatorname{Hom}_{\mathfrak{g}}(I(L), I(L'))$ such that $\psi(\Phi(L)) \subset \Phi(L')$. By Lemma 5.9 $\Phi(L)$ and $\Phi(L')$ generate respectively L and L'. Hence $\psi(L) \subset L'$. Thus, we obtain that the homomorphism

$$\Phi(L, L') : \operatorname{Hom}_{\mathfrak{g}}(L, L') \to \operatorname{Hom}_{\mathfrak{g}_c}(\Phi(L), \Phi(L'))$$

is also surjective.

To prove that Φ is essentially surjective, we use again Corollary 5.12 (b). We note that any $L \in \mathbb{T}_{\mathfrak{g}}$ is isomorphic to the kernel of $\varphi \in \operatorname{Hom}(T(V, W)^{\oplus k}, T(V, W)^{\oplus l})$ for some k and l and then apply Corollary 5.12 (a). □

Observe that Lemma 5.13 implies that

$$\Phi : \mathbb{T}_{\mathfrak{g}} \to \mathbb{T}_{\mathfrak{g}_c}$$

an equivalence of the abelian categories $\mathbb{T}_{\mathfrak{g}}$ and $\mathbb{T}_{\mathfrak{g}_c}$. To prove Theorem 5.5 it remains to check that Φ is an equivalence of monoidal categories. We therefore prove the following.

Lemma 5.14. *If $L, N \in \mathbb{T}_{\mathfrak{g}}$, then $\Phi(L \otimes N) \simeq \Phi(L) \otimes \Phi(N)$.*

Proof. We just consider the case $\mathfrak{sl}(V, W)$ as the orthogonal and symplectic cases are very similar. Let $\mathfrak{k} = \mathfrak{sl}(W_f^{\perp}, V_f^{\perp})$ for some finite-dimensional nondegenerate pair $V_f \subset V$, $W_f \subset W$. We claim that

$$(L \otimes N)^{\mathfrak{k}} = L^{\mathfrak{k}} \otimes N^{\mathfrak{k}}.$$

Indeed, using Lemma 4.2 one can easily show that

$$(T^{m,n})^{\mathfrak{k}} = V_f^{\otimes m} \otimes W_f^{\otimes n},$$

which implies the statement in the case when L and N are injective. For arbitrary L and N consider embeddings $L \hookrightarrow I$ and $N \hookrightarrow J$ for some injective $I, J \in \mathbb{T}_{\mathfrak{g}}$. Then

$$(L \otimes N)^{\mathfrak{k}} = (L \otimes N) \cap (I \otimes J)^{\mathfrak{k}} = (L \otimes N) \cap (I^{\mathfrak{k}} \otimes J^{\mathfrak{k}}) = L^{\mathfrak{k}} \otimes N^{\mathfrak{k}}.$$

Now we set $\mathfrak{k} = \mathfrak{sl}(W_i^{\perp}, V_i^{\perp})$ and finish the proof by passing to the direct limit. □

The proof of Theorem 5.5 is complete. □

6 Mackey Lie Algebras

Let $V \times W \to \mathbb{C}$ be a linear system. Then each of V and W can be considered as subspace of the dual of the other:

$$V \subset W^*, \quad W \subset V^*.$$

Let $\mathrm{End}_W(V)$ denote the algebra of endomorphisms $\varphi : V \to V$ such that $\varphi^*(W) \subset W$ where $\varphi^* : V^* \to V^*$ is the dual endomorphism. Clearly, there is a canonical anti-isomorphism of algebras

$$\mathrm{End}_W(V) \xrightarrow{\sim} \mathrm{End}_V(W), \quad \varphi \longmapsto \varphi^*_{|W}.$$

We call the Lie algebra associated with the associative algebra $\mathrm{End}_W(V)$ (or equivalently $\mathrm{End}_V(W)$) a *Mackey Lie algebra* and denote it by $\mathfrak{gl}^M(V, W)$.

Note that if V, W is a linear system, then for any subspaces $W' \subset V^*$ with $W \subset W'$, and $V' \subset W^*$ with $V \subset V'$, the pairs V, W' and V', W are linear systems. In particular V, V^* is a linear system and W^*, W is a linear system. Clearly, $\mathfrak{gl}^M(V, V^*)$ coincides with the Lie algebra of all endomorphisms of V (respectively, $\mathfrak{gl}^M(W^*, W)$ is the Lie algebra of all endomorphisms of W). Hence $\mathfrak{gl}^M(V, W) \subset \mathfrak{gl}^M(V, V^*)$, $\mathfrak{gl}^M(V, W) \subset \mathfrak{gl}^M(W^*, W)$. If V and $W = V_*$ are countable dimensional, the Lie algebra $\mathfrak{gl}^M(V, V_*)$ is identified with the Lie algebra of all matrices $X = (x_{ij})_{i\geq 1, j\geq 1}$ such that each row and each column of X have finitely many nonzero entries. The Mackey Lie algebra $\mathfrak{gl}^M(V, V^*)$ (for a countable dimensional space V) is identified with the Lie algebra of all matrices $X = (x_{ij})_{i\geq 1, j\geq 1}$ each column of which has finitely many nonzero entries. Alternatively, if a basis of V as above is enumerated by $\mathbb{Z}$ (i.e., we consider a basis $\{v_j\}_{j\in\mathbb{Z}}$ such that $V_* = \mathrm{span}\{v_j^*\}_{j\in\mathbb{Z}}$ where $v_j^*(v_i) = 0$ for $j \neq i$, $v_j^*(v_j) = 1$), then $\mathfrak{gl}^M(V, V_*)$ is identified with the Lie algebra of all matrices $(x_{ij})_{i,j\in\mathbb{Z}}$ whose rows and columns have finitely many nonzero entries, and $\mathfrak{gl}^M(V, V^*)$ is identified with the Lie algebra of all matrices $(x_{ij})_{i,j\in\mathbb{Z}}$ whose columns have finitely many nonzero entries.

Obviously V and W are $\mathfrak{gl}^M(V, W)$-modules. Moreover, V and W are not isomorphic as $\mathfrak{gl}^M(V, W)$-modules.

It is easy to see that $\mathfrak{gl}(V, W) = V \otimes W$ is the subalgebra of $\mathfrak{gl}^M(V, W)$ consisting of operators with finite-dimensional images in both V and W, and that it is an ideal in $\mathfrak{gl}^M(V, W)$. Furthermore, the Lie algebra $\mathfrak{gl}^M(V, W)$ has a 1-dimensional center consisting of the scalar operators $\mathbb{C}\mathrm{Id}$.

We now introduce the orthogonal and symplectic Mackey Lie algebras. Let V be a vector space endowed with a nondegenerate symmetric (respectively, antisymmetric) form, then $\mathfrak{o}^M(V)$ (respectively, $\mathfrak{sp}^M(V)$) is the Lie algebra

$$\{X \in \mathrm{End}(V) \mid (X \cdot v, w) + (v, X \cdot w) = 0 \ \forall v, w \in V\}.$$

If V is countable dimensional, there always is a basis $\{v_i, w_j\}_{i,j\in\mathbb{Z}}$ of V such that $\mathrm{span}\{v_i\}_{i\in\mathbb{Z}}$ and $\mathrm{span}\{w_j\}_{j\in\mathbb{Z}}$ are isotropic spaces and $(v_i, w_j) = 0$ for $i \neq j$, $(v_i, w_i) = 1$. The corresponding matrix form of $\mathfrak{o}^M(V)$ consists of all matrices

$$\left(\begin{array}{c|c} a_{ij} & b_{kl} \\ \hline c_{rs} & -a_{ji} \end{array}\right) \tag{9}$$

each row and column of which are finite and in addition $b_{kl} = -b_{lk}$, $c_{rs} = -c_{sr}$ where $i, j, k, l, r, s \in \mathbb{Z}$. The matrix form for $\mathfrak{sp}^M(V)$ is similar: here $b_{kl} = b_{lk}$, $c_{rs} = c_{sr}$.

It is clear that $\mathfrak{o}(V) \subset \mathfrak{o}^M(V)$ and $\mathfrak{sp}(V) \subset \mathfrak{sp}^M(V)$:

$$(v \wedge w) \cdot x = (v, x)w - (x, w)v \quad \text{for } v \wedge w \in \Lambda^2 V = \mathfrak{o}(V), \ x \in V$$

and

$$(vw) \cdot x = (v, x)w - (x, w)v \quad \text{for } vw \in S^2 V = \mathfrak{sp}(V), \ x \in V.$$

Moreover, $\mathfrak{o}(V)$ is an ideal in $\mathfrak{o}^M(V)$ and $\mathfrak{sp}(V)$ is an ideal in $\mathfrak{sp}^M(V)$, since both $\Lambda^2 V$ and $S^2 V$ consist of the respective operators with finite-dimensional image in V.

In this way we have the following exact sequences of Lie algebras:

$$0 \to \mathfrak{gl}(V, W) \to \mathfrak{gl}^M(V, W) \to \mathfrak{gl}^M(V, W)/\mathfrak{gl}(V, W) \to 0,$$

$$0 \to \mathfrak{o}(V) \to \mathfrak{o}^M(V) \to \mathfrak{o}^M(V)/\mathfrak{o}(V) \to 0,$$

$$0 \to \mathfrak{sp}(V) \to \mathfrak{sp}^M(V) \to \mathfrak{sp}^M(V)/\mathfrak{sp}(V) \to 0.$$

Lemma 6.1. $\mathfrak{sl}(V, W)$ *(respectively,* $\mathfrak{o}(V), \mathfrak{sp}(V)$*) is the unique simple ideal in* $\mathfrak{gl}^M(V, W)$ *(respectively,* $\mathfrak{o}^M(V), \mathfrak{sp}^M(V)$*).*

Proof. We will prove that if $I \neq \mathbb{C}\mathrm{Id}$ is a nonzero ideal in $\mathfrak{gl}^M(V, W)$, then I contains $\mathfrak{sl}(V, W)$. Indeed, assume that $X \in I$ and $X \neq c\mathrm{Id}$. Then one can find $v \in V$ and $w \in W$ such that $X \cdot v$ is not proportional to v and $X^* \cdot w$ is not proportional to w. Hence, $Z = [X, v \otimes w] = (X \cdot v) \otimes w - v \otimes (X \cdot w) \in \mathfrak{gl}(V, W) \cap I$ and $Z \neq 0$. Since $\mathfrak{sl}(V, W)$ is the unique simple ideal in $\mathfrak{gl}(V, W)$ and $\mathfrak{gl}(V, W) \cap I \neq 0$, we conclude that $\mathfrak{sl}(V, W) \subset I$.

The two other cases are similar and we leave them to the reader. □

Corollary 6.2. *(a) Two Lie algebras* $\mathfrak{gl}^M(V, W)$ *and* $\mathfrak{gl}^M(V', W')$ *are isomorphic if and only if the linear systems* $V \times W \to \mathbb{C}$ *and* $V' \times W' \to \mathbb{C}$ *are isomorphic.*

(b) Two Lie algebras $\mathfrak{o}^M(V)$ *and* $\mathfrak{o}^M(V')$ *(respectively,* $\mathfrak{sp}^M(V)$ *and* $\mathfrak{sp}^M(V')$*) are isomorphic if and only if there is an isomorphism of vector spaces* $V \simeq V'$ *transferring the form defining* $\mathfrak{o}^M(V)$ *(respectively* $\mathfrak{sp}^M(V)$*) into the form defining* $\mathfrak{o}^M(V')$ *(respectively,* $\mathfrak{sp}^M(V')$*).*

Proof. The statement follows from Proposition 1.1 and Lemma 6.1. □

The following is our main result about the structure of Mackey Lie algebras.

Theorem 6.3. *Let V be a countable-dimensional vector space.*

(a) $\mathfrak{gl}(V, V_*) \oplus \mathbb{C}\mathrm{Id}$ *is an ideal in* $\mathfrak{gl}^M(V, V_*)$ *and the quotient*

$$\mathfrak{gl}^M(V, V_*)/\left(\mathfrak{gl}(V, V_*) \oplus \mathbb{C}\mathrm{Id}\right)$$

is a simple Lie algebra.

(b) $\mathfrak{gl}(V, V^*) \oplus \mathbb{C}\mathrm{Id}$ *is an ideal in* $\mathrm{End}(V)$ *and the quotient* $\mathrm{End}(V)/\left(\mathfrak{gl}(V, V^*) \oplus \mathbb{C}\mathrm{Id}\right)$ *is a simple Lie algebra.*

(c) If V is equipped with a nondegenerate symmetric (respectively, antisymmetric) bilinear form, then $\mathfrak{o}^M(V)/\mathfrak{o}(V)$ *(respectively* $\mathfrak{sp}^M(V)/\mathfrak{sp}(V)$*) is a simple Lie algebra.*

Proof. The proof is subdivided into lemmas and corollaries.

Note that $\mathfrak{gl}(V, V_*) \subset \mathfrak{gl}(V, V^*) \subset \mathfrak{gl}^M(V, V^*) = \mathrm{End}(V)$. In what follows we fix a basis $\{v_i\}_{i\geq 1}$ in V and use the respective identification of $\mathfrak{gl}(V, V_*)$, $\mathfrak{gl}^M(V, V_*)$ and $\mathfrak{gl}^M(V, V^*) = \mathrm{End}(V)$ with infinite matrices. By E_{ij} we denote the elementary matrix whose only nonzero entry is 1 at position i, j.

Lemma 6.4. *Let* $\mathfrak{g}^M = \mathfrak{gl}^M(V, V_*)$, $\mathrm{End}(V)$. *Assume that an ideal* $I \subset \mathfrak{g}^M$ *contains a diagonal matrix* $D \notin \mathfrak{gl}(V, V_*) \oplus \mathbb{C}\mathrm{Id}$. *Then* $I = \mathfrak{g}^M$.

Proof. We first assume that $D = \sum_{i\geq 1} d_i E_{ii}$ satisfies $d_i \neq d_j$ for all $i \neq j$. Then $[D, \mathfrak{g}^M] = \mathfrak{g}_0^M$, where $\mathfrak{g}_0^M$ is the space of all matrices in $\mathfrak{g}^M$ with zeroes on the diagonal. Consequently, $\mathfrak{g}_0^M \subset I$. Furthermore, any diagonal matrix $\sum_i s_i E_{ii}$ can be written as the commutator

$$\left[\sum_{i\geq 1} E_{i\,i+1}, \sum_{j\geq 1} t_j E_{j+1\,j}\right]$$

with $t_j = \sum_{i=1}^{j} s_i$. Hence, $I = \mathfrak{g}^M$.

We now consider the case of an arbitrary $D \in I$. After permuting the basis elements of V, we can assume that $D = \sum_{i\geq 1} d_i E_{ii}$ with $d_{2m-1} \neq 0$ and $d_{2m-1} \neq d_{2m}$ for all $m > 0$. Let

$$X := \sum_{m=1}^{\infty} \frac{1}{d_{2m} - d_{2m-1}} E_{2m\,2m-1}, \quad Y := \sum_{m=1}^{\infty} s_m E_{2m-1\,2m},$$

where $s_m \neq \pm s_l$ for $m \neq l$. Then $[Y, [X, D]] = s_1 E_{11} - s_1 E_{22} + s_2 E_{33} - s_2 E_{44} + \cdots \in I$, and we reduce this case to the previous one. □

Lemma 6.5. *Let* $y = (y_{ij}) \in \mathfrak{gl}(n)$ *be a nonscalar matrix. There exist* $u, v, w \in \mathfrak{gl}(n)$ *such that* $[u, [v, [w, y]]]$ *is a nonzero diagonal matrix.*

Proof. If y is not diagonal, pick $i \neq j$ such that $y_{ij} \neq 0$. Set $w = E_{ii}, v = E_{jj}, u = E_{ji}$. If y is diagonal, pick $i \neq j$ such that $y_{ii} \neq y_{jj}$ and set $w = E_{ij}, v = E_{ii}, u = E_{ji}$. □

Corollary 6.6. *Let* $\prod_i \mathfrak{gl}(n_i)$ *for* $n_i \geq 2$ *be a block subalgebra of* $\mathfrak{g}^M$. *Suppose that* $X \in \left(\prod_i \mathfrak{gl}(n_i)\right) \cap I$ *for some ideal* $I \subset \mathfrak{g}^M$ *and that* $X \notin \mathfrak{gl}(V, V_*) \oplus \mathbb{C}\mathrm{Id}$. *Then* $I = \mathfrak{g}^M$.

Proof. Let $X = \prod_i X_i$, where $X_i \in \mathfrak{gl}(n_i)$. Without loss of generality we may assume that infinitely many X_i are not diagonal, as otherwise X is diagonal modulo $\mathfrak{gl}(V, V_*)$ and the result follows from Lemma 6.4. Now pick $u_i, v_i, w_i \in \mathfrak{g}_i$ as in Lemma 6.5. Set $u = \prod_i u_i, v = \prod_i v_i, w = \prod_i w_i$. Then $Z = [u,[v,[w, X]]$ is diagonal. By normalizing u_i we can ensure that $Z \notin \mathbb{C}\mathrm{Id}$. Since $Z \in I$, the statement follows from Lemma 6.4. □

Lemma 6.7. *For any* $X = (x_{ij})_{i\geq 1, j\geq 1} \in \mathfrak{gl}^M(V, V_*)$ *there exists an increasing sequence* $i_1 < i_2 < \dots$ *such that* $x_{ij} = 0$ *unless* $i, j \in [i_k, i_{k+2} - 1]$ *for some* k.

Proof. Set $i_1 = 1$,

$$i_2 = \max\{j \mid x_{1j} \neq 0 \text{ or } x_{j1} \neq 0\} + 1,$$

and construct the sequence recursively by setting

$$i_k = \max\{j > i_{k-1} \mid x_{ij} \neq 0 \text{ or } x_{ji} \neq 0 \text{ for some } i_{k-2} \leq i < i_{k-1}\} + 1.$$

□

We are now ready to prove Theorem 6.3 (a).

Corollary 6.8 (Theorem 6.3 (a)). *Let an ideal* I *of* $\mathfrak{gl}(V, V_*)$ *be not contained in* $\mathfrak{gl}(V, V_*) \oplus \mathbb{C}\mathrm{Id}$. *Then* $I = \mathfrak{gl}^M(V, V_*)$.

Proof. Let $X \in I \backslash \{\mathfrak{gl}(V, V_*) \oplus \mathbb{C}\mathrm{Id}\}$. Pick $i_1 < i_2 < \dots$ as in Lemma 6.7 and set

$$D = \mathrm{diag}(\underbrace{1, \dots, 1}_{i_2 - 1}, \underbrace{2, \dots, 2}_{i_3 - i_2}, \underbrace{3, \dots, 3}_{i_4 - i_3}, \dots).$$

Then $X = X_{-1} + X_0 + X_1$ where $[D, X_i] = iX_i$. If $X_0 \notin \mathfrak{gl}(V, V_*) \oplus \mathbb{C}\mathrm{Id}$ we are done by Corollary 6.6 as X_0 is a block matrix. Otherwise, at least one of X_1 and X_{-1} does not lie in $\mathfrak{gl}(V, V_*)$.

Assume for example that $X_1 = (x_{ij}) \notin \mathfrak{gl}(V, V_*)$. Then there exist infinite sequences $\{i_1 < i_2 < \dots\}$ and $\{j_1 < j_2 < \dots\}$ such that $x_{i_s j_s} \neq 0$. Moreover, we may assume that $\dots < i_s \leq j_s < i_{s+1} \leq j_{s+1} < \dots$. Set $Y = \sum_{s\geq 1} E_{j_s i_s}$. Then $[Y, X_1] \in I$ is a block matrix and we can again use Corollary 6.6. □

Next we prove Theorem 6.3 (b).

Let I be an ideal in $\mathrm{End}(V)$. Assume that I is not contained in $\mathfrak{gl}(V, V^*) \oplus \mathbb{C}\mathrm{Id}$. Let $X \in I \setminus \{\mathfrak{gl}(V, V^*) \oplus \mathbb{C}\mathrm{Id}\}$ and let $V_X \subset V$ denote the subspace of all X-finite vectors.

Assume first that $V_X \neq V$. Then there exists $v \in V$ such that $v, X \cdot v, X^2 \cdot v, \dots$ are linearly independent. Let $M = \mathrm{span}\{v, X \cdot v, X^2 \cdot v, \dots\}$ and let U be a subspace of V such that $V = M \oplus U$. Let π_M be the projector on M with kernel U. Then $Y := X + [X, \pi_M] \in I$. A simple calculation shows that both U and M are Y-stable and $Y|_M = X|_M$. Let $Z \in \mathrm{End}(M)$ be defined by $Z(U) = 0$, $Z(X^i \cdot v) = iX^{i-1} \cdot v$ for $i \geq 0$. Then $[Z, Y]$ is a diagonal matrix with infinitely many distinct entries. Hence $I = \mathrm{End}(V)$ by Lemma 6.4.

Now suppose that $V_X = V$. Then we have a decomposition $V = \bigoplus_\lambda V_\lambda$, where $V_\lambda := \bigcup_n \ker(X - \lambda\mathrm{Id})^n$ are generalized eigenspaces of X. First, we assume that for all λ there exists $n(\lambda)$ such that $V_\lambda = \ker(X - \lambda\mathrm{Id})^{n(\lambda)}$. In this case $V = \bigoplus_i V_i$ is a direct sum of X-stable finite-dimensional subspaces. Thus X is a block matrix and by Corollary 6.6 we obtain $I = \mathrm{End}(V)$. Next, we assume that for some λ the sequence $\ker(X-\lambda \mathrm{Id})^n$ does not stabilize. In this case there are linearly independent vectors $v_1, v_2, \ldots$ such that $(X-\lambda \mathrm{Id}) \cdot v_1 = 0$ and $(X-\lambda \mathrm{Id}) \cdot v_i = v_{i-1}$ for all $i > 1$. We repeat the argument from the previous paragraph. Set M to be the span of v_k, let $V = M \oplus U$ and define $Z \in \mathrm{End}(M)$ by setting $Z(U) = 0$, $Z(v_i) = i v_{i+1}$. Then $[Z, ([X, \pi_M] + X)] \in I$ is a diagonal matrix with infinitely many distinct entries. Hence $I = \mathrm{End}(V)$.

To complete the proof of Theorem 6.3 it remains to prove claim (c).

Lemma 6.9. *If $\mathfrak{g}^M = \mathfrak{o}^M(V)$ (respectively, $\mathfrak{sp}^M(V)$, then any nonzero proper ideal $I \subset \mathfrak{g}^M$ equals $\mathfrak{o}(V)$ (respectively, $\mathfrak{sp}(V)$).*

Proof. As follows from (9), one can define a $\mathbb{Z}$-grading $\mathfrak{g}^M = \mathfrak{g}^M_{-1} \oplus \mathfrak{g}^M_0 \oplus \mathfrak{g}^M_1$ such that $\mathfrak{g}^M_0 \simeq \mathfrak{gl}^M(V, V_*)$. This grading is defined by the matrix

$$D = \left(\begin{array}{c|c} \frac{1}{2}\mathrm{Id} & 0 \\ \hline 0 & -\frac{1}{2}\mathrm{Id} \end{array}\right),$$

i.e., $[D, X] = iX$ for $X \in \mathfrak{g}^M_i$. Since $D \in \mathfrak{g}^M$, any ideal $I \subset \mathfrak{g}^M$ is homogeneous in this grading. Note that the ideal generated by D equals the entire Lie algebra $\mathfrak{g}^M$. Hence we may assume that $D \notin I$, and thus that $I_0 := I \cap \mathfrak{g}^M_{-1}$ is a proper ideal in $\mathfrak{g}^M_0$.

Assume first that $I_1 := I \cap \mathfrak{g}^M_1$ is not contained in $\mathfrak{o}(V)$ (respectively, $\mathfrak{sp}(V)$) and let $X \in I_1 \setminus \mathfrak{o}(V)$ (respectively, $X \in I_1 \setminus \mathfrak{sp}(V)$). By an argument similar to the one at the end of the proof of Corollary 6.8, there exists $Y \in \mathfrak{g}^M_{-1}$ such that $[Y, X] \notin \mathfrak{gl}(V, V_*) \oplus \mathbb{C}D$. Therefore by Corollary 6.8 we obtain a contradiction with our assumption that I_0 is a proper ideal in $\mathfrak{g}^M_0$.

Thus, we have proved that $I_1 \subset \mathfrak{o}(V)$ (respectively, $\mathfrak{sp}(V)$) and, similarly, $I_{-1} := I \cap \mathfrak{g}^M_{-1} \subset \mathfrak{o}(V)$ (respectively, $\mathfrak{sp}(V)$). Moreover, $I_0 \subset \mathfrak{gl}(V, V_*)$ by Corollary 6.8. But then I is a nonzero ideal in $\mathfrak{o}(V)$ (respectively, $\mathfrak{sp}(V)$). Since both $\mathfrak{o}(V)$ and $\mathfrak{sp}(V)$ are simple, the statement follows. □

The proof of Theorem 6.3 is complete. □

Theorem 6.3 (a) gives a complete list of ideals in $\mathfrak{gl}^M(V, V_*)$ for a countable-dimensional V. Indeed, since $\mathfrak{sl}(V, V_*)$ is a simple Lie algebra, we obtain that all proper nonzero ideals in $\mathfrak{gl}^M(V, V_*)$ are $\mathfrak{gl}(V, V_*)$, $\mathfrak{sl}(V, V_*)$, $\mathbb{C}\mathrm{Id}$, $\mathfrak{sl}(V, V_*) \oplus \mathbb{C}\mathrm{Id}$ and $\mathfrak{gl}(V, V_*) \oplus \mathbb{C}\mathrm{Id}$. In the same way the Lie algebra $\mathrm{End}(V)$ also has five proper nonzero ideals.

Note that if V is not countable-dimensional, then $\mathfrak{gl}^M(V, V_*)$, $\mathrm{End}(V)$ and $\mathfrak{o}^M(V)$ (respectively, $\mathfrak{sp}^M(V)$) have the following ideal:

$$\{X \mid \dim(X \cdot V) \text{ is finite or countable}\}.$$

Hence, Theorem 6.3 does not hold in this case.

7 Dense Subalgebras

7.1 Definition and General Results

Definition 7.1. Let $\mathfrak{l}$ be a Lie algebra, R be an $\mathfrak{l}$-module, $\mathfrak{k} \subset \mathfrak{l}$ be a Lie subalgebra. We say that $\mathfrak{k}$ acts *densely* on R if for any finite set of vectors $r_1, \dots, r_n \in R$ and any $l \in \mathfrak{l}$ there is $k \in \mathfrak{k}$ such that $k \cdot r_i = l \cdot r_i$ for $i = 1, \dots, n$.

Lemma 7.2. *Let $\mathfrak{k} \subset \mathfrak{l}$ and let R, N be two $\mathfrak{l}$-modules such that $\mathfrak{k}$ acts densely on $R \oplus N$. Then* $\mathrm{Hom}_{\mathfrak{l}}(R, N) = \mathrm{Hom}_{\mathfrak{k}}(R, N)$.

Proof. There is an obvious inclusion $\mathrm{Hom}_{\mathfrak{l}}(R, N) \subset \mathrm{Hom}_{\mathfrak{k}}(R, N)$. Suppose there exists $\varphi \in \mathrm{Hom}_{\mathfrak{k}}(R, N) \setminus \mathrm{Hom}_{\mathfrak{l}}(R, N)$. Then one can find $r \in R, l \in \mathfrak{l}$ such that $\varphi(l \cdot r) \neq l \cdot \varphi(r)$. Since $\mathfrak{k}$ acts densely on $R \oplus N$, there exists $k \in \mathfrak{k}$ such that $k \cdot r = l \cdot r$ and $k \cdot \varphi(r) = l \cdot \varphi(r)$. Therefore we have

$$\varphi(l \cdot r) = \varphi(k \cdot r) = k \cdot \varphi(r) = l \cdot \varphi(r).$$

Contradiction. □

Lemma 7.3. *Let $\mathfrak{k} \subset \mathfrak{l}$ and R be an $\mathfrak{l}$-module on which $\mathfrak{k}$ acts densely. Then*

(a) $\mathfrak{k}$ acts densely on any $\mathfrak{l}$–subquotient of R;
(b) $\mathfrak{k}$ acts densely on $R^{\otimes n}$ for $n \geq 1$;
(c) $\mathfrak{k}$ acts densely on $R^{\oplus n}$ for $n \geq 1$;
(d) $\mathfrak{k}$ acts densely on $T(R)^{\oplus n}$ for $n \geq 1$.

Proof. (a) Let N be an $\mathfrak{l}$-submodule of R. It follows immediately from the definition that $\mathfrak{k}$ acts densely on N and on R/N. That implies the statement.
(b) Let $r_1, \dots, r_q \in R^{\otimes n}$. Write

$$r_i = \sum_{j=1}^{s(i)} m^i_{j1} \otimes \cdots \otimes m^i_{jn}$$

for some $m^i_{jp} \in R$. For any $l \in \mathfrak{l}$ there exists $k \in \mathfrak{k}$ such that $k \cdot m^i_{jp} = l \cdot m^i_{jp}$ for all $i \leq r$, $p \leq n$ and $j \leq s(i)$. Then $k \cdot r_i = l \cdot r_i$ for all $i \leq q$.
Proving (c) and (d) is similar to proving (b) and we leave it to the reader. □

Lemma 7.4. *Let $\mathfrak{k}$, $\mathfrak{l}$ and R be as in Lemma 7.3. Then a $\mathfrak{k}$-submodule of R is $\mathfrak{l}$-stable. Hence any $\mathfrak{k}$-subquotient of R has a natural structure of $\mathfrak{l}$-module.*

Proof. Straightforward from the definition. □

Theorem 7.5. *Let $C_{\mathfrak{l}}$ be a full abelian subcategory of* $\mathfrak{l}$-mod *such that $\mathfrak{k}$ acts densely on any object in C. Let* Res : $\mathfrak{l}$ − mod → $\mathfrak{k}$ − mod *be the functor of restriction. Let $C_{\mathfrak{k}}$ be the image of $C_{\mathfrak{l}}$ under* Res. *Then $C_{\mathfrak{k}}$ is a full abelian subcategory of $\mathfrak{k}$* − mod *and* Res *induces an equivalence of $C_{\mathfrak{k}}$ and $C_{\mathfrak{l}}$.*

Proof. The first assertion follows from Lemma 7.2. It also follows from the same lemma that $\mathrm{Res}(R) \simeq \mathrm{Res}(N)$ implies $R \simeq N$. Thus, every object in $\mathcal{C}_{\mathfrak{k}}$ has a unique (up to isomorphism) structure of $\mathfrak{l}$-module. This provides a quasi-inverse of Res. Hence the second assertion holds. □

Let R be an $\mathfrak{l}$-module. Denote by $\mathbb{T}_{\mathfrak{l}}^{R}$ the full subcategory of $\mathfrak{l}$-mod consisting of all finite length subquotients of finite direct sums $T(R)^{\oplus n}$ for $n \geq 1$.

Proposition 7.6. *Let $\mathfrak{k}$, $\mathfrak{l}$ and R be as in Lemma 8.2. Then the restriction functor*

$$\mathrm{Res} : \mathbb{T}_{\mathfrak{l}}^{R} \rightsquigarrow \mathbb{T}_{\mathfrak{k}}^{R}$$

is an equivalence of monoidal categories.

Proof. By Lemma 7.3, $\mathrm{Res}(\mathbb{T}_{\mathfrak{l}}^{R}) = \mathbb{T}_{\mathfrak{k}}^{R}$. Thus Res is an equivalence of $\mathbb{T}_{\mathfrak{l}}^{R}$ and $\mathbb{T}_{\mathfrak{k}}^{R}$ by Theorem 7.5. In addition, Res clearly commutes with $\otimes$, hence the statement. □

7.2 Dense Subalgebras of Mackey Lie Algebras

Now let $\mathfrak{g}^M$ denote one of the Lie algebras $\mathfrak{gl}^M(V, W)$, $\mathfrak{o}^M(V)$, $\mathfrak{sp}^M(V)$, and $\mathfrak{g}$ denote respectively the subalgebra $\mathfrak{gl}(V, W)$, $\mathfrak{o}(V)$, $\mathfrak{sp}(V)$. By R we denote the $\mathfrak{g}^M$-module $V \oplus W$ (respectively, V).

In what follows we call a Lie subalgebra $\mathfrak{a} \subset \mathfrak{g}^M$ *dense* if it acts densely on R. It is easy to see that $\mathfrak{g}$ is a dense subalgebra of $\mathfrak{g}^M$.

Here are further examples of dense subalgebras of $\mathfrak{gl}^M(V, V_*)$ for a countable-dimensional space V. We identify $\mathfrak{gl}^M(V, V_*)$ with the Lie algebra of matrices $(x_{ij})_{i\geq 1, j\geq 1}$ each row and column of which are finite.

1. The Lie algebra $\mathfrak{j}(V, V_*)$ consisting of matrices $J = (x_{ij})_{i\geq 1, j\geq 1}$ such that $x_{ij} = 0$ when $|i - j| > m_J$ for some $m_J \in \mathbb{Z}_{>0}$ (generalized Jacobi matrices), is dense in $\mathfrak{gl}^M(V, V_*)$.
2. The subalgebra $\mathfrak{lj}(V, V_*) \subset \mathfrak{gl}^M(V, V_*)$ consisting of matrices $X = (x_{ij})_{i\geq 1, j\geq 1}$ satisfying the condition $x_{ij} = 0$ when $i - j > c_X j$ for some $c_X \in \mathbb{Z}_{>0}$, is dense in $\mathfrak{gl}^M(V, V_*)$.
3. The subalgebra $\mathfrak{pj}(V, V_*)$ of matrices $Y = (x_{ij})_{i\geq 1, j\geq 1}$ satisfying the condition $x_{ij} = 0$ when $i - j > p_Y(j)$ for some polynomial $p_Y(t) \in \mathbb{Z}_{\geq 0}[t]$, is dense in $\mathfrak{gl}^M(V, V_*)$.
4. Let $\mathfrak{g}$ be a countable-dimensional diagonal Lie algebra. If $\mathfrak{g}$ is of type $\mathfrak{sl}$, fix a chain (1) of diagonal embeddings where $\mathfrak{g}_i \simeq \mathfrak{sl}(n_i)$. Observe that given a chain (3), we can always choose a chain

$$V_{\mathfrak{g}_1}^* \overset{\mu_1}{\hookrightarrow} V_{\mathfrak{g}_2}^* \overset{\mu_2}{\hookrightarrow} \cdots \hookrightarrow V_{\mathfrak{g}_i}^* \overset{\mu_i}{\hookrightarrow} V_{\mathfrak{g}_{i+1}}^* \hookrightarrow \cdots$$

so that the nondegenerate pairing $V_{\mathfrak{g}_{i+1}} \times V^*_{\mathfrak{g}_{i+1}} \to \mathbb{C}$ restricts to a nondegenerate pairing $\kappa_i(V_{\mathfrak{g}_i}) \times \mu_i(V^*_{\mathfrak{g}_i}) \to \mathbb{C}$. Therefore, by multiplying μ_i by a suitable constant, we can assume that κ_i and μ_i preserve the natural pairings $V_{\mathfrak{g}_i} \times V^*_{\mathfrak{g}_i} \to \mathbb{C}$. This shows that, given a natural representation V of $\mathfrak{g}$, there always is a natural representation V_* such that there is a nondegenerate $\mathfrak{g}$-invariant pairing $V \times V_* \to \mathbb{C}$. This gives an embedding of $\mathfrak{g}$ as a dense subalgebra in $\mathfrak{gl}^M(V, V_*)$

If $\mathfrak{g}$ is of type $\mathfrak{o}$ or $\mathfrak{sp}$, then a natural representation V of $\mathfrak{g}$ is defined again by a chain of embeddings (3). Moreover, V always carries a respective nondegenerate symmetric or symplectic form. Therefore $\mathfrak{g}$ can be embedded as a dense subalgebra in $\mathfrak{o}^M(V)$, or respectively in $\mathfrak{sp}^M(V)$.

The following statement is a particular case of Proposition 7.6.

Corollary 7.7. *Let $\mathfrak{a}$ be a dense subalgebra in $\mathfrak{g}^M$. Then the monoidal categories $\mathbb{T}^R_{\mathfrak{g}^M}$ and $\mathbb{T}^R_{\mathfrak{a}}$ are equivalent.*

7.3 Finite Corank Subalgebras of $\mathfrak{g}^M$ and the Category $\mathbb{T}_{\mathfrak{g}^M}$

We now generalize the notion of finite corank subalgebra to Mackey Lie algebras.

Let $V_f \subset V$, $W_f \subset W$ be a nondegenerate pair of finite-dimensional subspaces. Then $\mathfrak{gl}(W_f^\perp, V_f^\perp)$ is a subalgebra of $\mathfrak{gl}^M(W_f^\perp, V_f^\perp)$ and also a subalgebra of $\mathfrak{gl}^M(V, W)$. Moreover, the following important relation holds

$$\begin{aligned}\mathfrak{sl}(V, W)/\mathfrak{sl}(W_f^\perp, V_f^\perp) &= \mathfrak{gl}(V_f, W_f) \oplus (V_f \otimes V_f^\perp) \oplus (W_f^\perp \otimes W_f)\\ &= \mathfrak{gl}^M(V, W)/\mathfrak{gl}^M(W_f^\perp, V_f^\perp). \end{aligned} \tag{10}$$

We call a subalgebra $\mathfrak{k} \subset \mathfrak{gl}^M(V, W)$ a *finite corank subalgebra* if it contains $\mathfrak{gl}^M(W_f^\perp, V_f^\perp)$ for some nondegenerate pair $V_f \subset V$, $W_f \subset W$.

Similarly, let V be a vector space equipped with a symmetric (respectively, skew-symmetric) nondegenerate form and V_f be a nondegenerate finite-dimensional subspace. We have a well-defined subalgebra $\mathfrak{o}^M(V_f^\perp) \subset \mathfrak{o}^M(V)$ (respectively, $\mathfrak{sp}^M(V_f^\perp) \subset \mathfrak{sp}^M(V)$). Furthermore,

$$\begin{aligned}\mathfrak{o}(V)/\mathfrak{o}(V_f^\perp) &= \mathfrak{o}(V_f) \oplus (V_f \otimes V_f^\perp) = \mathfrak{o}^M(V)/\mathfrak{o}^M(V_f^\perp),\\ \mathfrak{sp}(V)/\mathfrak{sp}(V_f^\perp) &= \mathfrak{sp}(V_f) \oplus (V_f \otimes V_f^\perp) = \mathfrak{sp}^M(V)/\mathfrak{sp}^M(V_f^\perp). \end{aligned} \tag{11}$$

We call $\mathfrak{k} \subset \mathfrak{o}^M(V)$ (respectively, $\mathfrak{sp}^M(V)$) a *finite corank subalgebra* if it contains $\mathfrak{o}^M(V_f^\perp)$ (respectively, $\mathfrak{sp}^M(V_f^\perp)$) for some V_f as above.

Next, we say that $\mathfrak{g}^M$-module L *satisfies the large annihilator condition* if the annihilator in $\mathfrak{g}^M$ of any $l \in L$ contains a finite corank subalgebra. It follows immediately from the definition that if L_1 and L_2 satisfy the large annihilator condition, then the same is true for $L_1 \oplus L_2$ and $L_1 \otimes L_2$.

Lemma 7.8. *Let L be a $\mathfrak{g}^M$-module which is integrable as a $\mathfrak{g}$-module. If L satisfies the large annihilator condition (as a $\mathfrak{g}^M$-module), then $\mathfrak{g}$ acts densely on L.*

Proof. Since L satisfies the large annihilator condition as a $\mathfrak{g}^M$-module, so does also $L^{\oplus n}$. It suffices to show that for all $n \in \mathbb{Z}_{\geq 1}$ and all $l \in L^{\oplus n}$ we have

$$\mathfrak{g} \cdot l = \mathfrak{g}^M \cdot l. \tag{12}$$

However, as l is annihilated by $\mathfrak{gl}^M(W_f^{\perp}, V_f^{\perp})$ for an appropriate finite-dimensional nondegenerate pair $V_f \subset V, W_f \subset W$ in the case $\mathfrak{g} = \mathfrak{sl}(V, W)$ (respectively, by $\mathfrak{o}^M(V_f^{\perp}), \mathfrak{sp}^M(V_f^{\perp})$ in the case $\mathfrak{g} = \mathfrak{o}(V), \mathfrak{sp}(V)$), (12) follows from (10), (respectively, from (11)). □

Lemma 7.9. *Let L be a $\mathfrak{g}$-module satisfying the large annihilator condition. Then the $\mathfrak{g}$-module structure on L extends in a unique way to a $\mathfrak{g}^M$-module structure such that L satisfies the large annihilator condition as a $\mathfrak{g}^M$-module.*

Proof. Consider the case $\mathfrak{g} = \mathfrak{sl}(V, W)$. Any $l \in L$ is annihilated by $\mathfrak{sl}(W_f^{\perp}, V_f^{\perp})$ for an appropriate finite-dimensional nondegenerate pair $V_f \subset V, W_f \subset W$. Let $x \in \mathfrak{gl}^M(V, W)$. By (10) there exists $y \in \mathfrak{sl}(V, W)$ such that $x + \mathfrak{gl}^M(W_f^{\perp}, V_f^{\perp}) = y + \mathfrak{sl}(W_f^{\perp}, V_f^{\perp})$. Moreover, y is unique modulo $\mathfrak{sl}(W_f^{\perp}, V_f^{\perp})$. Thus we can set $x \cdot l := y \cdot l$. It is an easy check that this yields a well-defined $\mathfrak{gl}^M(V, W)$-module structure on L compatible with the $\mathfrak{sl}(V, W)$-module structure on L.

For $\mathfrak{g} = \mathfrak{o}(V), \mathfrak{sp}(V)$ one uses (11) instead of (10). □

We can now define the category $\mathbb{T}_{\mathfrak{g}^M}$ as an analogue of the category $\mathbb{T}_{\mathfrak{g}}$. More precisely, the category $\mathbb{T}_{\mathfrak{g}^M}$ is the full subcategory of $\mathfrak{g}^M$-mod consisting of all modules of finite length, integrable over $\mathfrak{g}$ and satisfying the large annihilator condition.

The following is our main result in Sect. 7.

Theorem 7.10. *(a) $\mathbb{T}_{\mathfrak{g}^M} = \mathbb{T}^R_{\mathfrak{g}^M}$, where $R = V \oplus W$ for $\mathfrak{g} = \mathfrak{sl}(V, W)$ and $R = V$ for $\mathfrak{g} = \mathfrak{o}(V), \mathfrak{sp}(V)$.*
(b) The functor Res $: \mathbb{T}_{\mathfrak{g}^M} \rightsquigarrow \mathbb{T}_{\mathfrak{g}}$ *is an equivalence of monoidal categories.*

Proof. It is clear that $\mathbb{T}^R_{\mathfrak{g}^M}$ is a full subcategory of $\mathbb{T}_{\mathfrak{g}^M}$. We need to show only that any $L \in \mathbb{T}_{\mathfrak{g}^M}$ is isomorphic to a subquotient of $T(R)^{\oplus n}$ for some n. Obviously, L satisfies the large annihilator condition as a $\mathfrak{g}$-module. Furthermore, by Lemma 7.9 (a), $\mathfrak{g}$ acts densely on L, hence L has finite length as a $\mathfrak{g}$-module. By Corollary 5.12 (b), L is isomorphic to a $\mathfrak{g}$subquotient of $T(R)^{\oplus n}$ for some n, and by Proposition 7.6 L is the restriction to $\mathfrak{g}$ of some $\mathfrak{g}^M$-subquotient L' of $T(R)^{\oplus n}$. However, since L' satisfies the large annihilator condition, Lemma 7.8 implies that there is an isomorphism of $\mathfrak{g}^M$-modules $L \simeq L'$. This proves (a).

(b) follows from (a) and Proposition 7.6. □

The following diagram summarizes the equivalences of monoidal categories established in this paper:

$$\mathbb{T}_{\mathfrak{a}} \overset{\mathrm{Res}}{\leftsquigarrow} \mathbb{T}_{\mathfrak{g}^M} = \mathbb{T}^R_{\mathfrak{g}^M} \overset{\mathrm{Res}}{\rightsquigarrow} \mathbb{T}_{\mathfrak{g}} \overset{\Phi}{\rightsquigarrow} \mathbb{T}_{\mathfrak{g}_c}.$$

Here $\mathfrak{a}$ is any dense subalgebra of $\mathfrak{g}^M$ and $R = V \oplus W$ for $\mathfrak{g} = \mathfrak{sl}(V, W)$, $R = V$ for $\mathfrak{g} = \mathfrak{o}(V), \mathfrak{sp}(V)$. In particular, when $\mathfrak{g} = \mathfrak{sl}(V, V_*)$ for countable-dimensional V and V_*, $\mathfrak{a}$ can be chosen as the Lie algebra $\mathfrak{j}(V, V_*)$ or as any countable-dimensional diagonal Lie algebra.

8 Further Results and Open Problems

Theorem 7.10 (a) can be considered an analogue of Theorem 5.1 and Corollary 5.12 (b) as it provides two equivalent descriptions of the category $\mathbb{T}_{\mathfrak{g}^M}$. It is interesting to have a longer list of such equivalent descriptions.

The following proposition provides another equivalent condition characterizing the objects of $\mathbb{T}_{\mathfrak{g}^M}$ under the additional assumption that $\mathfrak{g} = \mathfrak{sl}(V, V_*), \mathfrak{o}(V), \mathfrak{sp}(V)$ is countable dimensional.

Proposition 8.1. *Let $\mathfrak{g}^M = \mathfrak{gl}^M(V, V_*), \mathfrak{o}^M(V), \mathfrak{sp}^M(V)$ for a countable dimensional V, and let L be a $\mathfrak{g}^M$-module of finite length which is integrable as a $\mathfrak{g}$-module. Then L is an object of $\mathbb{T}_{\mathfrak{g}^M}$ if and only if $\mathfrak{g}$ acts densely on L.*

We first need a lemma.

Lemma 8.2. *Let $\mathfrak{g}^M = \mathfrak{gl}^M(V, V_*), \mathfrak{o}^M(V), \mathfrak{sp}^M(V)$ for a countable-dimensional V, and let L and L' be $\mathfrak{g}^M$-modules. Assume that L and L' have finite length as $\mathfrak{g}$-modules. Then*

$$\mathrm{Hom}_{\mathfrak{g}}(L, L') = \mathrm{Hom}_{\mathfrak{g}^M}(L, L').$$

In particular, if L and L' are isomorphic as $\mathfrak{g}$-modules, then L and L' are isomorphic as $\mathfrak{g}^M$-modules.

Proof. Observe that $\mathrm{Hom}_{\mathbb{C}}(L, L')$ has a natural structure of $\mathfrak{g}^M$-module defined by

$$(X \cdot \varphi)(l) := X \cdot \varphi(l) - \varphi(X \cdot l) \text{ for } X \in \mathfrak{g}^M, \varphi \in \mathrm{Hom}_{\mathbb{C}}(L, L'), l \in L. \quad (13)$$

Since $\mathfrak{g}$ is an ideal in $\mathfrak{g}^M$, $\mathrm{Hom}_{\mathfrak{g}}(L, L')$ is a $\mathfrak{g}^M$-submodule in $\mathrm{Hom}_{\mathbb{C}}(L, L')$. Moreover, $\mathrm{Hom}_{\mathfrak{g}}(L, L')$ is finite dimensional as L and L' have finite length over $\mathfrak{g}$. On the other hand, Theorem 6.3 implies that $\mathfrak{g}^M$ does not have proper ideals of finite codimension, hence any finite-dimensional $\mathfrak{g}^M$-module is trivial. Therefore (13) defines a trivial $\mathfrak{g}^M$-module structure of $\mathrm{Hom}_{\mathfrak{g}^M}(L, L')$, which means that any $\varphi \in \mathrm{Hom}_{\mathfrak{g}}(L, L')$ belongs to $\mathrm{Hom}_{\mathfrak{g}^M}(L, L')$. This shows that $\mathrm{Hom}_{\mathfrak{g}}(L, L') = \mathrm{Hom}_{\mathfrak{g}^M}(L, L')$. The second assertion follows immediately. □

Proof of Proposition 8.1. If $L \in \mathbb{T}_{\mathfrak{g}^M}$, then $\mathfrak{g}$ acts densely on L by Lemma 7.9.

Now let $\mathfrak{g}$ act densely on L. We first prove that L satisfies the large annihilator condition as a $\mathfrak{g}$-module. Assume that $\mathfrak{g}$ acts densely on L but L does not satisfy the large annihilator condition as a $\mathfrak{g}$-module. Using the matrix realizations of $\mathfrak{g}$ and $\mathfrak{g}^M$ one can show that there exists $l \in L$ and a sequence $\{X_i\}_{i \in \mathbb{Z}_{\geq 1}}$ of commuting linearly independent elements $X_i \in \mathfrak{g}$ which don't belong to the annihilator of l. Furthermore, one can find an infinite subsequence $\{Y_j = X_{i_j}\}$ such that each Y_j lies in an $\mathfrak{sl}(2)$-subalgebra $\mathfrak{s}_j \subset \mathfrak{g}$ with the condition $[\mathfrak{s}_j, \mathfrak{s}_s] = 0$ for $j \neq s$. Then $\prod_j \mathfrak{s}_j$ is a Lie subalgebra in $\mathfrak{g}^M$, and let $\mathfrak{s}$ be the diagonal subalgebra in $\prod_j \mathfrak{s}_j$. If $x \in \mathfrak{s}$, we denote by x_j its component in $\mathfrak{s}_j$.

Since $\mathfrak{g}$ acts densely on L, there exists a linear map $\theta : \mathfrak{s} \to \mathfrak{g}$ such that $\theta(y) \cdot l = y \cdot l$ for all $y \in \mathfrak{s}$. On the other hand, there exists $n \in \mathbb{Z}_{\geq 1}$ such that $[\theta(y), x_j] = 0$ for all $y, x \in \mathfrak{s}$ and $j > n$. Let $d_y := y - \theta(y)$. Then $d_y \cdot l = 0$ and

$$[d_y, x_j] = [y, x_j] = [y_j, x_j] \quad \text{for all } x, y \in \mathfrak{s} \text{ and } j > n. \tag{14}$$

Set $L_j := U(\mathfrak{s}_j) \cdot l$. Then (14) implies $d_y \cdot L_j \subset L_j$ for all $j > n$. Moreover, $\psi_y := d_y - y_j$ commutes with $\mathfrak{s}_j$, hence $\psi_y \in \mathrm{End}_{\mathfrak{s}_j}(L_j)$. Considering $y_j + \psi_y$ as an element of $\mathrm{End}_{\mathbb{C}}(L_j)$, we obtain in addition that $l \in \ker(y_j + \psi_y)$ for all $y \in \mathfrak{s}$ and all $j > n$.

Choose a standard basis $E, H, F \in \mathfrak{s}$. Since L_j is a finite-dimensional $\beta_j \simeq \mathfrak{sl}(2)$-module, we obtain easily

$$\ker(E_j + \psi_E) \cap L_j = L_j^{E_j}, \; \ker(F_j + \psi_F) \cap L_j = L_j^{F_j}.$$

Since

$$l \in \ker(E_j + \psi_E) \cap \ker(F_j + \psi_F) \cap L_j = L_j^{\mathfrak{s}_j},$$

we conclude that L_j is a trivial $\mathfrak{s}_j$-module for all $j > n$, which contradicts our original assumption that $Y_j \cdot l \neq 0$. Thus, L satisfies the large annihilator condition as a $\mathfrak{g}$-module.

Note that as $\mathfrak{g}$ acts densely on L, the length of L as a $\mathfrak{g}$-module is the same as the length of L as a $\mathfrak{g}^M$-module. Since L satisfies the large annihilator condition for $\mathfrak{g}$ and has finite length as a $\mathfrak{g}$-module, we conclude that $L_{\downarrow \mathfrak{g}}$ is a tensor module, i.e., an object of $\mathbb{T}_{\mathfrak{g}}$. By Theorem 7.10 (b), $L_{\downarrow \mathfrak{g}} = L'_{\downarrow \mathfrak{g}}$ for some $L' \in \mathbb{T}_{\mathfrak{g}^M}$. Finally, Lemma 8.2 implies that the $\mathfrak{g}^M$ modules L' and L are isomorphic, i.e., $L \in \mathbb{T}_{\mathfrak{g}^M}$. □

Next, under the assumption that V is countable dimensional, consider maximal subalgebras $\mathfrak{h}^M$ of $\mathfrak{g}^M$ which act semisimply on V and V_* (respectively only on V for $\mathfrak{g} = \mathfrak{o}(V), \mathfrak{sp}(V)$). It is straightforward to show that the centralizer in $\mathfrak{g}^M$ of any local Cartan subalgebra $\mathfrak{h}$ of $\mathfrak{g}$ is such a subalgebra of $\mathfrak{g}^M$. If $\mathfrak{g}^M = \mathfrak{gl}^M(V, V_*)$ is realized as the Lie algebra of matrices $X = (x_{ij})_{i,j \in \mathbb{Z}}$ with finite rows and columns, then $\mathfrak{h}^M$ can be chosen as the subalgebra of diagonal matrices.

The following statement looks plausible to us.

Conjecture 8.3. *Let* $\mathfrak{g} = \mathfrak{sl}(V, V_*), \mathfrak{o}(V), \mathfrak{sp}(V)$ *for a countable-dimensional* V. *Let* M *be a finite length* $\mathfrak{g}^M$*-module which is integrable as a* $\mathfrak{g}$*-module. The following conditions on* M *are equivalent:*

(a) $M \in \mathbb{T}_{\mathfrak{g}^M}$*;*
(b) M *is countable dimensional;*
(c) M *is a semisimple* $\mathfrak{h}^M$*-module for some subalgebra* $\mathfrak{h}^M \subset \mathfrak{g}^M$*;*
(d) M *is a semisimple* $\mathfrak{h}^M$*-module for any subalgebra* $\mathfrak{h}^M \subset \mathfrak{g}^M$.

Consider now the inclusion of Lie algebras

$$\mathfrak{g} = \mathfrak{sl}(V, V_*) \subset \mathfrak{gl}^M(V, V^*) = \mathrm{End}(V)$$

where V is an arbitrary vector space. The subalgebra $\mathfrak{g}$ is not dense in $\mathrm{End}(V)$, nevertheless the monoidal categories $\mathbb{T}_{\mathfrak{g}}$ and $\mathbb{T}_{\mathrm{End}(V)}$ are equivalent by Theorems 5.1 and 7.10. Here is a functor which most likely also provides such an equivalence. Let $M \in \mathbb{T}_{\mathrm{End}(V)}$. Set

$$\Gamma^{\mathrm{wt}}_{\mathfrak{g}}(M) := \cap_{\mathfrak{h} \subset \mathfrak{g}} \Gamma^{\mathrm{wt}}_{\mathfrak{h}}(M)$$

where $\mathfrak{h}$ runs over all local Cartan subalgebras of $\mathfrak{g}$.

Conjecture 8.4. $\Gamma^{\mathrm{wt}}_{\mathfrak{g}} : \mathbb{T}_{\mathrm{End}(V)} \rightsquigarrow \mathbb{T}_{\mathfrak{g}}$ *is an equivalence of monoidal categories.*

If V is countable dimensional, it is easy to check that V^*/V_* is a simple $\mathfrak{g}^M = \mathfrak{gl}^M(V, V_*)$-module. Hence V^* is a $\mathfrak{g}^M$-module of length 2. This raises the natural question of whether the entire category $\mathbb{T}_{\mathrm{End}(V)}$ consists of $\mathfrak{g}^M$-modules of finite length. A further problem is to compute the socle filtration as a $\mathfrak{g}^M$-module of a simple $\mathrm{End}(V)$-module in $\mathbb{T}_{\mathrm{End}(V)}$.

Another open question is whether there is an analogue of the category $\widetilde{\mathrm{Tens}}_{\mathfrak{g}}$ when we replace $\mathfrak{g}$ by $\mathfrak{g}^M$. More precisely, what can be said about the abelian monoidal category of $\mathfrak{g}^M$-modules obtained from $\mathbb{T}_{\mathfrak{g}^M}$ by iterated dualization in addition to taking submodules, quotients and applying $\otimes$? In particular, the adjoint representation, and therefore the coadjoint representation are objects of $\widetilde{\mathrm{Tens}}_{\mathfrak{g}^M}$. How can one describe the coadjoint representation $(\mathfrak{g}^M)^*$ of $\mathfrak{g}^M$?

Note added in proof: While this paper was under review, Alexandru Chirvasitu gave a proof of Conjecture 8.4 and computed the $\mathfrak{g}^M$-module socle filtration of any simple module in $\mathbb{T}_{\mathrm{End}(V)}$. His results appear in the article [C] in the present volume.

Acknowledgements Both authors thank the Max Planck Institute for Mathematics in Bonn where a preliminary draft of this paper was written in 2012. We also thank Jacobs University for its hospitality.

We thank the referee for several thoughtful suggestions, and in particular for showing us the module $\Lambda^{[\frac{\infty}{2}]}V$ from Sect. 3.1.

References

[Ba1] A. A. Baranov, Complex finitary simple Lie algebras, *Arch. Math.* **71** (1998), 1–6.

[Ba2] A. A. Baranov, Finitary simple Lie algebras, *Journ. Algebra* **219** (1999), 299–329.

[BaZh] A. A. Baranov, A. G. Zhilinskii, Diagonal direct limits of simple Lie algebras, *Comm. Algebra* **27** (1998), 2749–2766.

[BB] Y. Bakhturin, G. Benkart, Weight modules of direct limit Lie algebras, *Comm. Algebra* **27** (1998), 2249–2766.

[BS] A. A. Baranov, H. Strade, Finitary Lie algebras, *Journ. Algebra* **254** (2002), 173–211.

[BhS] Y. Bakhturin, H. Strade, Some examples of locally finite simple Lie algebras, *Arch. Math.* **65** (1995), 23–26.

[Bou] N. Bourbaki, *Groupes et Algèbres de Lie*, Hermann, Paris, 1975.

[C] A. Chirvasitu, Three results on representations of Mackey Lie algebras. This volume.

[DiP1] I. Dimitrov, I. Penkov, Weight modules of direct limit Lie algebras, *IMRN* 1999, no. 5, 223–249.

[DiP2] I. Dimitrov, I. Penkov, Locally semisimple and maximal subalgebras of the finitary Lie algebras $\mathfrak{gl}(\infty)$, $\mathfrak{sl}(\infty)$, $\mathfrak{so}(\infty)$ and $\mathfrak{sp}(\infty)$, *Journ. Algebra* **322** (2009), 2069–2081.

[DiP3] I. Dimitrov, I. Penkov, Borel subalgebras of $\mathfrak{gl}(\infty)$, *Resenhas* **6** (2004), no. 2–3, 153–163.

[DPS] E. Dan-Cohen, I. Penkov, V. Serganova, A Koszul category of representations of finitary Lie algebras, preprint 2011, arXiv:1105.3407.

[DPSn] E. Dan-Cohen, I. Penkov, N. Snyder, Cartan subalgebras of root-reductive Lie algebras, *Journ. Algebra* **308** (2007), 583–611.

[DPW] I. Dimitrov, I. Penkov, J. A. Wolf, A Bott-Borel-Weil theory for direct limits of algebraic groups, *Amer. Journ. Math.* **124** (2002), 955–998.

[DaPW] E. Dan-Cohen, I. Penkov, J. A. Wolf, Parabolic Subgroups of Real Direct Limit Lie Groups, *Contemporary Mathematics* **499**, AMS, 2009, pp. 47–59.

[FT] B. L. Feigin, B. L. Tsygan, Cohomologies of Lie algebras of generalized Jacobi matrices (Russian), *Funktsional. Anal. i Prilozhen.* **17** (1983), no. 2, 86–87.

[M] G. Mackey, On infinite dimensional linear spaces, *Trans. AMS* **57** (1945), 155–207.

[Ma] S. Markouski, Locally simple subalgebras of diagonal Lie algebras, *Journ. Algebra* **327** (2011), 186–207.

[N] K. -H. Neeb, Holomorphic highest weight representations of infinite-dimensional complex classical groups. *Journ. Reine Angew. Math.* **497** (1998), 171–222.

[Na] L. Natarajan, Unitary highest weight-modules of inductive limit Lie algebras and groups. *Journ. Algebra* **167** (1994), 9–28.

[NP] K. -H. Neeb, I. Penkov, Cartan subalgebras of $\mathfrak{gl}(\infty)$, *Canad. Math. Bull.* **46** (2003), 597–616.

[NS] K. -H. Neeb, N. Stumme, The classification of locally finite split simple Lie algebras. *Journ. Reine Angew. Math.* **533** (2001), 25–53.

[O] G. Olshanskii, Unitary representations of the group $SO_0(\infty, \infty)$ as limits of unitary representations of the groups $SO_0(n, \infty)$ as $n \to \infty$ (Russian), *Funktsional. Anal. i Prilozhen.* **20** (1986), no. 4, 46–57.

[PP] I. Penkov, A. Petukhov, On ideals in the enveloping algebra of a locally simple Lie algebra, *IMRN*, doi:10.1093/imzn/rnu085, 2014.

[PS] I. Penkov, V. Serganova, Categories of integrable $sl(\infty)$-, $o(\infty)$-, $sp(\infty)$-modules, in "Representation Theory and Mathematical Physics", *Contemporary Mathematics* 557 (2011), pp. 335–357.

[PStr] I. Penkov, H. Strade, Locally finite Lie algebras with root decomposition, *Archiv Math.* **80** (2003), 478–485.

[PStyr] I. Penkov, K. Styrkas, Tensor representations of infinite-dimensional root-reductive Lie algebras, in *Developments and Trends in Infinite-Dimensional Lie Theory*, Progress in Mathematics 288, Birkhäuser, 2011, pp. 127–150.

[PZ] I. Penkov, G. Zuckerman, A construction of generalized Harish-Chandra modules for locally reductive Lie algebras, *Transformation Groups* **13**, 2008, 799–817.

[S] V. Serganova, Classical Lie superalgebras at infinity, in *Advances in Lie superalgebras*, Springer, 2013, pp. 181–201.

[SS] S. Sam, A. Snowden, GL-equivariant modules over polynomial rings in infinitely many variables, arXiv:1206.2233.

[SSW] S. Sam, A. Snowden, J. Weyman, Homology of Littlewood complexes, arXiv:1209.3509.

[Z] G. Zuckerman, Generalized Harish-Chandra Modules, in *Highlights in Lie Algebraic Methods*, Progress in Mathematics 295, Birkhäuser, 2012, pp. 123–143.

[Zh1] A. G. Zhilinskii, Coherent systems of representations of inductive families of simple complex Lie algebras, (Russian) preprint of Academy of Belarussian SSR, ser. 38(438), Minsk, 1990.

[Zh2] A. G. Zhilinskii, Coherent finite-type systems of inductive families of non-diagonal inclusions (Russian), *Dokl. Acad. Nauk Belarusi* **36:1**(1992), 9–13, 92.

Algebraic Methods in the Theory of Generalized Harish-Chandra Modules

Ivan Penkov and Gregg Zuckerman

Abstract This paper is a review of results on generalized Harish-Chandra modules in the framework of cohomological induction. The main results, obtained during the last 10 years, concern the structure of the fundamental series of $(\mathfrak{g}, \mathfrak{k})$-modules, where $\mathfrak{g}$ is a semisimple Lie algebra and $\mathfrak{k}$ is an arbitrary algebraic reductive in $\mathfrak{g}$ subalgebra. These results lead to a classification of simple $(\mathfrak{g}, \mathfrak{k})$-modules of finite type with generic minimal $\mathfrak{k}$-types, which we state. We establish a new result about the Fernando–Kac subalgebra of a fundamental series module. In addition, we pay special attention to the case when $\mathfrak{k}$ is an eligible r-subalgebra (see the definition in Sect. 4) in which we prove stronger versions of our main results. If $\mathfrak{k}$ is eligible, the fundamental series of $(\mathfrak{g}, \mathfrak{k})$-modules yields a natural algebraic generalization of Harish-Chandra's discrete series modules.

Key words Generalized Harish-Chandra module • $(\mathfrak{g}, \mathfrak{k})$-module of finite type • Minimal $\mathfrak{k}$-type • Fernando–Kac subalgebra • Eligible subalgebra

Mathematics Subject Classification (2010): 17B10, 17B55.

Both authors have been partially supported by the DFG through Priority Program 1388 "Representation theory."

I. Penkov (✉)
Jacobs University Bremen, Campus Ring 1,
28759, Bremen, Germany
e-mail: i.penkov@jacobs-university.de

G. Zuckerman
Department of Mathematics, Yale University, 10 Hillhouse Avenue,
P.O. Box 208283, New Haven, CT 06520-8283, USA
e-mail: gregg.zuckerman@yale.edu

G. Mason et al. (eds.), *Developments and Retrospectives in Lie Theory: Algebraic Methods*, Developments in Mathematics 38,
DOI 10.1007/978-3-319-09804-3_15

Introduction

Generalized Harish-Chandra modules have now been actively studied for more than 10 years. A *generalized Harish-Chandra module M* over a finite-dimensional reductive Lie algebra $\mathfrak{g}$ is a $\mathfrak{g}$-module M for which there is a reductive in $\mathfrak{g}$ subalgebra $\mathfrak{k}$ such that as a $\mathfrak{k}$-module, M is the direct sum of finite-dimensional generalized $\mathfrak{k}$-isotypic components. If M is irreducible, $\mathfrak{k}$ acts necessarily semisimply on M, and in what follows we restrict ourselves to the study of generalized Harish-Chandra modules on which $\mathfrak{k}$ acts semisimply; see [Z] for an introduction to the topic.

In this paper we present a brief review of results obtained in the past 10 years in the framework of algebraic representation theory, more specifically in the framework of cohomological induction, see [KV] and [Z]. In fact, generalized Harish-Chandra modules have been studied also with geometric methods, see for instance [PSZ] and [PS1, PS2, PS3, Pe], but the geometric point of view remains beyond the scope of the current review. In addition, we restrict ourselves to finite-dimensional Lie algebras $\mathfrak{g}$ and do not review the paper [PZ4], which deals with the case of locally finite Lie algebras. We omit the proofs of most results which have already appeared.

The cornerstone of the algebraic theory of generalized Harish-Chandra modules so far is our work [PZ2]. In this work we define the notion of simple generalized Harish-Chandra modules with generic minimal $\mathfrak{k}$-type and provide a classification of such modules. The result extends in part the Vogan–Zuckerman classification of simple Harish-Chandra modules. It leaves open the questions of existence and classification of simple $(\mathfrak{g}, \mathfrak{k})$-modules of finite type whose minimal $\mathfrak{k}$-types are not generic. While the classification of such modules presents a major open problem in the theory of generalized Harish-Chandra modules, in the note [PZ3] we establish the existence of simple $(\mathfrak{g}, \mathfrak{k})$-modules with arbitrary given minimal $\mathfrak{k}$-type.

In the paper [PZ5] we establish another general result, namely the fact that each module in the fundamental series of generalized Harish-Chandra modules has finite length. We then consider in detail the case when $\mathfrak{k} = \mathfrak{sl}(2)$. In this case the highest weights of $\mathfrak{k}$-types are just nonnegative integers μ, and the genericity condition is the inequality $\mu \geq \Gamma$, Γ being a bound depending on the pair $(\mathfrak{g}, \mathfrak{k})$. In [PZ5] we improve the bound Γ to an, in general, much lower bound Λ. Moreover, we show that in a number of low-dimensional examples the bound Λ is sharp in the sense that the our classification results do not hold for simple $(\mathfrak{g}, \mathfrak{k})$-modules with minimal $\mathfrak{k}$-type $V(\mu)$ for μ lower than Λ. In [PZ5] we also conjecture that the Zuckerman functor establishes an equivalence of a certain subcategory of the thickening of category O and a subcategory of the category of $(\mathfrak{g}, \mathfrak{k} \simeq \mathfrak{sl}(2))$-modules.

Sections 2 and 3 of the present paper are devoted to a brief review of the above results. We also establish some new results in terms of the algebra $\tilde{\mathfrak{k}} := \mathfrak{k} + C(\mathfrak{k})$ (where $C(\cdot)$ stands for centralizer in $\mathfrak{g}$). A notable such result is Corollary 2.10 which gives a sufficient condition on a simple $(\mathfrak{g}, \mathfrak{k})$-module M for $\tilde{\mathfrak{k}}$ to be a maximal reductive subalgebra of $\mathfrak{g}$ which acts locally finitely on M.

The idea of bringing $\tilde{\mathfrak{k}}$ into the picture leads naturally to considering a preferred class of reductive subalgebras $\mathfrak{k}$ which we call eligible: they satisfy the condition $C(\mathfrak{t}) = \mathfrak{t} + C(\mathfrak{k})$ where $\mathfrak{t}$ is Cartan subalgebra of $\mathfrak{k}$. In Sect. 5 we study a natural generalization of Harish-Chandra's discrete series to the case of an eligible subalgebra $\mathfrak{k}$. A key statement here is that under the assumption of eligibility of $\mathfrak{k}$, the isotypic component of the minimal $\mathfrak{k}$-type of a generalized discrete series module is an irreducible $\tilde{\mathfrak{k}}$-module (Theorem 5.1).

1 Notation and Preliminary Results

We start by recalling the setup of [PZ2] and [PZ5].

1.1 Conventions

The ground field is $\mathbb{C}$, and if not explicitly stated otherwise, all vector spaces and Lie algebras are defined over $\mathbb{C}$. The sign $\otimes$ denotes tensor product over $\mathbb{C}$. The superscript * indicates dual space. The sign $\ni\!\!\!+$ stands for semidirect sum of Lie algebras (if $\mathfrak{l} = \mathfrak{l}' \ni\!\!\!+ \mathfrak{l}''$, then $\mathfrak{l}'$ is an ideal in $\mathfrak{l}$ and $\mathfrak{l}'' \cong \mathfrak{l}/\mathfrak{l}'$). $H^{\cdot}(\mathfrak{l}, M)$ stands for the cohomology of a Lie algebra $\mathfrak{l}$ with coefficients in an $\mathfrak{l}$-module M, and $M^{\mathfrak{l}} = H^0(\mathfrak{l}, M)$ stands for space of $\mathfrak{l}$-invariants of M. By $Z(\mathfrak{l})$ we denote the center of $\mathfrak{l}$, and by $\mathfrak{l}_{ss}$ we denote the semisimple part of $\mathfrak{l}$ when $\mathfrak{l}$ is reductive. $\Lambda^{\cdot}(\cdot)$ and $S^{\cdot}(\cdot)$ denote respectively the exterior and symmetric algebra.

If $\mathfrak{l}$ is a Lie algebra, then $U(\mathfrak{l})$ stands for the enveloping algebra of $\mathfrak{l}$ and $Z_{U(\mathfrak{l})}$ denotes the center of $U(\mathfrak{l})$. We identify $\mathfrak{l}$-modules with $U(\mathfrak{l})$-modules. It is well known that if $\mathfrak{l}$ is finite dimensional and M is a simple $\mathfrak{l}$-module (or equivalently a simple $U(\mathfrak{l})$-module), then $Z_{U(\mathfrak{l})}$ acts on M via a $Z_{U(\mathfrak{l})}$-*character*, i.e., via an algebra homomorphism $\theta_M : Z_{U(\mathfrak{l})} \to \mathbb{C}$, see Proposition 2.6.8 in [Dix].

We say that an $\mathfrak{l}$-module M is *generated* by a subspace $M' \subseteq M$ if $U(\mathfrak{l})\cdot M' = M$, and we say that M is *cogenerated* by $M' \subseteq M$, if for any nonzero homomorphism $\psi : M \to \bar{M}$, $M' \cap \ker\psi \neq \{0\}$.

By $\mathrm{Soc}M$ we denote the socle (i.e., the unique maximal semisimple submodule) of an $\mathfrak{l}$-module M. If $\omega \in \mathfrak{l}^*$, we put $M^{\omega} := \{m \in M \mid \mathfrak{l}\cdot m = \omega(\mathfrak{l})m \ \forall l \in \mathfrak{l}\}$. By $\mathrm{supp}_{\mathfrak{l}}M$ we denote the set $\{\omega \in \mathfrak{l}^* \mid M^{\omega} \neq 0\}$.

A finite *multiset* is a function f from a finite set D into $\mathbb{N}$. A *submultiset* of f is a multiset f' defined on the same domain D such that $f'(d) \leq f(d)$ for any $d \in D$. For any finite multiset f, defined on a subset D of a vector space, we put $\rho_f := \frac{1}{2}\sum_{d\in D} f(d)d$.

If $\dim M < \infty$ and $M = \bigoplus_{\omega\in\mathfrak{l}^*} M^{\omega}$, then M determines the finite multiset $\mathrm{char}_{\mathfrak{l}}\, M$ which is the function $\omega \mapsto \dim M^{\omega}$ defined on $\mathrm{supp}_{\mathfrak{l}}M$.

1.2 Reductive Subalgebras, Compatible Parabolics and Generic $\mathfrak{k}$-Types

Let $\mathfrak{g}$ be a finite-dimensional semisimple Lie algebra. By $\mathfrak{g}$-mod we denote the category of $\mathfrak{g}$-modules. Let $\mathfrak{k} \subset \mathfrak{g}$ be an algebraic subalgebra which is reductive in $\mathfrak{g}$. We set $\tilde{\mathfrak{k}} = \mathfrak{k} + C(\mathfrak{k})$ and note that $\tilde{\mathfrak{k}} = \mathfrak{k}_{ss} \oplus C(\mathfrak{k})$ where $C(\cdot)$ stands for centralizer in $\mathfrak{g}$. We fix a Cartan subalgebra $\mathfrak{t}$ of $\mathfrak{k}$ and let $\mathfrak{h}$ denote an as yet unspecified Cartan subalgebra of $\mathfrak{g}$. Everywhere, except in Sect. 1.3 below, we assume that $\mathfrak{t} \subseteq \mathfrak{h}$, and hence that $\mathfrak{h} \subseteq C(\mathfrak{t})$. By Δ we denote the set of $\mathfrak{h}$-roots of $\mathfrak{g}$, i.e., $\Delta = \{\mathrm{supp}_{\mathfrak{h}}\mathfrak{g}\}\backslash\{0\}$. Note that since $\mathfrak{k}$ is reductive in $\mathfrak{g}$, $\mathfrak{g}$ is a $\mathfrak{t}$-weight module, i.e., $\mathfrak{g} = \bigoplus_{\eta\in\mathfrak{t}^*} \mathfrak{g}^{\eta}$. We set $\Delta_{\mathfrak{t}} := \{\mathrm{supp}_{\mathfrak{t}}\mathfrak{g}\} \setminus \{0\}$. Note also that the $\mathbb{R}$-span of the roots of $\mathfrak{h}$ in $\mathfrak{g}$ fixes a real structure on $\mathfrak{h}^*$, whose projection onto $\mathfrak{t}^*$ is a well-defined real structure on $\mathfrak{t}^*$. In what follows, we denote by $\mathrm{Re}\eta$ the real part of an element $\eta \in \mathfrak{t}^*$. We fix also a Borel subalgebra $\mathfrak{b}_{\mathfrak{k}} \subseteq \mathfrak{k}$ with $\mathfrak{b}_{\mathfrak{k}} \supseteq \mathfrak{t}$. Then $\mathfrak{b}_{\mathfrak{k}} = \mathfrak{t} \niplus \mathfrak{n}_{\mathfrak{k}}$, where $\mathfrak{n}_{\mathfrak{k}}$ is the nilradical of $\mathfrak{b}_{\mathfrak{k}}$. We set $\rho := \rho_{\mathrm{char}_{\mathfrak{t}}\,\mathfrak{n}_{\mathfrak{k}}}$. The quartet $\mathfrak{g}, \mathfrak{k}, \mathfrak{b}_{\mathfrak{k}}, \mathfrak{t}$ will be fixed throughout the paper. By W we denote the Weyl group of $\mathfrak{g}$.

As usual, we parametrize the characters of $Z_{U(\mathfrak{g})}$ via the Harish-Chandra homomorphism. More precisely, if $\mathfrak{b}$ is a given Borel subalgebra of $\mathfrak{g}$ with $\mathfrak{b} \supset \mathfrak{h}$ ($\mathfrak{b}$ will be specified below), the $Z_{U(\mathfrak{g})}$-character corresponding to $\zeta \in \mathfrak{h}^*$ via the Harish-Chandra homomorphism defined by $\mathfrak{b}$ is denoted by θ_ζ ($\theta_{\rho_{\mathrm{char}_{\mathfrak{h}}\,\mathfrak{b}}}$ is the trivial $Z_{U(\mathfrak{g})}$-character). Sometimes we consider a reductive subalgebra $\mathfrak{l} \subset \mathfrak{g}$ instead of $\mathfrak{g}$ and apply this convention to the characters of $Z_{U(\mathfrak{l})}$. In this case we write $\theta_\zeta^{\mathfrak{l}}$ for $\zeta \in \mathfrak{h}_{\mathfrak{l}}^*$, where $\mathfrak{h}_{\mathfrak{l}}$ is a Cartan subalgebra of $\mathfrak{l}$.

By $\langle\cdot\ \cdot\rangle$ we denote the unique $\mathfrak{g}$-invariant symmetric bilinear form on $\mathfrak{g}^*$ such that $\langle\alpha, \alpha\rangle = 2$ for any long root of $\mathfrak{g}$. The form $\langle\cdot\,,\ \cdot\rangle$ enables us to identify $\mathfrak{g}$ with $\mathfrak{g}^*$. Then $\mathfrak{h}$ is identified with $\mathfrak{h}^*$, and $\mathfrak{k}$ is identified with $\mathfrak{k}^*$. We sometimes consider $\langle\cdot\,,\ \cdot\rangle$ as a form on $\mathfrak{g}$. The superscript $\perp$ indicates orthogonal space. Note that there is a canonical $\mathfrak{k}$-module decomposition $\mathfrak{g} = \mathfrak{k} \oplus \mathfrak{k}^{\perp}$ and a canonical decomposition $\mathfrak{h} = \mathfrak{t} \oplus \mathfrak{t}^{\perp}$ with $\mathfrak{t}^{\perp} \subseteq \mathfrak{k}^{\perp}$. We also set $\parallel \zeta \parallel^2 := \langle\zeta, \zeta\rangle$ for any $\zeta \in \mathfrak{h}^*$.

We say that an element $\eta \in \mathfrak{t}^*$ is $(\mathfrak{g}, \mathfrak{k})$-*regular* if $\langle\mathrm{Re}\eta, \sigma\rangle \neq 0$ for all $\sigma \in \Delta_{\mathfrak{t}}$. To any $\eta \in \mathfrak{t}^*$ we associate the following parabolic subalgebra $\mathfrak{p}_\eta$ of $\mathfrak{g}$:

$$\mathfrak{p}_\eta = \mathfrak{h} \oplus (\bigoplus_{\alpha\in\Delta_\eta} \mathfrak{g}^{\alpha}),$$

where $\Delta_\eta := \{\alpha \in \Delta \mid \langle\mathrm{Re}\eta, \alpha\rangle \geq 0\}$. By $\mathfrak{m}_\eta$ and $\mathfrak{n}_\eta$ we denote respectively the reductive part of $\mathfrak{p}$ (containing $\mathfrak{h}$) and the nilradical of $\mathfrak{p}$. In particular $\mathfrak{p}_\eta = \mathfrak{m}_\eta \niplus \mathfrak{n}_\eta$, and if η is $\mathfrak{b}_{\mathfrak{k}}$-dominant, then $\mathfrak{p}_\eta \cap \mathfrak{k} = \mathfrak{b}_{\mathfrak{k}}$. We call $\mathfrak{p}_\eta$ a $\mathfrak{t}$-*compatible parabolic subalgebra*. Note that

$$\mathfrak{p}_\eta = C(\mathfrak{t}) \oplus (\bigoplus_{\beta\in\Delta^+_{\mathfrak{t},\eta}} \mathfrak{g}^{\beta}),$$

where $\Delta^+_{\mathfrak{t},\eta} := \{\beta \in \Delta_{\mathfrak{t}} \mid \langle\mathrm{Re}\eta, \beta\rangle \geq 0\}$. Hence $\mathfrak{p}_\eta$ depends upon our choice of $\mathfrak{t}$ and η, but not upon the choice of $\mathfrak{h}$.

A $\mathfrak{t}$-compatible parabolic subalgebra $\mathfrak{p} = \mathfrak{m} \ni \mathfrak{n}$ (i.e., $\mathfrak{p} = \mathfrak{p}_\eta$ for some $\eta \in \mathfrak{t}^*$) is $\mathfrak{t}$-*minimal* (or simply *minimal*) if it does not properly contain another $\mathfrak{t}$-compatible parabolic subalgebra. It is an important observation that if $\mathfrak{p} = \mathfrak{m} \ni \mathfrak{n}$ is minimal, then $\mathfrak{t} \subseteq Z(\mathfrak{m})$. In fact, a $\mathfrak{t}$-compatible parabolic subalgebra $\mathfrak{p}$ is minimal if and only if $\mathfrak{m}$ equals the centralizer $C(\mathfrak{t})$ of $\mathfrak{t}$ in $\mathfrak{g}$, or equivalently, if and only if $\mathfrak{p} = \mathfrak{p}_\eta$ for a $(\mathfrak{g}, \mathfrak{k})$-regular $\eta \in \mathfrak{t}^*$. In this case $\mathfrak{n} \cap \mathfrak{k} = \mathfrak{n}_\mathfrak{k}$.

Any $\mathfrak{t}$-compatible parabolic subalgebra $\mathfrak{p} = \mathfrak{p}_\eta$ has a well-defined opposite parabolic subalgebra $\bar{\mathfrak{p}} := \mathfrak{p}_{-\eta}$; clearly $\mathfrak{p}$ is minimal if and only if $\bar{\mathfrak{p}}$ is minimal.

A $\mathfrak{k}$-*type* is by definition a simple finite-dimensional $\mathfrak{k}$-module. By $V(\mu)$ we denote a $\mathfrak{k}$-type with $\mathfrak{b}_\mathfrak{k}$-highest weight μ. The weight μ is then $\mathfrak{k}$-integral (or, equivalently, $\mathfrak{k}_{ss}$−integral) and $\mathfrak{b}_\mathfrak{k}$-dominant.

Let $V(\mu)$ be a $\mathfrak{k}$-type such that $\mu + 2\rho$ is $(\mathfrak{g}, \mathfrak{k})$-regular, and let $\mathfrak{p} = \mathfrak{m} \ni \mathfrak{n}$ be the minimal compatible parabolic subalgebra $\mathfrak{p}_{\mu+2\rho}$. Put $\tilde{\rho}_\mathfrak{n} := \rho_{\mathrm{char}_\mathfrak{h} \mathfrak{n}}$ and $\rho_\mathfrak{n} := \rho_{\mathrm{char}_\mathfrak{t} \mathfrak{n}}$. Clearly $\rho_\mathfrak{n} = \tilde{\rho}_\mathfrak{n}|_\mathfrak{t}$. We define $V(\mu)$ to be *generic* if the following two conditions hold:

1. $\langle \mathrm{Re}\mu + 2\rho - \rho_\mathfrak{n}, \alpha \rangle \geq 0 \; \forall \alpha \in \mathrm{supp}_\mathfrak{t} \mathfrak{n}_\mathfrak{k}$;
2. $\langle \mathrm{Re}\mu + 2\rho - \rho_S, \rho_S \rangle > 0$ for every submultiset S of $\mathrm{char}_\mathfrak{t}\, \mathfrak{n}$.

It is easy to show that there exists a positive constant C depending only on $\mathfrak{g}$, $\mathfrak{k}$ and $\mathfrak{p}$ such that $\langle \mathrm{Re}\mu + 2\rho, \alpha \rangle > C$ for every $\alpha \in \mathrm{supp}_\mathfrak{t} \mathfrak{n}$ implies $\mathfrak{p}_{\mu+2\rho} = \mathfrak{p}$ and that $V(\mu)$ is generic.

1.3 Generalities on $\mathfrak{g}$-Modules

Suppose M is a $\mathfrak{g}$-module and $\mathfrak{l}$ is a reductive subalgebra of $\mathfrak{g}$. M is *locally finite over* $Z_{U(\mathfrak{l})}$ if every vector in M generates a finite-dimensional $Z_{U(\mathfrak{l})}$-module. Denote by $\mathcal{M}(\mathfrak{g}, Z_{U(\mathfrak{l})})$ the full subcategory of $\mathfrak{g}$-modules which are locally finite over $Z_{U(\mathfrak{l})}$.

Suppose $M \in \mathcal{M}(\mathfrak{g}, Z_{U(\mathfrak{l})})$ and θ is a $Z_{U(\mathfrak{l})}$-character. Denote by $P(\mathfrak{l}, \theta)(M)$ the generalized θ-eigenspace of the restriction of M to $\mathfrak{l}$. The $Z_{U(\mathfrak{l})}$-*spectrum* of M is the set of characters θ of $Z_{U(\mathfrak{l})}$ such that $P(\mathfrak{l}, \theta)(M) \neq 0$. Denote the $Z_{U(\mathfrak{l})}$ spectrum of M by $\sigma(\mathfrak{l}, M)$. We say that θ is a *central character of* $\mathfrak{l}$ *in* M if $\theta \in \sigma(\mathfrak{l}, M)$. The following is a standard fact.

Lemma 1.1. *If $M \in \mathcal{M}(\mathfrak{g}, Z_{U(\mathfrak{l})})$, then*

$$M = \bigoplus_{\theta \in \sigma(\mathfrak{l}, M)} P(\mathfrak{l}, \theta)(M).$$

A $\mathfrak{g}$-module M is *locally Artinian over* $\mathfrak{l}$ if for every vector $v \in M$, $U(\mathfrak{l}) \cdot v$ is an $\mathfrak{l}$-module of finite length.

Lemma 1.2. *If M is locally Artinian over $\mathfrak{l}$, then $M \in \mathcal{M}(\mathfrak{g}, Z_{U(\mathfrak{l})})$.*

Proof. The statement follows from the fact that $Z_{U(\mathfrak{l})}$ acts via a character on any simple $\mathfrak{l}$-module. □

If $\mathfrak{p}$ is a parabolic subalgebra of $\mathfrak{g}$, by a $(\mathfrak{g}, \mathfrak{p})$-*module* M we mean a $\mathfrak{g}$-module M on which $\mathfrak{p}$ acts locally finitely. By $\mathcal{M}(\mathfrak{g}, \mathfrak{p})$ we denote the full subcategory of $\mathfrak{g}$-modules which are $(\mathfrak{g}, \mathfrak{p})$-modules.

In the remainder of this subsection we assume that $\mathfrak{h}$ is a Cartan subalgebra of $\mathfrak{g}$ such that $\mathfrak{h}_{\mathfrak{l}} := \mathfrak{h} \cap \mathfrak{l}$ is a Cartan subalgebra of $\mathfrak{l}$, and that $\mathfrak{p}$ is a parabolic subalgebra of $\mathfrak{g}$ such that $\mathfrak{h} \subset \mathfrak{p}$ and $\mathfrak{p} \cap \mathfrak{l}$ is a parabolic subalgebra of $\mathfrak{l}$. By M we denote a $\mathfrak{g}$-module from $\mathcal{M}(\mathfrak{g}, \mathfrak{p})$. Note that M is not necessarily semisimple as an $\mathfrak{h}$-module.

Lemma 1.3. *The set* $\mathrm{supp}_{\mathfrak{h}} M$ *is independent of the choice of* $\mathfrak{h} \subseteq \mathfrak{p}$, *i.e.* $\mathrm{supp}_{\mathfrak{h}} M$ *is equivariant with respect to inner automorphisms of* $\mathfrak{g}$ *preserving* $\mathfrak{p}$.

Proof. As $\mathfrak{p}$ acts locally finitely on M, the statement is an immediate consequence of the equivariance of the support (set of weights) of a finite-dimensional $\mathfrak{p}$−module. □

Proposition 1.4. *M is locally Artinian over $\mathfrak{l}$.*

Proof. We apply Proposition 7.6.1 in [Dix] to the pair $(\mathfrak{l}, \mathfrak{l} \cap \mathfrak{p})$. In particular, if $v \in M$, then $U(\mathfrak{l}) \cdot v$ has finite length as an $\mathfrak{l}$-module. □

Corollary 1.5. $M \in \mathcal{M}(\mathfrak{g}, Z_{U(\mathfrak{l})})$.

Lemma 1.6. $\sigma(\mathfrak{l}, M) \subseteq \{\theta^{\mathfrak{l}}_{(\eta|_{\mathfrak{h}_{\mathfrak{l}}})+\rho_{\mathfrak{l}}} \mid \eta \in \mathrm{supp}_{\mathfrak{h}} M\}$.

Proof. The simple $\mathfrak{l}$-subquotients of M are $(\mathfrak{l}, \mathfrak{l} \cap \mathfrak{p})$-modules, and our claim follows the well-known relationship between the highest weight of a highest weight module and its central character. □

Let N be a $\mathfrak{g}$-module, and let $\mathfrak{g}[N]$ be the set of elements $x \in \mathfrak{g}$ that act locally finitely in N. Then $\mathfrak{g}[N]$ is a Lie subalgebra of $\mathfrak{g}$, the *Fernando–Kac subalgebra associated to N*. The fact has been proved independently by V. Kac in [K] and by S. Fernando in [F].

Theorem 1.7. *Let M_1 be a nonzero subquotient of M. Assume that $\eta|_{\mathfrak{h}_{\mathfrak{l}}}$ is non-integral relative to $\mathfrak{l}$ for all $\eta \in \mathrm{supp}_{\mathfrak{h}} M$. Then $\mathfrak{l} \not\subseteq \mathfrak{g}[M_1]$.*

Proof. By Lemma 1.6, no central character of $\mathfrak{l}$ in M_1 is $\mathfrak{l}$-integral. Therefore, no nonzero $\mathfrak{l}$-submodule of M_1 is finite dimensional. But $M_1 \neq 0$. Hence, $\mathfrak{l} \not\subseteq \mathfrak{g}[M_1]$. □

In agreement with [PZ2], we define a $\mathfrak{g}$-module M to be a $(\mathfrak{g}, \mathfrak{k})$-*module* if M is isomorphic as a $\mathfrak{k}$-module to a direct sum of isotypic components of $\mathfrak{k}$-types. If M is a $(\mathfrak{g}, \mathfrak{k})$-module, we write $M[\mu]$ for the $V(\mu)$-isotypic component of M, and we say that $V(\mu)$ is a $\mathfrak{k}$-*type of M* if $M[\mu] \neq 0$. We say that a $(\mathfrak{g}, \mathfrak{k})$-module M is *of finite type* if $\dim M[\mu] \neq \infty$ for every $\mathfrak{k}$-type $V(\mu)$ of M. Sometimes, we also refer to $(\mathfrak{g}, \mathfrak{k})$-modules of finite type as *generalized Harish-Chandra modules*.

Note that for any $(\mathfrak{g}, \mathfrak{k})$-module of finite type M and any $\mathfrak{k}$-type $V(\sigma)$ of M, the finite-dimensional $\mathfrak{k}$-module $M[\sigma]$ is a $\tilde{\mathfrak{k}}$-module for $\tilde{\mathfrak{k}} = \mathfrak{k} + C(\mathfrak{k})$. In particular, M is a $(\mathfrak{g}, \tilde{\mathfrak{k}})$-module of finite type. We will write $M\langle\delta\rangle$ for the $\tilde{\mathfrak{k}}$-isotypic components of M where $\delta \in (\mathfrak{h} \cap \tilde{\mathfrak{k}})^*$.

If M is a module of finite length, a $\mathfrak{k}$-type $V(\mu)$ of M is *minimal* if the function $\mu' \mapsto \| \mathrm{Re}\mu' + 2\rho \|^2$ defined on the set $\{\mu' \in \mathfrak{t}^* \mid M[\mu'] \neq 0\}$ has a minimum at μ. Any nonzero $(\mathfrak{g}, \mathfrak{k})$-module M of finite length has a minimal $\mathfrak{k}$-type.

1.4 Generalities on the Zuckerman Functor

Recall that the *functor of* $\mathfrak{k}$*-finite vectors* $\Gamma^{\mathfrak{g},\mathfrak{t}}_{\mathfrak{g},\mathfrak{k}}$ is a well-defined left-exact functor on the category of $(\mathfrak{g}, \mathfrak{t})$-modules with values in $(\mathfrak{g}, \mathfrak{k})$-modules,

$$\Gamma^{\mathfrak{g},\mathfrak{t}}_{\mathfrak{g},\mathfrak{k}}(M) := \sum_{M' \subset M, \dim M' = 1, \dim U(\mathfrak{k})\cdot M' < \infty} M'.$$

By $R^{\cdot}\Gamma^{\mathfrak{g},\mathfrak{t}}_{\mathfrak{g},\mathfrak{k}} := \bigoplus_{i \geq 0} R^i\Gamma^{\mathfrak{g},\mathfrak{t}}_{\mathfrak{g},\mathfrak{k}}$ we denote as usual the total right derived functor of $\Gamma^{\mathfrak{g},\mathfrak{t}}_{\mathfrak{g},\mathfrak{k}}$, see [Z] and the references therein.

Proposition 1.8. *If* $\mathfrak{l}$ *is any reductive subalgebra of* $\mathfrak{g}$ *containing* $\mathfrak{k}$*, then there is a natural isomorphism of* $\mathfrak{l}$*-modules*

$$R^{\cdot}\Gamma^{\mathfrak{g},\mathfrak{t}}_{\mathfrak{g},\mathfrak{k}}(N) \cong R^{\cdot}\Gamma^{\mathfrak{l},\mathfrak{t}}_{\mathfrak{l},\mathfrak{k}}(N). \tag{1}$$

Proof. See Proposition 2.5 in [PZ4]. □

Proposition 1.9. *If* $\tilde{N} \in \mathcal{M}(\mathfrak{l}, \mathfrak{t}, Z_{U(\mathfrak{l})}) := \mathcal{M}(\mathfrak{l}, Z_{U(\mathfrak{l})}) \cap \mathcal{M}(\mathfrak{l}, \mathfrak{t})$*, then*

$$R^{\cdot}\Gamma^{\mathfrak{l},\mathfrak{t}}_{\mathfrak{l},\mathfrak{k}}(\tilde{N}) \in \mathcal{M}(\mathfrak{l}, \mathfrak{k}, Z_{U(\mathfrak{l})}).$$

Moreover,

$$\sigma(\mathfrak{l}, R^{\cdot}\Gamma^{\mathfrak{l},\mathfrak{t}}_{\mathfrak{l},\mathfrak{k}}(\tilde{N})) \subset \sigma(\mathfrak{l}, \tilde{N}).$$

Proof. See Proposition 2.12 and Corollary 2.8 in [Z]. □

Corollary 1.10. *If* $N \in \mathcal{M}(\mathfrak{g}, \mathfrak{t}, Z_{U(\mathfrak{l})}) := \mathcal{M}(\mathfrak{g}, Z_{U(\mathfrak{l})}) \cap \mathcal{M}(\mathfrak{g}, \mathfrak{t})$*, then*

$$R^{\cdot}\Gamma^{\mathfrak{g},\mathfrak{t}}_{\mathfrak{g},\mathfrak{k}}(N) \in \mathcal{M}(\mathfrak{g}, \mathfrak{k}, Z_{U(\mathfrak{l})}).$$

Moreover,

$$\sigma(\mathfrak{l}, R^{\cdot}\Gamma^{\mathfrak{g},\mathfrak{t}}_{\mathfrak{g},\mathfrak{k}}(N)) \subseteq \sigma(\mathfrak{l}, N).$$

Proof. Apply Propositions 1.8 and 1.9. □

Note that the isomorphism (1) enables us to write simply $\Gamma_{\mathfrak{k},\mathfrak{t}}$ instead of $\Gamma^{\mathfrak{g},\mathfrak{t}}_{\mathfrak{g},\mathfrak{k}}$.

For $\mathfrak{g} \supseteq \mathfrak{l} \supseteq \mathfrak{k} \supseteq \mathfrak{t}$ as above, let $\mathfrak{p}$ be a $\mathfrak{t}$-compatible parabolic subalgebra of $\mathfrak{g}$. Then $\mathfrak{l} \cap \mathfrak{p}$ is a $\mathfrak{t}$-compatible parabolic subalgebra of $\mathfrak{l}$. Let $\mathfrak{h}_{\mathfrak{l}} \subset \mathfrak{l} \cap \mathfrak{p}$ be a Cartan subalgebra of $\mathfrak{l}$ containing $\mathfrak{t}$, and let $\mathfrak{h} \subset \mathfrak{p}$ be a Cartan subalgebra of $\mathfrak{g}$ such that $\mathfrak{h}_{\mathfrak{l}} = \mathfrak{h} \cap \mathfrak{l}$. We have the following diagram of subalgebras:

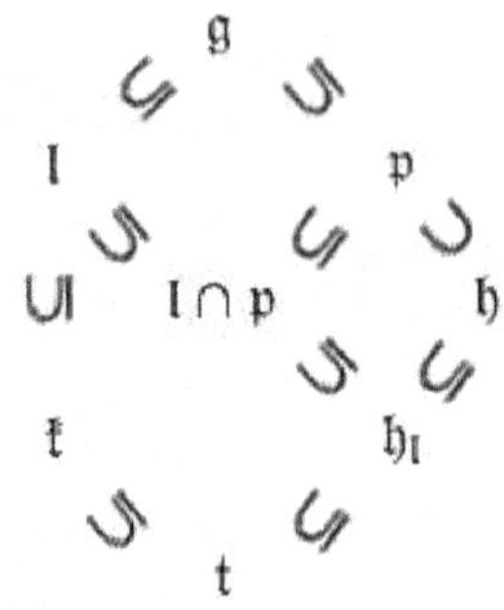

In this setup we have the following result.

Theorem 1.11. *Suppose $N \in \mathcal{M}(\mathfrak{g}, \mathfrak{p}) \cap \mathcal{M}(\mathfrak{g}, \mathfrak{t})$, M is a nonzero subquotient of $R^{\cdot}\Gamma_{\mathfrak{k},\mathfrak{t}}\ (N)$ and $\eta|_{\mathfrak{h}_{\mathfrak{l}}}$ is not $\mathfrak{l}$-integral for all $\eta \in \operatorname{supp}_{\mathfrak{h}} N$. Then $\mathfrak{l} \not\subseteq \mathfrak{g}[M]$.*

Proof. Every central character of $\mathfrak{l}$ in M is a central character of $\mathfrak{l}$ in N. This follows from Corollary 2.8 in [Z]. By our assumptions, no central character of $\mathfrak{l}$ in N is $\mathfrak{l}$-integral. Hence, no $\mathfrak{l}$-submodule of M is finite dimensional, and thus $\mathfrak{l} \not\subseteq \mathfrak{g}[M]$. □

2 The Fundamental Series: Main Results

We now introduce one of our main objects of study: the fundamental series of generalized Harish-Chandra modules.

We start by fixing some more notation: if $\mathfrak{q}$ is a subalgebra of $\mathfrak{g}$ and J is a $\mathfrak{q}$-module, we set $\operatorname{ind}_{\mathfrak{q}}^{\mathfrak{g}} J := U(\mathfrak{g}) \otimes_{U(\mathfrak{q})} J$ and $\operatorname{pro}_{\mathfrak{q}}^{\mathfrak{g}} J := \operatorname{Hom}_{U(\mathfrak{q})}(U(\mathfrak{g}), J)$. For a finite-dimensional $\mathfrak{p}$- or $\bar{\mathfrak{p}}$-module E, we set $N_{\mathfrak{p}}(E) := \Gamma_{\mathfrak{t},0}(\operatorname{pro}_{\mathfrak{p}}^{\mathfrak{g}}(E \otimes \Lambda^{\dim \mathfrak{n}}(\mathfrak{n})))$, $N_{\bar{\mathfrak{p}}}(E^*) := \Gamma_{\mathfrak{t},0}(\operatorname{pro}_{\bar{\mathfrak{p}}}^{\mathfrak{g}}(E^* \otimes \Lambda^{\dim \mathfrak{n}}(\mathfrak{n}^*)))$. One can show that both $N_{\mathfrak{p}}(E)$ and $N_{\bar{\mathfrak{p}}}(E^*)$ have simple socles as long as E itself is simple.

The *fundamental series* of $(\mathfrak{g}, \mathfrak{k})$-modules of finite type $F^{\cdot}(\mathfrak{k}, \mathfrak{p}, E)$ is defined as follows. Let $\mathfrak{p} = \mathfrak{m} \ni \mathfrak{n}$ be a minimal compatible parabolic subalgebra (recall that $\mathfrak{m} = C(\mathfrak{t})$), let E be a simple finite-dimensional $\mathfrak{p}$-module on which $\mathfrak{t}$ acts via a fixed weight $\omega \in \mathfrak{t}^*$, and $\mu := \omega + 2\rho_{\mathfrak{n}}^{\perp}$ where $\rho_{\mathfrak{n}}^{\perp} := \rho_{\mathfrak{n}} - \rho$. Set

$$F^{\cdot}(\mathfrak{k}, \mathfrak{p}, E) := R^{\cdot}\Gamma_{\mathfrak{k},\mathfrak{t}}(N_{\mathfrak{p}}(E)).$$

In the rest of the paper we assume that $\mathfrak{h} \cap \tilde{\mathfrak{k}}$ is a Cartan subalgebra of $\tilde{\mathfrak{k}}$.

Theorem 2.1. *(a) $F^{\cdot}(\mathfrak{k}, \mathfrak{p}, E)$ is a $(\mathfrak{g}, \mathfrak{k})$-module of finite type and $Z_{U(\mathfrak{g})}$ acts on $F^{\cdot}(\mathfrak{p}, E)$ via the $Z_{U(\mathfrak{g})}$-character $\theta_{\nu+\tilde{\rho}}$ where $\tilde{\rho} := \rho_{\operatorname{char}_{\mathfrak{h}} \mathfrak{b}}$ for some Borel subalgebra $\mathfrak{b}$ of $\mathfrak{g}$ with $\mathfrak{b} \supset \mathfrak{h}$, $\mathfrak{b} \subset \mathfrak{p}$ and $\mathfrak{b} \cap \mathfrak{k} = \mathfrak{b}_{\mathfrak{k}}$, and where ν is the $\mathfrak{b}$-highest weight of E (note that $\nu|_{\mathfrak{t}} = \omega$).*

(b) $F^{\cdot}(\mathfrak{k}, \mathfrak{p}, E)$ *is a* $(\mathfrak{g}, \mathfrak{k})$*-module of finite length.*
(c) There is a canonical isomorphism

$$F^{\cdot}(\mathfrak{k}, \mathfrak{p}, E) \simeq R^{\cdot}\Gamma_{\tilde{\mathfrak{k}}, \tilde{\mathfrak{k}}\cap\mathfrak{m}}(\Gamma_{\tilde{\mathfrak{k}}\cap\mathfrak{m},0}(\mathrm{pro}^{\mathfrak{g}}_{\mathfrak{p}}(E \otimes \Lambda^{\dim \mathfrak{n}}(\mathfrak{n})))). \quad (2)$$

Proof. Part (a) is a recollection of Theorem 1.11, (a) in [PZ2]. Part (b) is a recollection of Theorem 2.5 in [PZ5]. Part (c) follows from the comparison principle (Proposition 2.6) in [PZ4]. □

Corollary 2.2. $F^{\cdot}(\mathfrak{k}, \mathfrak{p}, E)$ *is a* $(\mathfrak{g}, \tilde{\mathfrak{k}})$*-module of finite type.*

Proof. As we observed in Sect. 1.3, every $(\mathfrak{g}, \mathfrak{k})$-module of finite type is a $(\mathfrak{g}, \tilde{\mathfrak{k}})$-module of finite type. □

Corollary 2.3. *Let* $\mathfrak{k}_1$ *and* $\mathfrak{k}_2$ *be two algebraic reductive in* $\mathfrak{g}$ *subalgebras such that* $\tilde{\mathfrak{k}}_1 = \tilde{\mathfrak{k}}_2$. *Suppose that* $\mathfrak{p}$ *is a parabolic subalgebra which is both* $\mathfrak{t}_1$*- and* $\mathfrak{t}_2$*-compatible and* $\mathfrak{t}_1$*- and* $\mathfrak{t}_2$*-minimal for some Cartan subalgebras* $\mathfrak{t}_1$ *of* $\mathfrak{k}_1$ *and* $\mathfrak{t}_2$ *of* $\mathfrak{k}_2$. *Then there exists a canonical isomorphism*

$$F^{\cdot}(\mathfrak{k}_1, \mathfrak{p}, E) \simeq F(\mathfrak{k}_2, \mathfrak{p}, E).$$

Proof. Consider the isomorphism (2) for $\mathfrak{k}_1$ and $\mathfrak{k}_2$, and notice that

$$R^{\cdot}\Gamma_{\tilde{\mathfrak{k}}, \tilde{\mathfrak{k}}\cap\mathfrak{m}}(\Gamma_{\tilde{\mathfrak{k}}\cap\mathfrak{m},0}(\mathrm{pro}^{\mathfrak{g}}_{\mathfrak{p}}(E \otimes \Lambda^{\dim \mathfrak{n}}(\mathfrak{n}))))$$

depends only on $\tilde{\mathfrak{k}}$ and $\mathfrak{p}$ but not on $\mathfrak{k}_1$ and $\mathfrak{k}_2$. □

Corollary 2.4. *Let* M *be any nonzero subquotient of* $F^{\cdot}(\mathfrak{k}, \mathfrak{p}, E)$. *If the* $\mathfrak{b}$*-highest weight* $\nu \in \mathfrak{h}^*$ *of* E *is non-integral after restriction to* $\mathfrak{h} \cap \mathfrak{l}$ *for any reductive subalgebra* $\mathfrak{l}$ *of* $\mathfrak{g}$ *such that* $\mathfrak{l} \supset \tilde{\mathfrak{k}}$, *then* $\tilde{\mathfrak{k}}$ *is a maximal reductive subalgebra of* $\mathfrak{g}[M]$.

Proof. Corollary 2.2 shows that $\tilde{\mathfrak{k}} \subseteq \mathfrak{g}[M]$. Theorem 1.11 shows that if $\mathfrak{l}$ is a reductive subalgebra of $\mathfrak{g}$ such that $\mathfrak{l}$ is strictly larger than $\tilde{\mathfrak{k}}$, then $\mathfrak{l} \not\subseteq \mathfrak{g}[M]$. The assumption on ν implies that all weights in $\mathrm{supp}_{\mathfrak{h}\cap\mathfrak{l}}(N_{\mathfrak{p}}(E))$ are non-integral with respect to $\mathfrak{l}$. □

Example. Here is an example to Corollary 2.4. Let $\mathfrak{g} = F_4$, $\mathfrak{k} \simeq \mathfrak{so}(3) \oplus \mathfrak{so}(6)$. Then $\mathfrak{k} = \tilde{\mathfrak{k}}$. By inspection, there is only one proper intermediate subalgebra $\mathfrak{l}$, $\tilde{\mathfrak{k}} \subset \mathfrak{l} \subset \mathfrak{g}$, and $\mathfrak{l}$ is isomorphic to $\mathfrak{so}(9)$. We have $\mathfrak{t} = \mathfrak{h}$, and $\varepsilon_1, \varepsilon_2, \varepsilon_3, \varepsilon_4$ is a standard basis of $\mathfrak{h}^*$, see [Bou]. A weight $\nu = \sum_{i=1}^{4} m_i \varepsilon_i$ is $\mathfrak{k}$-integral iff $m_1 \in \mathbb{Z}$ or $m_1 \in \mathbb{Z} + \frac{1}{2}$, and $(m_2, m_3, m_4) \in \mathbb{Z}^3$ or $(m_2, m_3, m_4) \in \mathbb{Z}^3 + (\frac{1}{2}, \frac{1}{2}, \frac{1}{2})$. On the other hand, ν is $\mathfrak{l}$-integral if $(m_1, m_2, m_3, m_4) \in \mathbb{Z}^4$ or $(m_1, m_2, m_3, m_4) \in \mathbb{Z}^4 + (\frac{1}{2}, \frac{1}{2}, \frac{1}{2}, \frac{1}{2})$. So if the $\mathfrak{b}$-highest weight ν of E is not $\mathfrak{l}$-integral, Corollary 2.4 implies that $\mathfrak{g}[M] = \tilde{\mathfrak{k}}$ for any simple subquotient M of $F^{\cdot}(\mathfrak{k}, \mathfrak{p}, E)$.

Remark. (a) In [PZ1] another method, based on the notion of a small subalgebra introduced by Willenbring and Zuckerman in [WZ], for computing maximal reductive subalgebras of the Fernando-Kac subalgebras associated to simple subquotients of $F^{\cdot}(\mathfrak{k}, \mathfrak{p}, E)$ is suggested. Note that the subalgebra $\mathfrak{k} \simeq \mathfrak{so}(3) \oplus \mathfrak{so}(6)$ of F_4 considered in the above example is not small in $\mathfrak{so}(9)$, so the above conclusion that $\mathfrak{g}[M] = \mathfrak{k}$ does not follow from [PZ1]. On the other hand, if one replaces $\mathfrak{k}$ in the example by $\mathfrak{k}' \simeq \mathfrak{so}(5) \oplus \mathfrak{so}(4)$, then a conclusion similar to that of the example can be reached both by the method of [PZ1] and by Corollary 2.4.

(b) There are pairs $(\mathfrak{g}, \mathfrak{k})$ to which neither the method of [PZ1] nor Corollary 2.4 apply. Such an example is a pair $(\mathfrak{g} = F_4, \mathfrak{k} \simeq \mathfrak{so}(8))$. The only proper intermediate subalgebra in this case is $\mathfrak{l} \simeq \mathfrak{so}(9)$; however $\mathfrak{so}(8)$ is not small in $\mathfrak{so}(9)$ and any $\mathfrak{k} = \tilde{\mathfrak{k}}$-integrable weight is also $\mathfrak{l}$-integrable.

If M is a $(\mathfrak{g}, \mathfrak{k})$-module of finite type, then $\Gamma_{\mathfrak{k},0}(M^*)$ is a well-defined $(\mathfrak{g}, \mathfrak{k})$-module of finite type and $\Gamma_{\mathfrak{k},0}(\cdot^*)$ is an involution on the category of $(\mathfrak{g}, \mathfrak{k})$-modules of finite type. We put $\Gamma_{\mathfrak{k},0}(M^*) := M^*_{\mathfrak{k}}$. There is an obvious $\mathfrak{g}$-invariant nondegenerate pairing $M \times M^*_{\mathfrak{k}} \to \mathbb{C}$.

The following five statements are recollections of the main results of [PZ2] (Theorem 2 through Corollary 4 in [PZ2]).

Theorem 2.5. *Assume that $V(\mu)$ is a generic $\mathfrak{k}$-type and that $\mathfrak{p} = \mathfrak{p}_{\mu+2\rho}$ (μ is necessarily $\mathfrak{b}_{\mathfrak{k}}$-dominant and $\mathfrak{k}$-integral).*

(a) $F^i(\mathfrak{k}, \mathfrak{p}, E) = 0$ for $i \neq s :=$ $\dim \mathfrak{n}_{\mathfrak{k}}$.

(b) There is a $\mathfrak{k}$-module isomorphism

$$F^s(\mathfrak{k}, \mathfrak{p}, E)[\mu] \cong \mathbb{C}^{\dim E} \otimes V(\mu),$$

and $V(\mu)$ is the unique minimal $\mathfrak{k}$-type of $F^s(\mathfrak{k}, \mathfrak{p}, E)$.

(c) Let $\bar{F}^s(\mathfrak{k}, \mathfrak{p}, E)$ be the $\mathfrak{g}$-submodule of $F^s(\mathfrak{k}, \mathfrak{p}, E)$ generated by $F^s(\mathfrak{k}, \mathfrak{p}, E)[\mu]$. Then $\bar{F}^s(\mathfrak{k}, \mathfrak{p}, E)$ is simple and $\bar{F}^s(\mathfrak{k}, \mathfrak{p}, E) = \mathrm{Soc} F^s(\mathfrak{k}, \mathfrak{p}, E)$. Moreover, $F^s(\mathfrak{k}, \mathfrak{p}, E)$ is cogenerated by $F^s(\mathfrak{k}, \mathfrak{p}, E)[\mu]$. This implies that $F^s(\mathfrak{k}, \mathfrak{p}, E)^*_{\mathfrak{t}}$ is generated by $F^s(\mathfrak{k}, \mathfrak{p}, E)^*_{\mathfrak{t}}[w_m(-\mu)]$, where $w_m \in W_{\mathfrak{k}}$ is the element of maximal length in the Weyl group $W_{\mathfrak{k}}$ of $\mathfrak{k}$.

(d) For any nonzero $\mathfrak{g}$-submodule M of $F^s(\mathfrak{k}, \mathfrak{p}, E)$ there is an isomorphism of $\mathfrak{m}$-modules

$$H^r(\mathfrak{n}, M)^{\omega} \cong E.$$

where $r :=$ $\dim(\mathfrak{n} \cap \mathfrak{k}^{\perp})$.

Theorem 2.6. *Let M be a simple $(\mathfrak{g}, \mathfrak{k})$-module of finite type with minimal $\mathfrak{k}$-type $V(\mu)$ which is generic. Then $\mathfrak{p} := \mathfrak{p}_{\mu+2\rho} = \mathfrak{m} \ni \mathfrak{n}$ is a minimal compatible parabolic subalgebra. Let $\omega := \mu - 2\rho_n^{\perp}$ (recall that $\rho_{\mathfrak{n}}^{\perp} = \rho_{\mathrm{ch}_{\mathfrak{t}}(\mathfrak{n} \cap \mathfrak{k}^{\perp})}$), and let*

E be the $\mathfrak{p}$-module $H^r(\mathfrak{n}, M)^\omega$ with trivial $\mathfrak{n}$-action, where $r = \dim(\mathfrak{n} \cap \mathfrak{k}^\perp)$. Then E is a simple $\mathfrak{p}$-module, the pair $(\mathfrak{p}, E)$ satisfies the hypotheses of Theorem 2.5, and M is canonically isomorphic to $\bar{F}^s(\mathfrak{p}, E)$ for $s = \dim(\mathfrak{n} \cap \mathfrak{k})$.

Corollary 2.7. *(Generic version of a theorem of Harish-Chandra). There exist at most finitely many simple $(\mathfrak{g}, \mathfrak{k})$-modules M of finite type with a fixed $Z_{U(\mathfrak{g})}$-character such that a minimal $\mathfrak{k}$-type of M is generic. (Moreover, each such M has a unique minimal $\mathfrak{k}$-type by Theorem 2.5 (b).)*

Proof. By Theorems 2.1 (a) and 2.6, if M is a simple $(\mathfrak{g}, \mathfrak{k})$-module of finite type with generic minimal $\mathfrak{k}$-type $V(\mu)$ for some μ, then the $Z_{U(\mathfrak{g})}$-character of M is $\theta_{\nu+\tilde{\rho}}$. There are finitely many Borel subalgebras $\mathfrak{b}$ as in Theorem 2.1 (a); thus, if $\theta_{\nu+\tilde{\rho}}$ is fixed, there are finitely many possibilities for the weight ν (as $\theta_{\nu+\tilde{\rho}}$ determines $\nu + \tilde{\rho}$ up to a finite choice). Hence, up to isomorphism, there are finitely many possibilities for the $\mathfrak{p}$-module E, and consequently, up to isomorphism, there are finitely many possibilities for M. □

Theorem 2.8. *Assume that the pair $(\mathfrak{g}, \mathfrak{k})$ is regular, i.e., $\mathfrak{t}$ contains a regular element of $\mathfrak{g}$. Let M be a simple $(\mathfrak{g}, \mathfrak{k})$-module (a priori of infinite type) with a minimal $\mathfrak{k}$-type $V(\mu)$ which is generic. Then M has finite type, and hence by Theorem 2.6, M is canonically isomorphic to $\bar{F}^s(\mathfrak{p}, E)$ (where $\mathfrak{p}$, E and s are as in Theorem 2.6).*

Corollary 2.9. *Let the pair $(\mathfrak{g}, \mathfrak{k})$ be regular.*

(a) There exist at most finitely many simple $(\mathfrak{g}, \mathfrak{k})$-modules M with a fixed $Z_{U(\mathfrak{g})}$-character, such that a minimal $\mathfrak{k}$-type of M is generic. All such M are of finite type (and have a unique minimal $\mathfrak{k}$-type by Theorem 2.5 (b)).
(b) (Generic version of Harish-Chandra's admissibility theorem). Every simple $(\mathfrak{g}, \mathfrak{k})$-module with a generic minimal $\mathfrak{k}$-type has finite type.

Proof. The proof of (a) is as the proof of Corollary 2.7 but uses Theorem 2.8 instead of Theorem 2.6, and (b) is a direct consequence of Theorem 2.8. □

The following statement follows from Corollary 2.4 and Theorem 2.6.

Corollary 2.10. *Let M be as in Theorem 2.6. If the $\mathfrak{b}$-highest weight of E is not $\mathfrak{l}$-integral for any reductive subalgebra $\mathfrak{l}$ with $\tilde{\mathfrak{k}} \subset \mathfrak{l} \subseteq \mathfrak{g}$, then $\tilde{\mathfrak{k}}$ is a maximal reductive subalgebra of $\mathfrak{g}[M]$.*

Definition 2.11. Let $\mathfrak{p} \supset \mathfrak{b}_{\mathfrak{k}}$ be a minimal $\mathfrak{t}$-compatible parabolic subalgebra and let E be a simple finite-dimensional $\mathfrak{p}$-module on which $\mathfrak{t}$ acts by ω. We say that the pair $(\mathfrak{p}, E)$ is *allowable* if $\mu = \omega + 2\rho_{\mathfrak{n}}^{\perp}$ is dominant integral for $\mathfrak{k}$, $\mathfrak{p}_{\mu+2\rho} = \mathfrak{p}$, and $V(\mu)$ is generic.

Theorem 2.6 provides a classification of simple $(\mathfrak{g}, \mathfrak{k})$-modules with generic minimal $\mathfrak{k}$-type in terms of allowable pairs. Note that for any minimal $\mathfrak{t}$-compatible parabolic subalgebra $\mathfrak{p} \supset \mathfrak{b}_{\mathfrak{k}}$, there exists a $\mathfrak{p}$-module E such that $(\mathfrak{p}, E)$ is allowable.

3 The Case $\mathfrak{k} \simeq \mathfrak{sl}(2)$

Let $\mathfrak{k} \simeq \mathfrak{sl}(2)$. In this case there is only one minimal $\mathfrak{t}$-compatible parabolic subalgebra $\mathfrak{p} = \mathfrak{m} \supsetplus \mathfrak{n}$ of $\mathfrak{g}$ which contains $\mathfrak{b}_{\mathfrak{k}}$. Furthermore, we can identify the elements of $\mathfrak{t}^*$ with complex numbers, and the $\mathfrak{b}_{\mathfrak{k}}$-dominant integral weights of $\mathfrak{t}$ in $\mathfrak{n} \cap \mathfrak{k}^{\perp}$ with nonnegative integers. It is shown in [PZ2] that in this case the genericity assumption on a $\mathfrak{k}$-type $V(\mu)$, $\mu \geq 0$, amounts to the condition $\mu \geq \Gamma := \tilde{\rho}(h) - 1$ where $h \in \mathfrak{h}$ is the semisimple element in a standard basis e, h, f of $\mathfrak{k} \simeq \mathfrak{sl}(2)$.

In our work [PZ5] we have proved a different sufficient condition for the main results of [PZ2] to hold when $\mathfrak{k} \simeq \mathfrak{sl}(2)$. Let λ_1 and λ_2 be the maximum and submaximum weights of $\mathfrak{t}$ in $\mathfrak{n} \cap \mathfrak{k}^{\perp}$ (if λ_1 has multiplicity at least two in $\mathfrak{n} \cap \mathfrak{k}^{\perp}$, then $\lambda_2 = \lambda_1$; if $\dim \mathfrak{n} \cap \mathfrak{k}^{\perp} = 1$, then $\lambda_2 = 0$). Set $\Lambda := \frac{\lambda_1+\lambda_2}{2}$.

Theorem 3.1. *If $\mathfrak{k} \simeq \mathfrak{sl}(2)$, all statements of Sect. 2 from Theorem 2.5 through Corollary 2.9 hold if we replace the assumption that μ is generic by the assumption $\mu \geq \Lambda$. As a consequence, the isomorphism classes of simple $(\mathfrak{g}, \mathfrak{k})$-modules whose minimal $\mathfrak{k}$-type is $V(\mu)$ with $\mu \geq \Lambda$ are parameterized by the isomorphism classes of simple $\mathfrak{p}$-modules E on which $\mathfrak{t}$ acts via $\mu - 2\rho_{\mathfrak{n}}^{\perp}$.*

The $\mathfrak{sl}(2)$-subalgebras of a simple Lie algebra are classified (up to conjugation) by Dynkin in [D]. We will now illustrate the computation of the bound Λ as well as the genericity condition on μ in examples.

We first consider three types of $\mathfrak{sl}(2)$-subalgebras of a simple Lie algebra: long root$-\mathfrak{sl}(2)$, short root$-\mathfrak{sl}(2)$ and principal $\mathfrak{sl}(2)$ (of course, there are short roots only for the series B, C and for G_2 and F_4). We compare the bounds Λ and Γ in the following table.

Table A

	Long root	Short root	Principal
$A_n, n \geq 2$	$\Gamma = n - 1 \geq 1 = \Lambda$	Not applicable	$\Gamma = \frac{n(n+1)(n+2)}{6} - 1 \geq 2n - 1 = \Lambda$
$B_n, n \geq 2$	$\Gamma = 2n - 3 \geq 1 = \Lambda$	$\Gamma = 2n - 2 \geq 2 = \Lambda$	$\Gamma = \frac{n(n+1)(4n-1)}{6} - 1 > 4n - 3 = \Lambda$
$C_n, n \geq 3$	$\Gamma = n - 1 > 1 = \Lambda$	$\Gamma = 2n - 2 > 2 = \Lambda$	$\Gamma = \frac{n(n+1)(2n+1)}{3} - 1 > 4n - 3 = \Lambda$
$D_n, n \geq 4$	$\Gamma = 2n - 4 > 1 = \Lambda$	Not applicable	$\Gamma = \frac{2(n-1)n(n+1)}{3} - 1 > 4n - 7 = \Lambda$
E_6	$\Gamma = 10 > 1 = \Lambda$	Not applicable	$\Gamma = 155 > 21 = \Lambda$
E_7	$\Gamma = 16 > 1 = \Lambda$	Not applicable	$\Gamma = 398 > 33 = \Lambda$
E_8	$\Gamma = 28 > 1 = \Lambda$	Not applicable	$\Gamma = 1{,}239 > 57 = \Lambda$
F_4	$\Gamma = 7 > 1 = \Lambda$	$\Gamma = 10 > 2 = \Lambda$	$\Gamma = 109 > 21 = \Lambda$
G_2	$\Gamma = 2 > 1 = \Lambda$	$\Gamma = 4 > 3 = \Lambda$	$\Gamma = 15 > 9 = \Lambda$

Let's discuss the case $\mathfrak{g} = F_4$ in more detail. Recall that the *Dynkin index* of a semisimple subalgebra $\mathfrak{s} \subset \mathfrak{g}$ is the quotient of the normalized $\mathfrak{g}$-invariant symmetric bilinear form on $\mathfrak{g}$ restricted to $\mathfrak{s}$ and the normalized $\mathfrak{s}$-invariant symmetric bilinear form on $\mathfrak{s}$, where for both $\mathfrak{g}$ and $\mathfrak{s}$ the square length of a long root equals 2. According to Dynkin [D], the conjugacy class of an $\mathfrak{sl}(2)$-subalgebra

$\mathfrak{k}$ of F_4 is determined by the Dynkin index of $\mathfrak{k}$ in F_4. Moreover, for $\mathfrak{g} = F_4$ the following integers are Dynkin indices of $\mathfrak{sl}(2)$-subalgebras: 1(long root), 2(short root), 3, 4, 6, 8, 9, 10, 11, 12, 28, 35, 36, 60, 156. The bounds Λ and Γ are given in the following table.

Table B

Dynkin index	1	2	3
	$\Gamma = 7 > 1 = \Lambda$	$\Gamma = 10 > 2 = \Lambda$	$\Gamma = 14 > 3 = \Lambda$
Dynkin index	4	6	8
	$\Gamma = 15 > 3 = \Lambda$	$\Gamma = 16 > 4 = \Lambda$	$\Gamma = 17 > 4 = \Lambda$
Dynkin index	9	10	11
	$\Gamma = 25 > 5 = \Lambda$	$\Gamma = 26 > 5 = \Lambda$	$\Gamma = 28 > 6 = \Lambda$
Dynkin index	12	28	35
	$\Gamma = 29 > 6 = \Lambda$	$\Gamma = 45 > 9 = \Lambda$	$\Gamma = 50 > 10 = \Lambda$
Dynkin index	36	60	156
	$\Gamma = 51 > 10 = \Lambda$	$\Gamma = 67 > 13 = \Lambda$	$\Gamma = 109 > 21 = \Lambda$

We conclude this section by recalling a conjecture from [PZ5]. Let $\mathcal{C}_{\bar{\mathfrak{p}},\mathfrak{t},n}$ denote the full subcategory of $\mathfrak{g}$-mod consisting of finite-length modules with simple subquotients which are $\bar{\mathfrak{p}}$-locally finite $(\mathfrak{g},\mathfrak{t})$-modules N whose $\mathfrak{t}$-weight spaces N^{β}, $\beta \in \mathbb{Z}$, satisfy $\beta \geq n$. Let $\mathcal{C}_{\mathfrak{k},n}$ be the full subcategory of $\mathfrak{g}$-mod consisting of finite length modules whose simple subquotients are $(\mathfrak{g},\mathfrak{k})$-modules with minimal $\mathfrak{k} \simeq \mathfrak{sl}(2)$-type $V(\mu)$ for $\mu \geq n$. We show in [PZ5] that the functor $R^1\Gamma_{\mathfrak{k},\mathfrak{t}}$ is a well-defined fully faithful functor from $\mathcal{C}_{\mathfrak{p},\mathfrak{t},n+2}$ to $\mathcal{C}_{\mathfrak{k},n}$ for $n \geq 0$. Moreover, we make the following conjecture.

Conjecture 3.2. Let $n \geq \Lambda$. Then $R^1\Gamma_{\mathfrak{k},\mathfrak{t}}$ is an equivalence between the categories $\mathcal{C}_{\bar{\mathfrak{p}},\mathfrak{t},n+2}$ and $\mathcal{C}_{\mathfrak{k},n}$.

We have proof of this conjecture for $\mathfrak{g} \simeq \mathfrak{sl}(2)$ and, jointly with V. Serganova, for $\mathfrak{g} \simeq \mathfrak{sl}(3)$.

4 Eligible Subalgebras

In what follows we adopt the following terminology. A *root subalgebra* of $\mathfrak{g}$ is a subalgebra which contains a Cartan subalgebra of $\mathfrak{g}$. An *r-subalgebra* of $\mathfrak{g}$ is a subalgebra $\mathfrak{l}$ whose root spaces (with respect to a Cartan subalgebra of $\mathfrak{l}$) are root spaces of $\mathfrak{g}$. The notion of r-subalgebra goes back to [D]. A root subalgebra is, of course, an r-subalgebra.

We now give the following key definition.

Definition 4.1. An algebraic reductive in $\mathfrak{g}$ subalgebra $\mathfrak{k}$ is eligible if $C(\mathfrak{t}) = \mathfrak{t} + C(\mathfrak{k})$.

Note that in the above definition one can replace $\mathfrak{t}$ with any Cartan subalgebra of $\mathfrak{k}$. Furthermore, if $\mathfrak{k}$ is eligible, then $\mathfrak{h} \subset C(\mathfrak{t}) = \mathfrak{t} + C(\mathfrak{k}) \subset \tilde{\mathfrak{k}} = \mathfrak{k} + C(\mathfrak{k})$, i.e., $\mathfrak{h}$ is a Cartan subalgebra of both $\tilde{\mathfrak{k}}$ and $\mathfrak{g}$. In particular, $\tilde{\mathfrak{k}}$ is a reductive root subalgebra of $\mathfrak{g}$. As $\mathfrak{k}$ is an ideal in $\tilde{\mathfrak{k}}$, $\mathfrak{k}$ is an r-subalgebra of $\mathfrak{g}$.

Proposition 4.2. *Assume $\mathfrak{k}$ is an r-subalgebra of $\mathfrak{g}$. The following three conditions are equivalent:*

(i) $\mathfrak{k}$ *is eligible;*
(ii) $C(\mathfrak{k})_{ss} = C(\mathfrak{t})_{ss}$*;*
(iii) $\dim C(\mathfrak{k})_{ss} = \dim C(\mathfrak{t})_{ss}$.

Proof. The implications (i)$\Rightarrow$(ii)$\Rightarrow$(iii) are obvious. To see that (iii) implies (i), observe that if $\mathfrak{k}$ is an r-subalgebra of $\mathfrak{g}$, then $\mathfrak{h} \subseteq \mathfrak{t} + C(\mathfrak{k}) \subseteq C(\mathfrak{t})$. Therefore the inclusion $\mathfrak{t} + C(\mathfrak{k}) \subseteq C(\mathfrak{t})$ is proper if and only if $\mathfrak{g}^{\pm\alpha} \in C(\mathfrak{t})\backslash C(\mathfrak{k})$ for some root $\alpha \in \Delta$, or equivalently, if the inclusion $C(\mathfrak{k})_{ss} \subseteq C(\mathfrak{t})_{ss}$ is proper. □

An algebraic, reductive in $\mathfrak{g}$, r-subalgebra $\mathfrak{k}$ may or may not be eligible. If $\mathfrak{k}$ is a root subalgebra, then $\mathfrak{k}$ is always eligible. If $\mathfrak{g}$ is simple of types A, C, D and $\mathfrak{k}$ is a semisimple r-subalgebra, then $\mathfrak{k}$ is necessarily eligible. In general, a semisimple r-subalgebra is eligible if and only if the roots of $\mathfrak{g}$ which vanish on $\mathfrak{t}$ are strongly orthogonal to the roots of $\mathfrak{k}$. For example, if $\mathfrak{g}$ is simple of type B and $\mathfrak{k}$ is a simple r-subalgebra of type B of rank less or equal than $\mathrm{rk}\mathfrak{g} - 2$, then $C(\mathfrak{k})_{ss}$ is simple of type D whereas $C(\mathfrak{t})_{ss}$ is simple of type B. Hence in this case $\mathfrak{k}$ is not eligible.

Note, however, that any semisimple r-subalgebra $\mathfrak{k}'$ can be extended to an eligible subalgebra $\mathfrak{k}$ just by setting $\mathfrak{k} := \mathfrak{k}' + \mathfrak{h}_{C(\mathfrak{k}')}$ where $\mathfrak{h}_{C(\mathfrak{k}')}$ is a Cartan subalgebra of $C(\mathfrak{k}')$. Finally, note that if x is any algebraic regular semisimple element of $C(\mathfrak{k}')$, then $\mathfrak{k} := \mathfrak{k}' \oplus Z(C(\mathfrak{k}')) + \mathbb{C}x$ is an eligible subalgebra of $\mathfrak{g}$. Indeed, if $\mathfrak{t}' \subseteq \mathfrak{k}'$ is a Cartan subalgebra of $\mathfrak{k}'$, and $\mathfrak{h}_{\mathfrak{k}} := \mathfrak{t}' \oplus Z(C(\mathfrak{k}')) + \mathbb{C}x$ is the corresponding Cartan subalgebra of $\mathfrak{k}$, then $C(\mathfrak{h}_{\mathfrak{k}})$ is a Cartan subalgebra of $\mathfrak{g}$. Hence,

$$C(\mathfrak{h}_{\mathfrak{k}}) = \mathfrak{h}_{\mathfrak{k}} + C(\mathfrak{k}) \tag{3}$$

as the right-hand side of (3) necessarily contains a Cartan subalgebra of $\mathfrak{g}$.

To any eligible subalgebra $\mathfrak{k}$ we assign a unique weight $\varkappa \in \mathfrak{h}^*$ (the "canonical weight associated with $\mathfrak{k}$"). It is defined by the conditions $\varkappa|_{(\mathfrak{h}\cap\mathfrak{k}_{ss})} = \rho$, $\varkappa|_{(\mathfrak{h}\cap C(\mathfrak{k}))} = 0$.

5 The Generalized Discrete Series

In what follows we assume that $\mathfrak{k}$ is eligible and $\mathfrak{h} \subset \tilde{\mathfrak{k}}$. In this case $\mathfrak{h}$ is a Cartan subalgebra both of $\tilde{\mathfrak{k}}$ and $\mathfrak{g}$. Let $\lambda \in \mathfrak{h}^*$ and set $\gamma := \lambda|_{\mathfrak{t}}$. Assume that $\mathfrak{m} := \mathfrak{m}_\gamma = C(\mathfrak{t})$. Assume furthermore that λ is $\mathfrak{m}$-integral and let E_λ be a simple finite-dimensional $\mathfrak{m}$-module with $\mathfrak{b}$-highest weight λ. Then

$$D(\mathfrak{k}, \lambda) := F^s(\mathfrak{k}, \mathfrak{p}_\gamma, E_\lambda \otimes \Lambda^{\dim \mathfrak{n}_\gamma}(\mathfrak{n}_\gamma^*))$$

is by definition a *generalized discrete series module.*

Note that since $D(\mathfrak{k}, \lambda)$ is a fundamental series module, Theorem 2.1 applies to $D(\mathfrak{k}, \lambda)$. In the case when $\mathfrak{k}$ is a root subalgebra and λ is regular, we have $\lambda = \gamma$ and $\mathfrak{p}_\gamma$ is a Borel subalgebra of $\mathfrak{g}$ which we denote by $\mathfrak{b}_\lambda$. Then $D(\mathfrak{k}, \lambda) = R^s \Gamma_{\mathfrak{k},\mathfrak{h}}(\Gamma_\mathfrak{h}(\mathrm{pro}^{\mathfrak{g}}_{\mathfrak{b}_\lambda} E_\lambda))$, i.e., $D(\mathfrak{k}, \lambda)$ is cohomologically co-induced from a 1-dimensional $\mathfrak{b}_\lambda$-module. If in addition, $\mathfrak{k}$ is a symmetric subalgebra, λ is $\mathfrak{k}$-integral, and $\lambda - \tilde{\rho}$ is $\mathfrak{b}_\lambda$-dominant regular, then $D(\mathfrak{k}, \lambda)$ is a $(\mathfrak{g}, \mathfrak{k})$-module in Harish-Chandra's discrete series, see [KV], Ch.XI.

Suppose $\mathfrak{k}$ is eligible but $\mathfrak{k}$ is not a root subalgebra. Suppose further that $\tilde{\mathfrak{k}}$ is symmetric. Any simple subquotient M of $D(\mathfrak{k}, \lambda)$ is a $(\mathfrak{g}, \tilde{\mathfrak{k}})$-module and thus a Harish-Chandra module for $(\mathfrak{g}, \tilde{\mathfrak{k}})$. However, M may or may not be in the discrete series of $(\mathfrak{g}, \tilde{\mathfrak{k}})$-modules. This becomes clear in Theorem 5.6 below.

Our first result is a sharper version of the main result of [PZ3] for an eligible $\mathfrak{k}$.

Theorem 5.1. *Let $\mathfrak{k} \subseteq \mathfrak{g}$ be eligible. Assume that $\lambda - 2\varkappa$ is $\tilde{\mathfrak{k}}$-integral and dominant. Then, $D(\mathfrak{k}, \lambda) \neq 0$. Moreover, if we set $\mu := (\lambda - 2\varkappa)|_\mathfrak{t}$, then $V(\mu)$ is the unique minimal $\mathfrak{k}$-type of $D(\mathfrak{k}, \lambda)$. Finally, there are isomorphisms of simple finite-dimensional $\tilde{\mathfrak{k}}$-modules*

$$D(\mathfrak{k}, \lambda)[\mu] \cong D(\mathfrak{k}, \lambda)\langle \lambda - 2\varkappa \rangle \simeq V_{\tilde{\mathfrak{k}}}(\lambda - 2\varkappa).$$

Proof. Note that $\mu = \gamma - 2\rho$. By Lemma 2 in [PZ3]

$$\dim \mathrm{Hom}_\mathfrak{k}(V(\mu), D(\mathfrak{k}, \lambda)) = \dim E_\lambda,$$

and hence $D(\mathfrak{k}, \lambda) \neq 0$. In addition, $V(\mu)$ is the unique minimal $\mathfrak{k}$-type of $D(\mathfrak{k}, \lambda)$. By construction, $D(\mathfrak{k}, \lambda)[\mu]$ is a finite-dimensional $\tilde{\mathfrak{k}}$-module. We will use Theorem 2.1 (c) to compute $D(\mathfrak{k}, \lambda)[\mu]$ as a $\tilde{\mathfrak{k}}$-module. Since $\mathfrak{k}$ is eligible, we have $\mathfrak{m} = \mathfrak{t} + C(\mathfrak{k})$. As $[\mathfrak{t}, C(\mathfrak{k})] = 0$ and $\mathfrak{t}$ is toral, the restriction of E_λ to $C(\mathfrak{k})$ is simple. We have

$$\tilde{\mathfrak{k}} = \mathfrak{k}_{ss} \oplus C(\mathfrak{k}),$$

and hence there is an isomorphism of $\tilde{\mathfrak{k}}$-modules

$$V_{\tilde{\mathfrak{k}}}(\lambda - 2\varkappa) \cong (V(\mu)|_{\mathfrak{k}_{ss}}) \boxtimes E_\lambda.$$

Consequently, we have isomorphisms of $C(\mathfrak{k})$-modules

$$\mathrm{Hom}_\mathfrak{k}(V(\mu), V_{\tilde{\mathfrak{k}}}(\lambda - 2\varkappa)) \cong \mathrm{Hom}_{\mathfrak{k}_{ss}}((V(\mu)|_{\mathfrak{k}_{ss}}), V_{\tilde{\mathfrak{k}}}(\lambda - 2\varkappa)) \cong E_\lambda. \quad (4)$$

Write $\mathfrak{p}_\gamma = \mathfrak{p}$ and note that $\tilde{\mathfrak{k}} \cap \mathfrak{m} = \mathfrak{m}$. By Theorem 2.1 (c), we have a canonical isomorphism

$$D(\mathfrak{k}, \lambda) \cong R^s \Gamma_{\tilde{\mathfrak{k}},\mathfrak{m}}(\Gamma_{\mathfrak{m},0}(\mathrm{pro}_{\mathfrak{p}}^{\mathfrak{g}} E_\lambda)).$$

According to the theory of the bottom layer [KV], Ch.V, Sec.6, $D(\mathfrak{k}, \lambda)$ contains the $\tilde{\mathfrak{k}}$-module

$$R^s \Gamma_{\tilde{\mathfrak{k}},\mathfrak{m}}(\Gamma_{\mathfrak{m},0}(\mathrm{pro}_{\tilde{\mathfrak{k}}\cap\mathfrak{p}}^{\tilde{\mathfrak{k}}} E_\lambda))$$

which is in turn isomorphic to $V_{\tilde{\mathfrak{k}}}(\lambda - 2\varkappa)$.

By the above argument, we have a sequence of injections

$$V_{\tilde{\mathfrak{k}}}(\lambda - 2\varkappa) \hookrightarrow D(\mathfrak{k}, \lambda)\langle\lambda - 2\varkappa\rangle \hookrightarrow D(\mathfrak{k}, \lambda)[\mu].$$

We conclude from (4) that the above sequence of injections is in fact a sequence of isomorphisms of simple $\tilde{\mathfrak{k}}$-modules. □

Corollary 5.2. *Under the assumptions of Theorem 5.1, there exists a simple* $(\mathfrak{g}, \mathfrak{k})$-*module M of finite type over $\mathfrak{k}$, such that if $V(\mu)$ is a minimal $\mathfrak{k}$-type of M, then $V(\mu)$ is the unique minimal $\mathfrak{k}$-type of M and there is an isomorphism of finite-dimensional $\tilde{\mathfrak{k}}$-modules*

$$M[\mu] \cong V_{\tilde{\mathfrak{k}}}(\lambda - 2\varkappa).$$

In particular, $M[\mu]$ is a simple $\tilde{\mathfrak{k}}$-submodule of M.

Proof. First we construct a module M as required. Let $\bar{D}(\mathfrak{k}, \lambda)$ be the $U(\mathfrak{g})$-submodule of $D(\mathfrak{k}, \lambda)$ generated by the $\tilde{\mathfrak{k}}$-isotypic component $D(\mathfrak{k}, \lambda)\langle\lambda - 2\varkappa\rangle$. Suppose N is a proper $\mathfrak{g}$-submodule of $\bar{D}(\mathfrak{k}, \lambda)$. Since $D(\mathfrak{k}, \lambda)\langle\lambda - 2\varkappa\rangle$ is simple over $\tilde{\mathfrak{k}}$,

$$N \cap (D(\mathfrak{k}, \lambda)\langle\lambda - 2\varkappa\rangle) = 0.$$

Thus, if $N(\mathfrak{k}, \lambda)$ is the maximum proper submodule of $\bar{D}(\mathfrak{k}, \lambda)$, the quotient module

$$M = \bar{D}(\mathfrak{k}, \lambda)/N(\mathfrak{k}, \lambda)$$

is a simple $(\mathfrak{g}, \tilde{\mathfrak{k}})$-module, and M has finite type over $\mathfrak{k}$. Theorem 5.1 implies now that $V(\mu)$ is the unique minimal $\mathfrak{k}$-type of M and that there is an isomorphism of finite-dimensional $\tilde{\mathfrak{k}}$-modules,

$$M[\mu] \cong V_{\tilde{\mathfrak{k}}}(\lambda - 2\varkappa).$$

□

If $\mathfrak{k}$ is symmetric (and hence $\tilde{\mathfrak{k}}$ is a root subalgebra due to the eligibility of $\mathfrak{k}$), Theorem 5.1 and Corollary 5.2 go back to [V] (where they are proven by a different method).

The following two statements are consequences of the main results of Sect. 2 and Theorem 5.1.

Corollary 5.3. *Let $\mathfrak{k}$ be eligible, $\lambda \in \mathfrak{h}^*$ be such that $\lambda - 2\varkappa$ is $\tilde{\mathfrak{k}}$-integral and $V(\mu)$ is generic for $\mu := \lambda|_{\mathfrak{t}} - 2\rho$.*

(a) $\mathrm{Soc}\, D(\mathfrak{k}, \lambda)$ is a simple $(\mathfrak{g}, \mathfrak{k})$-module with unique minimal $\mathfrak{k}$-type $V(\mu)$.
(b) There is a canonical isomorphism of $C(\mathfrak{k})$-modules

$$\mathrm{Hom}_{\mathfrak{k}}(V(\mu), \mathrm{Soc}\, D(\mathfrak{k}, \lambda)) \simeq E_\lambda.$$

(c) There is a canonical isomorphism of $\tilde{\mathfrak{k}}$-modules

$$V(\mu) \otimes \mathrm{Hom}_{\mathfrak{k}}(V(\mu), \mathrm{Soc}\, D(\mathfrak{k}, \lambda)) \simeq V_{\tilde{\mathfrak{k}}}(\lambda - 2\varkappa),$$

i.e. the $V(\mu)$-isotypic component of $\mathrm{Soc}D(\mathfrak{k}, \lambda)$ is a simple $\tilde{\mathfrak{k}}$-module isomorphic to $V_{\tilde{\mathfrak{k}}}(\lambda - 2\varkappa)$.
(d) If $\lambda - 2\varkappa$ is not $\mathfrak{l}$-integral for any reductive subalgebra $\mathfrak{l}$ such that $\tilde{\mathfrak{k}} \subset \mathfrak{l} \subseteq \mathfrak{g}$, then $\tilde{\mathfrak{k}}$ is a maximal reductive subalgebra of $\mathfrak{g}[M]$ for any subquotient M of $D(\mathfrak{k}, \lambda)$, in particular of $\mathrm{Soc}\, D(\mathfrak{k}, \lambda)$.

Proof. (a) Observe that $\mathfrak{p}_\gamma = \mathfrak{p}_{\mu+2\rho}$, and $D(\mathfrak{k}, \lambda) = F^s(\mathfrak{k}, \mathfrak{p}_{\mu+2\rho}, E_\lambda \otimes \Lambda^{\dim \mathfrak{n}}(\mathfrak{n}^*))$. So, (a) follows from Theorem 2.5 (c).
(b) By Theorem 2.5 (c), $\mathrm{Hom}_{\mathfrak{k}}(V(\mu), \mathrm{Soc}\, D(\mathfrak{k}, \lambda)) = \mathrm{Hom}_{\mathfrak{k}}(V(\mu), D(\mathfrak{k}, \lambda))$, which in turn is isomorphic to $\mathrm{Hom}_{\mathfrak{k}}(V(\mu), V_{\tilde{\mathfrak{k}}}(\lambda - 2\varkappa))$ by Theorem 5.1. The desired isomorphism follows now from (4).
(c) This follows from the isomorphism in (b) and the isomorphism $V(\mu) \otimes E_\lambda \cong V_{\tilde{\mathfrak{k}}}(\lambda - 2\varkappa)$ of $\tilde{\mathfrak{k}}$–modules.
(d) Follows from Corollary 2.4. Note that, since $\mathfrak{k}$ is eligible, $\tilde{\mathfrak{k}}$ is a root subalgebra and the condition that $\lambda - 2\varkappa$ be not $\mathfrak{l}$-integral involves only finitely many subalgebras $\mathfrak{l}$.

□

Corollary 5.4. *Let $\mathfrak{k}$ be eligible and let $V(\mu)$ be a generic $\mathfrak{k}$-type.*

(a) Let M be a simple $(\mathfrak{g}, \mathfrak{k})$-module of finite type with minimal $\mathfrak{k}$-type $V(\mu)$. Then $M[\mu]$ is a simple finite-dimensional $\tilde{\mathfrak{k}}$-module isomorphic to $V_{\tilde{\mathfrak{k}}}(\lambda)$ for some weight $\lambda \in \mathfrak{h}^*$ such that $\lambda|_{\mathfrak{t}} = \mu + 2\rho$ and $\mu - 2\varkappa$ is $\tilde{\mathfrak{k}}$-integral. Moreover,

$$M \cong \mathrm{Soc}\, D(\mathfrak{k}, \lambda).$$

If in addition λ is not $\mathfrak{l}$-integral for any reductive subalgebra $\mathfrak{l}$ with $\tilde{\mathfrak{k}} \subset \mathfrak{l} \subseteq \mathfrak{g}$, then $\tilde{\mathfrak{k}}$ is a unique maximal reductive subalgebra of $\mathfrak{g}[M]$.

(b) *If $\mathfrak{k}$ is regular in $\mathfrak{g}$, then (a) holds for any simple $(\mathfrak{g}, \mathfrak{k})$-module with generic minimal $\mathfrak{k}$-type $V(\mu)$. In particular M has finite type over $\mathfrak{k}$.*

Proof. (a) We apply Theorem 2.6. Since $V(\mu)$ is generic, $\mathfrak{p} = \mathfrak{p}_{\mu+2\rho} = \mathfrak{m} \ni \mathfrak{n}$ is a minimal $\mathfrak{t}$-compatible parabolic subalgebra. Let $\omega := \mu - 2\rho_{\mathfrak{n}}^{\perp}$ (recall that $\rho_{\mathfrak{n}}^{\perp} = \rho_{\mathfrak{n}} - \rho$) and let Q be the $\mathfrak{m}$-module $H^r(\mathfrak{n}, M)^{\omega}$ where $r = \dim(\mathfrak{k}^{\perp} \cap \mathfrak{n})$.
Observe that Q is a simple $\mathfrak{m}$-module and M is canonically isomorphic to $\bar{F}^s(\mathfrak{p}, Q) = \operatorname{Soc} F^s(\mathfrak{p}, Q)$. Let $\lambda \in \mathfrak{h}^*$ be so that $\lambda - 2\tilde{\rho}_{\mathfrak{n}}$ is an extreme weight of $\mathfrak{h}$ in Q. Thus, $F^s(\mathfrak{p}, Q) = F^s(\mathfrak{p}, E_\lambda \otimes \Lambda^{\dim \mathfrak{n}}(\mathfrak{n}^*)) = D(\mathfrak{k}, \lambda)$. Finally, $M \cong \operatorname{Soc} D(\mathfrak{k}, \lambda)$, and $\lambda|_{\mathfrak{t}} = \mu + 2\rho$. It follows that $\lambda - 2\varkappa$ is both $\mathfrak{k}$-integral and $C(\mathfrak{k})$-integral. Hence, the weight $\lambda - 2\varkappa$ is $\tilde{\mathfrak{k}}$-integral.
(b) We apply Theorem 2.8.

□

Corollary 5.5. *If $\mathfrak{k} \simeq \mathfrak{sl}(2)$, the genericity assumption on $V(\mu)$ in Corollaries 5.3 and 5.4 can be replaced by the assumption $\mu \geq \Lambda$.*

Proof. The statement follows directly from Theorem 3.1. □

We conclude this paper by discussing in more detail an example of an eligible $\mathfrak{sl}(2)$-subalgebra. Note first that if $\mathfrak{g}$ is any simple Lie algebra and $\mathfrak{k}$ is a long root $\mathfrak{sl}(2)$-subalgebra, then the pair $(\mathfrak{g}, \tilde{\mathfrak{k}})$ is a symmetric pair. This is a well-known fact and it implies in particular that any $(\mathfrak{g}, \mathfrak{k})$-module of finite type and of finite length is a Harish-Chandra module for the pair $(\mathfrak{g}, \tilde{\mathfrak{k}})$. The latter modules are classified under the assumption of simplicity, see [KV], Ch.XI; however, in general, it is an open problem to determine which simple $(\mathfrak{g}, \tilde{\mathfrak{k}})$-modules have finite type over $\mathfrak{k}$. Without having been explicitly stated, this problem has been discussed in the literature, see [GW, OW] and the references therein. On the other hand, in this case $\Lambda = 1$, hence Corollaries 5.4 and 5.5 provide a classification of simple $(\mathfrak{g}, \mathfrak{k})$-modules of finite type with minimal $\mathfrak{k}$-types $V(\mu)$ for $\mu \geq 1$. So the above problem reduces to matching the above two classifications in the case when $\mu \geq 1$, and finding all simple $(\mathfrak{g}, \mathfrak{k})$-modules of finite type whose minimal $\mathfrak{k}$-type equals $V(0)$ among the simple Harish-Chandra modules for the pair $(\mathfrak{g}, \tilde{\mathfrak{k}})$. We do this here in a special case.

Let $\mathfrak{g} = \mathfrak{sp}(2n+2)$ for $n \geq 2$. By assumption, $\mathfrak{k}$ is a long root $\mathfrak{sl}(2)$-subalgebra, and $\tilde{\mathfrak{k}} \simeq \mathfrak{sp}(2n) \oplus \mathfrak{k}$. Consider simple $(\mathfrak{g}, \tilde{\mathfrak{k}})$-modules with $Z_{U(\mathfrak{g})}$-character equal to the character of a trivial module. According to the Langlands classification, there are precisely $(n+1)^2$ pairwise non-isomorphic such modules, one of which is the trivial module. Following [Co] (see Figure 4.5 on page 93) we enumerate them as σ_t for $0 \leq t \leq n$ and σ_{ij} for $0 \leq i \leq n-1, 1 \leq j \leq 2n, i < j, i + j \leq 2n$. The modules σ_t are discrete series modules. The modules σ_{ij} are Langlands quotients of the principal series (all of them are proper quotients in this case).

We announce the following result which we intend to prove elsewhere.

Theorem 5.6. *Let $\mathfrak{g} = \mathfrak{sp}(2n+2)$ for $n \geq 2$ and $\mathfrak{k}$ be a long root $\mathfrak{sl}(2)$-subalgebra.*

(*a*) *Any simple* $(\mathfrak{g}, \mathfrak{k})$*-module of finite type is isomorphic to a subquotient of the generalized discrete series module* $D(\mathfrak{k}, \lambda)$ *for some* $\tilde{\mathfrak{k}} = \mathfrak{sp}(2n) \oplus \mathfrak{k}$*-integral weight* $\lambda - 2\varkappa$.

(*b*) *The modules* σ_0, σ_{0i} *for* $i = 1, \dots, 2n, \sigma_{12}$ *are, up to isomorphism, all of the simple* $(\mathfrak{g}, \mathfrak{k})$*-modules of finite type whose* $Z_{U(\mathfrak{g})}$*-character equals that of a trivial* $\mathfrak{g}$*-module. Moreover, their minimal* $\mathfrak{k}$*-types are as follows:*

Module	***Minimal*** $\mathfrak{k}$***−type***
σ_0	$V(2n)$
$\sigma_{0j}, n+1 \leq j \leq 2n$	$V(j-1)$
$\sigma_{0j}, 2 \leq j \leq n$	$V(j-2)$
σ_{01} *(trivial representation)*	$V(0)$
σ_{12}	$V(0)$

Acknowledgements I. Penkov thanks Yale University for its hospitality and partial financial support during the spring of 2012 when this paper was conceived, as well as the Max Planck Institute for Mathematics in Bonn where the work on the paper was continued. G. Zuckerman thanks Jacobs University for its hospitality.

References

[Bou] N. Bourbaki, *Groupes et Algèbres de Lie*, Ch.VI, Hermann, Paris, 1975.
[Co] D. H. Collingwood, *Representations of rank one Lie groups*, Pitman, Boston, 1985.
[Dix] J. Dixmier, *Enveloping algebras*, North Holland, Amsterdam, 1977.
[D] E. Dynkin, Semisimple subalgebras of semisimple Lie algebras, *Mat. Sbornik* (N.S.) **30** (72) (1952), 349–462 (Russian); English: Amer. Math. Soc. Transl. **6** (1957), 111–244.
[F] S. Fernando, *Lie algebra modules with finite-dimensional weight spaces I*, Transactions AMS Soc **332** (1990), 757–781.
[GW] B. Gross, N. Wallach, On quaternionic discrete series represenations and their continuations, *Journal reine angew. Math.* **481** (1996), 73–123.
[K] V. Kac, Constructing groups associated to infinite-dimensional Lie algebras, in: *Infinite-dimensional groups with applications* (Berkeley, 1984), Math. Sci. Res. Inst. Publ. 4, pp. 167–216.
[KV] A. Knapp, D. Vogan, *Cohomological Induction and Unitary Representations*, Princeton Mathematical Series, Princeton University Press, 1995.
[OW] B. Orsted, J.A. Wolf, Geometry of the Borel-de Siebenthal discrete series, *Journal of Lie Theory* **20** (2010), 175–212.
[PS1] I. Penkov, V. Serganova, Generalized Harish-Chandra modules, Moscow Math. *Journal* **2** (2002), 753–767.
[PS2] I. Penkov, V. Serganova, Bounded simple $(\mathfrak{g}, \mathfrak{sl}(2))$-modules for $\mathrm{rk}\mathfrak{g} = 2$, *Journal of Lie Theory* **20** (2010), 581–615.
[PS3] I. Penkov, V. Serganova, On bounded generalized Harish-Chandra modules, *annales de l'Institut Fourier* **62** (2012), 477–496.
[PSZ] I. Penkov, V. Serganova, G. Zuckerman, On the existence of $(\mathfrak{g}, \mathfrak{k})-$modules of finite type, Duke Math. *Journal* **125** (2004), 329–349.
[PZ1] I. Penkov, G. Zuckerman, Generalized Harish-Chandra modules: a new direction of the structure theory of representations, *Acta Applicandae Mathematicae* **81** (2004), 311–326.

[PZ2] I. Penkov, G. Zuckerman, Generalized Harish-Chandra modules with generic minimal $\mathfrak{k}$-type, *Asian Journal of Mathematics* **8** (2004), 795–812.
[PZ3] I. Penkov, G. Zuckerman, A construction of generalized Harish-Chandra modules with arbitrary minimal $\mathfrak{k}-$type, *Canad. Math. Bull.* **50** (2007), 603–609.

[PZ4] I. Penkov, G. Zuckerman, A construction of generalized Harish-Chandra modules for locally reductive Lie algebras, *Transformation Groups* **13** (2008), 799–817.
[PZ5] I. Penkov, G. Zuckerman, On the structure of the fundamental series of generalized Harish-Chandra modules, *Asian Journal of Mathematics* **16** (2012), 489–514.
[Pe] A. V. Petukhov, Bounded reductive subalgebras of $\mathfrak{sl}(n)$, *Transformation Groups* **16** (2011), 1173–1182.
[V] D. Vogan, The algebraic structure of the representations of semisimple Lie groups I, *Ann. of Math.* **109** (1979), 1–60.
[WZ] J. Willenbring, G. Zuckerman, Small semisimple subalgebras of semisimple Lie algebras, in: *Harmonic analysis, group representations, automorphic forms and invariant theory*, Lecture Notes Series, Institute Mathematical Sciences, National University Singapore, **12**, World Scientific, 2007, pp. 403–429.
[Z] G. Zuckerman, Generalized Harish-Chandra modules, in: *Highlights of Lie algebraic methods*, Progress in Mathematics 295, Birkhauser, 2012, pp.123–143.

On Exceptional Vertex Operator (Super) Algebras

Michael P. Tuite and Hoang Dinh Van

Abstract We consider exceptional vertex operator algebras and vertex operator superalgebras with the property that particular Casimir vectors constructed from the primary vectors of lowest conformal weight are Virasoro descendents of the vacuum. We show that the genus one partition function and characters for simple ordinary modules must satisfy modular linear differential equations. We show the rationality of the central charge and lowest weights of modules, modularity of solutions, the dimension of each graded space is a rational function of the central charge and that the lowest weight primaries generate the algebra. We also discuss conditions on the reducibility of the lowest weight primary vectors as a module for the automorphism group. Finally we analyse solutions for exceptional vertex operator algebras with primary vectors of lowest weight up to 9 and for vertex operator superalgebras with primary vectors of lowest weight up to 17/2. Most solutions can be identified with simple ordinary modules for known algebras but there are also four conjectured algebras generated by weight two primaries and three conjectured extremal vertex operator algebras generated by primaries of weight 3, 4 and 6, respectively.

Key words Vertex operator algebras • Vertex operator super algebras • Virasoro algebra • Group theory

Mathematics Subject Classification (2010): Primary 17B69, 17B68, 17B25, 17B67. Secondary 20E32.

---

*Research of Van was supported by Science Foundation Ireland Frontiers of Research Programme.

M.P. Tuite (✉) • H.D. Van
School of Mathematics, Statistics and Applied Mathematics, National University of Ireland Galway, University Road, Galway, Ireland
e-mail: michael.tuite@nuigalway.ie; vanhoangk15@gmail.com

G. Mason et al. (eds.), *Developments and Retrospectives in Lie Theory: Algebraic Methods*, Developments in Mathematics 38,
DOI 10.1007/978-3-319-09804-3_16

1 Introduction

Vertex Operator Algebras (VOAs) and Super Algebras (VOSAs) have deep connections to Lie algebras, number theory, group theory, combinatorics and Riemann surfaces (e.g., [FHL, FLM, Kac1, MN, MT]) and, of course, conformal field theory e.g., [DMS]. The classification of VOAs and VOSAs still seems to be a very difficult task, for example, there is no proof of the uniqueness of the Moonshine module [FLM]. Nevertheless, it would be very useful to be able to characterize VOA/VOSAs with interesting properties such as large automorphism groups (e.g., the Monster group for the Moonshine module), rational characters, generating vectors, etc. In [Mat], Matsuo introduced VOAs of class $\mathcal{S}^n$ with the defining property that the Virasoro vacuum descendents are the only Aut(V)-invariant vectors of weight $k \leq n$. Thus the Moonshine module [FLM] is of class $\mathcal{S}^{11}$, the Baby Monster VOA [Ho1] of class $\mathcal{S}^6$ and the level one Kac–Moody VOAs generated by Deligne's Exceptional Lie algebras $A_1, A_2, G_2, D_4, F_4, E_6, E_7, E_8$ [D] are of class $\mathcal{S}^4$.[1]

In this paper we consider a refinement and generalization of previous results in [T1, T2] concerning such exceptional VOAs. Assuming the VOA is simple and of strong CFT-type (e.g., [MT]) we consider quadratic Casimir vectors $\lambda^{(k)}$ of conformal weight $k = 0, 1, 2, \ldots$ constructed from the primary vectors of lowest conformal weight $l \in \mathbb{N}$. We say that a VOA is exceptional of lowest primary weight l if $\lambda^{(2l+2)}$ is a Virasoro vacuum descendent. Every VOA of class $\mathcal{S}^{2l+2}$ with lowest primary weight l is exceptional, but the converse is not known to be true. We show, using Zhu's theory for genus one correlation functions [Z], that for an Exceptional VOA of lowest primary weight l, the partition function and the characters for simple ordinary VOA modules satisfy a Modular Linear Differential Equation (MLDE) of order at most $l+1$. Given that order of the MLDE is exactly $l+1$ (which is verified for all $l \leq 9$) we show that the central charge c and module lowest weights h are rational, the MLDE solution space is modular invariant and the dimension of each VOA graded space is a rational function of c. Subject to a further indicial root condition (again verified for all $l \leq 9$) we show that an Exceptional VOA is generated by its primary vectors of lowest weight l.

We also consider other properties that arise from genus zero correlation functions for all l. Assuming the VOA is of class $\mathcal{S}^{2l+2}$ this leads to conditions on the reducibility of the lowest weight l primary space as a module for the VOA automorphism group.

A similar analysis is carried out for Exceptional VOSAs of lowest primary weight $l \in \mathbb{N} + \frac{1}{2}$ for which $\lambda^{(2l+1)}$ is a Virasoro vacuum descendent. Using a twisted version of Zhu theory [MTZ] we obtain a twisted MLDE of order at most $l+\frac{1}{2}$ which is satisfied by the partition function and simple ordinary VOA module characters. This differential equation leads to a similar set of general results to those for VOAs. Likewise, we can consider genus zero correlation functions for all $l \in \mathbb{N}+\frac{1}{2}$ leading

[1] In fact, the A_1 theory is of class $\mathcal{S}^\infty$ and the E_8 theory is of class $\mathcal{S}^6$.

to conditions on the reducibility of the space of the space of weight l primaries as a module for the VOSA automorphism group.

The paper also summarizes rational c, h solutions to the MLDE for all $l \leq 9$ and the twisted MLDE for all $l \leq \frac{17}{2}$. In most cases we can identify a VOA/VOSA with the requisite properties. These include a number of special VOA/VOSAs, some VOSAs obtained by commutant constructions, some simple current extensions of Virasoro minimal models and $\mathcal{W}$-algebras. We also present evidence for four candidate/conjectured VOAs with simple Griess algebras for $l = 2$ and three extremal VOAs for $l = 3, 4, 6$. All the VOSA solutions found can be identified with known theories.

2 Vertex Operator (Super) Algebras

We review some aspects of vertex operator super algebra theory (e.g., [FHL, FLM, Kac1, MN, MT]). A Vertex Operator Superalgebra (VOSA) is a quadruple $(V, Y(\cdot,\cdot), \mathbf{1}, \omega)$ with a $\mathbb{Z}_2$-graded vector space $V = V_{\bar{0}} \oplus V_{\bar{1}}$ with parity $p(u) = 0$ or 1 for $u \in V_{\bar{0}}$ or $V_{\bar{1}}$ respectively. $(V, Y(\cdot,\cdot), \mathbf{1}, \omega)$ is called a Vertex Operator Algebra (VOA) when $V_{\bar{1}} = 0$.

V also has a $\frac{1}{2}\mathbb{Z}$-grading with $V = \bigoplus_{r \in \frac{1}{2}\mathbb{Z}} V_r$ with $\dim V_r < \infty$. $\mathbf{1} \in V_0$ is the vacuum vector and $\omega \in V_2$ is called the conformal vector. Y is a linear map $Y : V \to \mathrm{End}(V)[[z, z^{-1}]]$ for formal variable z giving a vertex operator

$$Y(u, z) = \sum_{n \in \mathbb{Z}} u(n) z^{-n-1}, \tag{1}$$

for every $u \in V$. The linear operators (modes) $u(n) : V \to V$ satisfy creativity

$$Y(u, z)\mathbf{1} = u + O(z), \tag{2}$$

and lower truncation

$$u(n)v = 0, \tag{3}$$

for each $u, v \in V$ and $n \gg 0$. For the conformal vector ω

$$Y(\omega, z) = \sum_{n \in \mathbb{Z}} L(n) z^{-n-2}, \tag{4}$$

where $L(n)$ satisfies the Virasoro algebra for some central charge c

$$[L(m), L(n)] = (m - n)L(m + n) + \frac{c}{12}(m^3 - m)\delta_{m,-n}\, \mathrm{id}_V\,. \tag{5}$$

Each vertex operator satisfies the translation property

$$Y(L(-1)u, z) = \partial_z Y(u, z). \tag{6}$$

The Virasoro operator $L(0)$ provides the $\frac{1}{2}\mathbb{Z}$-grading with $L(0)u = \mathrm{wt}(u)u$ for $u \in V_r$ and with weight $\mathrm{wt}(u) = r \in \mathbb{Z} + \frac{1}{2}p(u)$. Finally, the vertex operators satisfy the Jacobi identity

$$z_0^{-1}\delta\left(\tfrac{z_1-z_2}{z_0}\right) Y(u, z_1)Y(v, z_2) - (-1)^{p(u)p(v)} z_0^{-1}\delta\left(\tfrac{z_2-z_1}{-z_0}\right) Y(v, z_2)Y(u, z_1)$$
$$= z_2^{-1}\delta\left(\tfrac{z_1-z_0}{z_2}\right) Y\left(Y(u, z_0)v, z_2\right),$$

with $\delta\left(\frac{x}{y}\right) = \sum_{r\in\mathbb{Z}} x^r y^{-r}$.

These axioms imply $u(n)V_r \subset V_{r-n+wt(u)-1}$ for u of weight $\mathrm{wt}(u)$. They also imply locality, skew-symmetry, associativity and commutativity:

$$(z_1 - z_2)^N Y(u, z_1)Y(v, z_2) = (-1)^{p(u)p(v)}(z_1 - z_2)^N Y(v, z_2)Y(u, z_1), \tag{7}$$

$$Y(u, z)v = (-1)^{p(u)p(v)} e^{zL(-1)} Y(v, -z)u, \tag{8}$$

$$(z_0 + z_2)^N Y(u, z_0 + z_2)Y(v, z_2)w = (z_0 + z_2)^N Y(Y(u, z_0)v, z_2)w, \tag{9}$$

$$u(k)Y(v, z) - (-1)^{p(u)p(v)} Y(v, z)u(k) = \sum_{j\geq 0} \binom{k}{j} Y(u(j)v, z)z^{k-j}, \tag{10}$$

for $u, v, w \in V$ and integers $N \gg 0$ [FHL, Kac1, MT].

We define an invariant symmetric bilinear form $\langle\ ,\ \rangle$ on V by

$$\left\langle Y\left(e^{zL(1)}\left(-z^{-2}\right)^{L(0)} w, z^{-1}\right)u, v\right\rangle = (-1)^{p(u)p(w)}\langle v, Y(w, z)v\rangle, \tag{11}$$

for all $u, v, w \in V$ [FHL]. V is said to be of *CFT-type* if $V_0 = \mathbb{C}\mathbf{1}$ and of *strong CFT-type* if additionally $L(1)V_1 = 0$ in which case $\langle\ ,\ \rangle$, with normalization $\langle \mathbf{1}, \mathbf{1}\rangle = 1$, is unique [Li]. Furthermore, $\langle\ ,\ \rangle$ is invertible if V is simple. All VOSAs in this paper are assumed to be of this type.

Every VOSA contains a subVOA,[2] which we denote by $V_{\langle\omega\rangle}$, generated by the Virasoro vector ω with Fock basis of vacuum descendents of the form

$$L(-n_1)L(-n_2)\ldots L(-n_k)\,\mathbf{1}, \tag{12}$$

[2]This subVOA is often denoted by M_c e.g., [FZ].

for $n_i \geq 2$. $\langle\ ,\ \rangle$ is singular on $\left(V_{\langle\omega\rangle}\right)_n$ iff the central charge is

$$c_{p,q} = 1 - 6\frac{(p-q)^2}{pq}, \tag{13}$$

for coprime integers $p, q \geq 2$ and $n \geq (p-1)(q-1)$ [Wa]. The Virasoro minimal model VOA $L(c_{p,q}, 0)$ is the quotient of $V_{\langle\omega\rangle}$ by the radical of $\langle\ ,\ \rangle$. $L(c_{p,q}, 0)$ has a finite number of simple ordinary V-modules $L(c_{p,q}, h_{r,s}) \cong L(c_{p,q}, h_{q-r,p-s})$ (e.g., [DMS]) with lowest weight

$$h_{r,s} = \frac{(pr-qs)^2 - (p-q)^2}{4pq}, \tag{14}$$

for $r = 1, \ldots, q-1$ and $s = 1, \ldots, p-1$.

3 Quadratic Casimirs and Genus One Zhu Theory

3.1 Quadratic Casimirs

Let $(V, Y(\cdot,\cdot), \mathbf{1}, \omega)$ be a simple VOA of strong CFT-type with unique invertible bilinear form $\langle\ ,\ \rangle$. Let Π_l denote the space of primary vectors of lowest weight $l \geq 1$, i.e., $L(n)u = 0$ for all $n > 0$ for $u \in \Pi_l$. Choose a Π_l-basis $\{u_i\}$ for $i = 1, \ldots, p_l = \dim \Pi_l$ with dual basis $\{\overline{u}_i\}$, i.e., $\langle u_i, \overline{u}_j\rangle = \delta_{ij}$. Define quadratic Casimir vectors $\lambda^{(n)}$ for $n \geq 0$ by [Mat, T1, T2]

$$\lambda^{(n)} = \sum_{i=1}^{p_l} u_i(2l-n-1)\overline{u}_i \in V_n. \tag{15}$$

In particular we find

$$\lambda^{(0)} = \sum_{i=1}^{p_l} u_i(2l-1)\overline{u}_i = (-1)^l \sum_{i=1}^{p_l} \langle u_i, \overline{u}_i\rangle\, \mathbf{1} = (-1)^l\, p_l\, \mathbf{1}.$$

Furthermore, if $l > 1$, then $\dim V_1 = 0$ and hence $\lambda^{(1)} = 0$, whereas for $l = 1$ the Jacobi identity implies $\lambda^{(1)} = \sum_{i=1}^{p_l} u_i(0)\overline{u}_i = -\sum_{i=1}^{p_l} \overline{u}_i(0)u_i = 0$ [T1]. Thus we find

Lemma 1. *$\lambda^{(0)} = (-1)^l\, p_l\, \mathbf{1}$ and $\lambda^{(1)} = 0$.*

Since the Π_l elements are primary then for all $m > 0$

$$L(m)\lambda^{(n)} = (n - m + l(m-1))\,\lambda^{(n-m)}. \tag{16}$$

Suppose that $\lambda^{(n)} \in V_{\langle\omega\rangle}$, then (16) implies that $\lambda^{(m)} \in V_{\langle\omega\rangle}$ for all $m \leq n$. Furthermore, since $\langle\, ,\, \rangle$ is invertible we have the following lemma

Lemma 2 (Matsuo [Mat]). *If* $\lambda^{(n)} \in V_{\langle\omega\rangle}$, *then* $\lambda^{(n)}$ *is uniquely determined.*

Thus if $\lambda^{(2)} \in V_{\langle\omega\rangle}$, then $\lambda^{(2)} = \kappa L(-2)\,\mathbf{1}$ for some κ so that $\langle L(-2)\,\mathbf{1}, \lambda^{(2)}\rangle = \kappa\frac{c}{2}$. But (11) and (16) imply $\langle L(-2)\,\mathbf{1}, \lambda^{(2)}\rangle = \langle \mathbf{1}, L(2)\lambda^{(2)}\rangle = (-1)^l\, p_l$, so that for $c \neq c_{2,3} = 0$ (cf. (13))

$$\lambda^{(2)} = p_l \frac{2\,(-1)^l\, l}{c} L(-2)\,\mathbf{1}\,. \tag{17}$$

Similarly, if $\lambda^{(4)} \in V_{\langle\omega\rangle}$ and $c \neq 0$ or $c \neq c_{2,5} = -22/5$, then [Mat, T1, T2]

$$\lambda^{(4)} = p_l \frac{2\,(-1)^l\, l\,(5l+1)}{c\,(5c+22)} L(-2)^2\,\mathbf{1} + p_l \frac{3\,(-1)^l\, l\,(c-2l+4)}{c\,(5c+22)} L(-4)\,\mathbf{1}\,. \tag{18}$$

These examples illustrate a general observation:

Lemma 3. *Each coefficient in the expansion of* $\lambda^{(n)} \in V_{\langle\omega\rangle}$ *in a basis of Virasoro Fock vectors is of the form* $p_l\, r(c)$ *for some rational function* $r(c)$ *of* c.

3.2 Genus One Constraints from Quadratic Casimirs

Define genus one partition and 1-point correlation functions for $u \in V$ by

$$Z_V(q) = \mathrm{Tr}_V\left(q^{L(0)-c/24}\right) = q^{-c/24}\sum_{n\geq 0} \dim V_n\, q^n, \tag{19}$$

$$Z_V(u,q) = \mathrm{Tr}_V\left(o(u) q^{L(0)-c/24}\right), \tag{20}$$

where q is a formal parameter and $o(u) = u(\mathrm{wt}\,(u)-1) : V_n \to V_n$ is the 'zero mode' for homogeneous u. By replacing V by a simple ordinary V-module N (on which $L(0)$ acts semisimply e.g., [FHL, MT]) these definitions may be extended to graded characters $Z_N(q)$ and 1-point functions $Z_N(u,q)$, e.g.,

$$Z_N(q) = \mathrm{Tr}_N\left(q^{L(0)-c/24}\right) = q^{h-c/24}\sum_{n\geq 0} \dim N_n\, q^n, \tag{21}$$

where h denotes the lowest weight of N. Zhu also introduced an isomorphic VOA $(V, Y[\cdot,\cdot], \mathbf{1}, \widetilde{\omega})$ with 'square bracket' vertex operators

$$Y[u,z] \equiv Y\left(e^{zL(0)}u, e^z - 1\right) = \sum_{n\in\mathbb{Z}} u[n] z^{-n-1}, \tag{22}$$

for Virasoro vector $\widetilde{\omega} = \omega - c/24\,\mathbf{1}$ with modes $\{L[n]\}$. $L[0]$ defines an alternative $\mathbb{Z}$ grading with $V = \bigoplus_{k\geq 0} V_{[k]}$ where $L[0]v = \mathrm{wt}[v]v$ for $\mathrm{wt}[v] = k$ for $v \in V_{[k]}$. Zhu obtained a reduction formula for the 2-point correlation function $Z_N(Y[u,z]v,q)$ for $u, v \in V$ in terms of the elliptic Weierstrass function

$$P_m(z) = \frac{1}{z^m} + (-1)^m \sum_{n\geq m} \binom{n-1}{m-1} E_n(q)\, z^{n-m}, \tag{23}$$

for $m \geq 1$ and with Eisenstein series $E_n(q) = 0$ for odd n and

$$E_n(q) = -\frac{B_n}{n!} + \frac{2}{(n-1)!} \sum_{k\geq 1} \frac{k^{n-1}q^k}{1-q^k}, \tag{24}$$

for even n with B_n the nth Bernoulli number. $P_m(z)$ converges absolutely and uniformly on compact subsets of the domain $|q| < |e^z| < 1$. $E_n(q)$ is a modular form of weight n for $n \geq 4$ and $E_2(q)$ is a quasi-modular form of weight 2, i.e., letting $q = \exp(2\pi i\tau)$ for $\tau \in \mathbb{H}_1$

$$E_n\left(\frac{\alpha\tau+\beta}{\gamma\tau+\delta}\right) = (\gamma\tau+\delta)^n E_n(\tau) - \frac{\gamma(\gamma\tau+\delta)}{2\pi i}\delta_{n2}, \tag{25}$$

for $\left(\begin{smallmatrix}\alpha & \beta\\ \gamma & \delta\end{smallmatrix}\right) \in \mathrm{SL}(2,\mathbb{Z})$ [Se]. We then have

Proposition 1 (Zhu [Z]). *Let N be a simple ordinary V-module.*

$$Z_N(Y[u,z]v,q) = \mathrm{Tr}_N\left(o(u)o(v)q^{L(0)-c/24}\right) + \sum_{m\geq 0} P_{m+1}(z) Z_N(u[m]v,q).$$

Taking $u = \widetilde{\omega}$ and noting that $o(\widetilde{\omega}) = L(0) - c/24$ we obtain:

Corollary 1. *The 1-point function of a Virasoro descendent $L[-k]v$ is*

$$Z_N(L[-k]v,q) = (-1)^k \sum_{r\geq 0} \binom{k+r-1}{k-2} E_{k+r}(q) Z_N(L[r]v,q),$$

for all $k \geq 3$, whereas for $k = 2$ we have

$$Z_N(L[-2]v,q) = \left(q\frac{\partial}{\partial q} + \mathrm{wt}[v]E_2(q)\right) Z_N(v,q) + \sum_{s\geq 1} E_{2s+2}(q) Z_N(L[2s]v,q).$$

Let us now consider a simple VOA V of strong CFT-type with lowest weight $l \geq 1$ Virasoro primary vectors Π_l so that

$$Z_V(q) = Z_{V_{\langle\omega\rangle}}(q) + O\left(q^{l+c/24}\right). \tag{26}$$

Let $\{u_i\}$ and $\{\overline{u}_i\}$ be a basis and dual basis for Π_l. Apply Proposition 1 to

$$Z_N\left(\sum_{i=1}^{p_l} Y[u_i, z]\overline{u}_i, q\right) = \sum_{n\geq 0} Z_N\left(\lambda^{[n]}, q\right) z^{n-2l}, \tag{27}$$

(for Casimir vector $\lambda^{[n]} \in V_{[n]}$ in square bracket modes) to find

$$\sum_{n\geq 0} Z_N\left(\lambda^{[n]}, q\right) z^{n-2l} = \mathrm{Tr}_N\left(\sum_{i=1}^{p_l} o(u_i)o(\overline{u}_i)q^{L(0)-c/24}\right) + \sum_{m=0}^{2l-1} P_{m+1}(z)Z_N\left(\lambda^{[2l-m-1]}, q\right). \tag{28}$$

Comparing the coefficients of z^{n-2l} for $n \geq 2l$ on both sides of this equality leads to a recursive identity between $Z_N\left(\lambda^{[n]}, q\right)$ and $Z_N\left(\lambda^{[m]}, q\right)$ for $m \leq n-2$. In particular, comparing the coefficients of z^2 we find

Proposition 2. *$Z_N\left(\lambda^{[2l+2]}, q\right)$ satisfies the recursive identity*

$$Z_N\left(\lambda^{[2l+2]}, q\right) = \sum_{r=0}^{l-1} \binom{2l-2k+1}{2} E_{2l-2k+2}(q)Z_N\left(\lambda^{[2k]}, q\right). \tag{29}$$

4 Exceptional VOAs

Consider a simple VOA of strong CFT-type with primary vectors of lowest weight $l \geq 1$ for which $\lambda^{(2l+2)} \in V_{\langle\omega\rangle}$ (or equivalently, $\lambda^{[2l+2]} \in V_{\langle\tilde{\omega}\rangle}$). We also assume that $\left(V_{\langle\omega\rangle}\right)_{2l+2}$ contains no Virasoro singular vector, i.e., $c \neq c_{p,q}$ for $(p-1)(q-1) \leq 2l+2$. We call such a VOA an *Exceptional VOA of lowest primary weight l*. (16) implies $\lambda^{(2k)} \in V_{\langle\omega\rangle}$ and $\lambda^{[2k]} \in V_{\langle\tilde{\omega}\rangle}$ for all $k \leq l$.

Proposition 3. *Let $\lambda^{[2k]} \in V_{\langle\tilde{\omega}\rangle}$. Then for a simple ordinary V-module N*

$$Z_N\left(\lambda^{[2k]}, q\right) = \sum_{m=0}^{k} f_{k-m}(q, c)D^m Z_N(q); \tag{30}$$

where D is the Serre modular derivative defined for $m \geq 0$ by

$$D^{m+1} Z_N(q) = \left(q\frac{\partial}{\partial q} + 2mE_2(q)\right) D^m Z_N(q); \tag{31}$$

$f_m(q,c)$ is a modular form of weight $2m$ whose coefficients over the ring of Eisenstein series are of the form $p_l\, r(c)$ for a rational function $r(c)$.

Proof. Equation (30) follows from Corollary 1 by induction in the number of Virasoro modes where the $D^k Z_N(q)$ term arises from a $L[-2]^k\mathbf{1}$ component in $\lambda^{[2k]}$. The coefficients of $f_m(q,c)$ over the ring of Eisenstein series are of the form $p_l\, r(c)$ for a rational function $r(c)$ from Lemma 3. □

Applying Proposition 3 to the recursive identity (29) implies $Z_N(q)$ satisfies a Modular Linear Differential Equation (MLDE) [Mas1].

Proposition 4. *Let V be an Exceptional VOA of lowest primary weight l. $Z_N(q)$ for each simple ordinary V–module N satisfies a MLDE of order $\le l+1$*

$$\sum_{m=0}^{l+1} g_{l+1-m}(q,c)D^m Z(q) = 0, \tag{32}$$

where $g_m(q,c)$ is a modular form of weight $2m$ whose coefficients over the ring of Eisenstein series are rational functions of c.

Remark 1. Proposition 4 states that each simple ordinary V-module N character $Z_N(q)$ satisfies the same MLDE (32). However, the MLDE may also have further solutions unrelated to module characters.

$g_0(q,c) = g_0(c)$ is independent of q since it is a modular form of weight 0. For $g_0(c) \neq 0$, the MLDE (32) is of order $l+1$ with a regular singular point at $q=0$ so that Frobenius–Fuchs theory concerning the $l+1$-dimensional solution space $\mathcal{F}$ applies, e.g., [Hi, I]. Any solution $Z(q) \in \mathcal{F}$ is holomorphic in q for $0 < |q| < 1$ since the MLDE coefficients $g_m(q,c)$ are holomorphic for $|q| < 1$. We may thus view each solution as a function of $\tau \in \mathbb{H}_1$ for $q = e^{2\pi i\tau}$.

Using the quasi-modularity of $E_2(\tau)$ and (31) with $q\frac{\partial}{\partial q} = \frac{1}{2\pi i}\frac{\partial}{\partial \tau}$, it follows that for all $\left(\begin{smallmatrix}\alpha & \beta\\ \gamma & \delta\end{smallmatrix}\right) \in \mathrm{SL}(2,\mathbb{Z})$, $Z\left(\frac{\alpha\tau+\beta}{\gamma\tau+\delta}\right)$ is also a solution of the MLDE since $g_{l+1-m}(q,c)$ is a modular form of weight $2l+2-2m$. Thus $T : \tau \to \tau+1$ has a natural action on $\mathcal{F} = \mathcal{F}_1 \oplus \ldots \oplus \mathcal{F}_r$ for distinct eigenspaces $\mathcal{F}_i$ with T monodromy eigenvalue $e^{2\pi i x_i}$ where x_i is a root of the indicial polynomial

$$\sum_{m=0}^{l+1} g_{l+1-m}(0,c)\prod_{s=0}^{m-1}\left(x - \frac{1}{6}s\right) = 0. \tag{33}$$

If $x_1 = x_2 \mod \mathbb{Z}$, for roots x_1, x_2, they determine the same monodromy eigenvalue. Let $\widehat{x_i}$ denote an indicial root with least real part for a given monodromy eigenvalue $e^{2\pi i x_i}$. Then $\mathcal{F}_i$ has a basis of the form [17, Hi]

$$f_i^n(\tau) = \phi_i^1(q) + \tau\phi_i^2(q) + \ldots + \tau^{n-1}\phi_i^n(q), \tag{34}$$

for $n = 1, \ldots, \dim \mathcal{F}_i$ and where each $\phi_i^n(q)$ is a q-series

$$\phi_i^n(q) = q^{\widehat{x}_i} \sum_{k \geq 0} a_{ik}^n q^k,$$

which is holomorphic on $0 < |q| < 1$. The solutions $f_i^n(\tau)$ for $n \geq 2$, which are referred to as logarithmic solutions (since they contain nonnegative integer powers of $\log q = 2\pi i\tau$), occur if the same indicial root occurs multiple times or, possibly, if two roots differ by an integer. However, every graded character $Z_N(q)$ for a simple ordinary module with lowest weight h has a pure q-series with indicial root $x = h - c/24$ from (21).

We now sketch a proof that the central charge c is rational following [AM] (which is extended to logarithmic solutions (34) in [Miy]). Suppose $c \notin \mathbb{Q}$ and consider $\phi \in \mathrm{Aut}(\mathbb{C})$ such that $\tilde{c} = \phi(c) \neq c$. Then $Z_V(\tau, \tilde{c})$ is a solution to the MLDE, found by replacing c by $\tilde{c}$ in (32). But since the coefficients in the q-expansion of $Z_V(\tau, c)$ are integral we have

$$Z_V(\tau, \tilde{c}) = q^{(\tilde{c}-c)/24} Z_V(\tau, c).$$

Applying the modular transformation $S : \tau \to -1/\tau$ we find

$$Z_V\left(-\frac{1}{\tau}, \tilde{c}\right) = \exp\left(-\frac{\pi i(\tilde{c} - c)}{12\tau}\right) Z_V\left(-\frac{1}{\tau}, c\right). \tag{35}$$

But $Z_V(-1/\tau, c)$ satisfies (32) and $Z_V(-1/\tau, \tilde{c})$ satisfies (32) with c replaced by $\tilde{c}$ and thus both are expressed in terms of the basis (34). Analysing (35) along rays $\tau = re^{i\theta}$ in the limit $r \to \infty$ with $0 < \theta < \pi$, a contradiction results unless $\tilde{c} = c$. Hence $c \in \mathbb{Q}$ [AM, Miy]. Similarly, the lowest conformal weight h of a simple ordinary module N is rational. Altogether we have

Proposition 5. *Let V be an Exceptional VOA of lowest primary weight $l \geq 1$ and central charge c and let N be a simple ordinary V-module of lowest weight h. Assuming $g_0(c) \neq 0$ in the MLDE (32), then*

(i) $Z_N(q)$ is holomorphic for $0 < |q| < 1$.
(ii) $Z_N\left(\frac{\alpha\tau+\beta}{\gamma\tau+\delta}\right)$ is a solution of the MLDE for all $\begin{pmatrix} \alpha & \beta \\ \gamma & \delta \end{pmatrix} \in \mathrm{SL}(2, \mathbb{Z})$ viewed as a function of $\tau \in \mathbb{H}_1$ for $q = e^{2\pi i\tau}$.
(iii) The central charge c and the lowest conformal weight h are rational.

Consider the general solution with indicial root $x = c/24$ of the form $Z(q) = q^{-c/24} \sum_{n \geq 0} a_n q^n$. Substituting into the MLDE, we obtain a linear equation in $a_0, \ldots, a_n$ for each n. This can be iteratively solved for a_n provided the coefficient of a_n is nonzero. This coefficient may vanish if $x = m - c/24$ is an indicial root for some integer $m > 0$. Hence we have

Proposition 6. *Let V be an Exceptional VOA of lowest primary weight $l \geq 1$ and central charge c. Suppose $g_0(c) \neq 0$ and that $m < l$ for any indicial root of the form $x = m - c/24$. Then*

(i) $Z_V(q)$ *is the unique q-series solution of the MLDE satisfying* (26).
(ii) $\dim V_n$ *is a rational function of c for each $n \geq 0$.*
(iii) V *is generated by the space of lowest weight primary vectors Π_l.*

Proof. (i) The $x = -c/24$ solution $Z(q) = q^{-c/24} \sum_{n\geq 0} a_n q^n$ is determined by a_0 and a_m for any indicial root(s) of the form $x = m - c/24$ for $m > 0$. Thus the partition function is uniquely determined by the l Virasoro leading terms (26) under the assumption that $m < l$.

(ii) The modular forms $g_m(q, c)$ of the MLDE of Proposition 4 have q-expansions whose coefficients are rational functions of c. Hence solving iteratively it follows that $a_n = \dim V_n$ is a rational function of c.

(iii) Let $V_{\langle \Pi_l \rangle} \subseteq V$ be the subalgebra generated by the lowest weight primary vectors Π_l. But $\omega \in V_{\langle \Pi_l \rangle}$ from (17) so that $V_{\langle \Pi_l \rangle}$ is a VOA of central charge c. Furthermore, since $\lambda^{(2l+2)} \in V_{\langle \Pi_l \rangle}$, the subVOA is an Exceptional VOA of lowest primary weight l. Hence $Z_{V_{\langle \Pi_l \rangle}}(q)$ satisfies the same MLDE as $Z_V(q)$. From (i) it follows that $Z_{V_{\langle \Pi_l \rangle}}(q) = Z_V(q)$ implying $V_{\langle \Pi_l \rangle} = V$. □

Remark 2. Note that $g_0(c) \neq 0$ provided $\lambda^{(2l+2)}$ contains an $L(-2)^{l+1}\mathbf{1}$ component. We conjecture that such a component exists for all l. We further conjecture that $m < l$ for any indicial root of the form $x = m - c/24$ for all l. These properties are verified for all $l \leq 9$ in Sect. 6.

4.1 Exceptional VOAs with $p_l = 1$

Let V be a simple VOA of strong CFT-type generated by one primary vector u of lowest weight l with dual $\overline{u} = u/\langle u, u\rangle$. Consider the commutator (10)

$$
\begin{aligned}
[u(m), Y(u, z)] &= \sum_{j\geq 0} \binom{m}{j} Y\left(u(j)u, z\right) z^{m-j} \\
&= \langle u, u\rangle \sum_{k=0}^{2l-1} \binom{m}{2l-k-1} Y\left(\lambda^{(k)}, z\right) z^{m+k+1-2l}, \qquad (36)
\end{aligned}
$$

using (15). Suppose that $\lambda^{(2l-1)} \in V_{\langle \omega \rangle}$ so that $\lambda^{(k)} \in V_{\langle \omega \rangle}$ for $0 \leq k \leq 2l-1$ which implies the RHS of (36) is expressed in terms of Virasoro modes. Thus (36) defines a $\mathcal{W}(l)$ algebra VOA with one primary vector u of weight l, e.g., [BFKNRV,F]. The further condition $\lambda^{(2l+2)} \in V_{\langle \omega \rangle}$ constrains c to specific rational values.

We consider two infinite families of Exceptional $\mathcal{W}(l)$-VOAs. One is of AD-type, from the ADE series of [CIZ], given by the simple current extension of a minimal model $L\left(c_{p,q},0\right)$ by an irreducible module $L\left(c_{p,q},l\right)$ with

$$l = h_{1,p-1} = \frac{1}{4}(p-2)(q-2) \in \mathbb{N}, \tag{37}$$

for $h_{r,s}$, of (14), i.e., for any coprime pair p,q such that p or $q = 2 \mod 4$. Then (36) is consistent with respect to the Virasoro fusion rule (e.g., [DMS])

$$L\left(c_{p,q},h_{1,p-1}\right) \times L\left(c_{p,q},h_{1,p-1}\right) = L\left(c_{p,q},0\right).$$

Furthermore, since

$$2l+2 = (p-1)(q-1) - \frac{1}{2}(pq-6) < (p-1)(q-1),$$

it follows that $\left(V_{\langle\omega\rangle}\right)_{2l+2}$ contains no Virasoro singular vectors. Hence

Proposition 7. *For a minimal model with $h_{1,p-1} \in \mathbb{N}$ there exists an Exceptional VOA with one primary vector of lowest weight $l = h_{1,p-1}$ of AD-type*

$$V = L\left(c_{p,q},0\right) \oplus L\left(c_{p,q},h_{1,p-1}\right). \tag{38}$$

A second infinite family of $\mathcal{W}(l)$-VOAs for $l = 3k$ for $k \geq 1$ is given in [BFKNRV, F]. A more complete VOA description of this construction will appear elsewhere [T3]. $\mathcal{W}(3k)$ is of central charge $c_k = 1-24k$ and contains a unique Virasoro primary vector of weight $h_n = (n^2-1)k$ for each $n \geq 1$. The corresponding Virasoro Verma module contains a unique singular vector of weight $h_n + n^2$ so that the partition function is [F]:

$$\begin{aligned} Z_{\mathcal{W}(3k)}(q) &= \sum_{n\geq 1} \frac{q^{-c_k/24}}{\prod_{m\geq 0}(1-q^m)}\left(q^{h_n} - q^{h_n+n^2}\right) \\ &= \frac{1}{2\eta(q)}\sum_{n\in\mathbb{Z}}\left(q^{n^2k} - q^{n^2(k+1)}\right). \end{aligned} \tag{39}$$

This VOA is generated by the lowest weight primary of weight $l = h_2 = 3k$ $\lambda^{(2l+2)} \in \left(V_{\langle\omega\rangle}\right)_{2l+2}$ and requires that $h_3 = 8k > 2l+2$ i.e., $k > 1$. Thus we find

Proposition 8. *For each $k \geq 2$ there exists an Exceptional VOA $\mathcal{W}(3k)$ with one primary vector of lowest weight $3k$ and central charge $c_k = 1-24k$.*

Remark 3. We conjecture that the two VOA series of Propositions 7 and 8 are the only Exceptional VOAs for which $p_l = 1$.

5 Genus Zero Constraints from Quadratic Casimirs

We next consider how an Exceptional VOA is also subject to local genus zero constraints following an approach originally described for $l = 1, 2$ in [T1, T2]. Let V be a simple VOA of strong CFT-type of central charge c with lowest primary weight $l \geq 1$. Let Π_l be the vector space of p_l primary vectors of weight l with basis $\{u_i\}$ and dual basis $\{\overline{u}_i\}$. Define the genus zero correlation function

$$F(a, b; x, y) = \left\langle a, \sum_{i=1}^{p_l} Y(u_i, x)Y(\overline{u}_i, y)b \right\rangle, \tag{40}$$

for $a, b \in \Pi_l$. $F(a, b; x, y)$ is linearly dependent on a and b and is constructed locally from Π_l alone.

Locality (7), associativity (9) and lower truncation (3) give

Proposition 9. *$F(a, b; x, y)$ is determined by a rational function*

$$F(a, b; x, y) = \frac{G(a, b; x, y)}{x^{2l}y^{2l}(x - y)^{2l}}, \tag{41}$$

for $G(a, b; x, y)$, a symmetric homogeneous polynomial in x, y of degree $4l$.

$F(a, b; x, y)$ can be considered as a rational function on the genus zero Riemann sphere and expanded in various domains to obtain the $2l+1$ independent parameters determining $G(a, b; x, y) = \sum_{r=0}^{4l} A_r x^{4l-r} y^r$ where $A_r = A_{4l-r}$. In particular, we expand in $\xi = -y/(x - y)$ using skew-symmetry (8), translation (6) and invariance of $\langle\ ,\ \rangle$ to find that

$$\begin{aligned} y^{2l}F(a, b; x, y) &= y^{2l} \sum_{i=1}^{p_l} \left\langle a, Y(u_i, x)e^{yL_{-1}}Y(b, -y)\overline{u}_i \right\rangle \\ &= y^{2l} \sum_{i=1}^{p_l} \left\langle a, e^{yL_{-1}}Y(u_i, x - y)Y(b, -y)\overline{u}_i \right\rangle \\ &= y^{2l} \sum_{i=1}^{p_l} \langle a, Y(u_i, x - y)Y(b, -y)\overline{u}_i \rangle \\ &= \sum_{m\geq 0} C_m \xi^m, \end{aligned} \tag{42}$$

for $C_m = \sum_{i=1}^{p_l} \langle a, u_i(m-1)b(2l-m-1)\overline{u}_i \rangle$. Since l is the lowest primary weight, we have $b(2l - m - 1)\overline{u}_i \in V_m = \left(V_{\langle\omega\rangle}\right)_m$ for $0 \leq m < l$ which determines the coefficients $C_0, \ldots, C_{l-1}$. This follows by writing $b(2l - m - 1)\overline{u}_i$

in a Virasoro basis with coefficients computed in a similar way as for the Casimir vectors in Lemma 2. On the other hand, from (41) we find using $y = -\xi x/(1-\xi)$ that

$$\begin{aligned} y^{2l} F(a,b;x,y) &= g\left(-\frac{\xi}{1-\xi}\right)(1-\xi)^{2l} \\ &= A_0 - (2lA_0 + A_1)\xi + O(\xi^2), \end{aligned}$$

for $g(y) = G(a,b;1,y) = \sum_{r=0}^{4l} A_r y^r$. Hence the coefficients $C_0, \ldots, C_{l-1}$ determine $A_0, \ldots, A_{l-1}$. For example, using $b(2l-1)\overline{u}_i = (-1)^l \langle b, \overline{u}_i\rangle \mathbf{1}$, we have

$$A_0 = C_0 = \sum_{i=1}^{p_l} \langle a, u_i(-1)b(2l-1)\overline{u}_i\rangle = (-1)^l \sum_{i=1}^{p_l} \langle a, u_i\rangle\langle b, \overline{u}_i\rangle = (-1)^l \langle a, b\rangle.$$

In general, $A_k = \langle a, b\rangle a_k(c)$ for $k = 0, \ldots, l-1$ where $a_k(c)$ is a rational function of c.

The other $l+1$ coefficients of $g(y)$ (recalling $A_r = A_{4l-r}$) are determined by using associativity (9) and expanding in $\zeta = (x-y)/y$ as follows:

$$\begin{aligned} (x-y)^{2l} F(a,b;x,y) &= \sum_{m\in\mathbb{Z}} \sum_{i=1}^{p_l} \langle a, Y(u_i(m)\overline{u}_i, y)b\rangle (x-y)^{2l-m-1} \\ &= \sum_{n\geq 0} B_n \zeta^n, \end{aligned} \tag{43}$$

for $B_n = \langle a, o(\lambda^{(n)})b\rangle$ for $n \geq 0$ and recalling $o(\lambda^{(n)}) = \lambda^{(n)}(n-1)$.

Lemma 4. *The leading coefficients of* (43) *are* $B_0 = (-1)^l p_l \langle a, b\rangle$ *and* $B_1 = 0$. *For* $k \geq 1$, *the odd labelled coefficients* B_{2k+1} *satisfy*

$$B_{2k+1} = \frac{1}{2}\sum_{r=2}^{2k} \binom{-r}{2k+1-r}(-1)^r B_r, \tag{44}$$

i.e., B_{2k+1} *is determined by the lower even labelled coefficients* $B_2, \ldots, B_{2k}$. *The even labelled coefficients for* $k \geq 0$ *are given by*

$$B_{2k} = A_{2l}\delta_{k,0} + \sum_{m=1}^{2l}\left[\binom{m}{2k} + \binom{-m}{2k}\right] A_{2l-m}. \tag{45}$$

Proof. From Lemma 1 we have $\lambda^{(0)} = (-1)^l p_l \mathbf{1}$ and $\lambda^{(1)} = 0$ so that $B_0 = (-1)^l p_l \langle a, b\rangle$ and $B_1 = 0$. Comparing (43) to (41) we find that

$$\sum_{n\geq 0} B_n \zeta^n = g\left(\frac{1}{1+\zeta}\right)(1+\zeta)^{2l} = g(1+\zeta)(1+\zeta)^{-2l},$$

since $G(a, b; x, y)$ is symmetric and homogeneous. Thus

$$\sum_{n\geq 0} B_n \zeta^n = \sum_{n\geq 0} B_n \left(\frac{-\zeta}{1+\zeta}\right)^n.$$

This implies $B_n = \sum_{r=0}^{n} \binom{-r}{n-r}(-1)^r B_r$. Taking $n = 2k+1$ leads to (44). (45) follows from the identity

$$\sum_{n\geq 0} B_n \zeta^n = A_{2l} + \sum_{m=1}^{2l} A_{2l-m}\left[(1+\zeta)^m + (1+\zeta)^{-m}\right].$$

□

We next assume that $\lambda^{(n)} \in V_{\langle\omega\rangle}$ for even $n \leq 2l$ giving $B_{2k} = \langle a, o(\lambda^{(2k)})b\rangle = p_l\langle a, b\rangle b_{2k}(c)$ for $k = 1, \ldots, l$ for some rational functions $b_{2k}(c)$ via Lemma 3. Note that we are not (yet) assuming $\lambda^{(2l+2)} \in V_{\langle\omega\rangle}$. $G(a, b; x, y)$ is uniquely determined provided we can invert (45) to solve for $A_l, \ldots, A_{2l}$. Define the $l \times l$ matrix

$$M_{mk} = \binom{m}{2k} + \binom{-m}{2k}, \tag{46}$$

of coefficients for A_{2l-m} of B_{2k} in (45), where $m, k = 1, \ldots, l$.

Lemma 5. *M is invertible with* $\det M = 1$.

Proof. Define unit diagonal lower and upper triangular matrices L and U by

$$L_{ij} = \begin{cases} \binom{2i-j-1}{j-1} & \text{for } i \leq j, \\ 0 & \text{for } i > j, \end{cases} \qquad U_{jk} = \begin{cases} \frac{k}{j}\binom{j+k-1}{2j-1} & \text{for } j \leq k, \\ 0 & \text{for } j > k. \end{cases}$$

By induction in k, one can show that $M_{ik} = (L\,U)_{ik}$ and so $\det M = 1$. □

Thus it follows that $A_{2l-m} = \sum_{k=1}^{l} B_{2k}(M^{-1})_{km}$ for $m = 1, \ldots, l$. Altogether, we have therefore shown the following.

Proposition 10. *Let V be a simple VOA of strong CFT–type of central charge c with lowest primary weight $l \geq 1$. Suppose that $\lambda^{(n)} \in V_{\langle\omega\rangle}$ for all even $n \leq 2l$. Then the genus zero correlation function is uniquely determined with*

$$F(a, b; x, y) = \frac{1}{x^{2l}y^{2l}(x-y)^{2l}} \sum_{r=0}^{2l} A_r \left(x^{4l-r}y^r + x^r y^{4l-r}\right),$$

where

$$A_k = \begin{cases} \langle a,b\rangle a_k(c), & k=0,\ldots,l-1,\\ p_l\langle a,b\rangle a_k(c), & k=l,\ldots,2l, \end{cases}$$

for $2l+1$ *specific rational functions* $a_0(c),\ldots,a_{2l}(c)$.

Next we assume $\lambda^{(2l+2)} \in V_{\langle\omega\rangle}$ so that V is an Exceptional VOA. This implies $B_{2l+2} = \langle a, o(\lambda^{(2l+2)})b\rangle = p_l\langle a,b\rangle b_{2l+2}(c)$ for some rational function $b_{2l+2}(c)$. But B_{2l+2} is already determined from (45) in terms of $A_1,\ldots,A_{2l}$ from Proposition 10. Hence we have

Proposition 11. *Let* V *be an Exceptional VOA with lowest primary weight* l. *Then the genus zero correlation function* $F(a,b;x,y)$ *is uniquely determined and* $p_l = p_l(c)$, *a specific rational function of* c.

For $l = 1,2$ we may use $F(a,b;x,y)$ to understand many properties of the corresponding VOA (as briefly reviewed below) [T1, T2]. We already know from Proposition 6(ii) that $p_l = \dim V_l - \dim \left(V_{\langle\omega\rangle}\right)_l$ is a rational function of c. In principle, the specific rational expressions for p_l may differ but, in practice, the same expression is observed to arise for all $l \le 9$. A more significant point is that the argument leading to Proposition 11 may be adopted to understanding some automorphism group properties of V.

5.1 Exceptional VOAs of Class $\mathcal{S}^{2l+2}$

Let $G = \mathrm{Aut}(V)$ denote the automorphism group of a VOA V and let V^G denote the sub-VOA fixed by G. Since the Virasoro vector is G invariant it follows that $V_{\langle\omega\rangle} \subseteq V^G$. V is said to be of Class $\mathcal{S}^n$ if $V_k^G = \left(V_{\langle\omega\rangle}\right)_k$ for all $k \le n$ [Mat]. (The related notion of conformal t-designs is described in [Ho2].) In particular, the quadratic Casimir (15) is G-invariant so it follows that a VOA V with lowest primary weight l of class $\mathcal{S}^{2l+2}$ is an Exceptional VOA. It is not known if every Exceptional VOA is of class $\mathcal{S}^{2l+2}$.

The primary vector space Π_l is a finite-dimensional G-module. Assuming Π_l is a completely reducible G-module (e.g., for G linearly reductive [Sp]) we have

Proposition 12. *Let* V *be an Exceptional VOA of class* $\mathcal{S}^{2l+2}$ *with primaries* Π_l *of lowest weight* l. *If* Π_l *is a completely reducible* G-*module, then it is either an irreducible* G-*module or the direct sum of two isomorphic irreducible* G-*modules.*

Remark 4. For odd p_l it follows that Π_l must be an irreducible G-module.

Proof. Let ρ be a G-irreducible component of Π_l and let $\overline{\rho}$ denote the $\langle\ ,\ \rangle$ dual vector space. $\overline{\rho}$ and ρ are isomorphic as G-modules. Define

$$R = \begin{cases} \rho & \text{if } \rho = \overline{\rho},\\ \rho\oplus\overline{\rho} & \text{if } \rho \ne \overline{\rho}. \end{cases}$$

Clearly $R \subseteq \Pi_l$ is a self-dual vector space. We next repeat the Casimir construction and analysis that lead up to Proposition 11. Choose an R-basis $\{v_i : i = 1, \ldots, \dim R\}$ and dual basis $\{\overline{v}_i\}$ and define Casimir vectors

$$\lambda_R^{(n)} = \sum_{i=1}^{\dim R} v_i(2l - n - 1)\overline{v}_i \in V_n, \quad n \geq 0. \tag{47}$$

But $\lambda_R^{(n)}$ is G-invariant and since V is of class $\mathcal{S}^{2l+2}$, it follows that $\lambda_R^{(n)} \in V_{\langle \omega \rangle}$ for all $n \leq 2l + 2$. We define a genus zero correlation function constructed from the vector space R

$$F_R(a, b; x, y) = \sum_{i=1}^{\dim R} \langle a, Y(v_i, x)Y(\overline{v}_i, y)b \rangle, \tag{48}$$

for all $a, b \in R$. We then repeat the earlier arguments to conclude that Proposition 11 also holds for $F_R(a, b; x, y)$ where, in particular, $\dim R = p_l(c)$, for the **same** rational function. Thus $\dim R = p_l$ and the result follows. □

6 Exceptional VOAs of Lowest Primary Weight $l \leq 9$

We now consider Exceptional VOAs of lowest primary weight $l \leq 9$. We denote by $E_n = E_n(q)$ the Eisenstein series of weight n appearing in the MLDE (32). For $l \leq 4$ we describe all the rational values for c, h, whereas for $5 \leq l \leq 9$ we give all rational values for c, h for which $p_l = \dim \Pi_l \leq 500{,}000$, found by computer algebra techniques. We also consider conjectured extremal self-dual VOAs with $c = 24(l - 1)$ [Ho1, Wi]. Any MLDE solution for rational h for which there is no graded character $Z_N(q)$ is marked with an asterisk. We obtain many examples of known Exceptional VOAs such as Deligne's exceptional series of Lie algebras, the Moonshine and Baby Monster modules. There are also a number of candidate solutions for which no construction yet exists indicated by question marks.

[$l = 1$]. This is discussed in much greater detail in [T1, T2]. Propositions 4–6 imply that $Z_N(q)$ satisfies the following 2nd order MLDE [T2]:

$$D^2 Z - \frac{5}{4}c\,(c + 4)\, E_4 Z = 0.$$

This MLDE has also appeared in [MatMS, KZ, Mas2, KKS, Kaw]. The indicial roots $x_1 = -c/24, x_2 = (c + 4)/24$ are exchanged under the MLDE symmetry $c \leftrightarrow -c - 24$. Solving iteratively for the partition function

$$Z_V(q) = q^{-c/24}\left(1 + p_1 q + (1 + p_1 + p_2)q^2 + (1 + 2p_1 + p_2 + p_3)q^3 + \ldots\right),$$

where $p_n = \dim \Pi_n$, for weight n primary vector space Π_n, we have

$$p_1 = \frac{c(5c+22)}{10-c}, \quad p_2 = \frac{5(5c+22)(c-1)(c+2)^2}{2(c-10)(c-22)},$$
$$p_3 = -\frac{5c(5c+22)(c-1)(c+5)(5c^2+268)}{6(c-10)(c-22)(c-34)}, \ldots.$$

For $c = 10 \mod 12$, the indicial roots differ by an integer leading to denominator zeros for all p_n.

By Proposition 6, V is generated by V_1 which defines a Lie algebra $\mathfrak{g}$. $F(a, b; x, y)$ from Proposition 11 determines the Killing form which can be used to show that $\mathfrak{g}$ is simple with dual Coxeter number [T1, MT]

$$h^\vee = 6k\frac{2+c}{10-c},$$

for some real level k. Thus $V = V_{\mathfrak{g}}(k)$, a level k Kac–Moody VOA.

The indicial root x_2 of the MLDE gives the lowest weight $h = (c+2)/12$ of any independent irreducible V-module(s) N. Therefore $V_{\mathfrak{g}}(k)$ has at most two independent irreducible characters so that the level k must be a positive integer [Kac2]. Comparing p_1 and $h^\vee$ to Cartan's list of simple Lie algebras shows that in fact $k = 1$ with $c = 1, 2, \frac{14}{5}, 4, \frac{26}{5}, 6, 7, 8$ with $\mathfrak{g} = A_1, A_2, G_2, D_4, F_4, E_6, E_7, E_8$, respectively, known as Deligne Exceptional Series [D, DdeM, MarMS, T2]. In summary, we have

$c > 0$	p_1	p_2	p_3	VOA	$h \in \mathbb{Q}$
1	3	0	0	$V_{A_1}(1)$	$0, \frac{1}{4}$
2	8	8	21	$V_{A_2}(1)$	$0, \frac{1}{3}$
$\frac{14}{5}$	14	27	84	$V_{G_2}(1)$	$0, \frac{2}{5}$
4	28	105	406	$V_{D_4}(1)$	$0, \frac{1}{2}$
$\frac{26}{5}$	52	324	1,547	$V_{F_4}(1)$	$0, \frac{3}{5}$
6	78	650	3,575	$V_{E_6}(1)$	$0, \frac{2}{3}$
7	133	1,539	10,108	$V_{E_7}(1)$	$0, \frac{3}{4}$
8	248	3,875	30,380	$V_{E_8}(1)$	$0, \frac{5}{6}*$

The table also shows h for a possible irreducible V-module(s). For $c = 2$ and 4 there are 2 independent irreducible modules but which share the same character (due to $\mathfrak{g}$

outer automorphisms). $V_{E_8}(1)$ is self-dual so that the MLDE solution with $h = \frac{5}{6}$ does not correspond to a graded character $Z_N(q)$.

[$l = 2$]. This case is also discussed in detail in [Mat, T1, T2]. Propositions 4–6 imply that $Z_N(q)$ satisfies the following 3rd order MLDE [T2]

$$D^3 Z - \frac{5}{124}\left(704 + 240c + 21c^2\right) E_4\, DZ - \frac{35}{248} c\left(144 + 66c + 5c^2\right) E_6\, Z = 0,$$

with indicial equation (33)

$$(x - x_1)\left(x^2 - \left(\frac{1}{2} + x_1\right)x + \frac{20x_1^2 - 11x_1 + 1}{62}\right) = 0,$$

for $x_1 = -c/24$. Solving iteratively for the partition function ($x = x_1$)

$$Z_V(q) = q^{-c/24}(1 + (1 + p_2)q^2 + (1 + p_2 + p_3)q^3 + \ldots),$$

where $p_n = \dim \Pi_n$, for weight n primary vector space Π_n, we find that

$$p_2 = \frac{(7c + 68)(2c - 1)(5c + 22)}{2(c^2 - 55c + 748)}, \quad p_3 = \frac{31c(7c + 68)(2c - 1)(5c + 44)(5c + 22)}{6(c^2 - 55c + 748)(c^2 - 86c + 1{,}864)}.$$

From Proposition 6, the Griess algebra generates V and from Proposition 11 the Griess algebra is simple [T1]. The solutions for $c, h \in \mathbb{Q}$ with positive p_3 and possible Exceptional VOAs are listed as follows:

c	p_2	p_3	VOA	$h \in \mathbb{Q}$
$-\frac{44}{5}$	1	0	$L\left(c_{3,10}, 0\right) \oplus L\left(c_{3,10}, 2\right)$	$0, -\frac{1}{5}, -\frac{2}{5}$
8	155	868	$V^+_{\sqrt{2}E_8}$	$0, \frac{1}{2}, 1$
16	2,295	63,240	$V^+_{\mathrm{BW}_{16}}$	$0, 1, \frac{3}{2}$
$\frac{47}{2}$	96,255	9,550,635	$\mathrm{VB}^\natural_{\mathbb{Z}}$	$0, \frac{3}{2}, \frac{31}{16}$
24	196,883	21,296,876	$V^\natural$	0
32	$3.7^2.13.73$	$2^4.3.7^2.13.31.73$	?? $V^+_L \oplus (V_L)^+_T$; L extremal S-D	0
$\frac{164}{5}$	$3^2.17.19.31$	$2.5.13.17.19.31.41$	??	$0, \frac{11}{5}, \frac{12}{5}$
$\frac{236}{7}$	$5.19.23.29$	$2.19.23.29.31.59$	??	$0, \frac{16}{7}, \frac{17}{7}$
40	$3^2.29.79$	$2^2.5.29.31.61.79$	?? $V^+_L \oplus (V_L)^+_T$; L extremal S-D	0

The list includes the famous Moonshine Module $V^\natural$ [FLM], the Baby Monster VOA $\mathrm{VB}^\natural_{\mathbb{Z}}$ [Ho1], V^+_L for $L = \sqrt{2}E_8$ [G] and the rank 16 Barnes–Wall lattice $L = \mathrm{BW}_{16}$ [Sh], and a minimal model simple current extension AD-type as in Proposition 7.

The value(s) of $h = x_i + c/24$ for the lowest weight(s) agree with those for the irreducible V-modules as do the corresponding MLDE solutions for the characters in each case. There are also four other possible candidates. For $c = 32$ and 40 one can construct a self-dual VOA from an extremal even self-dual lattice L (with no vectors of squared length 2). However, such lattices are not unique and it is not known which, if any, gives rise to a VOA satisfying the exceptional conditions. There are no known candidate constructions for $c = \frac{164}{5}$ and $\frac{236}{7}$.

Note that $p_2 = \dim \Pi_2$ is odd in every case and Proposition 12 implies that if Π_2 is completely Aut V-reducible, then it is irreducible. This is indeed the case in the first five known cases for $c \leq 24$ [Atlas]. Π_3 is also an Aut V-module whose dimension p_3 is given. The MLDE solutions (with positive coprime integer coefficients) for $c = 164/5$ with $h = 11/5, 12/5$ and for $c = 236/7$ with $h = 16/7, 17/7$ have respective leading q-expansions:

$$Z_{11/5}(q) = q^{5/6}\left(2^3.31.41 + 5.11.31.41.53\,q + O(q^2)\right),$$

$$Z_{12/5}(q) = q^{31/30}\left(2^2.11^2.31.41 + 2^5.11^2.31^2.41\,q + O(q^2)\right),$$

$$Z_{16/7}(q) = q^{37/42}\left(17.23.31 + 2^5.7.17.31.37\,q + O(q^2)\right),$$

$$Z_{17/7}(q) = q^{43/42}\left(2^4.29.31.59 + 2.3.17.29.31.43.59\,q + O(q^2)\right).$$

These coefficients constrain the possible structure of Aut V further.

[$l = 3$]. $Z_N(q)$ satisfies the 4th order MLDE:

$$\begin{aligned}
&(578c-7)\,D^4 Z - \frac{5}{2}\left(168c^3 + 2{,}979c^2 + 15{,}884c - 4{,}936\right) E_4\, D^2 Z \\
&-\frac{35}{2}\left(25c^4 + 661c^3 + 4{,}368c^2 + 10{,}852c + 1{,}144\right) E_6\, DZ \\
&-\frac{75}{16}c\left(14c^4 + 425c^3 + 3{,}672c^2 + 5{,}568c + 9{,}216\right) E_4^2\, Z = 0.
\end{aligned}$$

Solving iteratively for the partition function we find [T2]:

$$\begin{aligned}
Z_V &= q^{-c/24}(1 + q^2 + (1+p_3)q^3 + (2+p_3+p_4)q^3 + \ldots),\\
p_3 &= -\frac{(5c+22)(3c+46)(2c-1)(5c+3)(7c+68)}{5c^4 - 703c^3 + 32{,}992c^2 - 517{,}172c + 3{,}984},
\end{aligned}$$

and $p_4 = \frac{r(c)}{s(c)}$ for

$$r(c) = -\frac{1}{2}\,(2c-1)\,(3c+46)\,(5c-4)\,(7c+68)\,(5c+3)\,(7c+114)$$

$$\cdot\left(55c^3 - 5{,}148c^2 - 11{,}980c - 36{,}528\right),$$

$$s(c) = \left(5c^4 - 703c^3 + 32{,}992c^2 - 517{,}172c + 3{,}984\right)$$

$$\cdot\left(5c^4 - 964c^3 + 62{,}392c^2 - 1{,}355{,}672c + 13{,}344\right).$$

The $c, h \in \mathbb{Q}$ solutions for positive integer p_3 with possible VOAs are

c	p_3	p_4	VOA	$h \in \mathbb{Q}$
$-\frac{114}{7}$	1	0	$L(c_{3,14}, 0) \oplus L(c_{3,14}, 3)$	$0, -\frac{3}{7}, -\frac{4}{7}, -\frac{5}{7}$
$\frac{4}{5}$	1	0	$L(c_{5,6}, 0) \oplus L(c_{5,6}, 3)$	$0, \frac{1}{15}, \frac{2}{5}, \frac{2}{3}$
48	$3^2.19^2.101.131$	$5^6.19^2.71.101$	?? Höhn Extremal VOA	0

The Höhn Extremal VOA is a conjectural self-dual VOA [Ho1]. If Π_3 is a completely reducible Aut(V)-module, then it must be irreducible excluding Witten's suggestion that Aut(V) $= \mathbb{M}$, the Monster group [Wi].

[l = 4]. Proposition 4 implies $Z_N(q)$ satisfies the 5rd order MLDE:

$$(317c + 3)\ D^5 Z - \frac{5}{7}\left(297c^3 + 6{,}746c^2 + 53{,}133c + 4{,}644\right) E_4\, D^3 Z$$

$$-\frac{25}{8}\left(77c^4 + 3{,}057c^3 + 31{,}506c^2 + 129{,}736c - 24{,}096\right) E_6\, D^2 Z$$

$$-\frac{25}{112}\left(231c^5 + 12{,}117c^4 + 194{,}916c^3 + 843{,}728c^2 + 1{,}652{,}288c - 718{,}080\right) E_4{}^2\, DZ$$

$$-\frac{25}{32}c\,(c + 24)\left(15c^4 + 527c^3 + 5{,}786c^2 + 528c + 25{,}344\right) E_4 E_6\, Z = 0.$$

Solving iteratively for the partition function we find

$$Z_V = q^{-c/24}(1 + q^2 + q^3 + (2 + p_4)q^4 + (3 + p_4 + p_5)q^5 + \ldots),$$

$$p_4 = \frac{5(3c + 46)(2c - 1)(11c + 232)(7c + 68)(5c + 3)(c + 10)}{2(5c^4 - 1{,}006c^3 + 67{,}966c^2 - 1{,}542{,}764c - 12{,}576)(c - 67)},$$

and $p_5 = \frac{r(c)}{s(c)}$ where

$$r(c) = 3(c - 1)(5c + 22)(3c + 46)(2c - 1)(11c + 232)(7c + 68)(5c + 3)(c + 24)$$

$$.(59c^3 - 13{,}554c^2 + 788{,}182c - 398{,}640),$$

$$s(c) = 2(c - 67)(5c^4 - 1{,}006c^3 + 67{,}966c^2 - 1{,}542{,}764c - 12{,}576)$$

$$.(5c^5 - 1{,}713c^4 + 221{,}398c^3 - 12{,}792{,}006c^2 + 278{,}704{,}260c + 2{,}426{,}976).$$

The $c, h \in \mathbb{Q}$ solutions for $p_4 \leq 500{,}000$ and $c = 48$ with possible VOAs are

c	p_4	p_5	VOA	$h \in \mathbb{Q}$
1	2	0	V_L^+ for $L = 2\sqrt{2}\mathbb{Z}$	$0, \frac{1}{16}, \frac{1}{4}, \frac{9}{16}, 1$
72	$2^3.11^4.13^2.131$	$2.11^4.13^2.103.131.191$	?? Höhn Extremal VOA	0

The Höhn Extremal VOA is a conjectured self-dual VOA [Ho1, Wi]. If Π_4 is a completely reducible Aut(V)-module, then by Proposition 12, either p_4 or $\frac{1}{2}p_4$ is the dimension of an irreducible Aut(V)-module.

[$l = 5$]. Z_V satisfies a 6th order MLDE with $p_5 = \frac{r(c)}{s(c)}$ for

$$
\begin{aligned}
r(c) &= -(13c+350)(7c+25)(5c+126)(11c+232)\\
&\quad .(2c-1)(3c+46)(68+7c)(5c+3)(10c-7),\\
s(c) &= 1{,}750c^8 - 760{,}575c^7 + 132{,}180{,}881c^6 - 11{,}429{,}170{,}478c^5\\
&\quad +484{,}484{,}459{,}322c^4 - 7{,}407{,}871{,}790{,}404c^3 - 37{,}323{,}519{,}053{,}016c^2\\
&\quad +25{,}483{,}483{,}057{,}200c - 363{,}772{,}080{,}000.
\end{aligned}
$$

The $c, h \in \mathbb{Q}$ solutions for $p_5 \leq 500{,}000$ with possible VOAs are

c	p_5	VOA	$h \in \mathbb{Q}$
$-\frac{350}{11}$	1	$L(c_{3,22}, 0) \oplus L(c_{3,22}, 5)$	$0, -\frac{8}{11}, -\frac{10}{11}, -\frac{13}{11}, -\frac{14}{11}, -\frac{15}{11}$
$\frac{6}{7}$	1	$L(c_{6,7}, 0) \oplus L(c_{6,7}, 5)$	$0, \frac{1}{21}, \frac{1}{7}, \frac{10}{21}, \frac{5}{7}, \frac{4}{3}$

Witten's conjectured Extremal VOA for $c = 4.24 = 96$ does not appear [Wi].

[$l = 6$]. Z_V satisfies a 7th order MLDE with $p_6 = \frac{r(c)}{s(c)}$ for

$$
\begin{aligned}
r(c) &= \frac{7}{2}(13c+350)(5c+164)(7c+25)(11c+232)(3c+46)\\
&\quad .(4c+21)(5c+3)(10c-7)(5c^2+316c+3{,}600),\\
s(c) &= 1{,}750c^9 - 1{,}119{,}950c^8 + 297{,}661{,}895c^7 - 41{,}808{,}629{,}963c^6\\
&\quad +3{,}225{,}664{,}221{,}176c^5 - 123{,}384{,}054{,}679{,}580c^4 + 1{,}266{,}443{,}996{,}541{,}232c^3\\
&\quad +29{,}763{,}510{,}364{,}647{,}840c^2 + 96{,}385{,}155{,}929{,}078{,}400c + 7{,}743{,}915{,}615{,}744{,}000.
\end{aligned}
$$

The $c, h \in \mathbb{Q}$ solutions for $p_6 \leq 500{,}000$ with possible VOAs are

c	p_6	VOA	$h \in \mathbb{Q}$
$-\frac{516}{13}$	1	$L\left(c_{3,26},0\right) \oplus L\left(c_{3,26},6\right)$	$0,-\frac{10}{13},-\frac{15}{13},-\frac{17}{13}$ $-\frac{20}{13},-\frac{21}{13},-\frac{22}{13}$
-47	1	$\mathcal{W}(6)$	$0,-\frac{5}{4},-\frac{3}{2},-\frac{5}{3}$ $-\frac{15}{8},-\frac{23}{12},-2$
120	$2.7^2.11.29.43.67.97.191$	?? Witten Extremal VOA	0

$c = -47$ is first example of a $\mathcal{W}(3k)$-algebra of Proposition 8. The irreducible lowest weight h values and character solutions agree with [F]. Witten's conjecture Extremal VOA for $c = 5.24 = 120$ appears [Wi] where either p_6 or $\frac{1}{2}p_6$ is the dimension of an irreducible Aut(V)–module.

[$l = 7$]. Z_V satisfies an 8th order MLDE where $p_7 = \frac{r(c)}{s(c)}$ for

$$r(c) = -5\,(13c+350)\,(5c+164)\,(7c+25)\,(11c+232)\,(3c+46)\,(17c+658)$$
$$.\,(4c+21)\,(5c+3)\,(10c-7)\left(35c^3+3{,}750c^2+76{,}744c-32{,}640\right),$$

$$s(c) = 61{,}250c^{11} - 54{,}725{,}125c^{10} + 20{,}922{,}774{,}275c^9 - 4{,}421{,}902{,}106{,}730c^8$$
$$+553{,}932{,}117{,}001{,}488c^7 - 40{,}395{,}124{,}111{,}104{,}312c^6 + 1{,}491{,}080{,}056{,}338{,}817{,}984c^5$$
$$-12{,}528{,}046{,}696{,}953{,}576{,}896c^4 - 483{,}238{,}055{,}074{,}755{,}678{,}656c^3$$
$$-1{,}702{,}959{,}754{,}355{,}175{,}160{,}320c^2 + 249{,}488{,}376{,}255{,}167{,}616{,}000c$$
$$+362{,}620{,}505{,}915{,}136{,}000{,}000.$$

There are no rational c solutions for $p_7 \leq 500{,}000$.

[$l = 8$]. $Z_V(q)$ satisfies a 9th order MLDE with one $c, h \in \mathbb{Q}$ solution for $p_8 \leq 500{,}000$

c	p_8	VOA	$h \in \mathbb{Q}$
$-\frac{944}{17}$	1	$L\left(c_{3,34},0\right) \oplus L\left(c_{3,34},8\right)$	$0,-\frac{14}{17},-\frac{25}{17},-\frac{28}{17},-\frac{33}{17}$ $-\frac{35}{17},-\frac{38}{17},-\frac{39}{17},-\frac{40}{17}$

[$l = 9$]. $Z_V(q)$ satisfies a 10th order MLDE with $c, h \in \mathbb{Q}$ solutions for $p_9 \leq 500{,}000$

c	p_9	VOA	$h \in \mathbb{Q}$
$-\frac{1{,}206}{19}$	1	$L\left(c_{3,38},0\right) \oplus L\left(c_{3,38},9\right)$	$0, -\frac{16}{19}, -\frac{29}{19}, -\frac{36}{19}, -\frac{44}{19}$
			$-\frac{46}{19}, -\frac{49}{19}, -\frac{50}{19}, -\frac{51}{19}$
$-\frac{208}{35}$	1	$L\left(c_{5,14},0\right) \oplus L\left(c_{5,14},9\right)$	$0, -\frac{2}{7}, -\frac{9}{35}, -\frac{4}{35}$
			$\frac{1}{7}, \frac{11}{35}, \frac{9}{7}, \frac{8}{5}$
$-\frac{14}{11}$	1	$L\left(c_{6,11},0\right) \oplus L\left(c_{6,11},9\right)$	$0, -\frac{1}{11}, -\frac{2}{33}, \frac{1}{11}, \frac{7}{33}$
			$\frac{6}{11}, \frac{25}{33}, \frac{14}{11}, \frac{52}{33}, \frac{8}{3}$
-71	1	$\mathcal{W}(9)$	$0, -2, -\frac{9}{4}, -\frac{39}{16}, -\frac{8}{3}$
			$-\frac{11}{4}, -\frac{35}{12}, -\frac{47}{16}, -3, -\frac{9}{8}*$

For $c = -71$, the MLDE solutions agree with all the graded characters for $\mathcal{W}(9)$ except for $h = -\frac{9}{8}$ [F].

7 Exceptional VOSAs

7.1 VOSA Quadratic Casimirs and Zhu Theory

We now give an analysis for Vertex Operator Superalgebras (VOSAs). Many of the results are similar but there are also significant differences, e.g., here the MLDEs involve twisted Eisenstein series. Let V be a simple VOSA of strong CFT-type with unique invertible bilinear form $\langle\ ,\ \rangle$. Let Π_l denote the space of Virasoro primary vectors of *lowest half integer weight* $l \in \mathbb{N} + \frac{1}{2}$, i.e., Π_l is of odd parity and $V_k = (V_{\langle\omega\rangle})_k$ for all $k \leq l - \frac{1}{2}$. We construct quadratic Casimir vectors $\lambda^{(n)}$ as in Sect. 3.1 (from the odd parity space Π_l) which enjoy the same properties as VOA Casimir vectors.

Define the genus one partition function of a VOSA V by

$$Z_V(q) = \mathrm{Tr}_V\left(\sigma\, q^{L(0)-c/24}\right) = q^{-c/24}\sum_{n\geq 0}(-1)^{2n}\dim V_n\, q^n, \tag{49}$$

for fermion number operator σ where $\sigma u = (-1)^{p(u)}u$ for u of parity $p(u)$ and with a corresponding definition for a simple ordinary V-module N. We also define the genus one 1-point correlation function

$$Z_N(u,q) = \mathrm{Tr}_N\left(\sigma\, o(u) q^{L(0)-c/24}\right). \tag{50}$$

In [MTZ] a Zhu reduction formula for the 2-point correlation function $Z_N(Y[u,z]v,q)$ for $u, v \in V$ is found expressed in terms of *twisted elliptic Weierstrass functions* parameterized by $\theta, \phi \in \{\pm 1\}$. Let $\phi = e^{2\pi i\kappa}$ for $\kappa \in \{0, \frac{1}{2}\}$. Then (23) and (24) are generalized to [MTZ]

$$P_m \begin{bmatrix} \theta \\ \phi \end{bmatrix} (z) = \frac{1}{z^m} + (-1)^m \sum_{n \geq m} \binom{n-1}{m-1} E_n \begin{bmatrix} \theta \\ \phi \end{bmatrix} (q) z^{n-m}, \tag{51}$$

for twisted Eisenstein series $E_n \left[\begin{smallmatrix} \theta \\ \phi \end{smallmatrix} \right](q) = 0$ for n odd, and for n even

$$E_n \begin{bmatrix} \theta \\ \phi \end{bmatrix} (q) = -\frac{B_n(\kappa)}{n!} + \frac{2}{(n-1)!} \sum_{k \in \mathbb{N}+\kappa} \frac{k^{n-1}\theta q^k}{1-\theta q^k}, \tag{52}$$

and where $B_n(\kappa)$ is the Bernoulli polynomial defined by

$$\frac{e^{z\kappa}}{e^z - 1} = \frac{1}{z} + \sum_{n \geq 1} \frac{B_n(\kappa)}{n!} z^{n-1}. \tag{53}$$

(51) and (52) agree with (23) and (24) respectively for $(\theta, \phi) = (1, 1)$. $P_m \left[\begin{smallmatrix} \theta \\ \phi \end{smallmatrix} \right](z)$ converges absolutely and uniformly on compact subsets of the domain $|q| < |e^z| < 1$ and $E_n \left[\begin{smallmatrix} \theta \\ \phi \end{smallmatrix} \right](q)$ is a holomorphic function of $q^{\frac{1}{2}}$ for $|q| < 1$. For $(\theta, \phi) \neq (1, 1)$, $E_n \left[\begin{smallmatrix} \theta \\ \phi \end{smallmatrix} \right]$ is modular of weight n in the sense that

$$E_n \begin{bmatrix} \theta^\alpha \phi^\beta \\ \theta^\gamma \phi^\delta \end{bmatrix} \left(\frac{\alpha\tau + \beta}{\gamma\tau + \delta} \right) = (\gamma\tau + \delta)^n E_n \begin{bmatrix} \theta \\ \phi \end{bmatrix} (\tau), \tag{54}$$

for $\left(\begin{smallmatrix} \alpha & \beta \\ \gamma & \delta \end{smallmatrix} \right) \in \mathrm{SL}(2, \mathbb{Z})$. The Zhu reduction formula of Proposition 1 has been generalized in [MTZ] as follows

Proposition 13. *Let N be a simple ordinary V-module for a VOSA V. For u of parity $p(u)$ and for all v we have*

$$\begin{aligned} Z_N(Y[u,z]v,q) &= \mathrm{Tr}_N \left(\sigma\, o(u) o(v) q^{L(0)-c/24} \right) \delta_{p(u)1} \\ &\quad + \sum_{m \geq 0} P_{m+1} \begin{bmatrix} 1 \\ p(u) \end{bmatrix} (z) Z_N(u[m]v, q). \end{aligned}$$

For even parity u this agrees with Proposition 1. In particular, Corollary (1) concerning Virasoro vacuum descendents holds. Much as before, Proposition 13 implies that the Casimir vectors $\lambda^{[n]} \in V_{[n]}$ satisfy

$$\sum_{n\geq 0} Z_N\left(\lambda^{[n]},q\right)z^{n-2l} = \sum_{m=0}^{2l-1} P_{m+1}\left[\begin{matrix}1\\-1\end{matrix}\right](z)Z_N\left(\lambda^{[2l-m-1]},q\right). \quad (55)$$

Equating the z coefficients implies the following variant of Proposition 2.

Proposition 14. *$Z_N\left(\lambda^{[2l+1]},q\right)$ satisfies the recursive identity*

$$Z_N\left(\lambda^{[2l+1]},q\right) = -2\sum_{r=0}^{l-\frac{1}{2}}(l-r)E_{2(l-r)+1}\left[\begin{matrix}1\\-1\end{matrix}\right](q)Z_N\left(\lambda^{[2r]},q\right). \quad (56)$$

7.2 Exceptional VOSAs

Let V be a simple VOSA of strong CFT-type with primary vectors of lowest weight $l \in \mathbb{N}+\frac{1}{2}$ for which $\lambda^{(2l+1)} \in V_{\langle\omega\rangle}$. We further assume that $\left(V_{\langle\omega\rangle}\right)_{2l+1}$ contains no Virasoro singular vectors. We call V an *Exceptional VOSA of Odd Parity Lowest Primary Weight l*. Proposition 3 implies

Proposition 15. *Let V be an Exceptional VOSA of lowest weight $l \in \mathbb{N}+\frac{1}{2}$ and central charge c. Then $Z_N(q)$ for a simple ordinary V-module N satisfies a Twisted Modular Linear Differential Equation (TMLDE)*

$$\sum_{m=0}^{l+\frac{1}{2}} g_{l+\frac{1}{2}-m}\left[\begin{matrix}1\\-1\end{matrix}\right](q,c)\, D^m Z(q) = 0, \quad (57)$$

where $g_k\left[\begin{smallmatrix}1\\-1\end{smallmatrix}\right](q,c)$ is a twisted modular form of weight $2k$ whose coefficients over the ring of twisted Eisenstein series (52) *are rational functions of c.*

The TMLDE (57) is of order $l+\frac{1}{2}$ with a regular singular point at $q=0$ provided $g_0\left[\begin{smallmatrix}1\\-1\end{smallmatrix}\right](q,c) = g_0(c) \neq 0$ so that Frobenius–Fuchs theory implies that its solutions are holomorphic in $q^{\frac{1}{2}}$ for $0<|q|<1$. Furthermore, from (54), $\widehat{Z}_N = Z_N\left(\frac{\alpha\tau+\beta}{\gamma\tau+\delta}\right)$ is a solution of the TMLDE

$$\sum_{m=0}^{l+\frac{1}{2}} g_{l+\frac{1}{2}-m}\left[\begin{matrix}(-1)^{\beta}\\(-1)^{\gamma}\end{matrix}\right](q,c)\, D^m \widehat{Z}(q) = 0, \quad (58)$$

which is again of regular singular type provided $g_0(c) \neq 0$. We can repeat the results of Sect. 4 concerning TMLDE $q^{\frac{1}{2}}$-series solutions and the rationality of c and h noting that $Z_V(-1/\tau,c)$ (cf. (35)) satisfies (58) for $\left(\begin{smallmatrix}\alpha&\beta\\\gamma&\delta\end{smallmatrix}\right) = \left(\begin{smallmatrix}0&1\\-1&0\end{smallmatrix}\right)$. We therefore find the VOSA analogues of Propositions 5 and 6.

Proposition 16. *Let V be an Exceptional VOSA of lowest primary weight $l \in \mathbb{N}+\frac{1}{2}$ and central charge c and let N be a simple ordinary V-module of lowest weight h. Assuming $g_0(c) \neq 0$ in the TMLDE* (57) *then*

(i) $Z_N(q)$ *is holomorphic in* $q^{\frac{1}{2}}$ *for* $0 < |q| < 1$.
(ii) $Z_N\left(\frac{\alpha\tau+\beta}{\gamma\tau+\delta}\right)$ *is a solution of the TMLDE* (58) *for all* $\left(\begin{smallmatrix}\alpha & \beta \\ \gamma & \delta\end{smallmatrix}\right) \in \mathrm{SL}(2,\mathbb{Z})$.
(iii) The central charge c and the lowest conformal weight h are rational.

Proposition 17. *Let V be an Exceptional VOSA of lowest primary weight $l \in \mathbb{N}+\frac{1}{2}$ and central charge c. Assuming that $g_0(c) \neq 0$ and that $m \leq l - \frac{1}{2}$ for any indicial root of the form $x = m - c/24$. We then find*

(i) $Z_V(q)$ *is the unique* $q^{\frac{1}{2}}$*–series solution of the TMLDE with leading expansion* $Z_V(q) = Z_{V_{\langle\omega\rangle}}(q) + O\left(q^{l-c/24}\right)$.
(ii) $\dim V_r$ *is a rational function of c for each* $r \in \frac{1}{2}\mathbb{N}$.
(iii) V is generated by the space of lowest weight primary vectors Π_l.

We verify below for $l \leq 17/2$ that $g_0(c) \neq 0$ and that $m \leq l - \frac{1}{2}$ for any indicial root $x = m - c/24$. We conjecture these conditions hold in general.

We can construct two infinite series of $p_l = 1$ Exceptional VOSAs which we conjecture are the only examples.

Proposition 18. *For each Virasoro minimal model with $h_{1,p-1} \in \mathbb{N}+\frac{1}{2}$ there exists an Exceptional VOSA with one odd parity primary vector of lowest weight $l = h_{1,p-1}$ of AD-type*

$$V = L\left(c_{p,q}, 0\right) \oplus L\left(c_{p,q}, h_{1,p-1}\right). \tag{59}$$

Proposition 19. *For each $k \in \mathbb{N} + \frac{1}{2}$ for $k \geq \frac{3}{2}$ there exists an Exceptional VOSA $\mathcal{W}(3k)$ with one odd parity primary vector of lowest weight $3k$ and central charge $c_k = 1 - 24k$.*

Finally, similarly to Sect. 5, with $G = \mathrm{Aut}(V)$ we have

Proposition 20. *Let V be an Exceptional VOSA of class $\mathcal{S}^{2l+1}$ with primaries Π_l of lowest weight $l \in \mathbb{N}+\frac{1}{2}$. If Π_l is a completely reducible G-module then it is either an irreducible G-module or the direct sum of two isomorphic irreducible G-modules.*

8 Exceptional SVOAs with Lowest Primary Weight with $l \in \mathbb{N} + \frac{1}{2}$ for $l \leq \frac{17}{2}$

We now consider examples of Exceptional VOSAs of lowest primary weight $l \leq \frac{17}{2}$. We denote by $E_n = E_n(q)$ the Eisenstein series and $F_n = E_n\left[\begin{smallmatrix}-1 \\ 1\end{smallmatrix}\right](q)$ the twisted Eisenstein series of weight n appearing in the order $l + \frac{1}{2}$ TMLDE (57). For $l \leq \frac{3}{2}$ we find all $c, h \in \mathbb{Q}$, whereas for $\frac{5}{2} \leq l \leq \frac{17}{2}$ we find all $c, h \in \mathbb{Q}$ for which

$p_l = \dim \Pi_l \leq 500{,}000$ found by computer algebra techniques. We obtain many examples of known exceptional VOAs such as the free fermion VOSAs and the Baby Monster VOSA $\mathrm{VB}^\natural = \mathrm{Com}(V^\natural, \omega_{\frac{1}{2}})$, the commutant of $V^\natural$ with respect to a Virasoro vector of central charge $\frac{1}{2}$ [Ho1]. Some other such commutant theories also arise.

[$l = \frac{1}{2}$]. Propositions 15–17 imply that $Z(q)$ satisfies the 1st order TMLDE

$$DZ + cF_2 Z = 0.$$

But $F_2(q) = \frac{1}{24} + 2\sum_{r\in\mathbb{N}+\frac{1}{2}} \frac{rq^r}{1-q^r}$ so that $Z(q) = \left(\frac{\eta(\tau/2)}{\eta(\tau)}\right)^{2c}$ with $p_{1/2} = 2c$. An Exceptional VOSA exists for all $p_{1/2} = m \geq 1$ given by the tensor product of m copies of the free fermion VOSA $L\left(\frac{1}{2}, 0\right) \oplus L\left(\frac{1}{2}, \frac{1}{2}\right)$ of central charge $c = \frac{m}{2}$.

[$l = \frac{3}{2}$]. $Z(q)$ satisfies a 2nd order TMLDE:

$$D^2 Z + \frac{2}{17} F_2(5c+22)\, DZ + \frac{1}{34} c\left(4(5c+22)F_4 + 17E_4\right)\, Z = 0,$$

with indicial roots $x_1 = -c/24$, $x_2 = (7c+24)/408$ with iterative solution

$$Z_V(q) = q^{-c/24}(1 - p_{3/2}q^{3/2} + (1+p_2)q^2 - (p_{3/2}+p_{5/2})q^{5/2} \ldots),$$

$$p_{3/2} = \frac{8c(5c+22)}{3(2c-49)}, \quad p_2 = \frac{(5c+22)(4c+21)(10c-7)}{2(c-33)(2c-49)},$$

$$p_{5/2} = -\frac{136c(5c+22)(4c+21)(10c-7)}{15(2c-83)(c-33)(2c-49)}.$$

For $2c = -2 \mod 17$, the indicial roots differ by an integer leading to denominator zeros for p_n. The $c, h \in \mathbb{Q}$ solutions with possible VOAs are

c	$p_{3/2}$	p_2	$p_{5/2}$	VOSA	$h \in \mathbb{Q}$
$-\frac{21}{4}$	1	0	0	$L\left(c_{3,8}, 0\right) \oplus L\left(c_{3,8}, \frac{3}{2}\right)$	$0, -\frac{1}{4}$
$\frac{7}{10}$	1	0	0	$L\left(c_{4,5}, 0\right) \oplus L\left(c_{4,5}, \frac{3}{2}\right)$	$0, \frac{1}{10}$
$\frac{15}{2}$	35	119	238	$\mathrm{Com}\left(V^+_{\sqrt{2}E_8}, \omega_{\frac{1}{2}}\right)$	$0, \frac{1}{2}$
16	256	2,295	13,056	$V^+_{\mathrm{BW}_{16}} \oplus \left(V^+_{\mathrm{BW}_{16}}\right)_{3/2}$	0, 1
$\frac{114}{5}$	2,432	48,620	537,472	$\mathrm{Com}\left(\mathrm{VB}^\natural, \omega_{\frac{7}{10}}\right)$	$0, \frac{7}{5}$
$\frac{47}{2}$	4,371	96,255	1,139,374	$\mathrm{VB}^\natural$	$0, \frac{49}{34}*$

The $c = \frac{15}{2} = 8 - \frac{1}{2}$ VOSA is the commutant of $V^{+}_{\sqrt{2}E_8}$ with respect to a Virasoro vector of central charge $\frac{1}{2}$ with $\mathrm{Aut}(V) = S_8(2)$ [LSY] and $VB^{\natural}$ is the Baby Monster VOSA with $\mathrm{Aut}(VB^{\natural}) = \mathbb{B}$ [Ho1]. In both cases, $p_{3/2}$ is odd and $\Pi_{3/2}$ is $\mathrm{Aut}(V)$-irreducible in agreement with Proposition 20 [Atlas]. The $c = \frac{15}{2}$ VOSA is the simple current extension of the Barnes–Wall Exceptional VOA by its $h = \frac{3}{2}$ module. The $c = \frac{114}{5} = \frac{47}{2} - \frac{7}{10}$ VOSA is the commutant of $VB^{\natural}$ with respect to a Virasoro vector of central charge $\frac{7}{10}$ [HLY, Y]. In the later case, we expect $\mathrm{Aut}(V) = 2.^2E_6(2):2$, the maximal subgroup of $\mathbb{B}$, which has a 2,432 dimensional irreducible representation [Atlas]. $VB^{\natural}$ is self-dual so that the $h = \frac{49}{34}$ TMLDE solution does not correspond to a graded character $Z_N(q)$.

[$l = \frac{5}{2}$]. $Z(q)$ satisfies a 3rd order TMLDE:

$$
\begin{aligned}
&(734c+49)\, D^3 Z + 27(2c-1)(7c+68) F_2\, D^2 Z \\
&+\left(6(7c+68)(2c-1)(5c+22)F_4 + \frac{1}{2}(2{,}634c^2+1{,}739c-29{,}348)E_4\right) DZ \\
&+\Big(2c(7c+68)(2c-1)(5c+22)F_6 + \frac{27}{2}c(2c-1)(7c+68)E_4F_2 \\
&+5c(36c^2+622c-2{,}413)E_6\Big)\, Z = 0,
\end{aligned}
$$

where

$$
p_{5/2} = \frac{8\,(7c+68)\,(2c+5)\,(2c-1)\,(5c+22)}{5(8c^3-716c^2+16{,}102c+239)}.
$$

There is one $c, h \in \mathbb{Q}$ solution with possible VOSA for $p_{5/2} \le 500{,}000$

c	$p_{5/2}$	VOSA	$h \in \mathbb{Q}$
$-\frac{13}{14}$	1	$L\left(c_{4,7}, 0\right) \oplus L\left(c_{4,7}, \frac{5}{2}\right)$	$0, -\frac{1}{14}, \frac{1}{7}$

[$l = \frac{7}{2}$]. $Z_V(q)$ satisfies a 4th order TMLDE where $p_{7/2} = \frac{r(c)}{s(c)}$ for

$$
\begin{aligned}
r(c) &= 128(5c+22)(3c+46)(2c-1)(14+c)(5c+3)(7c+68), \\
s(c) &= 7(160c^5 - 31{,}176c^4 + 2{,}015{,}748c^3 - 41{,}830{,}202c^2 \\
&\qquad -92{,}625{,}711c + 1{,}017{,}681).
\end{aligned}
$$

The $c, h \in \mathbb{Q}$ solutions with possible VOSA for $p_{7/2} \le 500{,}000$ are

c	$p_{7/2}$	VOSA	$h \in \mathbb{Q}$
$-\frac{161}{8}$	1	$L\left(c_{3,16},0\right) \oplus L\left(c_{3,16},\frac{7}{2}\right)$	$0,-\frac{5}{8},-\frac{3}{4},-\frac{7}{8}$
$-\frac{19}{6}$	1	$L\left(c_{4,9},0\right) \oplus L\left(c_{4,9},\frac{7}{2}\right)$	$0,-\frac{1}{9},-\frac{1}{6},\frac{1}{6}$

$[l = \frac{9}{2}]$. $Z_V(q)$ satisfies a 5th order TMLDE where $p_{9/2} = \frac{r(c)}{s(c)}$ for

$$r(c) = 160(3c+46)(2c-1)(5c+3)(11c+232)(68+7c)(40c^2+1{,}778c+11{,}025),$$
$$s(c) = 9(3{,}200c^6 - 1{,}096{,}320c^5 + 140{,}381{,}096c^4 - 7{,}850{,}716{,}276c^3 + 149{,}541{,}921{,}538c^2$$
$$+829{,}856{,}821{,}745c + 7{,}484{,}560{,}125).$$

The $c, h \in \mathbb{Q}$ solutions with possible VOSA for $p_{9/2} \leq 500{,}000$ are

c	$p_{9/2}$	VOSA	$h \in \mathbb{Q}$
$-\frac{279}{10}$	1	$L\left(c_{3,20},0\right) \oplus L\left(c_{3,20},\frac{9}{2}\right)$	$0,-\frac{7}{10},-1,-\frac{11}{10},-\frac{6}{5}$
$-\frac{125}{22}$	1	$L\left(c_{4,11},0\right) \oplus L\left(c_{4,11},\frac{9}{2}\right)$	$0,-\frac{3}{22},-\frac{5}{22},-\frac{3}{11},\frac{2}{11}$
$-\frac{7}{20}$	1	$L\left(c_{5,8},0\right) \oplus L\left(c_{5,8},\frac{9}{2}\right)$	$0,-\frac{1}{20},\frac{1}{4},\frac{7}{10},\frac{891}{1{,}850}*$
-35	1	$\mathcal{W}(\frac{9}{2})$	$0,-\frac{11}{10},-\frac{4}{3},-\frac{7}{5},-\frac{3}{2}$

The $c = -\frac{7}{20}$, $h = \frac{891}{1{,}850}$ TMLDE solution does not correspond to a graded character $Z_N(q)$.

$[l = \frac{11}{2}]$. $Z_V(q)$ satisfies a 6th order TMLDE where $p_{11/2} = \frac{r(c)}{s(c)}$ for

$$r(c) = -640(13c+350)(7c+25)(11c+232)(2c-1)(3c+46)(68+7c)$$
$$.(5c+3)(10c-7)(40c^2+3{,}586c+50{,}743),$$
$$s(c) = 11(2{,}240{,}000c^9 - 1{,}185{,}856{,}000c^8 + 249{,}718{,}385{,}120c^7 - 25{,}848{,}494{,}429{,}040c^6$$
$$+1{,}266{,}635{,}173{,}648{,}176c^5 - 18{,}264{,}666{,}939{,}042{,}072c^4 - 336{,}264{,}778{,}062{,}263{,}522c^3$$
$$-861{,}021{,}133{,}326{,}393{,}167c^2 + 653{,}498{,}177{,}653{,}904{,}632c - 9{,}760{,}778{,}116{,}675{,}215).$$

The $c, h \in \mathbb{Q}$ solutions with possible VOSA for $p_{11/2} \leq 500{,}000$ are

c	$p_{11/2}$	VOSA	$h \in \mathbb{Q}$
$-\frac{217}{26}$	1	$L\left(c_{4,13},0\right) \oplus L\left(c_{4,13},\frac{11}{2}\right)$	$0,-\frac{2}{13},-\frac{7}{26},-\frac{9}{26},-\frac{5}{13},\frac{5}{26}$

[$l = \frac{13}{2}$]. $Z_V(q)$ satisfies a 7th order TMLDE with $p_{13/2} = \frac{r(c)}{s(c)}$ for

$$r(c) = 4{,}480(13c + 350)(5c + 164)(7c + 25)(11c + 232)(3c + 46)(4c + 21)$$
$$(5c + 3)(10c - 7)(1{,}120c^4 + 187{,}160c^3 + 6{,}889{,}980c^2 + 58{,}079{,}018c - 24{,}165{,}453),$$
$$s(c) = 13(125{,}440{,}000c^{11} - 94{,}806{,}656{,}000c^{10} + 29{,}650{,}660{,}755{,}200c^9$$
$$-4{,}865{,}828{,}683{,}343{,}040c^8 + 431{,}531{,}398{,}085{,}049{,}664c^7 - 18{,}001{,}596{,}789{,}986{,}119{,}984c^6$$
$$+107{,}049{,}283{,}968{,}364{,}390{,}448c^5 + 9{,}359{,}034{,}900{,}957{,}509{,}468{,}076c^4$$
$$+76{,}817{,}948{,}684{,}836{,}018{,}331{,}724c^3 + 155{,}170{,}276{,}090{,}966{,}927{,}173{,}843c^2$$
$$-81{,}951{,}451{,}902{,}336{,}562{,}695{,}126c - 7{,}944{,}030{,}229{,}978{,}323{,}194{,}805).$$

The $c, h \in \mathbb{Q}$ solutions with possible VOSA for $p_{13/2} \leq 500{,}000$ are

c	$p_{13/2}$	VOSA	$h \in \mathbb{Q}$
$-\frac{611}{14}$	1	$L\left(c_{3,28}, 0\right) \oplus L\left(c_{3,28}, \frac{13}{2}\right)$	$0, -\frac{11}{14}, -\frac{19}{14}, -\frac{3}{2}, -\frac{12}{7}, -\frac{25}{14}, -\frac{13}{7}$
$-\frac{111}{10}$	1	$L\left(c_{4,15}, 0\right) \oplus L\left(c_{4,15}, \frac{13}{2}\right)$	$0, -\frac{1}{6}, -\frac{3}{10}, -\frac{2}{5}, -\frac{7}{15}, -\frac{1}{2}, \frac{1}{5}$

[$l = \frac{15}{2}$]. $Z_V(q)$ satisfies an 8th order TMLDE where $p_{15/2} = \frac{r(c)}{s(c)}$ for

$$r(c) = -28{,}672(13c + 350)(5c + 164)(7c + 25)(11c + 232)$$
$$.(3c + 46)(17c + 658)(4c + 21)(5c + 3)(10c - 7)$$
$$.(560c^4 + 146{,}584c^3 + 9{,}082{,}444c^2 + 133{,}381{,}952c - 27{,}346{,}605),$$
$$s(c) = 21{,}073{,}920{,}000c^{12} - 21{,}694{,}120{,}448{,}000c^{11} + 9{,}524{,}271{,}218{,}201{,}600c^{10}$$
$$-2{,}298{,}054{,}501{,}201{,}632{,}000c^9 + 325{,}029{,}065{,}007{,}052{,}546{,}624c^8$$
$$-26{,}081{,}744{,}761{,}028{,}079{,}338{,}944c^7 + 968{,}808{,}700{,}001{,}847{,}281{,}619{,}664c^6$$
$$+787{,}299{,}295{,}625{,}321{,}246{,}276{,}560c^5 - 696{,}312{,}046{,}814{,}218{,}010{,}729{,}784{,}676c^4$$
$$-7{,}887{,}852{,}431{,}045{,}609{,}558{,}472{,}152{,}948c^3 - 21{,}020{,}840{,}196{,}255{,}652{,}876{,}820{,}528{,}205c^2$$
$$+3{,}455{,}907{,}491{,}220{,}404{,}701{,}398{,}711{,}750c + 4{,}568{,}101{,}033{,}862{,}110{,}116{,}156{,}159{,}375.$$

The $c, h \in \mathbb{Q}$ solutions with possible VOSA for $p_{15/2} \leq 500{,}000$ are

c	$p_{15/2}$	VOSA	$h \in \mathbb{Q}$
$-\frac{825}{16}$	1	$L\left(c_{3,32},0\right)\oplus L\left(c_{3,32},\frac{15}{2}\right)$	$0,-\frac{13}{16},-\frac{23}{16},-\frac{7}{4},$ $-\frac{15}{8},-\frac{33}{16},-\frac{17}{8},-\frac{35}{16}$
$-\frac{473}{34}$	1	$L\left(c_{4,17},0\right)\oplus L\left(c_{4,17},\frac{15}{2}\right)$	$0,-\frac{3}{17},-\frac{11}{34},-\frac{15}{34},$ $-\frac{9}{17},-\frac{10}{17},-\frac{21}{34},\frac{7}{34}$
$-\frac{39}{10}$	1	$L\left(c_{5,12},0\right)\oplus L\left(c_{5,12},\frac{15}{2}\right)$	$0,\frac{1}{2},\frac{13}{10},-\frac{1}{6},-\frac{1}{5},\frac{2}{15}$
$\frac{25}{28}$	1	$L\left(c_{7,8},0\right)\oplus L\left(c_{7,8},\frac{15}{2}\right)$	$0,\frac{1}{28},\frac{3}{28},\frac{5}{14},\frac{3}{4},\frac{9}{7}$
-59	1	$\mathcal{W}(\frac{15}{2})$	$0,-\frac{13}{7},-\frac{21}{10},-\frac{31}{14},$ $-\frac{12}{5},-\frac{17}{7},-\frac{5}{2},-\frac{67}{62}*$

The $c=-59$, $h=-\frac{67}{62}$ TMLDE solution does not correspond to a graded character $Z_N(q)$ [F].

[$l=\frac{17}{2}$]. $Z_V(q)$ satisfies a 9th order TMLDE. The only $c,h\in\mathbb{Q}$ solution with possible VOSA for $p_{17/2}\leq 500{,}000$ is

c	$p_{17/2}$	VOSA	$h \in \mathbb{Q}$
$-\frac{637}{38}$	1	$L\left(c_{4,19},0\right)\oplus L\left(c_{4,19},\frac{17}{2}\right)$	$0,-\frac{7}{38},-\frac{13}{38},-\frac{9}{19},-\frac{11}{19},$ $-\frac{25}{38},-\frac{27}{38},-\frac{14}{19},\frac{4}{19}$

Acknowledgements The authors are particularly grateful to Geoffrey Mason for very useful comments over many discussions about this work. The authors also thank Terry Gannon, Atsushi Matsuo and Ching Hung Lam for their comments.

References

[AM] G. Anderson and G. Moore, *Rationality in conformal field theory*, Comm. Math. Phys. **117** (1988), 441–450.

[Atlas] J.H. Conway, R.T. Curtis, S.P. Norton, R.A. Parker and R.A. Wilson, *Atlas of Finite Groups*, Oxford, Clarendon, 1985.

[BFKNRV] R. Blumenhagen, M. Flohr, A. Kliem, W. Nahm, A. Recknagel and R. Varnhagen, *W-algebras with two and three generators*, Nucl. Phys. **B361** (1991), 255–289.

[CIZ] A. Cappelli, C. Itzykson and J. B. Zuber, *The A-D-E Classification of Minimal and $A_1^{(1)}$ Conformal Invariant Theories*, Commun. Math. Phys. **113** (1987), 1–26.

[D] P. Deligne, *La série exceptionnelle de groupes de Lie (The exceptional series of Lie groups)*, C.R.Acad.Sci.Paris Sér. I Math. **322** (1996), 321–326.

[DdeM] P. Deligne and R. de Man, *La série exceptionnelle de groupes de Lie II (The exceptional series of Lie groups II)*, C.R.Acad.Sci.Paris Sér. I Math. **323** (1996), 577–582.

[DMS] P. Di Francesco, P. Mathieu, and D. Senechal, *Conformal Field Theory*, Springer Graduate Texts in Contemporary Physics, Springer-Verlag, New York, 1997.

[F] M. Flohr, *$\mathcal{W}$-algebras, new rational models and completeness of the $c = 1$ classification*, Commun.Math.Phys. **157** (1993), 179–212.

[FHL] I. Frenkel, Y-Z. Huang and J. Lepowsky, *On axiomatic approaches to vertex operator algebras and modules*, Mem. Amer. Math. Soc. **104** (1993), no. 494.

[FLM] I. Frenkel, J. Lepowsky and A. Meurman, *Vertex Operator Algebras and the Monster*, New York, Academic Press, 1988.

[FZ] I. Frenkel and Y. Zhu, *Vertex operator algebras associated to representations of affine and Virasoro algebras*, Duke. Math. J. **66**, (1992), 123–168.

[G] R.L. Griess, *The vertex operator algebra related to E_8 with automorphism group* $O^+(10, 2)$, in The Monster and Lie algebras, Columbus, Ohio, 1996, Ohio State University Math. Res. Inst. Public. **7**, Berlin, de Gruyter, 1998.

[Hi] E. Hille, *Ordinary Differential Equations in the Complex domain*, Dover, New York, 1956.

[HLY] G. Höhn, C.H. Lam and H. Yamauchi, *McKay's E7 observation on the Baby Monster*, Int. Math. Res. Notices **2012**, (2012), 166–212.

[Ho1] G. Höhn, *Selbstduale Vertexoperatorsuperalgebren und das Babymonster*, Ph.D. thesis, Bonn. Math. Sch., **286**, (1996), 1–85.

[Ho2] G. Höhn, *Conformal designs based on vertex operator algebras*, Adv. Math. **217** (2008), 2301–2335.

[I] E. Ince, *Ordinary Differential Equations*, Dover, New York, 1976.

[Kac1] V. Kac, *Vertex Operator Algebras for Beginners*, University Lecture Series, Vol. 10, Boston, AMS, 1998.

[Kac2] V. Kac, *Infinite Dimensional Lie Algebras*, Cambridge, CUP, 1995.

[KKS] M. Kaneko, K. Nagatomo and Y. Sakai, *Modular forms and second order ordinary differential equations: applications to vertex operator algebras*, Lett. Math. Phys. **103** (2013), 439–453.

[KZ] M. Kaneko and D. Zagier, *Supersingular j -invariants, hypergeometric series, and Atkin's orthogonal polynomials*, AMS/IP Stud. Adv. Math. **7** (1998), 97–126.

[Kaw] K. Kawasetsu, *The intermediate vertex subalgebras of the lattice vertex operator algebras*, preprint: arXiv:1305.6463.

[Li] H. Li, *Symmetric invariant bilinear forms on vertex operator algebras*, J. Pure. Appl. Alg. **96** (1994), 279–297.

[LSY] C.H. Lam, S. Sakuma and H. Yamauchi, *Ising vectors and automorphism groups of commutant subalgebras related to root systems*, Math. Z. **255** (2007), 597–626.

[MarMS] H. Maruoka, A. Matsuo and H. Shimakura, *Trace formulas for representations of simple Lie algebras via vertex operator algebras*, unpublished preprint, (2005).

[Mas1] G. Mason, *Vector-valued modular forms and linear differential operators*, Int. J. Num. Th. **3** (2007), 377–390.

[Mas2] G. Mason, *2-Dimensional vector-valued modular forms*, Ramanujan J. **17** (2008), 405–427.

[MatMS] S.D. Mathur, S. Mukhi and A. Sen, *On the classification of rational conformal field theories*, Phys. Lett. **B213** (1988), 303–308.

[Mat] A. Matsuo, *Norton's trace formula for the Griess algebra of a vertex operator algebra with large symmetry*, Commun. Math. Phys. **224** (2001), 565–591.

[MN] A. Matsuo and K. Nagatomo, *Axioms for a vertex algebra and the locality of quantum fields*, Math. Soc. Jap. Mem. **4** (1999).

[MT] G. Mason and M.P. Tuite, *Vertex operators and modular forms*, MSRI Publications **57** (2010) 183–278 , 'A Window into Zeta and Modular Physics', eds. K. Kirsten and F. Williams, Cambridge University Press, Cambridge, 2010.

[MTZ] G. Mason, M.P. Tuite and A. Zuevsky, *Torus chiral n-point functions for* $\mathbb{R}$ *graded vertex operator superalgebras and continuous fermion orbifolds*, Commun. Math. Phys. **283** (2008), 305–342.

[Miy] M. Miyamoto, *Modular invariance of vertex operator algebras satisfying* $C-2$*–cofiniteness* Duke Math. J. **122** (2004), 51–91.

[Se] J.-P. Serre, *A Course in Arithmetic*, Springer-Verlag, Berlin 1978.

[Sh] H. Shimakura, *The automorphism group of the vertex operator algebra* V_L^+ *for an even lattice L without roots*, J. Alg. **280** (2004), 29–57.

[Sp] T.A. Springer, *Linear Algebraic Groups*, Volume 9, Progress in Mathematics, Birkhäuser, Boston 1998.

[T1] M.P. Tuite, *The Virasoro algebra and some exceptional Lie and finite groups*, SIGMA **3** (2007), 008.

[T2] M.P. Tuite, *Exceptional vertex operator algebras and the Virasoro algebra*, Contemp. Math. **497** (2009), 213–225.

[T3] M.P. Tuite, *On the construction of some* $\mathcal{W}$ *algebras*, to appear.

[Y] H. Yamauchi, *Extended Griess algebras and Matsuo-Norton trace formulae*, arXiv:1206.3380.

[Wa] W. Wang, *Rationality of Virasoro vertex operator algebras*, Int. Math. Res. Notices. **1993** (1993), 197–211.

[Wi] E. Witten, *Three-dimensional gravity revisited*, preprint, arXiv:0706.3359.

[Z] Y. Zhu, *Modular invariance of characters of vertex operator algebras*, J. Amer. Math. Soc. **9** (1996), 237–302.

The Cubic, the Quartic, and the Exceptional Group G_2

Anthony van Groningen and Jeb F. Willenbring

Abstract We study an example first addressed in a 1949 paper of J. A. Todd, in which the author obtains a complete system of generators for the covariants in the polynomial functions on the eight-dimensional space of the double binary form of degree (3,1), under the action of $SL_2 \times SL_2$. We reconsider Todd's result by examining the complexified Cartan complement corresponding to the maximal compact subgroup of simply connected split G_2. A result of this analysis involves a connection with the branching rule from the rank two complex symplectic Lie algebra to a principally embedded $\mathfrak{sl}_2$-subalgebra. Special cases of this branching rule are related to covariants for cubic and quartic binary forms.

Key words Binary form • Branching rule • Double binary form • G_2 • Harmonic polynomials • Principal $\mathfrak{sl}_2$ • Symmetric space

Mathematics Subject Classification (2010): 20G05, 22E45, 17B10.

1 Introduction

Let F^k denote the irreducible $k+1$-dimensional representation of SL_2 over the field $\mathbb{C}$. If V is a complex vector space, then we will denote the complex algebra of polynomial functions on V by $\mathbb{C}[V]$, which is graded by degree. That is, for a nonnegative integer k denote

This research was supported by the National Security Agency grant # H98230-09-0054.

A. van Groningen
Milwaukee School of Engineering, 1025 North Broadway, Milwaukee, WI 53202-3109, USA
e-mail: vangroningen@msoe.edu

J.F. Willenbring (✉)
Department of Mathematical Sciences, University of Wisconsin-Milwaukee, EMS Building, Room E403, 3200 N Cramer Street, Milwaukee, WI 53211-3029, USA
e-mail: jw@uwm.edu

G. Mason et al. (eds.), *Developments and Retrospectives in Lie Theory: Algebraic Methods*, Developments in Mathematics 38,
DOI 10.1007/978-3-319-09804-3_17

$$\mathbb{C}[V]_k = \left\{ f \in \mathbb{C}[V] | f(tv) = t^k f(v) \text{ for all } t \in \mathbb{C},\, v \in V \right\}.$$

Thus, $\mathbb{C}[V] = \bigoplus \mathbb{C}[V]_k$. We will identify $F^k = \mathbb{C}[V]_k$ where $V = \mathbb{C}^2$ is the defining representation of SL_2 with the usual action defined by $g \cdot f(v) = f(vg)$ for $g \in \mathrm{SL}_2$, $v \in V$ and $f \in F^k$.

Let x and y denote the standard coordinate functionals on $\mathbb{C}^2$, so that $\mathbb{C}[V]$ may be identified with the polynomial algebra $\mathbb{C}[x, y]$. Thus, $F^k = \mathbb{C}[x, y]_k$ for each nonnegative integer k. In particular, we have

$$F^2 = \left\{ ax^2 + bxy + cy^2 | a, b, c \in \mathbb{C} \right\}.$$

The SL_2-invariant subalgebra of $\mathbb{C}[F^2]$ is generated by $\delta = b^2 - 4ac$ (i.e., the discriminant), which defines a nondegenerate symmetric bilinear form on F^2. The image of $\mathfrak{sl}_2$ in $\mathrm{End}(F^2)$ is contained in $\mathfrak{so}_3$ with respect to this symmetric form. Define

$$\Delta = \frac{\partial^2}{\partial b^2} - 4\frac{\partial^2}{\partial a \partial c},$$

which is in turn a generator of the constant coefficient SL_2-invariant differential operators on $\mathbb{C}[F^2]$. The spherical harmonic polynomials are denoted by

$$\ker \Delta = \left\{ f \in \mathbb{C}[F^2] | \Delta f = 0 \right\}.$$

It is well known that we have a "separation of variables", $\mathbb{C}[F^2] = \mathbb{C}[\delta] \otimes \ker \Delta$. Furthermore, as a representation of SL_2, the graded components of $\ker \Delta$ are irreducible SL_2-representations. Specifically, the degree k harmonic polynomials is equivalent to F^{2k}. Thus, the theory of spherical harmonics provides a graded decomposition of $\mathbb{C}[F^2]$ into irreducible representations of SL_2.

One desires a similar decomposition of $\mathbb{C}[F^k]$ for $k > 2$ (see [7]), which we present for $d \le 4$. Specifically, our goal in this article is to point out a connection between the (infinite-dimensional) representation theory of the simply connected split real form of the exceptional group G_2 with the invariant theory for the cubic ($k = 3$, see Corollary 1) and quartic ($k = 4$, see Corollary 2) binary forms.

An important ingredient to the background of this paper concerns the various embedding of $\mathfrak{sl}_2$ into a given Lie algebra. Specifically, a subalgebra $\mathfrak{l}$ of a semisimple Lie algebra $\mathfrak{g}$ is said to be a *principal* $\mathfrak{sl}_2$*-subalgebra* if $\mathfrak{l} \cong \mathfrak{sl}_2$ and contains a regular nilpotent element of $\mathfrak{g}$ [4, 6, 9]. These subalgebras are conjugate, so we sometimes speak of "the" principal $\mathfrak{sl}_2$-subalgebra. There is a beautiful connection between the principal $\mathfrak{sl}_2$-subalgebra and the cohomology of the corresponding simply connected compact Lie group (see [9]). There is a nice discussion of this theory in [3].

A principal $\mathfrak{sl}_2$ subalgebra of $\mathfrak{sl}_n$ is given by the image of $\mathfrak{sl}_2$ in the n-dimensional irreducible representation of $\mathfrak{sl}_2$ (over $\mathbb{C}$). For $n \geq 4$ these are not maximal. Rather, for even n the image is contained in the standard symplectic subalgebra of $\mathfrak{sl}_n$, while for odd n the image is contained in the standard orthogonal subalgebra of $\mathfrak{sl}_n$.

Since we are interested in the cubic (n=4) and quartic (n=5), it is worthwhile to recall the Lie algebra isomorphism $\mathfrak{sp}_4 \cong \mathfrak{so}_5$. As we shall see, we will exploit an instance of Howe duality for $\mathfrak{sp}_4$. Nonetheless, an exposition of the results presented here could equally well be cast for $\mathfrak{so}_5$. Along these lines, the referee has pointed out to us that another proof of the main theorem may be obtained from the theory of spinors (following [14]).

Central to our treatment is the observation that the complex rank two symplectic Lie algebra $\mathfrak{sp}_4(\cong \mathfrak{so}_5)$ acts on the cubic and quartic forms. In particular, these are the two fundamental representations of $\mathfrak{sp}_4$ of dimension four and five respectively. From this point of view the problem of decomposing finite-dimensional representations of $\mathfrak{sp}_4$ when restricted to the principally embedded $\mathfrak{sl}_2$-subalgebra generalizes the invariant theory of the cubic and quartic.

The unitary dual of simply connected split G_2 was determined in [18]. This important example was an outgrowth of decades of work emerging from the study of Harish-Chandra modules. More recently, several authors have taken on the task of classification of the *generalized Harish-Chandra modules*. Specifically, we find motivation from [13].

Since $F^k \cong (F^k)^*$ as an SL_2-representation, we have an SL_2-invariant form in $\mathrm{End}(F^k)$. For even k, this invariant is symmetric and for odd k the form is skew-symmetric. The cubic forms, F^3, define a four-dimensional representation of $\mathfrak{sl}_2$, which is symplectic. That is,

$$\mathfrak{sl}_2 \hookrightarrow \mathfrak{sp}_4 \hookrightarrow \mathrm{End} F^3,$$

where $\mathfrak{sp}_4$ is the subalgebra of $\mathrm{End}(F^3)$ preserving the degree-two skew-symmetric form.

On the other hand, the quartic forms, F^4, define a five-dimensional representation of $\mathfrak{sl}_2$, which is orthogonal. That is,

$$\mathfrak{sl}_2 \hookrightarrow \mathfrak{so}_5 \hookrightarrow \mathrm{End} F^4,$$

where $\mathfrak{so}_5$ is the subalgebra of $\mathrm{End} F^4$ preserving the degree-two symmetric form.

Upon a fixed choice of Cartan subalgebra $\mathfrak{h} \subset \mathfrak{sp}_4$, and Borel subalgebra $\mathfrak{h} \subset \mathfrak{b} \subset \mathfrak{sp}_4$ we obtain a choice of root system, Φ with positive roots, Φ^+, and simple roots denoted Π. Let ω_1 and ω_2 denote the corresponding fundamental weights, and denote the irreducible finite-dimensional representation indexed by $\lambda = a\omega_1 + b\omega_2 \in \mathfrak{h}^*$ for nonnegative integer a and b by

$$L(\lambda) = L(a, b),$$

where we let $\dim L(1, 0) = 4$ and $\dim L(0, 1) = 5$.

The key point here is that upon restriction to $\mathfrak{sl}_2$, the space of the $\mathfrak{sp}_4$-representation $L(1,0)$ is the space of cubic forms, while $L(0,1)$ is the space of quartic forms. This unification of cubic and quartic forms was first systematically studied in [11] and [12].

In general, if $\mathcal{V}$ denotes a representation of a group (resp. Lie algebra) and $\mathcal{W}$ denotes an irreducible representation of a subgroup (resp. subalgebra) H, then upon restriction to H, let

$$[\mathcal{W}, \mathcal{V}] = \dim \mathrm{Hom}_{\mathrm{H}}(\mathcal{W}, \mathcal{V})$$

denote the multiplicity.

The present article concerns, in part, the function, $b(k,l,m) = [F^k, L(l,m)]$, for a principally embedded $\mathfrak{sl}_2$ in $\mathfrak{sp}_4$. It is important to note that the numbers $b(k,l,m)$ are not difficult to compute nor are they theoretically mysterious. The point is the relationship to the group G_2, as we shall see.

We relate this situation to the problem of decomposing harmonic polynomials, in the sense of B. Kostant and S. Rallis [10]. Specifically, we let K denote a symmetrically embedded subgroup of the complex exceptional group G_2, such that the Lie algebra, $\mathfrak{k}$, of K is isomorphic to $\mathfrak{sl}_2 \oplus \mathfrak{sl}_2$. Let $\mathfrak{p}$ denote the complexified Cartan complement of $\mathfrak{p}$ in the Lie algebra of G_2. The group K acts on $\mathfrak{p}$ by restricting the adjoint representation. From the general theory of Kostant and Rallis, we know that the K-invariant subalgebra, $\mathbb{C}[\mathfrak{p}]^{\mathrm{K}}$, is a polynomial ring and $\mathbb{C}[\mathfrak{p}]$ is a free module over $\mathbb{C}[\mathfrak{p}]^{\mathrm{K}}$.

In the instance where $\mathfrak{g}$ is the Lie algebra of G_2 and $\mathrm{K} = \mathrm{SL}_2 \times \mathrm{SL}_2$, we have $\mathbb{C}[\mathfrak{p}]^{\mathrm{K}} = \mathbb{C}[\delta_1, \delta_2]$ with δ_i degree 2 and 6 respectively. Let Δ_i $(i = 1, 2)$ denote the corresponding constant coefficient K-invariant differential operators and set

$$\mathcal{H}_{\mathfrak{p}} = \{ f \in \mathbb{C}[\mathfrak{p}] | \Delta_i(f) = 0 \text{ for } i = 1, 2 \} .$$

to be the harmonic polynomials together with their gradation by degree. For each $d = 0, 1, 2, \ldots$ the group K acts linearly on $\mathcal{H}_{\mathfrak{p}}^d$. See [19] for general results on the decomposition of $\mathcal{H}_{\mathfrak{p}}^d$ into irreducible K-representations.

Motivation for considering $\mathcal{H}_{\mathfrak{p}}$ for the symmetric pair corresponding to split G_2 comes from considering the (underlying Harish-Chandra module) of spherical principal series, which are isomorphic to $\mathcal{H}_{\mathfrak{p}}$ as representations of K.

The finite-dimensional irreducible representations of $\mathrm{K} \cong \mathrm{SL}_2 \times \mathrm{SL}_2$ with polynomial matrix coefficients are of the form $F^k \otimes F^l$. In the G_2 case when $\mathrm{K} = \mathrm{SL}_2 \times \mathrm{SL}_2$, we have $\mathfrak{p} = F^3 \otimes F^1$. We organize the graded K-multiplicities in a formal power series in q defined by

$$p_{k,l}(q) = \sum_{d=0}^{\infty} [F^k \otimes F^l, \mathcal{H}_{\mathfrak{p}}^d] q^d .$$

We set $p(k,l;d) = [F^k \otimes F^l, \mathcal{H}_{\mathfrak{p}}^d]$ for nonnegative integers k, l and d.

It turns out that $p_{k,l}(q)$ is a polynomial in q. That is, K-irreps occur with finite multiplicity in $\mathcal{H}_{\mathfrak{p}}$. In the last section we recall the results from [10] concerning this multiplicity. In the special case of G_2 falls into the literature on double binary (3,1) forms studied by J. A. Todd (see [16]).

From our point of view, we find the unification of the cubic and quartic seen in [12] via $\mathfrak{sp}_4$-representations in parallel with the work presented in [16] on (3,1)-forms. Neither of these references mention the group G_2. Since the group G_2 plays an important role in invariant theory (see [1]), we would like to point out:

Theorem 1. *For nonnegative integers k, l and m we have*

$$b(k,l,m) = \sum_{j \geq 0} p(k,l;2m + l - 6j).$$

We prove this theorem in the next section, and point out special cases, including the decomposition of the polynomial functions on the cubic and quartic. The proof uses an instance of R. Howe's theory of dual pairs (see [5]) applied to the problem of computing branching multiplicities (see [8]). In the final section we recall the result of Todd, and provide a picture of the Brion polytope (see [2, 15]) associated with this example.

2 Proof of the Main Theorem

If $\lambda = (\lambda_1 \geq \lambda_2 \geq \cdots \geq \lambda_l > 0)$ is an integer partition, then let F_n^λ denote the irreducible finite-dimensional representation of GL_n (resp. $\mathfrak{gl}_n$) with highest weight indexed by λ as in [5]. Note that these representations restrict irreducibly to SL_n (resp. $\mathfrak{sl}_n$). In the case $n = 2$ we will use the notation $F^d = F_2^{(d)}$.

For positive integers r and c, let $M_{r,c}$ denote the $r \times c$ complex matrices. Throughout this section we set $\mathfrak{p} = F^3 \otimes F^1$, which as a vector space may be identified with $M_{4,2}$. The group $GL_4 \times GL_2$ acts on $\mathbb{C}[M_{4,2}]$ by the action $(g_1, g_2) \cdot f(X) = f(g_1^T X g_2)$ for $g_1 \in GL_4$, $g_2 \in GL_2$, $f \in \mathbb{C}[M_{4,2}]$, and $X \in M_{4,2}$. (Here g^T denotes the transpose of a matrix g.) Under this action, we have a multiplicity free decomposition

$$\mathbb{C}[M_{4,2}]_d \cong \bigoplus F_4^\mu \otimes F_2^\mu,$$

where the sum is over all partitions $\mu = (\mu_1 \geq \mu_2 \geq 0)$ with at most two parts and of size d (i.e., $\mu_1 + \mu_2 = d$).

This, multiplicity free decomposition of $\mathbb{C}[M_{4,2}]_d$ into irreducible $GL_4 \times GL_2$-representations is the starting point. We will proceed as follows: (1) restrict to the group $Sp_4 \times GL_2 \subset GL_4 \times GL_2$ and decompose by using $(Sp_4, \mathfrak{so}_4)$-Howe duality, then (2) restrict further to the principally embedded SL_2 in Sp_4 on the left, and the

$SL_2 \subset GL_2$ on the right. Lastly, we compare this decomposition into irreducibles as predicted by Kostant–Rallis theory.

With this plan in mind, we will set

$$H(k,l) = \sum_{d=0}^{\infty} \left[F^k \otimes F^l, \mathbb{C}[M_{4,2}]_d \right] q^d$$

for nonnegative integers k and l. We will compute two expressions for $H(k,l)$. For the first, we start with (GL_4, GL_2)-Howe duality,

$$H(k,l) = \sum_{\lambda_1 \geq \lambda_2 \geq 0} \left[F^k \otimes F^l, F_4^\lambda \otimes F_2^\lambda \right] q^{\lambda_1 + \lambda_2},$$

where we restrict the GL_4 and GL_2 irreps to (the principally embedded) SL_2. The key idea is to first restrict the left acting GL_4 to Sp_4, then restrict from Sp_4 to SL_2.

2.1 Symplectic-Orthogonal Howe Duality

The $(Sp_{2k}, \mathfrak{so}(2n))$ instance of Howe duality concerns two commuting actions on $\mathbb{C}[M_{2k,n}]$. The first is given by left multiplication of Sp_{2k}, while the second is given by a Lie algebra of Sp_{2k}-invariant polynomial coefficient differential operators. The operators in this second action are generated by a set spanning a Lie algebra isomorphic to $\mathfrak{so}(2n)$ (with respect to the commutator bracket). We have a multiplicity free decomposition:

Theorem 2. *Given integers k and n,*

$$\mathbb{C}[M_{2k,n}] \cong \bigoplus V_{2k}^{\mu} \otimes V_{\mu}^{2n}$$

where the sum is over $\mu = (\mu_1 \geq \cdots \geq \mu_l > 0)$—a nonnegative integer partition with $l \leq \min(k,n)$.

The modules V_{2k}^{μ} are finite-dimensional irreducible representations of Sp_{2k}, while V_{μ}^{2n} are, in general ($n > 1$), infinite-dimensional irreducible highest weight representations of the Lie algebra $\mathfrak{so}(2n)$.

Using the results of [8] one is able to obtain, in a stable range ($k \geq n$), a concrete combinatorial description of the branching rule from SL_{2n} to the symmetric subgroup Sp_{2n}. These results come from translating the branching problem to one of decomposing the $\mathfrak{so}_{2n}$ irreps under the actin of $\mathfrak{gl}_n$.

In the stable range (see [8]), the $\mathfrak{so}_{2n}$-representations V_μ^{2n} are (generalized) Verma modules. This fact implies that upon restriction to the symmetrically embedded $\mathfrak{gl}_n$ subalgebra we have

$$V_\mu^{2n} = \mathbb{C}[\wedge^2\mathbb{C}^n] \otimes F_n^\mu,$$

as a $\mathfrak{gl}_n$-representation.

If $n = 2$, then $\wedge^2\mathbb{C}^2$ is invariant for SL_2. Thus, it is very easy to see the decomposition of these $\mathfrak{so}_4$ irreps into $\mathfrak{sl}_2$-irreps. Consequently, in the very special case of decomposing SL_4 representations into Sp_4-irreps, we obtain a multiplicity free branching rule. This is no surprise since SL_4 is isomorphic to Spin(6) and Sp_4 is isomorphic to Spin(5), and the (Spin(6), Spin(5)) branching rule is well known to be multiplicity free (see [5]).

In any case, we have that, upon restriction, the GL_4-representations,

$$F_4^\mu \cong V_4^\mu \oplus V_4^{\mu-(1,1)} \oplus V_4^{\mu-(2,2)} \oplus \cdots \oplus V^{(\mu_1-\mu_2,0)},$$

as Sp_4-representations.

The fact that the partition (1, 1) has size 2 means that there is a degree 2 $Sp_4 \times SL_2$-invariant, f_1, in $\mathbb{C}[M_{4,2}]$. This polynomial, obviously, remains invariant under the subgroup $SL_2 \times SL_2$. Moreover, the algebra $\mathbb{C}[M_{4,2}]$ is free as a module over $\mathbb{C}[f_1]$. Since the degree of f_1 is two, we have the following:

$$H(k,l) = \frac{1}{1-q^2}\sum_\mu \left[F^k \otimes F^l, V^\mu \otimes F^{\mu_1-\mu_2}\right] q^{\mu_1+\mu_2},$$

where the sum is over all $\mu = (\mu_1 \geq \mu_2 \geq 0)$ with $|\mu| = \mu_1 + \mu_2$. Thus, $l = \mu_1 - \mu_2$.

Let $m = \mu_2$ so that $\mu_1 + \mu_2 = 2m + l$. That is, $V^\mu = L(l, m)$ and we have

$$b(k,l,m) = \left[F^k, L(l,m)\right].$$

Restricting from Sp_4 to the principally embedded SL_2 we obtain

$$H(k,l) = \frac{1}{1-q^2}\sum_{m=0}^{\infty} b(k,l,m) q^{2m+l}. \tag{1}$$

2.2 *The Symmetric Pair* (G_2, $\mathfrak{sl}_2 \oplus \mathfrak{sl}_2$)

An ordered pair of groups, (G, K), is said to be a *symmetric pair* if G is a connected reductive linear algebraic group (over $\mathbb{C}$), and K is an open subgroup of the fixed points of a regular involution on G. The Lie algebra of G, denoted $\mathfrak{g}$, contains the Lie algebra of K, denoted $\mathfrak{k}$. The differential of the involution, denoted θ has -1 eigenspace

$$\mathfrak{p} = \{X \in \mathfrak{g} | \theta(X) = -X\}.$$

Let $\mathfrak{a}$ denote a maximal toral[1] subalgebra contained in $\mathfrak{p}$. Let M denote the centralizer of $\mathfrak{a}$ in K.

Upon restricting the adjoint representation of G on $\mathfrak{g}$, we obtain a linear group action of K on $\mathfrak{p}$. From [10], we know that $\mathbb{C}[\mathfrak{p}]^K$ is freely generated by dim$\mathfrak{a}$ elements, as an algebra. The same is true about the constant coefficient K-invariant differential operators, denoted $\mathcal{D}(\mathfrak{p})^{\mathrm{K}}$. Define

$$\mathcal{H}_{\mathfrak{p}} = \{f \in \mathbb{C}[\mathfrak{p}] | \partial f = 0 \text{ for all } \partial \in \mathcal{D}(\mathfrak{p})^{\mathrm{K}}\}.$$

These are the K-harmonic polynomials on $\mathfrak{p}$. Set $\mathcal{H}_{\mathfrak{p}}^d = \mathcal{H}_{\mathfrak{p}} \cap \mathbb{C}[\mathfrak{p}]_d$. As a K-representation we have a gradation, $\mathcal{H}_{\mathfrak{p}} = \bigoplus_{d=0}^{\infty} \mathcal{H}_{\mathfrak{p}}^d$.

From [10], $\mathcal{H}_{\mathfrak{p}} \cong \mathrm{Ind}_M^K 1$. That is, as a representation of K,

$$\mathcal{H} \cong \{f \in \mathbb{C}[K] | f(mk) = f(k) \text{ for all } m \in \mathrm{M} \text{ and } k \in \mathrm{K}\}.$$

By Frobenius reciprocity, each irreducible K-representation, F, occurs with multiplicity $\dim F^M$ in $\mathcal{H}_{\mathfrak{p}}$.

In the present article, we consider the case when $\mathfrak{g}$ is the Lie algebra of G_2 and $\mathfrak{k} = \mathfrak{sl}_2 \oplus \mathfrak{sl}_2$ in detail. We have remarked that $\mathfrak{p} \cong F^3 \otimes F^1$. The group M is isomorphic to the eight element quaternion group,

$$M \cong \{\pm 1, \pm \mathrm{i}, \pm \mathrm{j}, \pm \mathrm{k}\}.$$

An embedding of M into K is given by

$$i \mapsto \left(\begin{bmatrix} 0 & 1 \\ -1 & 0 \end{bmatrix}, \begin{bmatrix} 0 & 1 \\ -1 & 0 \end{bmatrix} \right)$$

$$j \mapsto \left(\begin{bmatrix} -\mathrm{i} & 0 \\ 0 & \mathrm{i} \end{bmatrix}, \begin{bmatrix} -\mathrm{i} & 0 \\ 0 & \mathrm{i} \end{bmatrix} \right)$$

$$k \mapsto \left(\begin{bmatrix} 0 & \mathrm{i} \\ \mathrm{i} & 0 \end{bmatrix}, \begin{bmatrix} 0 & \mathrm{i} \\ \mathrm{i} & 0 \end{bmatrix} \right).$$

[1] A *toral* subalgebra is an abelian subalgebra consisting of semisimple elements.

In [5], this example is treated in Chapter 12. It is pointed out that $\mathbb{C}[\mathfrak{p}]^K = \mathbb{C}[f_1, f_2]$ where deg $f_1 = 2$ and deg $f_2 = 6$. The fact that $\mathbb{C}[\mathfrak{p}]$ is free over $\mathbb{C}[f_1, f_2]$ leads to

$$H(k,l) = \frac{p_{k,l}(q)}{(1-q^2)(1-q^6)} = \frac{1}{(1-q^2)(1-q^6)} \sum_{d=0}^{\infty} p(k,l;d)q^d$$

$$= \frac{1}{1-q^2} \sum_{d=0}^{\infty} \sum_{j=0}^{\infty} p(k,l;d)q^{d+6j}.$$

We introduce a new parameter m, and re-index with $d + 6j = l + 2m$ and we obtain

$$H(k,l) = \frac{1}{1-q^2} \sum_{m=0}^{\infty} \sum_{j=0}^{\infty} p(k,l;l+2m-6j)q^{l+2m}. \tag{2}$$

The main result follows by comparing Eqs. 1 and 2.

2.3 Special Cases

The insolvability of the quintic motivates special attention to forms of degree less than five. The invariant theory of quadratic forms reduces to the theory of spherical harmonics (i.e., Kostant–Rallis theory for the pair (SO(4),SO(3)).) mentioned at the beginning of this article.

The structure of G_2 is intimately related to the cubic (see [1]). The complexified Cartan decomposition of split G_2 has $\mathfrak{p}$ being two copies of the cubic. However, there is no symmetric pair in which $\mathfrak{p}$ consists of exactly on copy of the cubic form. Nonetheless, the algebra of polynomial functions on the cubic forms is free over the invariant subalgebra (see Chapter 12 of [5]).

From our point of view, the multiplicity of an irreducible SL_2-representation in the polynomial function on the cubic is given by reduces to:

Corollary 1 (The Cubic).

$$\left[F^k, \mathbb{C}[F^3]_d\right] = \sum_{j \geq 0} p(k,d;d-6j).$$

Proof. The irreducible representation of $\mathfrak{sp}_4$ of the form $L[d,0] \cong \mathbb{C}[V]_d$ where $V = \mathbb{C}^4$ is the defining representation. Upon restriction to a principal $\mathfrak{sl}_2$ subalgebra $V \cong F^3$. The result follows by taking $m = 0$ in the main theorem.

The polynomial function on the quartic forms are a free module over the invariants. This fact is a special case of Kostant–Rallis theory for (SL_3, SO_3), where $\mathfrak{p}$ consists of the five-dimensional space of trace zero 3×3 symmetric matrices, denoted $SM(3)_0$. As a representation of SL_2 (locally SO_3), $\mathfrak{p} = SM(3)_0$ is the quartic under the conjugation action of SO_3.

The relationship between the quartic and G_2 is less transparent. In a nutshell, since Sp_4 is isomorphic to Spin(5), and the five-dimensional defining representation of Spin(5) restricts irreducibly to the quartic with respect to the principally embedded SL_2.

However, it follows from the split G_2 example by specializing the main theorem that:

Corollary 2 (The Quartic). *Let $\mathcal{H}(F^4)$ denote the space of spherical harmonics with respect to the orthogonal representation, F^4. For each d, we have $L(0, d) = \mathcal{H}^d(F^4)$, and*

$$\left[F^k, \mathcal{H}^d(F^4)\right] = \sum_{j \geq 0} p(k, 0; 2d - 6j).$$

Proof. Let $l = 0$ from the main theorem.

There is a third obvious special case describing the $\mathfrak{sl}_2$-invariants in an arbitrary $\mathfrak{sp}_4$-representation with respect to the principal embedding. That is:

Corollary 3 (principal invariants).

$$\dim\left(L(l, m)\right)^{SL_2} = \sum_{j \geq 0} p(0, l; 2m - 6j)$$

Proof. Let $k = 0$ in the main theorem.

The pair $(\mathfrak{sp}_4, \mathfrak{sl}_2)$ is not a symmetric pair. In particular, there is no instance of the Cartan–Helgason theorem describing $\mathfrak{sp}_4$-irreps. with a $\mathfrak{sl}_2$-invariant vector. In fact, the $\mathfrak{sl}_2$-invariant subspace does not have to be one dimensional.

We mention this special case since it suggests a relationship between the $(\mathfrak{sp}_4, \mathfrak{sl}_2)$ generalized Harish-Chandra modules and the unitary dual of split G_2.

3 Todd's Covariants for Double Binary (3,1) Forms, and the Brion Polytope

We conclude this article with the remark that the $(G_2, \mathfrak{sl}_2 \oplus \mathfrak{sl}_2)$ instance of Kostant–Rallis theory is seen in the double binary (3,1) forms, as studied by J. A. Todd in 1949. Although there is no mention of the group G_2, a generating set of covariants is provided. That is, for each $d = 1, 2, 3, 4, 5, 6, 7, 8, 9, 10, 11, 12$ a list of highest weight vectors (given as polynomials) is provided. The weights are summarized in the following.

3.1 *Table:* $SL_2 \times SL_2$*-Covariants in* $\mathbb{C}[F^3 \otimes F^1]$

The $SL_2 \times SL_2$ highest weight vectors in $\mathbb{C}[F^3 \otimes F^1]$ span a graded subalgebra – the covariants. In the table below, we denote Todd's generator for the highest weight of $F^k \otimes F^l$ by $[k, l]$. The degrees are indicated.

degree	weight			
1	[3, 1]			
2	[4, 0]	[2, 2]	[0, 0]	
3	[5, 1]	[3, 3]	[1, 1]	
4	[4, 0]	[2, 2]	[0, 4]	
5	[3, 1]	[1, 3]		
6	[6, 0]	[4, 2]	[2, 4]	[0, 0]
7	[1, 3]			
8	[0, 4]			
9	[1, 5]			
12	[0, 6]			

Not included in [16] are the *relations* between these covariants, nor is there any information about the vector space dimension of their span in each multi-graded component. Note that, modulo the degree 2 and 6 invariants, the multiplicity of a given covariant may be computed by the dimension of the M-fixed vectors. Thus, implicitly the eight element quaternion group plays a role in this example.

A convenient way to "visualize" this example comes from the theory of the *Brion polytope* as described in [15] (see also [2]). That is, for each occurrence of $F^k \otimes F^l$ in $\mathcal{H}_{\mathfrak{p}}^d$, with $d > 0$, plot the point

$$\left(\frac{k}{d}, \frac{l}{d}\right)$$

on the xy-plane. The closure of these points is a polytope. In the special case corresponding to G_2, this polytope is two dimensional and can be visualized as

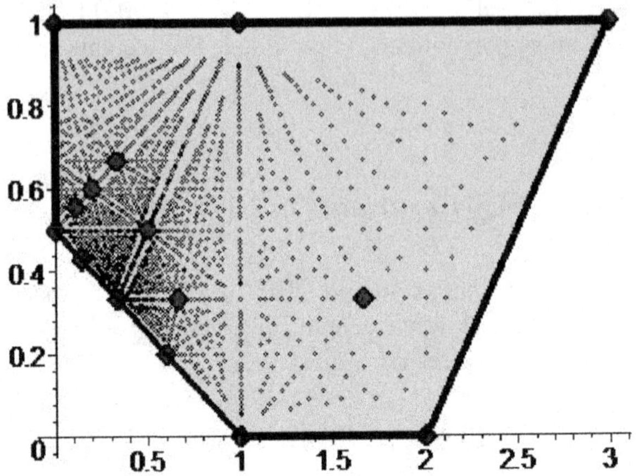

The large dots in the above picture are the Todd covariants. The smaller dots correspond to specific K-types. We refer the reader to [17] for a detailed account of how this polytope relates to the $(\mathfrak{sp}_4, \mathfrak{sl}_2)$ branching rule, asymptotically.

Acknowledgements We thank Allen Bell, Nolan Wallach and Gregg Zuckerman for helpful conversations about the results presented here. The first author's Ph.D. thesis [17] contains a more thorough treatment of the graded K-multiplicities associated with the symmetric pair $(G_2, \mathfrak{sl}_2 \oplus \mathfrak{sl}_2)$, which was jointly directed by Allen Bell and the second author. The problem concerning the graded decomposition of $\mathcal{H}_\mathfrak{p}$ as a K-representation was suggested by Nolan Wallach (see [19]), while the problem of studying the restriction of a finite-dimensional representation to a principally embedded $\mathfrak{sl}_2$-subalgebra was suggested by Gregg Zuckerman (see [20]).

Finally, we would like to thank the referee for many helpful suggestions, and corrections.

References

[1] Ilka Agricola, *Old and new on the exceptional group G_2*, Notices Amer. Math. Soc. **55** (2008), no. 8, 922–929.

[2] Michel Brion, *On the general faces of the moment polytope*, Internat. Math. Res. Notices **4** (1999), 185–201, DOI 10.1155/S1073792899000094.

[3] David H. Collingwood and William M. McGovern, *Nilpotent orbits in semisimple Lie algebras*, Van Nostrand Reinhold Mathematics Series, Van Nostrand Reinhold Co., New York, 1993.

[4] E. B. Dynkin, *Semisimple subalgebras of semisimple lie algebras*, Mat. Sbornik N.S. **30(72)** (1952), 349–462 (3 plates).

[5] Roe Goodman and Nolan R. Wallach, *Symmetry, Representations, and Invariants*, Graduate Texts in Mathematics, Vol. 255, Springer, New York, 2009.

[6] È. B. Vinberg, V. V. Gorbatsevich, and A. L. Onishchik, *Structure of Lie groups and Lie algebras*, Current problems in mathematics. Fundamental directions, Vol. 41 (Russian), Itogi Nauki i Tekhniki, Akad. Nauk SSSR, Vsesoyuz. Inst. Nauchn. i Tekhn. Inform., Moscow, 1990, pp. 5–259 (Russian).

[7] Roger Howe, *The classical groups and invariants of binary forms*, The mathematical heritage of Hermann Weyl (Durham, NC, 1987), Proc. Sympos. Pure Math., Vol. 48, Amer. Math. Soc., Providence, RI, 1988, pp. 133–166.

[8] Roger Howe, Eng-Chye Tan, and Jeb F. Willenbring, *Stable branching rules for clas- sical symmetric pairs*, Trans. Amer. Math. Soc. **357** (2005), no. 4, 1601–1626, DOI 10.1090/S0002-9947-04-03722-5.

[9] Bertram Kostant, *The principal three-dimensional subgroup and the Betti numbers of a complex simple Lie group*, Amer. J. Math. **81** (1959), 973–1032.

[10] B. Kostant and S. Rallis, *Orbits and representations associated with symmetric spaces*, Amer. J. Math. **93** (1971), 753–809.

[11] Yannis Yorgos Papageorgiou, *SL(2)(C), the cubic and the quartic*, ProQuest LLC, Ann Arbor, MI, 1996. Thesis (Ph.D.), Yale University.

[12] Yannis Y. Papageorgiou, sl_2, *the cubic and the quartic*, Ann. Inst. Fourier (Grenoble) **48** (1998), no. 1, 29–71 (English, with English and French summaries).

[13] Ivan Penkov and Vera Serganova, *Bounded simple* $(\mathfrak{g}, \mathrm{sl}(2))$*-modules for* $\mathrm{rk}\,\mathfrak{g} = 2$. J. Lie Theory, **20** (2010), no. 3, 581–615.

[14] Roger Penrose and Wolfgang Rindler, *Spinors and space-time. Vol. 1*, Cambridge Monographs on Mathematical Physics, Cambridge University Press, Cambridge, 1987. Two-spinor calculus and relativistic fields.

[15] A. V. Smirnov, *Decomposition of symmetric powers of irreducible representations of semisimple lie algebras, and the brion polytope*, Tr. Mosk. Mat. Obs. **65** (2004), 230–252. (Russian, with Russian summary); English transl., Trans. Moscow Math. Soc. (2004), 213–234.

[16] J. A. Todd, *The geometry of the binary* $(3, 1)$ *form*, Proc. London Math. Soc. (2) **50** (1949), 430–437.

[17] Anthony Paul van Groningen, *Graded multiplicities of the nullcone for the algebraic symmetric pair of type G*. ProQuest LLC, Ann Arbor, MI, 2007. Thesis (Ph.D.), The University of Wisconsin, Milwaukee.

[18] David A. Vogan Jr., *The unitary dual of* G_2, Invent. Math. **116** (1994), no. 1–3, 677–791.

[19] N. R. Wallach and J. Willenbring, *On some* q*-analogs of a theorem of kostant-rallis*, Canad. J. Math. **52** (2000), no. 2, 438–448, DOI 10.4153/CJM-2000-020-0.

[20] Jeb F. Willenbring and Gregg J. Zuckerman, *Small semisimple subalgebras of semisimple lie algebras*, Harmonic analysis, group representations, automorphic forms and invariant theory, Lect. Notes Ser. Inst. Math. Sci. Natl. Univ. Singap., vol. 12, World Sci. Publ., Hackensack, NJ, 2007, pp. 403–429.

Printed in the United States
By Bookmasters